World Sustainability Series

Series editor

Walter Leal Filho, HAW Hamburg, Hamburg, Germany

Due to its scope and nature, sustainable development is a matter which is very interdisciplinary, and draws from knowledge and inputs from the social sciences and environmental sciences on the one hand, but also from physical sciences and arts on the other. As such, there is a perceived need to foster integrative approaches, whereby the combination of inputs from various fields may contribute to a better understanding of what sustainability is, and means to people. But despite the need for and the relevance of integrative approaches towards sustainable development, there is a paucity of literature which address matters related to sustainability in an integrated way.

More information about this series at http://www.springer.com/series/13384

Walter Leal Filho
Editor

Social Responsibility and Sustainability

How Businesses and Organizations Can Operate in a Sustainable and Socially Responsible Way

Springer

Editor
Walter Leal Filho
FTZ-NK
HAW Hamburg
Hamburg, Germany

ISSN 2199-7373 ISSN 2199-7381 (electronic)
World Sustainability Series
ISBN 978-3-030-03561-7 ISBN 978-3-030-03562-4 (eBook)
https://doi.org/10.1007/978-3-030-03562-4

Library of Congress Control Number: 2018960219

This Springer imprint is published by the registered company Springer Nature Switzerland AG
The registered company address is: Gewerbestrasse 11, 6330 Cham, Switzerland

Preface

Social responsibility and sustainable development are two different concepts, whose integration over the years has led to significant progresses in the way enterprises see and perceive their operations. It is not only about policies or steps taken to meet legal requirements, but it also about social equality and environmental accountability, also bearing in mind the links with eco-efficiency, innovation, and the health and well-being of workers.

According to ISO 26000, Social responsibility is defined as the responsibility of an organisation for the impacts of its decisions and activities on society and the environment, through transparent and ethical behaviour that:

- contributes to sustainable development, including health and the welfare of society
- takes into account the expectations of stakeholders
- is in compliance with applicable law and consistent with international norms of behaviour
- is integrated throughout the organisation and practised in its relationships.

But even though the relations between social responsibility and sustainability are strong, it is still necessary to encourage organisations to ideally adhere to or at least follow the principles of sustainable development in their operations, giving something back to the community.

There is thus a perceived need for a better understanding of how social responsibility is related to sustainable development and for the identification of processes, methods, and tools which may help the integration of these two important elements. There is also a real need to showcase successful examples of how to structure behaviour and institutional practice in line with the sustainability challenges we face today.

It is against this background that this book has been produced. It contains papers presented at the "World Symposium on Social Responsibility and Sustainability" organised by the University of Edinburgh, the Research and Transfer Centre "Sustainable Development and Climate Change Management" of the Hamburg

University of Applied Sciences, Germany, and the Inter-University Sustainable Development Research Programme (IUSDRP), as well as by other contributors.

The book is a truly interdisciplinary piece, with inputs from scholars, representatives from social movements, practitioners and members of governmental agencies, undertaking research and/or executing projects focusing on social responsibility and sustainability from across the world.

The book serves the purpose of showcasing experiences from research, field projects, and best practice to foster social responsibility and sustainability, which may be useful or implemented elsewhere.

Enjoy your reading.

Hamburg, Germany Walter Leal Filho

Contents

Part I
Conceptual Frameworks

Reviewing the Stakeholder Value Creation Literature: Towards a Sustainability Approach

Riikka Tapaninaho and Johanna Kujala

Abstract The purpose of this study is to examine distinctive narratives of stakeholder value creation and discuss how they consider sustainability. Based on an extensive research review spanning over three decades of material, we present four categories of the stakeholder value creation literature: (1) a focal firm orientation with an economic value perspective, (2) a stakeholder orientation with an economic value perspective, (3) a focal firm orientation with a multiple value perspective, and (4) a stakeholder orientation with a multiple value perspective. In each of these categories, we identified several narratives of stakeholders, value creation, and sustainability. This study reveals an increased interest in sustainability issues and their more coherent incorporation into stakeholder research in recent years. We suggest that, with respect to sustainability, future research should consider the dynamic, systemic, and multilevel nature of stakeholder relationships and collaboration. Additionally, a more versatile understanding of value and value creation, as well as a broader comprehension of stakeholders and their needs, should be incorporated into future research. Finally, conceptualising sustainability and the normative core of sustainable business, as well as elaborating on the purpose and role of business regarding sustainability, are interesting focus areas for future research.

Keywords Stakeholder theory · Value creation · Sustainability · Research review

1 Introduction

Stakeholder theory is one of the main management frameworks used to discuss social responsibility issues in business. As stakeholder theory perceives stakeholders broadly, referring to those who can affect or are affected by a firm's operations (Freeman 1984), the theory has acquired a prominent place as a framework that addresses social responsibility issues as a natural part of business. Moreover, stakeholder theory

---

R. Tapaninaho (✉) · J. Kujala
Faculty of Management, University of Tampere, 33014 Tampere, Finland
e-mail: riikka.tapaninaho@uta.fi

W. Leal Filho (ed.), *Social Responsibility and Sustainability*, World Sustainability Series, https://doi.org/10.1007/978-3-030-03562-4_1

has been described as a management theory of the 21st century suitable for understanding and redefining the role of business and value creation in society (Freeman 2010). Freeman et al. (2010) argue that stakeholder theory is fundamentally 'about value creation and trade and how to manage a business effectively. "Effective" can be seen as "create as much value as possible".' However, academics and business practitioners have given sustainability increased attention, leading to calls for management scholars to rethink extant management theories and their underlying assumptions in the context of sustainability (e.g., Derry 2012; Starik and Kanashiro 2013). Stakeholder theory has also been advanced directly regarding sustainability. In effect, Hörisch et al. (2014) examined the applicability of stakeholder theory in sustainability management and developed a conceptual framework to enhance the theory's application in the context of sustainability. Scholars have particularly emphasised the need to incorporate sustainability management into conventional management theories to have an impact on more sustainable business decisions instead of building distinctive theories and new languages to learn for this purpose (ibid.).

Stakeholder theory has served to analyse and understand multiple phenomena in various fields of the organisational sciences, such as strategic management (e.g., Freeman 1984; Haksever et al. 2004; Strand and Freeman 2015), corporate responsibility (e.g., Brower and Mahajan 2013; Sachs and Maurer 2009; Smith and Rönnegard 2016; Strand et al. 2015), business ethics (e.g., Phillips 1997; Purnell and Freeman 2012; Wicks 1996), and international business (e.g., Ansari et al. 2013; Christmann 2004; Lehtimaki and Kujala 2015). Moreover, stakeholder value creation has been used in many recent studies to better understand stakeholder concerns and cooperation (e.g., Garcia-Castro and Aguilera 2015; Garriga 2014; Harrison and Wicks 2013; Rühli et al. 2017; Schneider and Sachs 2015; Tantalo and Priem 2016). Therefore, we need a comprehensive understanding of what stakeholder value creation means in the organisational sciences and what value means in the stakeholder literature.

While stakeholder theory was not originally developed to address complex sustainability issues, its potential for further development has been acknowledged and acted on (Freeman 1994; Hörisch et al. 2014), and it has served as a traditional management theory in research on corporate sustainability to some extent (Montiel and Delgado-Ceballos 2014). However, stakeholder theory, like other management theories, has been criticised quite heavily for being too focused on economic, firm-centric value creation with anthropocentric premises in the context of sustainability (e.g., Banerjee 2000, 2001; Clifton and Amran 2011). Starik and Kanashiro (2013) have also criticised management theorists as lacking a systematic examination of sustainable development, which has been the case in stakeholder research. In essence, calls have been made for researchers to pay attention to the premises of their research, including research questions, assumptions, and broader paradigms followed in their studies (e.g., Derry 2012).

The purpose of this study is to examine distinctive narratives of stakeholder value creation and discuss how they consider sustainability. Analysing the stakeholder value creation literature from the sustainability viewpoint allows researchers and practitioners to become more aware of the various uses of the concepts and of the impacts different research questions, settings, and ontological and epistemological assumptions have on research findings. Hence, the findings of this study help us better understand how the chosen approaches influence the development of management theories and business practices.

Sustainability, or sustainable development, is understood in this study via its most profound definition, which refers to 'development which meets the needs of the present without compromising the ability of future generations to meet their own needs' and consists of economic, social, and environmental dimensions (United Nations World Commission on Environment and Development 1987). Sustainability has been regarded as somewhat difficult to translate and implement in traditional businesses (e.g., Gallo and Christensen 2011), and compromises have often been made between the economic, social, and environmental dimensions (Gallo and Christensen 2011). For true sustainability, though, all three dimensions should be considered equally and acted upon (Bansal 2005). Studies on strong sustainability take the issue even further and suggest that concerns related to the natural environment should be considered as an elementary part of all studies related to management and organisations (Heikkinen et al. 2018; Heikkurinen 2017). Relating to the idea of strong sustainability, this chapter discusses how the studies on stakeholder value creation contribute to sustainability when considered an important issue affecting organisations and society today.

The remainder of the chapter is organised as follows. In the next section, we will explain the collection and analysis of studies on stakeholder value creation. Then, based on an inductive categorisation, we will present the findings of our review in four categories according to their orientation toward stakeholders (i.e., focal firm or stakeholder orientation) and value (i.e., economic or multiple value perspective). In each of these categories, we identify several narratives of stakeholders, value creation, and sustainability. The chapter is concluded by discussing the stakeholder value creation narratives and their relation to sustainability.

2 Collection and Analysis of Studies on Stakeholder Value Creation

To examine the distinctive narratives of stakeholder value creation, we reviewed articles published in leading management journals from 1985 to 2015 and found 210 related scholarly articles for our review. We followed the advice given by Short (2009) and focused our search on eleven high-quality management journals including both top journals, such as *Academy of Management Review* and *Academy of Management Journal*, and specialty journals, such as *Business & Society* and *Journal of Business Ethics*. The journal choice was made with an objective to allow discussion with the mainstream management theories instead of concentrating on more conditioned journals, where the importance of sustainability issues would be more acknowledged (ibid.). In essence, we are participating in efforts to incorporate sustainability into conventional business thinking and language.

The research process was iterative in nature. It started by searching for and identifying relevant articles in the chosen journals. Altogether, three search rounds were conducted: in November–December of 2015, February–March of 2016, and September–November of 2016. Each search round was followed by a close reading and analysis of the articles, as well as the development and refinement of the inductive categorisation and distinctive narratives and documenting the findings. The analysis was finalised in 2018 after crosschecking and fine-tuning the findings, resulting in the final version of the article.

To identify and select relevant articles, we read the titles and abstracts of articles and, when necessary, their introductions and conclusions. We based the selection of articles on the following criteria: they had to examine value creation in a way that emphasised stakeholders and stakeholder relationships and needed to recognise that value could be created for one or more stakeholders other than shareholders. We chose these criteria based on the premises and main thesis of the stakeholder theory, which considers cooperation between various actors to be important (e.g., Freeman et al. 2007). We also chose these criteria based on the assumption that sustainability views value creation as extending beyond economic returns and shareholder value maximisation to achieving environmental and social benefits (e.g., Starik and Kanashiro 2013). Table 1 presents a summary of the selected articles.

After collecting the articles, we began our analysis by closely reading articles published in the top management journals. Based on our reading, we performed an inductive categorisation by paying attention to the perception of value in the articles and how each of the articles approached stakeholder issues. Consequently, we positioned the articles on two different dimensions, which ranged from a single value perspective to a multiple value perspective and from a focal firm orientation to a stakeholder orientation. We then extended our analysis to articles published in specialty journals. These articles confirmed our inductive categorisation and convinced

Table 1 Summary of the stakeholder value creation articles in this review

Journal	No.
Academy of Management Review	36
Academy of Management Journal	18
Strategic Management Journal	25
Journal of Management	5
Administrative Science Quarterly	4
Journal of Management Studies	14
Organization Science	12
Journal of Business Ethics	46
Business & Society	12
Business Ethics Quarterly	18
Organization & Environment	20
Total	210

Table 2 Content classification system for coding and analysis

Code	Description
Year	Publication year
Authors	Name of the authors
Article	Title of the article
Journal	Name of the journal
Research questions	Explicitly stated research questions
Theories	Theories used within the study
Empirical methods and setting	Empirical method, data collection, and description of data source
Key findings	Explicitly stated key findings in the article
View on stakeholders	Focal firm orientation or stakeholder orientation
View on value	A single value perspective or a multiple value perspective
View on sustainability	Explicit or implicit reference and view on sustainability

us to proceed accordingly. As a result, we divided all 210 articles into four categories along with the dimensions of a focal firm versus stakeholder orientation and an economic value versus multiple value perspective. To conduct a systematic examination of the articles, we used qualitative content analysis as a method of investigation (Krippendorff 2004; Weber 1990). This methodology allowed us to inductively develop a unified coding system and systematically analyse and extract relevant information from the articles. Table 2 presents the content classification system for coding and analysis.

Finally, we paid attention to the different views of sustainability in each of these categories. In general, we noted the clear (i.e., the article mentioned sustainability or sustainable development explicitly) and embedded sustainability references (i.e., the article did not use sustainability or sustainable development as a concept directly but used the social or environmental dimension implicitly) in the articles. In the next section, we will explain the results of our analysis in more detail and depict the four categories and their different narratives of stakeholders, value creation, and sustainability in each category.

3 Narratives of Stakeholder Value Creation

Our examination shows that authors have studied stakeholder value creation quite extensively and that the literature has widely acknowledged the responsibilities of business and the creation of versatile value beyond economic measures. We divided the previous studies on stakeholder value creation into four categories: (1) a focal firm orientation with an economic value perspective (25 articles), (2) a stakeholder orientation with an economic value perspective (20 articles), (3) a focal firm orienta-

	Focal firm orientation	Stakeholder orientation
Multiple value perspective	Focal firm orientation with a multiple value perspective (84 articles)	Stakeholder orientation with a multiple value perspective (81 articles)
Economic value perspective	Focal firm orientation with an economic value perspective (25 articles)	Stakeholder orientation with an economic value perspective (20 articles)

Fig. 1 Categories of stakeholder value creation articles

tion with a multiple value perspective (84 articles), and (4) a stakeholder orientation with a multiple value perspective (81 articles) (Fig. 1).

Our examination indicates that interest in sustainability issues and theory development within stakeholder value creation studies has increased recently. The fourth category, stakeholder orientation with a multiple value perspective, consisted of articles that widely acknowledge the responsibilities of businesses in society, the importance of stakeholder cooperation, and collaboration for versatile value creation and value beyond economic measures. Therefore, the fourth category shows the most potential to address the question of how stakeholder value creation relates to sustainability. We will now present the different narratives of stakeholder value creation within in each category in more detail. A more detailed description of the literature related to each category and narrative is presented in the Appendix.

3.1 Focal Firm Orientation and the Economic Value Perspective

The first category represents an instrumental view of stakeholder value creation, placing the focal firm and business performance in the centre of the study. Although over half of the articles referred to sustainability issues implicitly or explicitly—mainly in the form of common social performance measures, such as KLD or sustainability ratings—sustainability issues and measurements were treated as subordinate to traditional strategic issues and performance measures. Studies in this category criticise the stakeholder approach to value creation and capture for its lack of guidance for managers in situations in which trade-offs need to be made between stakeholders. Instead, some argued that the single objective function, with the primacy of the firm's long-term value maximisation, should always guide managerial decision-making (e.g., Jensen 2002). In a similar fashion, Sundaram and Inkpen (2004a) defended shareholder value maximisation as the primary corporate objective function, arguing that it is the only objective that will profit all stakeholders in the end. Hence, this category instrumentally investigates whether and how sustainability or stakeholder issues should be dealt with and follows the prevailing economic paradigm. Furthermore, scholars emphasised that stakeholder theory has not provided enough empirical evidence for its stakeholder value maximisation claim (Sundaram and Inkpen 2004b). The three stakeholder value creation narratives within this category were (1) the narrative of stakeholder identification, (2) the narrative of stakeholder management influencing firm performance, and (3) the narrative of value creation and capture. In the following paragraphs, we will explain them in more detail.

3.1.1 The Narrative of Stakeholder Identification

Studies using this narrative show interest in the identification of stakeholders and analysing how and why companies respond to pressures from different stakeholders. Firm-stakeholder relationships were examined from the managerial or organisational perspective, as well as through understanding the accrued effects of the networks in which firms participate. Factors affecting salience and decisions regarding different stakeholder issues were linked to, for example, stakeholder status (Perrault 2017), directors' personal values and roles (Adams et al. 2011), and the organisational life cycle (Jawahar and McLaughlin 2001). Rowley (1997), however, built on social network analysis and stated that firms resist stakeholder demands based on the simultaneous effects of the stakeholder network density and the firm's centrality within this network.

3.1.2 The Narrative of Stakeholder Management Influencing Firm Performance

The studies using this narrative are interested in the relationship between stakeholder management and firm performance. They showed contradictory results concerning whether attending stakeholder concerns can be beneficial for a firm and what should come first in the managerial agenda. To start with, Berman et al. (1999) argued that firms address stakeholder concerns when they expect positive effects on financial performance. The studies that showed evidence of positive effects examined impacts on shareholder value (Hillman and Keim 2001), financial performance (Choi and Wang 2009), and long-term shareholder value (Garcia-Castro et al. 2011). Furthermore, Ogden and Watson (1999) examined whether a firm is able to improve the interests of shareholders and stakeholders simultaneously and showed that it is possible to align the interests of different stakeholders, at least to some extent, without compromising shareholder returns, although Garcia-Castro and Francoeur (2016) found that additional investments in stakeholders do not necessarily benefit firms linearly and can also become costly. Additionally, Wang et al. (2008) depicted the relationship between corporate philanthropy and financial performance in the form of an inverse U-shape, showing positive effects on financial performance in the beginning but negative effects after a certain point.

Innovations and temporal aspects were also taken into consideration within the studies trying to show the link between stakeholder management and competitive advantage. For example, the interconnections between innovations, stakeholder relationships, and competitive advantage were examined, underlining the importance of cultural and industry contexts when choosing the most efficient stakeholder management approach to create a competitive advantage through innovations (Harting et al. 2006). It was also argued that achieving a competitive advantage depends on a firm's ability to adapt and transform its stakeholder management practices over time (Verbeke and Tung 2013).

Finally, the studies regarding CEOs, stakeholder management, and business performance revealed, for instance, how it might be disadvantageous for a CEO to pursue stakeholder-related initiatives, as they can have negative effects on CEO compensation, even if there are financial improvements (Coombs and Gilley 2005). A newly appointed CEO may sacrifice long-term stakeholder value, such as pension funds, research and development (R&D) investments, and capital equipment investments, while attempting to create short-term profits in their self-interest (Harrison and Fiet 1999). It also seemed that the economic performance of a firm has a moderating effect on the CEO's stakeholder concerns (Dooley and Lerner 1994).

3.1.3 The Narrative of Dynamics in Value Creation

The dynamics of value creation were of interest in quite a few studies. For example, Bridoux and Stoelhorst (2014) showed that the heterogeneity of stakeholders allows firms to create sustained market value with both fairness and an arms-length approach to stakeholder management. Additionally, the roles of different stakeholders and environments in economic value creation were focused on, for example, through exploring the strategies to manage consumer benefited experiences (Priem 2007), examining the firm's political environment (Oliver and Holzinger 2008), and analysing the secondary stakeholders in the socially complex cases of product diversification (Su and Tsang 2015). Some of the studies approached value creation through the examination of who finally captures the value. As an example, Blyler and Coff (2003) suggested that, in the context of dynamic capability, internal stakeholders with social capital may capture the economic rents for their personal gain before they appear in traditional performance measures. To improve the management of stakeholder claims and value capture, Crane et al. (2015) argued that stakeholder accounting and financialising stakeholder claims could be helpful.

3.2 Stakeholder Orientation and the Economic Value Perspective

The second category, along with focusing on economic or business value, drew attention to the importance of stakeholder relationships and cooperation for business success and value creation. This view is also instrumental in nature; for example, the studies on instrumental stakeholder theory are situated within this category. These studies examined concepts such as trust in cooperation networks, the consequences of blurring organisational boundaries, and value creation and capture logics in multiple stakeholder settings. As the economic value creation function of firms was emphasised in this category, only a few articles either implicitly or explicitly referred to sustainability. Three stakeholder value creation narratives within this category were (1) trust in cooperation networks, (2) blurring organisational boundaries, and (3) value creation and capture logics. In the following paragraphs, we will explain them in more detail.

3.2.1 The Narrative of Trust in Cooperation Networks

An overriding issue that many of the articles in this category discussed was trust. In effect, several articles based on instrumental stakeholder theory emphasised the role of trust and cooperation in creating organisational wealth and a competitive advantage (e.g., Jones 1995; Preston and Donaldson 1999). The influence of trust in stakeholder relationships was elaborated on even further by Wicks et al. (1999), who stated that the optimal level of trust in stakeholder relationships positively affects firm performance. Furthermore, Pirson and Malhotra (2011) contended that different organisational stakeholders appreciate different dimensions of trustworthiness. The studies on strategic partnerships (Ireland et al. 2002), innovation networks (Dhanaraj and Parkhe 2006), and interorganisational relationships (e.g., Barringer and Harrison 2000; Connelly et al. 2015) highlighted the fact that productive and cost-efficient cooperation and value creation require building trust into these relationships.

3.2.2 The Narrative of Blurring Organisational Boundaries

Another major narrative concerns the consequences of blurring organisational boundaries and roles of different actors in multi-stakeholder networks. Due to the blurring of organisational boundaries, understanding the role of different stakeholders in value creation was regarded as important. As an example, Korschun (2015) investigated the important role of employees and concluded that a strong identification with the company leads to adversarial stakeholder relationships, while a collectivistic organisational identity and seeing stakeholders as organisational members supports a cooperative approach. Kochan and Rubinstein (2000) emphasised the importance of employees, too, by giving employees a prominent place in value creation and improving the organisational performance of American organisations. The cognitive side of value creation and construction of a competitive advantage were also focused on, as using and exchanging resources within firm-constituent interactions would require 'communication about and interpretations of those exchanges' (Rindova and Fombrun 1999). Other contexts in which different stakeholder groups were examined included, for example, leaders and stakeholder management in radix organisations (Schneider 2002), external stakeholder engagement in creating sustainable shareholder value (Henisz et al. 2014), and cooperation between isolated firms and stakeholders with the help of government support institutions in the case of product upgrading in emerging markets (McDermott et al. 2009).

3.2.3 The Narrative of Value Creation and Capture Logics

Again, value creation and capture logics emerged as an important narrative. Garcia-Castro and Aguilera (2015) presented a conceptual framework of incremental value creation and appropriation, which expands the concept of value and value capture to consider all stakeholders of the firm. The scholars argued that value creation and appropriation should be viewed dynamically, as multiple stakeholders participate in value creation processes with their resources and capabilities, meriting their proportion of the economic rents created (ibid.). Priem et al. (2013) emphasised the role of consumers and stated that it is necessary to pay attention to value creation opportunities for consumers and corresponding business models and ecosystems.

However, some of the studies showed how value creation in stakeholder relationships may harm the value capture of a firm or an individual stakeholder group. For instance, Coff (2010) examined how different stakeholders participating in capability development may use their bargaining power for rent appropriation in different stages of the capability life cycle, causing direct effects on firm performance. Kivleniece and Quelin (2012) examined value creation and capture in public-private collaboration and stated that private actors' involvement might be jeopardised if public partner opportunism or external stakeholder activism restrained private actors' value capture. The empirical results of Poulain-Rehm and Lepers' (2013) study did not support the hypothesis that employee share ownership plans and employees' growing role in company decision-making would enhance value creation and capture for either shareholders or stakeholders. Finally, Bridoux et al. (2011) emphasised that a firm should adapt its motivational system to the heterogeneous motives of different employees to enhance collective value creation and interfirm performance.

3.3 Focal Firm Orientation and the Multiple Value Perspective

The third category approaches value creation mainly from the focal firm perspective but recognises the social or environmental responsibilities of companies in addition to economic value creation. Most of the studies in this category recognised the need to expand the view of stakeholder value creation further and challenged the current paradigm to develop more sustainable business practices. In effect, most of the articles in this category had either explicit or implicit sustainability references. However, the conceptions of sustainability-related values were not coherent, as some of the studies solely examined environmental value, while others more broadly discussed social or stakeholder value considerations but without further specifying value conception. The common themes emerged around five narratives: (1) challenging the traditional corporate social responsibility (CSR) approach, (2) stakeholder identification and salience, (3) stakeholder management practices, (4) an expanded view of value creation, (5) environmental and sustainability management. In the following paragraphs, we will explain these five narratives in more detail.

3.3.1 The Narrative of Challenging the Traditional CSR Approach

The first narrative in this category criticises the traditional CSR and corporate social performance (CSP) approaches by aiming to understand responsible business practices through stakeholder theory (e.g., Clarkson 1995; Jamali 2008; Rowley and Berman 2000; Sachs and Maurer 2009; Smith and Rönnegard 2016). Sachs and Maurer (2009), for instance, argued that CSR research should move toward dynamic corporate stakeholder responsibility and should not artificially distinguish between economic and social responsibilities. Smith and Rönnegard (2016) even implied that stakeholder theory could be set as a paradigm for business and business responsibilities in the future. To challenge the traditional view of CSR, O'Riordan and Fairbrass (2014) provided a practical model for organisational decision-makers to conceive their firms as inherently linked with society and to address collective value creation for all stakeholders within the value chain. Furthermore, Margolis and Walsh (2003) encouraged researchers to set aside persistent attempts to explain the relationship between a firm's social and economic performance and concentrate instead on the question of when firm activities could be beneficial to society. In addition, Kroeger and Weber (2014) introduced a conceptual framework to measure the benefits of social value creation.

Other CSR- and CSP-related studies concentrated on the effects of good CSP on accessing finance (Cheng et al. 2014), the stakeholder landscape and its impacts on the breadth of corporate social performance (Brower and Mahajan 2013), and differences in firms' CSR responses to institutional pressures (Crilly et al. 2012). Moreover, some scholars studied firms' identity orientation toward stakeholders in explaining CSP activity (Bingham et al. 2011), the effects of the sociocognitive factors of the top management team and organisational decision-making structures on corporate social performance (Wong et al. 2011), and the impacts of changes in CSP on financial performance (Ruf et al. 2001). Additionally, some articles scrutinised stakeholder reactions and stakeholder relationships. Crilly et al. (2016) examined stakeholder evaluations and reactions to the social responsibility activities of multinational corporations, while Madsen and Rodgers (2015) investigated stakeholder attention to firm CSR activities and its effects on corporate financial performance, and Bendheim et al. (1998) concentrated on the best practices in firm-stakeholder relationships.

3.3.2 The Narrative of Stakeholder Identification and Salience

Identifying relevant stakeholders is a special narrative within the stakeholder literature. Probably the best-known model of stakeholder identification and salience was presented by Mitchell et al. (1997), who defined three relationship attributes (i.e., power, legitimacy, and urgency) as relevant in defining stakeholder salience through managerial perception. To support this theory, Agle et al. (1999) found evidence for the attribute-salience relationships, while Bundy et al. (2013) examined how organisational identity and strategic frames guide managerial interpretations of issue salience. Whereas the studies mainly focused on manager-led processes of stakeholder identification, Tashman and Raelin (2013) suggested that stakeholder salience should move beyond managerial perceptions, as they might ignore important stakeholders due to market frictions. Other studies examined stakeholder identification based on the social identity of stakeholders (Crane and Ruebottom 2011), stakeholder salience in family business settings (Mitchell et al. 2011), stakeholder legitimacy (Phillips 2003), and the role of stakeholder culture in stakeholder-related decisions (Jones et al. 2007). Moreover, scholars have investigated firms' responses to conflicting institutional demands (Pache and Santos 2010) and secondary stakeholder action (Eesley and Lenox 2006), instrumental and normative perspectives on understanding why firms respond to stakeholders (Welcomer et al. 2003), and stakeholder orientations of boards of directors (Wang and Dewhirst 1992).

3.3.3 The Narrative of Stakeholder Management Practices

In this narrative, attention is drawn to the stakeholder management practices of focal firms. Managerial cognition and its effects on stakeholder management was identified as a research gap in the stakeholder literature (Laplume et al. 2008) and was examined by various scholars (e.g., Crilly and Sloan 2012). Moreover, De Luque et al. (2008) showed how managers' stakeholder and economic values relate to subordinates' perceptions of leadership and firm performance. Minoja (2012) called for an ambidextrous approach for stakeholder management to ensure stakeholder cooperation and long-term firm performance, while Kaufman (2002) argued that stakeholder management approaches should include a double fiduciary duty consisting of loyalty to corporate stakeholders, as well as loyalty to fair bargaining and freedom. Organisational factors such as organisational architecture (Crilly and Sloan 2013) and enterprise strategy (Crilly 2013) were also identified as influencing stakeholder management practices. Wheeler et al. (2002) highlighted the difficulties an organisation might face when developing more stakeholder-responsive orientations related to environmental and social issues throughout the organisation. Meanwhile, Winn (2001) examined what a multiple stakeholder decision-making model would look like. Some articles paid attention to stakeholder activism, for example, through the study of CEO ideology and its effects on social activism (Briscoe et al. 2014), as well as through the examination of differences in firms' responses to activism (Waldron et al. 2013).

3.3.4 The Narrative of an Expanded View on Value Creation

An expanded view of value creation was the fourth main narrative within this category. In this narrative, the traditional view of economic value creation is challenged, for instance, by arguing that value creation and capture, and what is of value are contingent and subjective, and these arguments should be considered in the research related to value creation and capture (Lepak et al. 2007). Haksever et al. (2004) showed how firms and their managers may, through their decisions, create or even destroy value for their stakeholders in different dimensions. The long-term success of the firms was sought by creating happiness and well-being for all stakeholders instead of following the objective function of shareholder wealth maximisation (e.g., Dierksmeier and Pirson 2009; Jones and Felps 2013a, b). Although the traditional view of the firm was challenged to a great extent, and business was suggested to be defined with regard to its ability to create common good (Shankman 1999), a strategic approach to social value creation was considered superior regarding social outcomes than a purely altruistic approach (Husted and de Jesus Salazar 2006).

Research on social entrepreneurship and social enterprises also took an extended view on value creation, by combining social problems with economic value creation. The studies investigated, for example, entrepreneurs' motivation to engage in social entrepreneurship (Fauchart and Gruber 2011; Miller et al. 2012). It was even argued that the role of entrepreneurship in society should be defined as naturally considering blended value creation, including financial, social, and environmental aspects (Zahra and Wright 2015). McMullen and Warnick (2015) regarded the blended value creation objective at its best as an ideal model, which should not be normative or a legal obligation for all new entrepreneurial ventures. In effect, the tensions between social missions and business objectives were recognised, and stakeholder theory was seen as a possible solution to manage them (Smith et al. 2013).

3.3.5 The Narrative of Environmental and Sustainability Management

Over the years, stakeholder theorists have been arguing whether the natural environment should have a stakeholder status. Although nature has been ascribed a stakeholder status (Starik 1995) or even given primacy in the stakeholder identification and salience processes based on its relationship attribute of proximity (Driscoll and Starik 2004), it has been argued that the environment does not need a stakeholder status as environmental issues are considered through other legitimate stakeholders. (e.g., Phillips and Reichart 2000). In either case, stakeholder value creation studies have been widely interested in expanding the value creation to also include environmental issues.

Many scholars have shown interest in what drives companies toward environmentally friendly practices and how environmental friendliness is reflected in stakeholder relationships. Companies were regarded as changing their behaviour mostly due to external pressures from their operating environment. For example, Fineman and Clarke (1996) identified campaigners and regulators as external stakeholders that manage to evoke pro-environmental responses within companies. As managers were accused of perceiving corporate environmentalism and their firm's relationship with the environment through an economic rationale that focuses on how environmental initiatives benefit the firm financially, regulatory forces and stakeholder activity were presented as central in advancing environmentally friendly activities (Banerjee 2001). Regarding climate change, the temporal orientations of managers were argued to be future-oriented but rely heavily on public policy development (Sarasini and Jacob 2014). It was even argued that companies would enter partnerships to address environmental problems (e.g., with the government) based on a threat or an opportunity and being dependent on a firm's resources and positioning (Lin 2014).

Generally, environmental management was examined from three different perspectives. Some of the studies investigated the influences of external stakeholders on environmental management practices, for instance, by studying the effect of shareholder activists' status and reputation on firm responsiveness to environmental issues (Perrault and Clark 2015), or more generally, stakeholder impacts on choosing environmental response patterns (Murillo-Luna et al. 2008). Meanwhile, other scholars focused on examining the internal factors affecting firm responses to environmental management. These studies shed light on a number of issues, such as entrepreneurs' disengagement with pro-environmental values (Shepherd et al. 2013), the effects a firm's size has on its stakeholder responsiveness and the adoption of proactive environmental strategies (Darnall et al. 2010), the determinants of multinationals' global environmental policies (Christmann 2004), stakeholder management and proactive environmental strategies (Buysse and Verbeke 2003), the ecological responsiveness model (Bansal and Roth 2000), the influence of supervisory support and environmental policies on employees' eco-initiatives (Ramus and Steger 2000), and the role of corporations in achieving ecological sustainability (Shrivastava 1995a, b). Finally, some studies examined environmental management and value creation from an institutional viewpoint by focusing on a number of topics, such as the role of national context in explaining how green innovation can enhance firm-level financial performance (Aguilera-Caracuel and Ortiz-de-Mandojana 2013), the legal environment and its effect on a firm's self-regulation (Short and Toffel 2010), the effects of public and private politics on corporate climate change strategies (Reid and Toffel 2009), and community and regulatory stakeholder pressures and the environmental performance of companies (Kassinis and Vafeas 2006).

Studies focusing on sustainability management followed similar kinds of patterns and research interests as environmentally oriented studies. For example, Sharma and Henriques (2005) argued that the resources of a firm's social, ecological, and economic stakeholders influence the adoption of sustainability practices. Hahn et al. (2014) were interested in how the cognitive frames of managers affect managerial sensemaking in sustainability issues, and Zollo et al. (2013) stated that sustainability research should direct attention toward companies' internal learning and change processes instead of concentrating on external stakeholders. Furthermore, Gallo and Christensen (2011) highlighted that firm size, ownership, and industry are related to behaviours firms adopt regarding sustainability, and a longitudinal study of corporate sustainable development conducted by

Bansal (2005) showed how both institutional and resource-based factors have influenced the adoption of firms' sustainability activities.

Stakeholder theory and management practices were also criticised regarding sustainability. Gladwin et al. (1995) stated that attempting to adapt to sustainability while relying on the current anthropocentric worldview, which is the basis for most management theories, including stakeholder theory, is insufficient. Instead, a shift is needed to sustain centrism, which considers both environmental and social sustainability as important. According to Clifton and Amran (2011), the stakeholder approach fails in advancing sustainability both in terms of its normative principles and in management practices. Banerjee (2000) also expressed a critical viewpoint on sustainability-related issues and posited that current stakeholder theories and management practices follow Western economic rationality, which leads to neglecting marginalised stakeholders and their needs.

3.4 Stakeholder Orientation and the Multiple Value Perspective

This category consists of traditional stakeholder studies, which are built around the narrative of cooperative and trusting relationships between firms and their stakeholders (e.g., Donaldson and Preston 1995; Freeman 1984; Jones and Wicks 1999) with a broad view of value (e.g., Harrison et al. 2010; Harrison and Wicks 2013). However, our analysis revealed that the original design of stakeholder theory does not address broader sustainability issues, although some scholars argue that the theory could potentially be expanded due to its normative core, its consideration of those who affect and are affected by business, and its pluralistic nature. In effect, stakeholder theory's applicability in sustainability management was advanced just recently (Hörisch et al. 2014). The sustainability-oriented approach is prominent in ecologically oriented studies, studies related to multi-stakeholder settings that address 'wicked' socioeconomic problems, and in more recent research streams, such as the development of sustainable business models.

Our analysis reveals that few articles explicitly discuss value considerations regarding sustainability. Although the researchers recognised the importance of stakeholder welfare and value creation beyond economic measures, the main focus was on those stakeholders who contribute to the value creation processes of organisations. This category consists of three narratives: (1) grounds for responsible business behaviour, (2) defining value, and (3) the principles and mechanisms of how value is created. In the following paragraphs, we will explain these three narratives in more detail.

3.4.1 The Narrative of Grounds for Responsible Business Behaviour

This narrative focuses on the intertwined connections between business and ethics and the role of business in society. While scholars argue for the grounds of responsible business from different perspectives, the primary focus is on determining why firms should engage in responsible business practices. For instance, Phillips (1997) relied on the principle of fairness, whereas Burton and Dunn (1996) built upon the principles of caring derived from feminist ethics. Various stakeholder theorists asserted that it is not meaningful to discuss business and ethics separately and that value creation and trade call for moral decision-making influenced by personal values (e.g., Freeman 2000; Harris and Freeman 2008; Purnell and Freeman 2012; Wicks 1996). Donaldson (1999) suggested that there are both ethical and economic reasons to address multiple stakeholder interests. Moreover, Harrison and Freeman (1999) argued that economic and social issues should be viewed jointly rather than separately, and Argandoña (1998) asserted that responsible business behaviour could rely on the objective of creating common good for all stakeholders.

Due to the environmental, social, and ethical challenges businesses face today, scholars also contend that it is necessary to expand management theory and business strategies to achieve more responsible business practices (Hahn et al. 2010; Strand and Freeman 2015). Gibson (2012) advocated promoting sustainability through stakeholder management and collaboration and perceived sustainability as the guiding principle in business. Indeed, scholars have noticed the growing interest in understanding sustainability and social responsibility within business contexts. Shrivastava (1995a, b) defined the main goals of ecocentric management as sustainability, quality of life, and stakeholder welfare. Additionally, Marcus et al. (2010) argued that, because of systemic limits and the existential dependency of business and society on nature, it is necessary to consider the natural environment in business-society relationships. However, only a few articles use sustainability objectives as justification for responsible behaviour. It is even argued that stakeholder theory connects to organisational ethics without intending to answer all moral questions, including those related to sustainability (Phillips et al. 2003). Nevertheless, stakeholder theory emphasises that business and ethics are inseparable, and many researchers have acknowledged the need for expanding the theory to include sustainability issues (e.g., Agle et al. 2008; Freeman 1994).

3.4.2 The Narrative of Defining Value

This narrative is devoted to defining what 'value' means within stakeholder value creation. Griesinger (1990) proposed a subjective viewpoint, arguing that individuals' motivations for participating in cooperative exchange within organisational relationships are for reasons other than economic interests, such as interpersonal resources and the expectation of personal betterment. The subjective nature of value was also supported by Ramirez (1999), who stated that value cannot be defined by a single metric. Furthermore, Harrison and Wicks (2013) defined value as the utility that stakeholders seek from a company, concluding that stakeholders' perceived utility consists of four factors, but these factors do not directly consider social or environmental concerns. Garriga (2014) viewed stakeholder utility and welfare through the lens of stakeholder capability and brought at least environmental concerns to direct attention through the capability of being green.

The special issue on stakeholder accounting in *Journal of Management Studies* published in 2015 further advanced stakeholder value considerations. In this issue, Hall et al. (2015) examined the use of social return on investment as an accounting methodology that allows managers to manage and communicate about the social value created for different stakeholders. Including different stakeholders in organisational decision-making was also considered important (ibid.). Mitchell et al. (2015) introduced a theory of value-creation stakeholder accounting, emphasising the role of stakeholder partnerships in value creation processes, as well as in sharing risks. Furthermore, Brown and Dillard (2015) presented dialogic accountings and related technologies as solutions to go beyond economically and managerially focused accounting practices to offer stakeholders all necessary information concerning organisational impacts related to environmental, political, cultural, economic, and financial value. Finally, it was posited that public accounting professionals should engage in the development of more responsible accounting practices that consider the variety of different stakeholders and provide them with information regarding their risk management needs (Harrison and Van der Laan Smith 2015).

3.4.3 The Narrative of Principles and Mechanisms of How Value Is Created

The most recent narrative of stakeholder value creation addresses the question of how value is created in cooperative relationships. Researchers have studied innovation in multi-stakeholder settings (Rühli et al. 2017), value creation in issue-based stakeholder networks (Schneider and Sachs 2015), collaboration among non-profit stakeholders (Butterfield et al. 2004), interdependencies of public and private interests (Mahoney et al. 2009), cross-sector partnerships (Dentoni et al. 2016; Koschmann et al. 2012), and value creation in public-private ventures (York et al. 2013). Moreover, scholars have studied the sensemaking of ethical complexity in a multi-stakeholder setting (Reinecke and Ansari 2015a); multi-stakeholder learning dialogues (Payne and Calton 2004); value creation at the intersection of markets and developments (Reinecke and Ansari 2015b); cooperation between nation-states, NGOs, and multinationals (Ansari et al. 2013); and multi-stakeholder partnerships in addressing climate change and sustainable development (Pinkse and Kolk 2012).

Recently, many scholars have focused on stakeholder value creation in multi-stakeholder settings that target wicked socioeconomic issues. For example, Rühli et al. (2017) found that the design of social interaction processes, such as participative stakeholder innovation in healthcare, is crucial for mutual value creation. Similarly, Schneider and Sachs (2015) proposed that the process of inductive identity salience supports cooperation and value creation within an issue-based stakeholder network.

Additionally, Dentoni et al. (2016) argued that the dynamic capabilities of sensing, interacting with, learning from, and changing based on stakeholders are beneficial in cross-sector partnerships, as they may help to solve wicked sustainability issues.

Traditionally, researchers have linked successful stakeholder value creation to creating and maintaining mutually trusting and cooperative stakeholder relationships (e.g., Jones and Wicks 1999). Instead of concentrating on trade-offs, value creation involves stakeholders being able to jointly satisfy their needs (Freeman 2010). This line of thinking relies strongly on the principles of stakeholder capitalism, such as stakeholder cooperation, stakeholder engagement, and continuous creation (Freeman et al. 2007), as well as freedom and voluntary action (Freeman and Phillips 2002; Freeman et al. 2004). Harrison et al. (2010) emphasised that managing stakeholder utility functions and allocating more value to legitimate stakeholders than necessary are essential to enhance value creation opportunities. Moreover, Tantalo and Priem (2016) posited that, by concentrating on stakeholder synergy and stakeholders' multi-attribute utility functions, novel and innovative value creation possibilities can be identified, and versatile value can be created for several stakeholders simultaneously. More recently, Mitchell et al. (2016) suggested the reconceptualisation of companies as multi-objective corporations in which managers can consider better social welfare when making decisions.

Various studies on stakeholder value creation emphasised the importance of justice and trust in stakeholder relationships (e.g., Bosse et al. 2009; Harrison et al. 2010; Myllykangas et al. 2011). Bosse et al. (2009) asserted that firms whose stakeholders perceive them as fair create more value based on reciprocal relationships and cooperation. Additionally, Myllykangas et al. (2011) found that trust, along with the potential of stakeholders to learn, the history of the relationships, the objectives of the stakeholders, and the amount of interaction and information sharing in the relationships, influence the dynamics of stakeholder relations and value creation. Researchers have also acknowledged the use of language in shaping stakeholder relationships and their dynamics (Lehtimaki and Kujala 2015; Prasad and Elmes 2005). Brickson (2005, 2007), in contrast, argued that one's orientation toward social value creation and stakeholder relations varies according to one's organisational identity orientation, resulting in an individualistic identity orientation with a motivational basis in organisational self-interest, a relational identity orientation with a motivational basis in another's benefits or a collectivistic identity orientation with a motivational basis in the welfare of a greater collective. The role of firms' internal change agents in enhancing social value creation (Sonenshein 2016) and the importance of incorporating responsibility and the creation of shared value to a firm's everyday operations (Szmigin and Rutherford 2013) have also been emphasised.

Studies connecting value creation directly to sustainability drew attention, for example, to multilevel and multi-systemic organisational relationships (Starik and Rands 1995), the importance of firms addressing the complex nature of climate change through interconnections and collaboration (Slawinski and Bansal 2015), the necessity of stakeholder relations management to tackle economic, social, and environmental stakeholder claims (Steurer et al. 2005), and responsible leadership understood as a relational and moral phenomenon in which leaders create mutually beneficial relationships with their stakeholders, enabling the creation of social capital that contributes to both business and the common good (Maak 2007). Furthermore, Hörisch et al. (2014) argued that companies need to concentrate on developing sustainability as a common value for their stakeholders.

As cooperation around sustainability often meets certain challenges and conflicts while also being beneficial for all parties, paying attention to firm-stakeholder relationships and analysing why and how these relationships change is regarded as important (Friedman and Miles 2002). The studies related to sustainable strategic management and sustainable business models present concrete approaches to sustainability, advancing management theory at both the strategic and operational levels of sustainability. Dyllick and Muff (2015) approached the strategic level of sustainable business by defining a truly sustainable business as a business that 'shifts its perspective from seeking to minimise its negative impacts to understanding how it can create a significant positive impact in critical and relevant areas for society and the planet'. Collaborative strategies, including social and environmental value considerations, are regarded as central for sustainable value creation at the strategic level (e.g., Tencati and Zsolnai 2009; Stead and Stead 2000). Furthermore, Stead and Stead (2013) defined sustainable strategic management activities to consider different needs in the global markets and emphasised the role of business ecosystems in creating social, environmental, and economic value within undeveloped, developing, and developed markets.

Stubbs and Cocklin (2008) wrote the seminal article on sustainable business model conceptualisation, arguing that organisations need to move beyond neoclassical economic assumptions and strive for sustainability and collaboration with key stakeholders by developing their internal structural and cultural capabilities. More recently, Upward and Jones (2015) defined an ontology for strongly sustainable business models that provides guidelines for the development of an entire enterprise aligned with the social and natural sciences. Business model transformations were called for, requiring changes in a firm's value proposition considering all stakeholders, as well as a firm's value creation and capture logics (Abdelkafi and Täuscher 2016; Schaltegger et al. 2016) Finally, it is argued that distinct, random solutions to sustainability are not enough; rather, a fundamental shift is necessary for defining the purpose of a business and its value creation logic, which would hence redefine the business model for sustainability (Gauthier and Gilomen 2016). Purpose-driven urban entrepreneurship aiming to enhance the quality of life of citizens locally (e.g., Cohen and Muñoz 2015) and hybrid organisations drawing attention from profits and growth to social and environmental systems (e.g., Haigh and Hoffman 2014) are examples of truly sustainable and collaborative business models.

On a meta-level, researchers have argued that responsible value creation with and for stakeholders requires changes in the mindsets of organisational and academic actors (e.g., Buchholz and Rosenthal, 2005; Derry 2012). Researchers argue that, by questioning the dominating stakeholder models and changing the language and narratives we use to describe business and stakeholders, it is possible to expand the view of different stakeholders and their needs (Derry 2012; Jensen and Sandström 2013). For example, Hummels (1998) criticised the domination of manager-oriented perspectives, stating that the primary job of managers is to facilitate debates between stakeholders with different positions and interests to obtain satisfying and sustainable outcomes. Buchholz and Rosenthal (2005) proposed that to create nurturing and harmonious relationships with stakeholders, organisations should not see themselves as separate or isolated from their stakeholders. Similarly, Derry (2012) called for scholars to carefully consider research models and questions in the context of sustainability to challenge the traditional firm-centric and manager-oriented perspectives. Finally, Waddock (2011) suggested that the Earth should be the focal entity that all living beings and ecosystems are dependent on, and she perceived them as the stakeholders of the Earth. She concluded that to advance sustainability, the interdependencies between all stakeholders and the Earth should be given much more emphasis and thought in business (ibid.).

4 Discussion

The presented categorisation and the related narratives on stakeholder value creation show that organisational scholars have paid a lot of attention to stakeholder relationships and collaboration and acknowledged the importance of these factors in creating economic, social, and environmental value. They have also examined multistakeholder settings and systems, especially within the sustainability-related studies. However, this study shows that, while researchers have approached stakeholder value creation from many different perspectives, there is no coherent conception of how sustainability issues should be defined and addressed when studying value creation in business contexts. Hence, the narratives of stakeholder value creation studies differ in both depth and breadth, especially in relation to sustainability. Therefore, we suggest that the presented narratives can offer scholars from different organisation research streams a possibility to become aware of the strengths and weaknesses of their own and others' approaches and theories and gain insights from different research streams regarding stakeholder value creation and sustainability.

This chapter indicates, first of all, that scholars can examine and conceive stakeholder value creation through either a strong focal firm orientation or a more stakeholder-orientated approach. The focal firm orientation emphasises the central role of a firm in managing stakeholder relationships to either benefit the firm and its economic commitments or create value for various stakeholders and contribute to social and environmental well-being. Stakeholder orientation, in turn, draws attention to the importance of stakeholder cooperation and relationship dynamics within value creation for either mainly economic reasons or broader value creation purposes. In either case, most of the studies on stakeholder value creation still follow the current economic paradigm and anthropocentric worldview, which do not question the dominating position of traditional economic performance assumptions behind business success and human-centric starting points.

As the presented narratives have illustrated, there is great variance in understanding and defining stakeholder value creation, as well as value itself, in the previous literature. Stakeholder value creation studies vary between narrow and broad conceptualisations of who creates value, what kind of value is created, and with and for whom it is created. For example, researchers regard value creation differently depending on whether the value benefits stakeholders who contribute to firm value creation processes or stakeholders who otherwise affect or are affected by a firm's operations. Financial measures can help define stakeholder value, but the definition of stakeholder value often considers nonfinancial values or even extends further to social, environmental, or sustainable value considerations.

Moreover, our analysis shows that there is no common conceptualisation of sustainability issues or social responsibilities of business within stakeholder value creation studies. The stakeholder approach, in its original form, did not address the complex sustainability issues but instead aimed to illustrate possibilities for effective, responsible management of companies beyond shareholder value maximisation. Hence, sustainability is not a matter of interest in many stakeholder studies. However, it is important to note that most stakeholder value creation studies refer to or incorporate sustainability or social responsibility to some extent. The continuously increasing interest in sustainability and the role of business in responding to sustainability challenges puts organisational scholars in a position where they need to embrace sustainability. Thus, scholars need to decide whether they want to participate in discussion and theory development regarding sustainable business.

If the variance between narratives within stakeholder value creation studies is taken into the context of sustainability (i.e., sustainability is considered as if it mattered) the following questions become crucial: What does "stakeholder value creation" mean in the context of sustainability? What does sustainable "value" mean, and whom does it benefit? How should we perceive and understand sustainability and sustainable value creation? Moreover, researchers should closely consider and explicate which value creation narrative and assumptions they relate and contribute to. As many sustainability-focused value creation studies highlight, the current Western, economic, and firm-centric paradigm serves sustainability-oriented value creation poorly. This problem creates a need to question current stakeholder value creation approaches and identify influences from, for example, studies concentrating on systemic approaches and sustainable business models. Such studies have managed to challenge the current economic paradigm by incorporating sustainability into the core of the business models and value creation, expanding the views of different actors in sustainable value creation, highlighting the importance of collective effort and collaboration, and accepting the complexities and contradictories inherent in sustainability to enhance their management.

Sustainability management may also require researchers to question and redefine the role and purpose of the firm and its dominating position within value creation research. This proposition creates a call for transformational thinking in both business practice and research. As this study shows, researchers do not define sustainability univocally within business studies. To develop management theories for true sustainability, which considers the complex social, environmental, and economic challenges touching both current and future generations locally and globally, we must see the variances in current definitions and strive for more common and shared definitions of the firm and its role in society.

Our study also has some managerial implications. First, the presented categorisation and narratives related to stakeholder value creation reveal the breadth of management theories and approaches that guide and influence business decision-makers in their strategic and operational-level business conduct. Stakeholder value creation studies highlight the possible unproductive effects of the shareholder maximisation objective and the firm-centric approach, and these studies show how to view business strategically from stakeholder perspectives, too. Moreover, most of the stakeholder value creation studies encourage managers to examine the moral value considerations inherent in business decisions and take a stand on what kind of role their business plays within stakeholder networks, in either a narrow or broad sense. Additionally, some studies provide managers with ideas on how to conceive value creation through stakeholder networks instead of traditional value chain thinking. Many studies also offer examples of how to promote and manage stakeholder cooperation within these networks.

Regarding sustainable business practices, our study shows how the realities and objectives of traditional business firms differ from the realities and objectives of those who promote, for example, environmental sustainability. Many studies offer insights and tools for managers to, first of all, examine their own behaviour inside their firms and participate more efficiently in discussions and activities to promote sustainability. At the strategic level, firms and their managers can choose how they want to participate in enhancing sustainability and organise firm operations accordingly. For example, research on sustainable business models has described how companies can transform their business models at both the strategic and operational levels. Furthermore, studies on the dynamics in multi-stakeholder settings give guidance on how to manage and promote cooperation between different actors within blurring organisational boundaries. In sum, firm managers can decide to participate in the long-term development of their business models to correspond with the objectives of sustainable development based on either moral or business reasons, or they may reap all the benefits now and let regulations and other external factors influence the future state of their business.

This study has at least the following limitations. First, as the objective was to conduct a research review, we have scrutinised a broad range of journal articles and showed example studies from the various narratives. This process has both advantages and disadvantages. It brings together a great deal of research around the selected theme, but it does not necessarily further deepen and advance the discussions. However, we have attempted to find the most topical issues from the vast amount of literature and elucidate directions for future research avenues accordingly. Moreover, we admit that there certainly are alternatives to our inductively developed categorisation and the narratives we have identified. Yet, our objective was not to make the categorisation normative or recommend its use in the future but to present the studies and their perspectives within this research theme for readers. Whether we have succeeded in achieving this objective remains up to the reader's justification. Finally, the task of analysing sustainability views in each study turned out to be quite challenging, which confirmed the arguments presented in earlier studies that the definition and understanding of business sustainability are not unequivocal, and therefore, more research needs to be done in this important field.

5 Conclusions

The purpose of this study was to examine distinctive narratives of stakeholder value creation and discuss how they consider sustainability. Based on an extensive research review spanning over three decades of material, we contribute to the stakeholder value creation literature and advancement of social responsibility and sustainability in business by showing how stakeholder theory as a management theory provides us with different approaches to value creation, depending on their orientation towards stakeholders (i.e., focal firm vs. stakeholder orientation), and value (i.e., the economic perspective vs. the multiple value perspective). While the study reveals an increased interest in sustainability issues and their more coherent incorporation into stakeholder research in recent years, the study also extends our knowledge of the existing variance within this research stream and the different narratives a chosen approach generates about stakeholders, value creation, and sustainability. The different approaches used in the research naturally lead to affecting the development of conventional management theory and business practices in different ways. Hence, with our research, we implicitly or explicitly participate in developing business language, which either increases or decreases business involvement in tackling social responsibility and sustainability issues.

As sustainability is the most important aspect affecting our society and planet today, it requires stronger involvement from businesses and organisations and positive solutions instead of trade-offs or minimising strategies. Therefore, we renew the calls to challenge existing stakeholder research to involve sustainability more consistently and reframe the purpose towards addressing sustainability objectives in value creation. We suggest that, with respect to sustainability, future research should consider the dynamic, systemic, and multilevel nature of stakeholder relationships and collaboration. Moreover, a more versatile understanding of value and value creation, as well as a broader understanding of stakeholders and their needs, should be incorporated into future research. To conclude, the conceptualisation of sustainability, the normative core of sustainable business, and elaboration on the purpose and role of business regarding sustainability serve as important and interesting focus areas for future research.

Acknowledgements The authors gratefully acknowledge the financial support from the Academy of Finland (Decision no. 298663).

Appendix

Focal firm orientation and the economic value perspective		
Narratives	Authors	
Stakeholder identification	Adams et al. (2011) Jawahar and McLaughlin (2001)	Perrault (2017) Rowley (1997)
Stakeholder management influencing firm performance	Berman et al. (1999) Choi and Wang (2009) Coombs and Gilley (2005) Dooley and Lerner (1994) Garcia-Castro et al. (2011) Garcia-Castro and Francoeur (2016)	Harrison and Fiet (1999) Harting et al. (2006) Hillman and Keim (2001) Ogden and Watson (1999) Verbeke and Tung (2013)
Dynamics in value creation	Blyler and Coff (2003) Bridoux and Stoelhorst (2014) Crane et al. (2015) Jensen (2002)	Oliver and Holzinger (2008) Priem (2007) Sundaram and Inkpen (2004a, b) Su and Tsang (2015)

Stakeholder orientation and the economic value perspective		
Narratives	Authors	
Trust in cooperation networks	Barringer and Harrison (2000) Connelly et al. (2015) Dhanaraj and Parkhe (2006) Ireland et al. (2002)	Jones (1995) Pirson and Malhotra (2011) Preston and Donaldson (1999) Wicks et al. (1999)
Blurring organisational boundaries	Henisz et al. (2014) Kochan and Rubinstein (2000) Korschun (2015)	McDermott et al. (2009) Rindova and Fombrun (1999) Schneider (2002)
Value creation and capture logics	Bridoux et al. (2011) Coff (2010) Garcia-Castro and Aguilera (2015)	Kivleniece and Quelin (2012) Poulain-Rehm and Lepers (2013) Priem et al. (2013)

Focal firm orientation and the multiple value perspective			
Narratives	Authors		
Challenging traditional CSR approach	Bendheim et al. (1998) Bingham et al. (2011) Brower and Mahajan (2013) Cheng et al. (2014) Clarkson (1995) Crilly et al. (2012)	Crilly et al. (2016) Jamali (2008) Kroeger and Weber (2014) Madsen and Rodgers (2015) Margolis and Walsh (2003) O'Riordan and Fairbrass (2014)	Rowley and Berman (2000) Ruf et al. (2001) Sachs and Maurer (2009) Smith and Rönnegard (2014) Wong et al. (2011)
Stakeholder identification and salience	Agle et al. (1999) Bundy et al. (2013) Crane and Ruebottom (2011) Eesley and Lenox (2006)	Jones et al. (2007) Mitchell et al. (1997) Mitchell et al. (2011) Pache and Santos (2010)	Phillips (2003) Tashman and Raelin (2013) Welcomer et al. (2003) Wang and Dewhirst (1992)
Stakeholder management practices	Briscoe et al. (2014) Crilly (2013) Crilly and Sloan (2012) Crilly and Sloan (2013)	de Luque et al. (2008) Hosseini and Brenner (1992) Kaufman (2002) Laplume et al. (2008)	Minoja (2012) Waldron et al. (2013) Wheeler et al. (2002) Winn (2001)
Expanded view on value creation	Dierksmeier and Pirson (2009) Fauchart and Gruber (2011) Haksever et al. (2004) Husted and de Jesus Salazar (2006)	Jones and Felps (2013a) Jones and Felps (2013b) Lepak et al. (2007) McMullen and Warnick (2015)	Miller et al. (2012) Shankman (1999) Smith et al. (2013) Zahra and Wright (2015)
Environmental and sustainability management	Aguilera-Caracuel and Ortiz-de-Mandojana (2013) Banerjee (2000) Banerjee (2001) Bansal (2005) Bansal and Roth (2000) Buysse and Verbeke (2003) Christmann (2004) Clifton and Amran (2011) Darnall et al. (2010) Driscoll and Starik (2004)	Egri and Herman (2000) Fineman and Clarke (1996) Gallo and Christensen (2011) Gladwin et al. (1995) Hahn et al. (2014) Henriques and Sadorsky (1999) Kassinis and Vafeas (2006) Lin (2014) Murillo-Luna et al. (2008) Perrault and Clark (2015)	Phillips and Reichart (2000) Ramus and Steger (2000) Reid and Toffel (2009) Sarasini and Jacob (2014) Sharma and Henriques (2005) Shepherd et al. (2013) Short and Toffel (2010) Shrivastava (1995a, b) Starik (1995) Zollo et al. (2013)

Stakeholder orientation and the multiple value perspective

Narratives	Authors		
Grounds for responsible business behaviour	Agle et al. (2008) Argandoña (1998) Burton and Dunn (1996) Donaldson (1999) Freeman (1994) Freeman (1999)	Freeman (2000) Gibson (2012) Hahn et al. (2010) Harris and Freeman (2008) Harrison and Freeman (1999) Phillips (1997)	Marcus et al. (2010) Phillips et al. (2003) Purnell and Freeman (2012) Shrivastava (1995a, b) Strand and Freeman (2015) Wicks (1996)
Defining value	Brown and Dillard (2015) Garriga (2014) Griesinger (1990)	Hall et al. (2015) Harrison and Wicks (2013) Harrison and Van der Laan Smith (2015)	Mitchell et al. (2015) Ramirez (1999) Strand et al. (2015)
Principles and mechanisms of how value is created	Abdelkafi and Täuscher (2016) Ansari et al. (2013) Bosse et al. (2009) Brickson (2005) Brickson (2007) Buchholz and Rosenthal (2005) Butterfield et al. (2004) Cohen and Muñoz (2015) Dentoni et al. (2016) Derry (2012) Donaldson and Preston (1995) Dyllick and Muff (2015) Freeman (2010) Freeman et al. (2007) Freeman et al. (2004) Freeman and Phillips (2002) Friedman and Miles (2002) Gauthier and Gilomen (2016)	Haigh and Hoffman (2014) Harrison et al. (2010) Hummels (1998) Hörisch et al. (2014) Jensen and Sandström (2013) Jones and Wicks (1999) Pinkse and Kolk (2012) Koschmann et al. (2012) Lehtimaki and Kujala 2015 Maak (2007) Mahoney et al. (2009) Mitchell et al. (2016) Myllykangas et al. (2011) Payne and Calton (2004) Prasad and Elmes (2005) Reinecke and Ansari (2015a) Reinecke and Ansari (2015b) Rühli et al. (2017)	Sachs et al. (2010) Schaltegger et al. (2016) Schneider and Sachs (2015) Slawinski and Bansal (2015) Sonenshein (2016) Starik and Rands (1995) Stead and Stead (2000) Stead and Stead (2013) Steurer et al. (2005) Stubbs and Cocklin (2008) Szmigin and Rutherford (2013) Tantalo and Priem (2016) Tencati and Zsolnai (2009) Upward and Jones (2015) York et al. (2013) Waddock (2011) Walls and Paquin (2015)

References

Abdelkafi N, Täuscher K (2016) Business models for sustainability from a system dynamics perspective. Organ Environ 29:74–96

Adams RB, Licht AN, Sagiv L (2011) Shareholders and stakeholders: how do directors decide? Strat Manag J 32(12):1331–1355

Agle BR, Mitchell RK, Sonnenfeld JA (1999) Who matters to CEOs? An investigation of stakeholder attributes and salience, corporate performance, and CEO values. Acad Manag J 42(5):507–525

Agle BR, Donaldson T, Freeman RE, Jensen MC, Mitchell RK, Wood DJ (2008) Dialogue: Toward superior stakeholder theory. Bus Ethics Quart 18(02):153–190

Aguilera-Caracuel J, Ortiz-de-Mandojana N (2013) Green innovation and financial performance an institutional approach. Organ Environ 26(4):365–385

Ansari S, Wijen F, Gray B (2013) Constructing a climate change logic: an institutional perspective on the 'tragedy of the commons'. Organ Sci 24(4):1014–1040

Argandoña A (1998) The stakeholder theory and the common good. J Bus Ethics 17(9–10):1093–1102

Banerjee SB (2000) Whose land is it anyway? National interest, indigenous stakeholders, and colonial discourses. The case of the Jabiluka uranium mine. Organ Environ 13(1):3–38

Banerjee SB (2001) Managerial perceptions of corporate environmentalism: interpretations from industry and strategic implications for organizations. J Manag Stud 38(4):489–513
Bansal P (2005) Evolving sustainably: a longitudinal study of corporate sustainable development. Strat Manag J 26(3):197–218
Bansal P, Roth K (2000) Why companies go green: a model of ecological responsiveness. Acad Manag J 43(4):717–736
Barringer BR, Harrison JS (2000) Walking a tightrope: creating value through interorganizational relationships. J Manag 26:367–403
Bendheim CL, Waddock SA, Graves SB (1998) Determining best practice in corporate-stakeholder relations using data envelopment analysis: an industry-level study. Bus Soc 37(3):306–338
Berman SL, Wicks AC, Kotha S, Jones TM (1999) Does stakeholder orientation matter? The relationship between stakeholder management models and firm financial performance. Acad Manag J 42(5):488–506
Bingham JB, Dyer WG Jr, Smith I, Adams GL (2011) A stakeholder identity orientation approach to corporate social performance in family firms. J Bus Ethics 99(4):565–585
Blyler M, Coff RW (2003) Dynamic capabilities, social capital, and rent appropriation: ties that split pies. Strat Manag J 24:677–686
Bosse DA, Phillips RA, Harrison JS (2009) Stakeholders, reciprocity, and firm performance. Strat Manag J 30(4):447–456
Brickson SL (2005) Organizational identity orientation: forging a link between organizational identity and organizations' relations with stakeholders. Adm Sci Q 50(4):576–609
Brickson SL (2007) Organizational identity orientation: the genesis of the role of the firm and distinct forms of social value. Acad Manag Rev 32(3):864–888
Bridoux F, Stoelhorst JW (2014) Microfoundations for stakeholder theory: managing stakeholders with heterogeneous motives. Strat Manag J 35(1):107–125
Bridoux F, Coeurderoy R, Durand R (2011) Heterogeneous motives and the collective creation of value. Acad Manag Rev 36(4):711–730
Briscoe F, Chin MK, Hambrick DC (2014) CEO ideology as an element of the corporate opportunity structure for social activists. Acad Manag J 57(6):1786–1809
Brower J, Mahajan V (2013) Driven to be good: a stakeholder theory perspective on the drivers of corporate social performance. J Bus Ethics 117(2):313–331
Brown J, Dillard J (2015) Dialogic accountings for stakeholders: on opening up and closing down participatory governance. J Manag Stud 52(7):961–985
Buchholz RA, Rosenthal SB (2005) Toward a contemporary conceptual framework for stakeholder theory. J Bus Ethics 58:137–148
Bundy J, Shropshire C, Buchholtz AK (2013) Strategic cognition and issue salience: toward an explanation of firm responsiveness to stakeholder concerns. Acad Manag Rev 38(3):352–376
Burton BK, Dunn CP (1996) Feminist ethics as moral grounding for stakeholder theory. Bus Ethics Q 6(02):133–147
Butterfield KD, Reed R, Lemak DJ (2004) An inductive model of collaboration from the stakeholder's perspective. Bus Soc 43(2):162–195
Buysse K, Verbeke A (2003) Proactive environmental strategies: a stakeholder management perspective. Strat Manag J 24:453–470
Cheng B, Ioannou I, Serafeim G (2014) Corporate social responsibility and access to finance. Strat Manag J 35(1):1–23
Choi J, Wang H (2009) Stakeholder relations and the persistence of corporate financial performance. Strat Manag J 30(8):895–907
Christmann P (2004) Multinational companies and the natural environment: determinants of global environmental policy. Acad Manag J 47(5):747–760
Clarkson ME (1995) A stakeholder framework for analyzing and evaluating corporate social performance. Acad Manag Rev 20(1):92–117
Clifton D, Amran A (2011) The stakeholder approach: a sustainability perspective. J Bus Ethics 98(1):121–136

Coff RW (2010) The coevolution of rent appropriation and capability development. Strat Manag J 31(7):711–733
Cohen B, Muñoz P (2015) Toward a theory of purpose-driven urban entrepreneurship. Organ Environ 28(3):264–285
Connelly BL, Crook TR, Combs JG, Ketchen DJ, Aguinis H (2015) Competence- and integrity-based trust in interorganizational relationships: which matters more? J Manag 44(3):919–945
Coombs JE, Gilley KM (2005) Stakeholder management as a predictor of CEO compensation: main effects and interactions with financial performance. Strat Manag J 26(9):827–840
Crane A, Ruebottom T (2011) Stakeholder theory and social identity: rethinking stakeholder identification. J Bus Ethics 102(1):77–87
Crane A, Graham C, Himick D (2015) Financializing stakeholder claims. J Manag Stud 52(7):878–906
Crilly D (2013) Recasting enterprise strategy: towards stakeholder research that matters to general managers. J Manag Stud 50(8):1427–1447
Crilly D, Sloan P (2012) Enterprise logic: explaining corporate attention to stakeholders from the 'inside-out'. Strat Manag J 33(10):1174–1193
Crilly D, Sloan P (2013) Autonomy or control? Organizational architecture and corporate attention to stakeholders. Organ Sci 25(2):339–355
Crilly D, Zollo M, Hansen MT (2012) Faking it or muddling through? Understanding decoupling in response to stakeholder pressures. Acad Manag J 55(6):1429–1448
Crilly D, Ni N, Jiang Y (2016) Do-no-harm versus do-good social responsibility: attributional thinking and the liability of foreignness. Strat Manag J 37:1316–1329
Darnall N, Henriques I, Sadorsky P (2010) Adopting proactive environmental strategy: the influence of stakeholders and firm size. J Manag Stud 47(6):1072–1094
de Luque MS, Washburn NT, Waldman DA, House RJ (2008) Unrequited profit: how stakeholder and economic values relate to subordinates' perceptions of leadership and firm performance. Adm Sci Q 53(4):626–654
Dentoni D, Bitzer V, Pascucci S (2016) Cross-sector partnerships and the co-creation of dynamic capabilities for stakeholder orientation. J Bus Ethics 135(1):35–53
Derry R (2012) Reclaiming marginalized stakeholders. J Bus Ethics 111:253–264
Dhanaraj C, Parkhe A (2006) Orchestrating innovation networks. Acad Manag Rev 31(3):659–669
Dierksmeier C, Pirson M (2009) Oikonomia versus chrematistike: learning from Aristotle about the future orientation of business management. J Bus Ethics 88(3):417–430
Donaldson T (1999) Making stakeholder theory whole. Acad Manag Rev 24(2):237–241
Donaldson T, Preston LE (1995) The stakeholder theory of the corporation: concepts, evidence, and implications. Acad Manag Rev 20(1):65–91
Dooley RS, Lerner LD (1994) Pollution, profits, and stakeholders: the constraining effect of economic performance on CEO concern with stakeholder expectations. J Bus Ethics 13(9):701–711
Driscoll C, Starik M (2004) The primordial stakeholder: advancing the conceptual consideration of stakeholder status for the natural environment. J Bus Ethics 49(1):55–73
Dyllick T, Muff K (2015) Clarifying the meaning of sustainable business introducing a typology from business-as-usual to true business sustainability. Organ Environ 29(2):156–174
Eesley C, Lenox MJ (2006) Firm responses to secondary stakeholder action. Strat Manag J 27(8):765–781
Egri CP, Herman S (2000) Leadership in the North American environmental sector: values, leadership styles, and contexts of environmental leaders and their organizations. Acad Manag J 43(4):571–604
Fauchart E, Gruber M (2011) Darwinians, communitarians, and missionaries: the role of founder identity in entrepreneurship. Acad Manag J 54(5):935–957
Fineman S, Clarke K (1996) Green stakeholders: industry interpretations and response. J Manag Stud 33(6):715–730
Freeman RE (1984) Strategic management: a stakeholder approach. Pitman, Marshfield, MA

Freeman RE (1994) The politics of stakeholder theory: some future directions. Bus Ethics Q 4(04):409–421
Freeman RE (1999) Divergent stakeholder theory. Acad Manag Rev 24(2):233–236
Freeman RE (2000) Business ethics at the millennium. Bus Ethics Q 10(01):169–180
Freeman RE (2010) Managing for stakeholders: trade-offs or value creation. J Bus Ethics 96:7–9
Freeman RE, Phillips RA (2002) Stakeholder theory: a libertarian defense. Bus Ethics Q 12(03):331–349
Freeman RE, Wicks AC, Parmar B (2004) Stakeholder theory and 'the corporate objective revisited'. Organ Sci 15(3):364–369
Freeman RE, Martin K, Parmar B (2007) Stakeholder capitalism. J Bus Ethics 74(4):303–314
Freeman E, Harrison JE, Wicks A, Parmar B, Colle S (2010) Stakeholder theory: the state of art. Cambridge, Cambridge University Press
Friedman AL, Miles S (2002) Developing stakeholder theory. J Manag Stud 39(1):1–21
Gallo PJ, Christensen LJ (2011) Firm size matters: an empirical investigation of organizational size and ownership on sustainability-related behaviors. Bus Soc 50(2):315–349
Garcia-Castro R, Aguilera RV (2015) Incremental value creation and appropriation in a world with multiple stakeholders. Strat Manag J 36(1):137–147
Garcia-Castro R, Francoeur C (2016) When more is not better: complementarities, costs and contingencies in stakeholder management. Strat Manag J 37(2):406–424
Garcia-Castro R, Ariño MA, Canela MA (2011) Over the long-run? Short-run impact and long-run consequences of stakeholder management. Bus Soc 50(3):428–455
Garriga E (2014) Beyond stakeholder utility function: stakeholder capability in the value creation process. J Bus Ethics 120(4):489–507
Gauthier C, Gilomen B (2016) Business models for sustainability energy efficiency in urban districts. Organ Environ 29:124–144
Gibson K (2012) Stakeholders and sustainability: an evolving theory. J Bus Ethics 109(1):15–25
Gladwin TN, Kennelly JJ, Krause TS (1995) Shifting paradigms for sustainable development: implications for management theory and research. Acad Manag Rev 20(4):874–907
Griesinger DW (1990) The human side of economic organization. Acad Manag Rev 15(3):478–499
Hahn T, Kolk A, Winn M (2010) A new future for business? Rethinking management theory and business strategy. Bus Soc 49(3):385–401
Hahn T, Preuss L, Pinkse J, Figge F (2014) Cognitive frames in corporate sustainability: managerial sensemaking with paradoxical and business case frames. Acad Manag Rev 39(4):463–487
Haigh N, Hoffman AJ (2014) The new heretics hybrid organizations and the challenges they present to corporate sustainability. Organ Environ 27(3):223–241
Haksever C, Chaganti R, Cook RG (2004) A model of value creation: strategic view. J Bus Ethics 49(3):295–307
Hall M, Millo Y, Barman E (2015) Who and what really counts? Stakeholder prioritization and accounting for social value. J Manag Stud 52(7):907–934
Harris JD, Freeman RE (2008) The impossibility of the separation thesis: a response to Joakim Sandberg. Bus Ethics Q 18(04):541–548
Harrison JS, Fiet JO (1999) New CEOs pursue their own self-interests by sacrificing stakeholder value. J Bus Ethics 19(3):301–308
Harrison JS, Freeman RE (1999) Stakeholders, social responsibility, and performance: empirical evidence and theoretical perspectives. Acad Manag J 42(5):479–485
Harrison JS, Van der Laan Smith J (2015) Responsible accounting for stakeholders. J Manag Stud 52(7):935–960
Harrison JS, Wicks AC (2013) Stakeholder theory, value, and firm performance. Bus Ethics Q 23(1):97–124
Harrison JS, Bosse DA, Phillips RA (2010) Managing for stakeholders, stakeholder utility function, and competitive advantage. Strat Manag J 31(1):58–74
Harting TR, Harmeling SS, Venkataraman S (2006) Innovative stakeholder relations: when 'ethics pays' (and when it doesn't). Bus Ethics Q 16(01):43–68

Heikkinen A, Kujala J, Mäkelä H, Nieminen J, Jokinen A, Rekola H (2018) Urban ecosystem services and stakeholders: towards a sustainable capability approach. In: Bonnedahl KJ, Heikkurinen P (eds) Strongly sustainable societies. Organising human activities on a hot and full Earth. Routledge, New York, NY, pp 116–133

Heikkurinen P (ed) (2017) Sustainability and peaceful coexistence for the Anthropocene. Routledge, New York, NY

Henisz WJ, Dorobantu S, Nartey LJ (2014) Spinning gold: the financial returns to stakeholder engagement. Strat Manag J 35(12):1727–1748

Henriques I, Sadorsky P (1999) The relationship between environmental commitment and managerial perceptions of stakeholder importance. Acad Manag J 42(1):87–99

Hillman AJ, Keim GD (2001) Shareholder value, stakeholder management, and social issues: what's the bottom line? Strat Manag J 22(2):125–139

Hörisch J, Freeman RE, Schaltegger S (2014) Applying stakeholder theory in sustainability management: links, similarities, dissimilarities, and a conceptual framework. Organ Environ 27(4):328–346

Hosseini JC, Brenner SN (1992) The stakeholder theory of the firm: a methodology to generate value matrix weights. Bus Ethics Q 2(02):99–119

Hummels H (1998) Organizing ethics: a stakeholder debate. J Bus Ethics 17(13):1403–1419

Husted BW, de Jesus Salazar J (2006) Taking Friedman seriously: maximizing profits and social performance. J Manag Stud 43(1):75–91

Ireland RD, Hitt MA, Vaidyanath D (2002) Alliance management as a source of competitive advantage. J Manag 28(3):413–446

Jamali D (2008) A stakeholder approach to corporate social responsibility: a fresh perspective into theory and practice. J Bus Ethics 82(1):213–231

Jawahar M, McLaughlin G (2001) Toward a descriptive stakeholder theory: an organizational life cycle approach. Acad Manag Rev 26(3):397–414

Jensen MC (2002) Value maximization, stakeholder theory, and the corporate objective function. Bus Ethics Q 12(2):235–256

Jensen T, Sandström J (2013) In defence of stakeholder pragmatism. J Bus Ethics 114(2):225–237

Jones TM (1995) Instrumental stakeholder theory: a synthesis of ethics and economics. Acad Manag Rev 20(2):404–437

Jones TM, Felps W (2013a) Shareholder wealth maximization and social welfare: a utilitarian critique. Bus Ethics Q 23:207–238

Jones TM, Felps W (2013b) Stakeholder happiness enhancement: a neo-utilitarian objective for the modern corporation. Bus Ethics Q 23(03):349–379

Jones TM, Wicks AC (1999) Convergent stakeholder theory. Acad Manag Rev 24(2):206–221

Jones TM, Felps W, Bigley GA (2007) Ethical theory and stakeholder-related decisions: the role of stakeholder culture. Acad Manag Rev 32(1):137–155

Kassinis G, Vafeas N (2006) Stakeholder pressures and environmental performance. Acad Manag J 49(1):145–159

Kaufman A (2002) Managers' double fiduciary duty: to stakeholders and to freedom. Bus Ethics Q 12(02):189–214

Kivleniece I, Quelin BV (2012) Creating and capturing value in public-private ties: a private actor's perspective. Acad Manag Rev 37(2):272–299

Kochan TA, Rubinstein SA (2000) Toward a stakeholder theory of the firm: the Saturn partnership. Organ Sci 11(4):367–386

Korschun D (2015) Boundary-spanning employees and relationships with external stakeholders: a social identity approach. Acad Manag Rev 40(4):611–629

Koschmann MA, Kuhn TR, Pfarrer MD (2012) A communicative framework of value in cross-sector partnerships. Acad Manag Rev 37(3):332–354

Krippendorff K (2004) Content analysis: an introduction to its methodology. Sage, Thousand Oaks, CA

Kroeger A, Weber C (2014) Developing a conceptual framework for comparing social value creation. Acad Manag Rev 39(4):513–540
Laplume AO, Sonpar K, Litz RA (2008) Stakeholder theory: reviewing a theory that moves us. J Manag 34(6):1152–1189
Lehtimaki H, Kujala J (2015) Framing dynamically changing firm–stakeholder relationships in an international dispute over a foreign investment. A discursive analysis approach. Bus Soc 56(3):487–523
Lepak DP, Smith KG, Taylor MS (2007) Value creation and value capture: a multilevel perspective. Acad Manag Rev 32(1):180–194
Lin H (2014) Government–business partnership formation for environmental improvements. Organ Environ 27(4):383–398
Maak T (2007) Responsible leadership, stakeholder engagement, and the emergence of social capital. J Bus Ethics 74(4):329–343
Madsen PM, Rodgers ZJ (2015) Looking good by doing good: the antecedents and consequences of stakeholder attention to corporate disaster relief. Strat Manag J 36(5):776–794
Mahoney JT, McGahan AM, Pitelis CN (2009) Perspective—the interdependence of private and public interests. Organ Sci 20(6):1034–1052
Marcus J, Kurucz EC, Colbert BA (2010) Conceptions of the business–society–nature interface: implications for management scholarship. Bus Soc 49(3):402–438
Margolis JD, Walsh JP (2003) Misery loves companies: rethinking social initiatives by business. Adm Sci Q 48:268–305
McDermott GA, Corredoira RA, Kruse G (2009) Public-private institutions as catalysts of upgrading in emerging market societies. Acad Manag J 52(6):1270–1296
McMullen JS, Warnick BJ (2015) Should we require every new venture to be a hybrid organization? J Manag Stud 53(4):630–662
Miller TL, Grimes MG, McMullen JS, Vogus TJ (2012) Venturing for others with heart and head: how compassion encourages social entrepreneurship. Acad Manag Rev 37(4):616–640
Minoja M (2012) Stakeholder management theory, firm strategy, and ambidexterity. J Bus Ethics 109(1):67–82
Mitchell RK, Agle BR, Wood DJ (1997) Toward a theory of stakeholder identification and salience: defining the principle of who and what really counts. Acad Manag Rev 22(4):853–886
Mitchell RK, Agle BR, Chrisman JJ, Spence LJ (2011) Toward a theory of stakeholder salience in family firms. Bus Ethics Q 21(02):235–255
Mitchell RK, Van Buren HJ, Greenwood M, Freeman RE (2015) Stakeholder inclusion and accounting for stakeholders. J Manag Stud 52(7):851–877
Mitchell RK, Weaver GR, Agle BR, Bailey AD, Carlson J (2016) Stakeholder agency and social welfare: pluralism and decision making in the multi-objective corporation. Acad Manag Rev 41(2):252–275
Montiel I, Delgado-Ceballos J (2014) Defining and measuring corporate sustainability: are we there yet? Organ Environ 27:113–139
Murillo-Luna JL, Garcés-Ayerbe C, Rivera-Torres P (2008) Why do patterns of environmental response differ? A stakeholders' pressure approach. Strat Manag J 29(11):1225–1240
Myllykangas P, Kujala J, Lehtimäki H (2011) Analyzing the essence of stakeholder relationships: what do we need in addition to power, legitimacy, and urgency? J Bus Ethics 96:65–72
O'Riordan L, Fairbrass J (2014) Managing CSR stakeholder engagement: a new conceptual framework. J Bus Ethics 125(1):121–145
Ogden S, Watson R (1999) Corporate performance and stakeholder management: balancing shareholder and customer interests in the UK privatized water industry. Acad Manag J 42(5):526–538
Oliver C, Holzinger I (2008) The effectiveness of strategic political management: a dynamic capabilities framework. Acad Manag Rev 33(2):496–520
Pache AC, Santos F (2010) When worlds collide: the internal dynamics of organizational responses to conflicting institutional demands. Acad Manag Rev 35(3):455–476

Payne SL, Calton JM (2004) Exploring research potentials and applications for multi-stakeholder learning dialogues. J Bus Ethics 55(1):71–78
Perrault EA (2017) 'Names-and-faces approach' to stakeholder identification and salience: a matter of status. J Bus Ethics 146(1):25–38
Perrault E, Clark C (2015) Environmental shareholder activism considering status and reputation in firm responsiveness. Organ Environ 29(2):194–211
Phillips RA (1997) Stakeholder theory and a principle of fairness. Bus Ethics Q 7(01):51–66
Phillips R (2003) Stakeholder legitimacy. Bus Ethics Q 13(01):25–41
Phillips RA, Reichart J (2000) The environment as a stakeholder? A fairness-based approach. J Bus Ethics 23(2):185–197
Phillips R, Freeman RE, Wicks AC (2003) What stakeholder theory is not. Bus Ethics Q 13(04):479–502
Pinkse J, Kolk A (2012) Addressing the climate change sustainable development nexus: the role of multistakeholder partnerships. Bus Soc 51(1):176–210
Pirson M, Malhotra D (2011) Foundations of organizational trust: what matters to different stakeholders? Organ Sci 22(4):1087–1104
Poulain-Rehm T, Lepers X (2013) Does employee ownership benefit value creation? The case of France (2001–2005). J Bus Ethics 112(2):325–340
Prasad P, Elmes M (2005) In the name of the practical: unearthing the hegemony of pragmatics in the discourse of environmental management. J Manag Stud 42(4):845–867
Preston LE, Donaldson T (1999) Dialogue. Acad Manag Rev 24(4):619–620
Priem RL (2007) A consumer perspective on value creation. Acad Manag Rev 32(1):219–235
Priem RL, Butler JE, Li S (2013) Toward reimagining strategy research: retrospection and prospection on the 2011 AMR decade award article. Acad Manag Rev 38(4):471–489
Purnell LS, Freeman RE (2012) Stakeholder theory, fact/value dichotomy, and the normative core: how Wall Street stops the ethics conversation. J Bus Ethics 109(1):109–116
Ramirez R (1999) Value co-production: intellectual origins and implications for practice and research. Strat Manag J 20(1):49–65
Ramus CA, Steger U (2000) The roles of supervisory support behaviors and environmental policy in employee 'ecoinitiatives' at leading-edge European companies. Acad Manag J 43(4):605–626
Reid EM, Toffel MW (2009) Responding to public and private politics: corporate disclosure of climate change strategies. Strat Manag J 30(11):1157–1178
Reinecke J, Ansari S (2015a) What is a 'fair' price? Ethics as sensemaking. Organ Sci 26(3):867–888
Reinecke J, Ansari S (2015b) When times collide: temporal brokerage at the intersection of markets and developments. Acad Manag J 58(2):618–648
Rindova VP, Fombrun CJ (1999) Constructing competitive advantage: the role of firm-constituent interactions. Strat Manag J 20(8):691–710
Rowley TJ (1997) Moving beyond dyadic ties: a network theory of stakeholder influences. Acad Manag Rev 22(4):887–910
Rowley T, Berman S (2000) A brand new brand of corporate social performance. Bus Soc 39(4):397–418
Ruf BM, Muralidhar K, Brown RM, Janney JJ, Paul K (2001) An empirical investigation of the relationship between change in corporate social performance and financial performance: a stakeholder theory perspective. J Bus Ethics 32(2):143–156
Rühli E, Sachs S, Schmitt R, Schneider T (2017) Innovation in multistakeholder settings: the case of a wicked issue in health care. J Bus Ethics 143(2):289–305
Sachs S, Maurer M (2009) Toward dynamic corporate stakeholder responsibility. J Bus Ethics 85(3):535–544
Sarasini S, Jacob M (2014) Past, present, or future? Managers' temporal orientations and corporate climate action in the Swedish electricity sector. Organ Environ 27(3):242–262
Schaltegger S, Hansen EG, Lüdeke-Freund F (2016) Business models for sustainability origins, present research, and future avenues. Organ Environ 29(1):3–10
Schneider M (2002) A stakeholder model of organizational leadership. Organ Sci 13(2):209–220

Schneider T, Sachs S (2015) The impact of stakeholder identities on value creation in issue-based stakeholder networks. J Bus Ethics 144(1):41–57
Shankman NA (1999) Reframing the debate between agency and stakeholder theories of the firm. J Bus Ethics 19(4):319–334
Sharma S, Henriques I (2005) Stakeholder influences on sustainability practices in the Canadian forest products industry. Strat Manag J 26(2):159–180
Shepherd DA, Patzelt H, Baron RA (2013) 'I care about nature, but…': disengaging values in assessing opportunities that cause harm. Acad Manag J 56(5):1251–1273
Short J (2009) The art of writing a review article. J Manag 35(6):1312–1317
Short JL, Toffel MW (2010) Making self-regulation more than merely symbolic: the critical role of the legal environment. Adm Sci Q 55(3):361–396
Shrivastava P (1995a) Ecocentric management for a risk society. Acad Manag Rev 20(1):118–137
Shrivastava P (1995b) The role of corporations in achieving ecological sustainability. Acad Manag Rev 20(4):936–960
Slawinski N, Bansal P (2015) Short on time: intertemporal tensions in business sustainability. Organ Sci 26(2):531–549
Smith NC, Rönnegard D (2016) Shareholder primacy, corporate social responsibility, and the role of business schools. J Bus Ethics 134(3):463–478
Smith WK, Gonin M, Besharov ML (2013) Managing social-business tensions: a review and research agenda for social enterprise. Bus Ethics Q 23(03):407–442
Sonenshein S (2016) How corporations overcome issue illegitimacy and issue equivocality to address social welfare: the role of the social change agent. Acad Manag Rev 41(2):349–366
Starik M (1995) Should trees have managerial standing? Toward stakeholder status for non-human nature. J Bus Ethics 14(3):207–217
Starik M, Kanashiro P (2013) Toward a theory of sustainability management: uncovering and integrating the nearly obvious. Organ Environ 26(1):7–30
Starik M, Rands GP (1995) Weaving an integrated web: Multilevel and multisystem perspectives of ecologically sustainable organizations. Acad Manag Rev 20(4):908–935
Stead JG, Stead E (2000) Eco-enterprise strategy: standing for sustainability. J Bus Ethics 24(4):313–329
Stead JG, Stead WE (2013) The coevolution of sustainable strategic management in the global marketplace. Organ Environ 26(2):162–183
Steurer R, Langer ME, Konrad A, Martinuzzi A (2005) Corporations, stakeholders and sustainable development I: a theoretical exploration of business–society relations. J Bus Ethics 61(3):263–281
Strand R, Freeman RE (2015) Scandinavian cooperative advantage: the theory and practice of stakeholder engagement in Scandinavia. J Bus Ethics 127(1):65–85
Strand R, Freeman RE, Hockerts K (2015) Corporate social responsibility and sustainability in Scandinavia: an overview. J Bus Ethics 127(1):1–15
Stubbs W, Cocklin C (2008) Conceptualizing a 'sustainability business model'. Organ Environ 21(2):103–127
Su W, Tsang EW (2015) Product diversification and financial performance: the moderating role of secondary stakeholders. Acad Manag J 58(4):1128–1148
Sundaram AK, Inkpen AC (2004a) The corporate objective revisited. Organ Sci 15(3):350–363
Sundaram AK, Inkpen AC (2004b) Stakeholder theory and 'The corporate objective revisited': a reply. Organ Sci 15(3):370–371
Szmigin I, Rutherford R (2013) Shared value and impartial spectator test. J Bus Ethics 114:171–182
Tantalo C, Priem RL (2016) Value creation through stakeholder synergy. Strat Manag J 37(2):314–329
Tashman P, Raelin J (2013) Who and what really matters to the firm: moving stakeholder salience beyond managerial perceptions. Bus Ethics Q 23(04):591–616
Tencati A, Zsolnai L (2009) The collaborative enterprise. J Bus Ethics 85(3):367–376
United Nations World Commission on Environment and Development (1987) Our common future. Oxford University Press, Oxford, England

Upward A, Jones P (2015) An ontology for strongly sustainable business models defining an enterprise framework compatible with natural and social science. Organ Environ 29(1):97–123
Verbeke A, Tung V (2013) The future of stakeholder management theory: a temporal perspective. J Bus Ethics 112(3):529–543
Waddock S (2011) We are all stakeholders of Gaia: a normative perspective on stakeholder thinking. Organ Environ 24(2):192–212
Waldron TL, Navis C, Fisher G (2013) Explaining differences in firms' responses to activism. Acad Manag Rev 38(3):397–417
Walls JL, Paquin RL (2015) Organizational perspectives of industrial symbiosis: a review and synthesis. Organ Environ 28:32–53
Wang J, Dewhirst HD (1992) Boards of directors and stakeholder orientation. J Bus Ethics 11(2):115–123
Wang H, Choi J, Li J (2008) Too little or too much? Untangling the relationship between corporate philanthropy and firm financial performance. Organ Sci 19(1):143–159
Weber RP (1990) Basic content analysis. Sage, Thousand Oaks, CA
Welcomer SA, Cochran PL, Rands G, Haggerty M (2003) Constructing a Web effects of power and social responsiveness on firm-stakeholder relationships. Bus Soc 42(1):43–82
Wheeler D, Fabig H, Boele R (2002) Paradoxes and dilemmas for stakeholder responsive firms in the extractive sector: lessons from the case of Shell and the Ogoni. J Bus Ethics 39(3):297–318
Wicks AC (1996) Overcoming the separation thesis. The need for a reconsideration of business and society research. Bus Soc 35(1):89–118
Wicks AC, Berman SL, Jones TM (1999) The structure of optimal trust: moral and strategic implications. Acad Manag Rev 24(1):99–116
Winn MI (2001) Building stakeholder theory with a decision modeling methodology. Bus Soc 40(2):133–166
Wong EM, Ormiston ME, Tetlock PE (2011) The effects of top management team integrative complexity and decentralised decision making on corporate social performance. Acad Manag J 54(6):1207–1228
York JG, Sarasvathy SD, Wicks AC (2013) An entrepreneurial perspective on value creation in public-private ventures. Acad Manag Rev 38(2):307–309
Zahra SA, Wright M (2015) Understanding the social role of entrepreneurship. J Manag Stud 53(4):610–629
Zollo M, Cennamo C, Neumann K (2013) Beyond what and why understanding organisational evolution towards sustainable enterprise models. Organ Environ 26(3):241–259

Innovative Approaches to Organisational Sustainability: State-of-the-Art and Conceptual Framework

Patrícia Tourais and Nuno Videira

Abstract Organisational Sustainability (OS) has been one of the key research topics in sustainability debates. Organisational approaches that support pathways towards a more sustainable society, such as environmental management and social responsibility instruments, have been widely adopted. However, there are critical gaps in understanding how companies may design strategic pathways, assess alternatives and implement sustainability transitions. Despite multiple theories and methods on how organisations integrate sustainability into their strategies and operations, their efforts to improve sustainability performance still lack effective and consistent results. The main goal of this study is thus to identify key research challenges related with organisational sustainability and develop a conceptual framework to support the implementation of sustainability transitions in organisations. Departing from an in-depth literature review, the proposed conceptual framework supports analysis and implementation of organisational pathways, bringing insights on the key challenges and enablers of innovative long-term solutions. Preliminary results from such analysis point out that organisational sustainability transitions ought to be built in close collaboration with stakeholders, in order to achieve a shared vision of sustainability for the organisation. The pathways towards this vision should then consider the dynamics of the organisational external context and internal factors, such as organisational capabilities, innovation and participatory and systemic assessments of the organisational performance.

Keywords Organisational sustainability · Corporate sustainability · Meta-review Conceptual framework

P. Tourais (✉) · N. Videira
CENSE - Center for Environmental and Sustainability Research, NOVA School of Science and Techonology, NOVA University Lisbon, Campus de Caparica, 2829-516 Caparica, Portugal
e-mail: ptf@fct.unl.pt

N. Videira
e-mail: nmvc@fct.unl.pt

W. Leal Filho (ed.), *Social Responsibility and Sustainability*, World Sustainability Series, https://doi.org/10.1007/978-3-030-03562-4_2

1 Introduction

The emergence of the sustainability paradigm brought voluntary sustainability tools to a higher level of importance to companies and become essential to assure the success of most companies. Organisational Sustainability (OS) arose as a globalised trend among organisations, influencing their competitiveness and survival regardless of the part of the globe where the company operates (Brones et al. 2017).

Despite their popularity and the increasing number of actors involved in the transition towards sustainability in organisations, the effectiveness of the transition is questioned and widely debated in literature due to lack of environmental and social results and a persistence or worsening of the problems (Lankoski 2016). The implementation of sustainability voluntary tools has proved to improve organisations' performance in some cases; however, some of the outcomes are intangible or associated with other organisational areas rather than environmental and social performance (Daddi et al. 2011; Iraldo et al. 2009). More sceptical authors argue organisations use of sustainability voluntary tools promotes greenwashing by allowing the achievement of competitive advantage through an environmental-friendly reputation, without effectively implementing sustainability practices (Clapp 1998, 2005).

Summarising, there is a gap between organisations' intentions and their actions and performance in relation to sustainability (Porter 2008). Further research on organisational sustainability, focusing on how to perform a transition towards sustainability, would be useful to close this gap. Organisations have been focusing on stakeholder involvement, sustainability management through voluntary-based tools and different approaches to each dimension of sustainability (Engert and Baumgartner 2016; Garcia et al. 2016; Zollo et al. 2016). Providing insights on how to adopt more dynamic, holistic and inclusive approaches to sustainability, would allow organisations to close the gap between their intentions and their actions and performance.

The current study aims to provide insights on the main key challenges emerging from literature and regarding the transition of organisations towards sustainability, through the development of a conceptual framework. The following section details the methods applied in the current study. The succeeding two sections refer to the results of the meta-review, firstly by focusing on OS from a conceptual perspective to identify both the theoretical landscape and the key elements of OS definition; secondly describing the emergent issues referring to the integration of sustainability in the organisational structure and functions. The final section reflects results from the meta-review summarised in a conceptual framework, informing on the key emerging challenges in OS research.

2 Methods

A priori issue on the literature about sustainability in organisations is the diversity of concepts applied in literature. An exploratory search revealed the use of diverse definitions, such as OS, Corporate Social Responsibility (CSR) and Corporate Sustainability (CS). Either these concepts are interchangeable or not is a controversial issue that is recommendable to define in each study (Chang et al. 2017; Montiel 2008). Since the main goal of the present study was set on the general idea of sustainability in organisations (either companies or other types of organisations, such as public institutions), CS and CSR are considered as similar concepts, despite their differences are acknowledged. CS is an evolution of the CSR concept, as presented by Chang et al. (2017). OS is a more recent concept, however used by researchers as an equivalent of CS, as in Batista and de Francisco (2018).

The identification of the key emerging challenges in OS research were identified through a meta-review of the literature. The search for scientific articles was limited to literature reviews, considering the amount of studies developed either from a thematic or methodological perspective and the need for a holistic perspective on OS.

Two sets of keywords were applied in the search for scientific articles on the Scopus database: (1) "organisational sustainability" and "review"; (2) "corporate sustainability" and "review". The search was limited to articles' title, abstract and keywords and to articles classified as reviews. Publication date was another limitation criteria, since were only considered literature reviews published between 1987 (since the CS concept emerged from the Brundtland report that was published in 1987) and February 2018.

The results of the Scopus searches are displayed in Fig. 1. A screening exercise on the title and abstract of the articles resulting from the search was performed, aimed

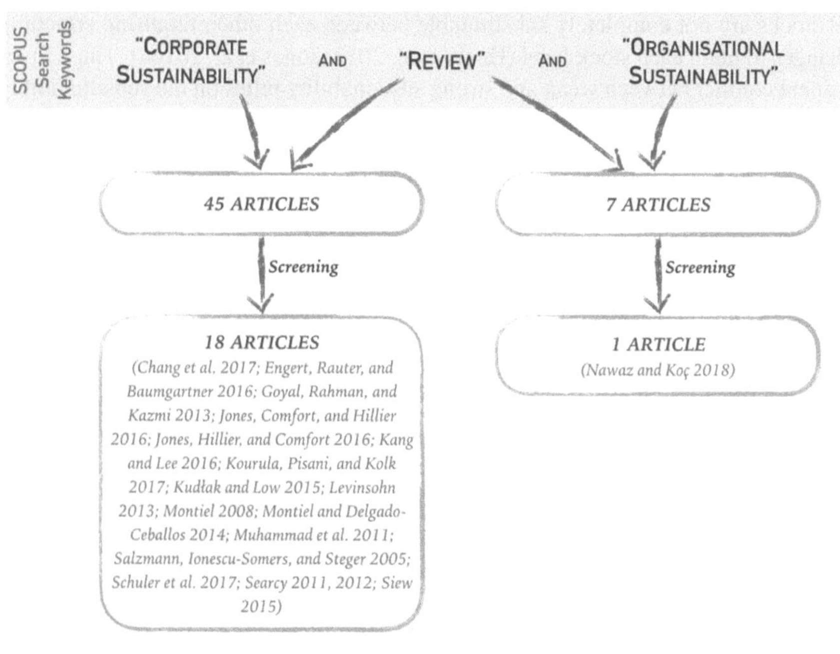

Fig. 1 Selection of scientific articles included in the meta-review, including the search and selection method

to identify those effectively focused on OS and conducting a systematic literature review.

The final set of 19 articles was analysed and provided the foundation for elaborating the conceptual framework and identifying the key emerging challenges in OS research presented below.

3 Organisational Sustainability

Sustainability is an evolving concept that has been defined by multiple authors in diverse variations of the most popular definition published in the Brundtland report in 1987. This definition relies on two basic assumptions: the idea of needs, focusing specially on those that live in poorer conditions; and the limits of the planet to meet both the present and future needs, taking in consideration technology and the society demand (Schuler et al. 2017). Simplifying, sustainability requires the modification of the production and consumption patterns, the end of poverty and the preservation of ecosystems, in the long-term (Chang et al. 2017).

Sustainability requires the maintenance of current capital stocks, that can be disaggregated into four main types: manufactured, human, social and natural. Based on this definition, authors distinguish between "weak" and "strong" sustainability. The former requires the maintenance of the total capital stock, regardless the partition between each type, representing incremental change. The latter implies that capital stocks are not completely substitutable between each other, requiring structural changes to main each stock level (Healy et al. 2013; Jones et al. 2016a). Thus, a permanent conflict between weak and strong sustainability relies on the substitutability between different capital stocks, implying different levels of change in society (Healy et al. 2013; Levinsohn 2013).

Sustainability is often characterised as vague and difficult to operationalise; however, organisations rely on this concept to create their own vision of sustainability. Due to the growing importance of sustainability in the organisational and international realms, multiple definitions of sustainability in organisations were developed recurring to diverse research fields and theories (Chang et al. 2017; Siew 2015).

3.1 Theoretical Landscape: Research Fields and Theories

The multidisciplinary approach to sustainability is supported in literature by the contribute of diverse research fields and theories to the development of knowledge on OS. Figure 2 summarises the main research fields and theories found in literature which contribute to the extensive debate on OS.

Apart from specialised journals and studies, sustainability in organisations is a recurring theme in management and business literature. The structural changes needed in organisations to promote an effective transition towards sustainability are

Fig. 2 Theoretical landscape on organisational sustainability

studied at different levels of organisational management: strategic, tactical and operational. The implications of sustainability integration at a strategic level have been addressed in a specific research field: strategic management (Engert et al. 2016).

Strategic management is focused on the relation between the organisational structure and strategy, implying that each company defines a unique position, differentiating itself from competitors by reducing costs or better service (Engert et al. 2016). External and internal factors (such as stakeholders) must be included in strategic management to guarantee long-term effectiveness (Engert et al. 2016; Muhammad et al. 2011).

Including sustainability as part of strategy formulation is a complex task, demanding multi-disciplinary knowledge, management and organisational learning (Engert et al. 2016). Still at this level, Kang and Lee (2016) developed a study focused on Corporate Environmental Strategy (CES) that is defined as the integration of environmental activities in the strategic planning process. The relation between CES and organisational performance is explored, taking into consideration relevant variables, such as stakeholder pressure, environmental standards and environmental technology.

The tactical level requires that sustainability is integrated in the designing of organisational structures, policies and processes, while considering stakeholders needs (identified at a strategic level). The development of a system allowing the evaluation of sustainability integration in structures, policies and processes should happen at this level, since the operational level embraces the implementation of the tasks defined at the tactical level, as well as the evaluation of processes and systems in place (Muhammad et al. 2011).

Thus, OS requires a meta-management approach, assuring the organisation is seen as a whole. Vertical and horizontal integration throughout the organisation support this approach. Vertical integration by guaranteeing that the main elements of the organisational strategy are integrated into the organisational goals, targets and process and consequently, the strategy is operationalised. Horizontal integration requires the connection of processes, chains, departments and positions within the organisation, including structures and competencies (Muhammad et al. 2011).

Business ethics also contributes to OS research through the enlightenment on the values behind OS and the implications of employees' ethics in the transition towards OS. Schuler et al. (2017) explores the implications of environmental ethics in the achievement of OS, balancing three main sustainability concepts [CSR, Environmental Management (EM) and Corporate Political Activity], with four environmental ethics orientations: sustainable resource use, conservation and preservation, rights-based perspectives, and deep ecology (progressive categories from an instrumental value to an intrinsic value perspective). Powell (2011) explores the role of employees ethics in the creation of an organisational identity and culture prone to sustainability.

Stakeholders and corporate identity are key elements in marketing research, since the creation of an organisational brand or image is essential to stablish a uniform organisational identity. Stakeholder identification is common to marketing and OS, however from the marketing perspective customers and channel members are the key stakeholders (Powell 2011).

Sustainability entrepreneurship refers to businesses adopting innovations with environmental or social benefits, often driven by values or causes linked to the entrepreneur. Businesses built onto sustainability values from the initial stages are distinguished from those establishing an environment business branch due to opportunity discovery, providing different insights on the needs to transition to OS (Levinsohn 2013).

The bias in OS studies towards the environment is also relevant in sustainability entrepreneurship research (Levinsohn 2013), reinforcing the importance of EM research. Schuler et al. (2017) points EM as one of the organisational areas of CS management, emphasising two major lines of research: the relationship between environmental and financial performance; and the ways to minimise environmental impacts. The former focus on efficiency improvements through lean manufacturing and waste reduction, innovation, risk reduction and dealing with stakeholders needs; while the latter studies the effects of regulation, self-regulation (e.g., ISO 14001 standard) and environmental ratings and rankings. An alternative line of research explores the time dimension in environmental changes (short vs. long term).

Organisational and other social systems change towards sustainability is the main focus of Transitions Management research, by interlinking systems and governance theories. Business strategies, innovation and the governance of processes assume a key role in the transition towards sustainability (Chang et al. 2017; Levinsohn 2013). A popular approach in the field of research is the Multi-Level Perspective developed by Geels and Schot (2007), by providing typologies of transition of the socio technical system. The transition results from a non-linear process resulting from the interconnections between three levels: niches, socio-technical regimes, and a socio-technical landscape (Chang et al. 2017).

As a complex and holistic concept, OS draws from diverse theories the needed foundation to conceptualise its main elements (Montiel and Delgado-Ceballos 2014; Muhammad et al. 2011). Stakeholder theory, institutional theory and Resource-Based View (RBV) are traditionally associated to OS research (Montiel and Delgado-Ceballos 2014); however other theories are mentioned in literature, as co-evolution, contingency, organisational, motivation, internationalisation, control and market, lean manufacturing, multi-level perspective, environmentalism and sustaincentrism (Chang et al. 2017; Kang and Lee 2016; Montiel 2008; Montiel and Delgado-Ceballos 2014; Schuler et al. 2017; Searcy 2011). As an example, co-evolution theory classifies companies and environments as co-evolving systems in the transition to OS (Chang et al. 2017). The next sub-sections focus on the contributions of traditional organisational theories to OS research.

3.1.1 Stakeholder Theory

Stakeholder theory was developed by Freeman in 1984 and is fundamental in OS research (Jones et al. 2016b; Montiel 2008; Searcy 2012). Its main argument reinforces stakeholders importance at a strategic level due to their role in the organisation's survival (Chang et al. 2017; Freeman 1984; Montiel and Delgado-Ceballos 2014). All types of stakeholders should be considered: the traditional (e.g. suppliers, customers and employees) or others as governments, non-governmental organisations (NGOs) and local communities (Chang et al. 2017). This approach diverges from the traditional shareholder supremacy, accountable for the exploitation of market failures in order to achieve short-term profitability, despite social and environmental impacts (Kudłak and Low 2015).

Primary and secondary stakeholders have different levels of influence in organisations, and therefore different strategies may be required to each of them (Chang et al. 2017; Muhammad et al. 2011; Searcy 2011). As an example, stakeholders' expectations on an effective transition to OS might motivate the inclusion of sustainability in the organisational strategy (Engert et al. 2016; Kang and Lee 2016).

Stakeholders legitimacy, urgency and power are dynamic variables influencing the priority level of sustainability issues (Muhammad et al. 2011). Monitoring this variation in time inform the sustainability performance measurement system (Searcy 2011).

Effective communication with stakeholders promotes vertical and horizontal integration of OS, through the interaction among different stakeholders, the development of competences and knowledge and the establishment of continuous improvement (Muhammad et al. 2011). Stakeholder engagement in the definition of the organisational strategy supports the achievement of long-term goals and the creation of shared value through sponsorships, strategic partnerships and research cooperation (Engert et al. 2016).

Despite the recognised benefits, stakeholders' identification and engagement are costly and time demanding tasks, not always fully supported by top management with human and financial resources. These limitations can compromise the process by the application of stakeholders' involvement formats that are not aligned with the vision, strategy and core competences of the organisation (Engert et al. 2016).

3.1.2 Triple Bottom Line

The Triple Bottom Line (TBL) is often pointed as the operationalisation of OS since it requires the evaluation of the economic, social and environmental performance in the organisation (Chang et al. 2017; Engert et al. 2016; Muhammad et al. 2011; Siew 2015). The balance of these three dimensions leads to the achievement of sustainability and social responsibility in the long-term (Goyal et al. 2013; Montiel 2008). Shell developed the 3 Ps terminology (profit, planet and people), often associated to TBL in literature (Levinsohn 2013; Montiel and Delgado-Ceballos 2014).

TBL is criticised because it promotes a utilitarian and anthropocentric implementation of sustainability, allowing organisations to maintain a profit oriented format (Chang et al. 2017; Schuler et al. 2017).

3.1.3 Institutional Theory

Institutional theory explains how sustainability and the associated practices spread among organisations and were accepted as common practice (Kang and Lee 2016; Montiel and Delgado-Ceballos 2014). Similar coercive forces on organisations lead to the implementation of also similar sustainability practices (Kang and Lee 2016). This is also a way that organisations found to legitimise their activities, particularly on foreign countries (Montiel and Delgado-Ceballos 2014). In these specific cases, the question debated in literature is whether sustainability practices adopted by Multi-National Corporations (MNCs) abroad are driven by the replication of the global organisational strategy or if MNCs tend to transfer non sustainable practices to the foreign units (Kudłak and Low 2015; Montiel and Delgado-Ceballos 2014).

3.1.4 Resource Based View

The RBV refers to the internal resources of an organisation as a source of intangible assets, providing the organisation a unique culture and format. Unique characteristics and intangible assets prevent replication and consequently guarantees competitive advantage and long-term performance. Thus, RBV theory relies on the relationships between resources, capabilities, competitive advantage and performance (Kang and Lee 2016). Since the internal resources and capabilities are the main factor of success, they affect strategic decisions (Engert et al. 2016).

Derived theories emerged from RBV, such as knowledge-based-view, capability-based-view and the most popular among them, the Natural Resource-Based View (NRBV) (Engert et al. 2016; Montiel and Delgado-Ceballos 2014). Sustainability competences depend on organisational capabilities, in special dynamic capabilities

(Kang and Lee 2016; Kudłak and Low 2015). Sustainable development is one of the key strategic capabilities of the NRBV, along with pollution prevention and product stewardship, to achieve superior social and environmental performance (Montiel and Delgado-Ceballos 2014; Schuler et al. 2017). This key strategic capability evolved into two research lines: clean technology and base of the pyramid strategies (Montiel and Delgado-Ceballos 2014).

3.2 *Conceptual Definition of Corporate and Organisational Sustainability*

How organisations conceptualise sustainability influences the identification and contextualisation of problems, as well as solutions (Schuler et al. 2017). OS, CS and CSR are popular conceptualisations of sustainability in organisations. The overlapping between CS and CSR is debated in literature, since some authors argue that the concepts are interchangeable, while others focus on the differences between them (Chang et al. 2017; Kourula et al. 2017; Montiel 2008).

CSR emerged from the proactive organisational approach towards environmental and social issues, firstly embedded in management literature and focused on the role of organisations as social agents, and then evolved into a specialised line of research more oriented towards the study of ethics and internal organisational mechanism, among others (Kudłak and Low 2015; Montiel 2008).

From a CSR perspective, the environment and local communities are legitimate stakeholders, essential for the survival of the organisation and consequently worthy of attention in the long-term (Chang et al. 2017; Schuler et al. 2017). Wider boundaries are criticised since this approach contradicts the traditional shareholders primacy (Chang et al. 2017); however, an economic approach to CSR allows organisations to prioritise value creation to shareholders, by limiting their responsibilities on the society and the environment only to those situations where shared value is possible. In this context, organisations only adopt CSR activities that aim to improve their competitiveness through value chain efficiency improvements or an enhanced market position (Schuler et al. 2017).

CS has its roots in the sustainable development concept and lacks a universal definition. Nevertheless, the recognition of the importance of stakeholders needs and the balance between social, environmental and economic performance in the long-term are common to most definitions (Chang et al. 2017; Montiel and Delgado-Ceballos 2014).

TBL is often associated with CS (Chang et al. 2017), despite in some literature the conceptualisation of CS is focused only on two (social and environmental) or one (environmental) dimensions of sustainability (Montiel 2008; Montiel and Delgado-Ceballos 2014). Scholars define CS more from a holistic, complex and philosophical perspective, while practitioners are focused on providing guidance (Montiel and Delgado-Ceballos 2014).

From an initial risk based compliance, CS becomes complex and an alternative to the traditional business models (Jones et al. 2016a). This shift was promoted by increasingly stricter environmental regulations, costly and scarce natural resources, stakeholder influence in organisational investments and a growth in media coverage of sustainability related issues and actions (Jones et al. 2016b).

Sustainable business models represent a possibility to achieve competitive advantage, through the improvement of products or services (Chang et al. 2017), since traditional business models fail to integrate sustainability at an operation level (Nawaz and Koç 2018). Sustainability reporting became a mainstream practice as a source of competitive advantage (Chang et al. 2017; Jones et al. 2016b), although this exposes the critical challenge of combining profit making with extended sustainability objectives (Chang et al. 2017).

CS and CSR main differences are: (i) the origin of the concepts (CSR in the 1970s and CS in 1990s with the publication in 1987 of the Brundtland report); (ii) research approaches on the economic, social and environmental dimensions; (iii) interconnection between all the dimensions; (iv) conceptualisation of the economic dimension; (v) instrumental versus intrinsic value of the environment. Despite the differences, the concepts have been converging and from an operational perspective are interchangeable (Montiel 2008), providing a definition of sustainability in organisations, as well as operational tips on the transition to OS (Chang et al. 2017).

3.3 *Sustainability Dynamics*

Time is an embedded variable of sustainability, considering the necessity to address the needs of the present and future generations, continuously interacting with multiple stakeholders (with variable legitimacy, urgency and power), constant innovation and an increasing complex business environment. Dealing with sustainability dynamics requires the development of organisational capabilities and knowledge (Muhammad et al. 2011; Searcy 2011), as well as a culture of organisational learning to address constant change (Nawaz and Koç 2018).

The dynamic nature of sustainability has implications at the multiple levels of organisational management, from the strategic planning to the evaluation of performance. A cyclical and systematic approach guarantees the flexibility and adaptive skills within organisations, as the example on the evaluation of performance provided by Searcy (2011), through the creation of a framework based on an iterative process.

3.3.1 Innovation

The transition towards OS requires not only the improvement of existing capabilities, but also innovation and organisational learning (Nawaz and Koç 2018). Employees' experiences and skills influence the integration of new knowledge into organisational processes, leading to innovation and increasing organisational resilience by providing the capabilities to address emerging challenges (Muhammad et al. 2011).

Another path towards sustainability innovations is the business model adopted by the organisation (Chang et al. 2017). Sustainability entrepreneurship can be defined as the individual or organisational capability to keep innovating and commercialise these innovations. At an organisational level, innovations vary on degree (incremental or radical innovation) and focus (social or technical), being influenced by diverse factors, such as size (Amui et al. 2017; Levinsohn 2013).

Large and Small and Medium Enterprises (SMEs) play a complementary role in the innovation process, since SMEs have an advantage in the development of innovations (due to a more flexible structure and business model and different competitive pressures), while large companies have the resources to commercialise sustainable innovations (Levinsohn 2013).

Despite the conditioning factors, both technical (process and product) and social innovation contribute to the achievement of strategic advantages, by promoting the inclusion of sustainability and stakeholders' interests in the core strategy. Summarising, innovation promotes the transition of organisations towards sustainability, and the integration of sustainability in the core organisational strategy supports continuously innovation (Engert, Rauter, and Baumgartner 2016).

3.4 Stakeholders and Global Governance

The globalisation of economic activities has led to the internationalisation of local value chains and to an increase in the volume and velocity of international trade and investment. Such evolution caused a new balance in global governance, where companies gained power over national states, but not always accommodated the associated responsibilities (Kourula et al. 2017; Kudłak and Low 2015).

In this context, governmental approaches became insufficient to address environmental and social issues, while companies adopted an active role in the creation and redesign of institutional order (Kourula et al. 2017; Kudłak and Low 2015). This shift of power propelled two lines of research: one oriented to the study of the role of enterprises as social agents leading change in society and the other focused on the integration of responsibility values in organisational structure and functions, where can be included in the study of sustainability in organisations (Kudłak and Low 2015).

Corporate Political Activity is an approach adopted in the study of the politicisation of companies, in which the main concern is how organisations maximize their influence over government to create policies that would benefit them (Kudłak and Low 2015; Schuler et al. 2017). However, business is a heterogeneous group and not all the organisations are able to afford the costs of influencing the public policy process, producing a biased influence in the government policy-making process (Schuler et al. 2017).

Companies are able to influence the live of citizens (mostly large companies) and becoming social agents. Their engagement in the development of solutions to social and environmental problems caused by their activities has been rising (Chang et al. 2017; Kudłak and Low 2015). Considering this demand, organisations adopted mainly four types of strategies: reaction, defence, accommodation and proactivity (Chang et al. 2017). Nevertheless, is not clear if responsible organisational behaviour produces effective positive impacts in society and environment, since the increased power of companies allowed them to self-regulate their activities, avoiding governmental supervision (Kudłak and Low 2015).

In these cases, NGOs appear as providers of supervision and alternative forms of regulation to fill the voids of legislation, either national or international (Kudłak and Low 2015). This is especially relevant in cases where companies move their socially irresponsible behaviour to foreign units, which is mostly the case of MNCs (Montiel and Delgado-Ceballos 2014).

MNCs constitute a popular example of organisational irresponsibility in literature, explained in detail by Kudłak and Low (2015). Some of these companies re-located their activities in developing countries, looking for cheap labour, weak governance structures and local natural resources, in order to improve their profit. Despite the potential to act as relevant development agents in these countries in collaboration with NGOs and national governments, some MNCs are often associated to the worsening of social and environmental problems, such as political corruption and environmental degradation. As a result of local stakeholders' pressure or to replicate the global trend to look for sustainability, MNCs implemented sustainability practices in local communities related to health care, education, human rights and environment, among others. However, these efforts have mostly failed to cause positive effects, since the measures adopted are in most cases displaced from the local reality, focused on the improvement of the corporate image and paternalist.

Thus, despite the potential for positive impacts, the shift of MNCs' manufacturing sites to developing countries have increased the gap between the prosperity of developed countries and the poverty of developing countries, reinforcing the negative effects of globalisation and leading environmental movements to encourage local value chains or the effective engagement of local communities (Levinsohn 2013; Schuler et al. 2017). Another possible approach is the development of performance evaluation systems (such as indicators) to the whole supply chain (Searcy 2012).

4 Sustainability Management in Organisations

Transitioning to sustainability requires a high effort from organisations, since the concept itself is complex, holistic and consequently difficult to grasp. Another difficulty is the continuous change in the environmental, social and economic aspects in which the organisation must focus and the need to adapt sustainability initiatives to local conditions and characteristics. This variability can be attributed to three main reasons: the influence of internal and external factors in organisational resources;

the change in legitimacy, power and urgency of stakeholders; and the increase of the complexity of business (Muhammad et al. 2011; Searcy 2012).

Thus, performing a transition towards sustainability is a high effort task for organisations. To be effectively successful they are required to introduce significant changes in the organisational structure and functions (Kudłak and Low 2015; Searcy 2012). In most cases, efforts to transitioning are supported by voluntary initiatives of self-regulation, which is the case of the implementation of management systems. The obstacles faced by organisations are driven by diverse factors, such as lack of education and training and difficulties in finding the way to integrate sustainability in organisational activities (Searcy 2012).

4.1 *Strategic Level*

The integration of sustainability values at a strategic level allows organisations to change from a reactive to a proactive approach to sustainability. The proactive approach presuppose that organisations go beyond environmental and social legal compliance and actively seek innovative solutions to improve products and organisational processes by sustainability criteria. As a result, organisations with a proactive approach towards sustainability have increased competitive advantage in relation to those that are reactive and do not integrate sustainability issues at a strategic level (Kang and Lee 2016).

Three factors are relevant in the integration of sustainability at a strategic level: manager attitudes and perceptions of sustainability, organisational culture and the need for more investments. Organisational culture refers to a group of assumptions that the members of an organisation share and influence their thinking and actions and when transitioning towards sustainability, its values must be embedded in the organisational culture in order to be successful. Organisational learning is a possible vehicle of change of organisational culture (Engert et al. 2016).

To monitor the effectiveness of sustainability integration and the results obtained at other levels, performance variables must be considered, such as environmental and financial performance (Kang and Lee 2016).

4.2 *Tactical Level*

At the management level, developing and maintaining a sustainability orientation represents a challenge to organisations, since a structural change in internal organisational attributes and functions is required (Kudłak and Low 2015). However, sustainability activities are a source of added value and in most cases also of cost reduction, contributing to the competitive advantage of organisations involved in the transition (Engert et al. 2016). In order to maintain the sustainability orientation in the long-term, organisations must consider the following factors: the perception of managers on the legitimacy of sustainability models, partnerships with diverse sectors, research and development and internationalisation (Kudłak and Low 2015).

The integration of sustainability in organisations also requires financial efforts in its initial stage, specifically in new technologies, certifications, communication and marketing and human resources. Consequently, the transition must be mostly supported by top management, since commonly the middle management tends to reproduce top management attitudes and behaviours (Engert et al. 2016).

The change process requires strong leadership, thus attitudes, perceptions and the behaviour of the manager responsible to lead the organisation in the transition towards sustainability is relevant. Stakeholder involvement is another essential factor in the organisational change (Engert et al. 2016; Nawaz and Koç 2018).

4.3 Operational Level

At an operational level, the transition towards sustainability, requires organisations to introduce significant modifications in organisational processes, such as communication and the management of human resources. For example, the involvement of employees is crucial in the integration of sustainability in processes; therefore, the internal communication must guarantee transparency and that employees are aware and fully understand the organisational sustainability goals, strategies and measures (Engert et al. 2016; Powell 2011). The change in organisational processes is also dependent on the employees' experiences, skills, knowledge and availability to learn (Muhammad et al. 2011), leading to the development of new organisational capabilities and innovation (Powell 2011). Employees' engagement is a key element in OS, thus understanding employees values and perceptions on sustainability is essential to align organisational and employees' ethics. A possible approach is employees' empowerment in OS activities (Powell 2011).

Kudłak and Low (2015) reinforces the trend to focus the study of sustainability integration in organisations on human resources issues, such as employees' well-being, job satisfaction and the development and maintenance of a responsibility culture in organisations. Labour ethical problems also are relevant in the transition to sustainability itself, such as working hours, work intensification and addiction. Also, the relations between managers and employees influence the success of sustainability integration.

Management control and the selection of performance indicators that are complementary and consistent with the conventional management tools is another key issue in the management of OS (Engert et al. 2016).

4.4 Performance Measurement and Tools

Organisational sustainability performance is a dynamic concept that requires measurement, including both internal and external factors promoting the system change. Therefore, a sustainability performance evaluation system must be balanced and take into consideration the singular characteristics and conditions of the organisation, include stakeholders' involvement and provide an effective feedback on how to address sustainability in the organisation, including its long-term approach (Searcy 2011).

Multiple reasons motivate organisations to develop a performance measurement system, such as the evaluation of the progress achieved considering defined goals, informing the decision-making process at a strategic and tactical level and enhance relationships with stakeholders (Searcy 2011).

How organisations design and implement the evaluation system is a recurrent issue analysed in literature, focusing on the development, selection criteria and design processes and how to use data in the evaluation systems. However, the relation between the development and implementation effectiveness of the evaluation system is still a gap in literature, as well as cumulative impacts assessment interlinked with the long-term perspective of sustainability (Searcy 2012).

The performance evaluation system itself must be assesses to allow constant updates, to ensure the cohesion of the system and the consideration of changes in stakeholders' legitimacy, power and urgency. The organisation should consider the possibility of benchmarking with other organisations from the same sector, as well as if the assessment should be conducted internally or externally. Three levels of the system should be included in the assessment: individual measures, the whole system and the relation between the system and the operating environment (Searcy 2011).

Using performance indicators associated with Environmental Management Systems (EMS), sustainability targets or sustainability reporting allows organisations to avoid that indicators became an administrative burden (Searcy 2012). Management systems provide a framework for organisations to manage their sustainability issues and stakeholders demand in a systematic way (Muhammad et al. 2011), while sustainability reporting aims to measure organisational performance in the sustainability field, by including a social, economic and environmental assessment (Siew 2015).

Summarising, a wide range of tools is available for organisations that aim to measure sustainability performance. The diversity of tools makes more difficult the comparison between organisations, however allow organisations to select the tools that better suit their goals and strategic vision of sustainability. Simplifying, sustainability tools can be divided into three categories: frameworks, standards and ratings and indices (Siew 2015).

The Global Reporting Initiative (GRI) constitutes a framework of sustainability performance assessment, providing guidance to organisations and enabling comparison through the standardisation of reports (Chang et al. 2017; Montiel and Delgado-Ceballos 2014). GRI suggests the division of sustainability into the three pillars of sustainability: in the economic pillar are included indirect impacts but not corporate governance; the social dimension is mostly focused on human rights and the impacts on society; the environmental component is exhaustive, including the diverse thematic areas, such as biodiversity, water, energy, transports and environmental compliance (Montiel and Delgado-Ceballos 2014). Nevertheless, GRI guidelines do not assure continuous performance improvement, focusing on the management dimension and possibly leading to a false sense of evolution (Siew 2015).

Diverse standards on management systems are available to organisations, such as ISO 14001 and EMAS. Both are popular standards to address the environmental dimension of sustainability (Montiel and Delgado-Ceballos 2014; Siew 2015), however Muhammad et al. (2011) argues that Integrated Management Systems (IMS) provide a potential framework for the integration of sustainability in organisations. Each individual management system allows to fill the needs of different stakeholders and when integrated under a sustainability approach provide a governance structure to promote the transition. From another perspective, Nawaz and Koç (2018) emphasise the voids in IMS left by the differences in design and scope of each management system and provides a sustainability management system developed as an whole.

The Dow Jones Sustainable Index (DJSI) is an example of an index commonly used in literature to identify sustainability frontrunners and in the study of the relationship between firms performance and sustainability. Organisations are selected to the DJSI based on a frontrunner approach, since only the organisations with best results in long-term economic, environmental and social criteria are included in the index. DJSI is also based on the three sustainability pillar: corporate governance is considered in the economic dimension; the social component is focused on employees' attraction and evolution, stakeholders engagement and reporting of social matter; and the environmental pillar includes product stewardship, despite not considering compliance (Montiel and Delgado-Ceballos 2014).

Notwithstanding, the positive aspects described in the application of these tools, corporate and organisational sustainability are also debated in literature as ways to decrease stakeholders pressure, without performing a real effort in the transition towards sustainability (Clapp 2005; Siew 2015). Stakeholders pressure often lead organisations to report scarce and limited sustainability information, since non-existent reporting would create a greater competitiveness loss than having a non-effective reporting system (Cho et al. 2015).

5 Conceptual Framework on OS

Sustainability is by definition a complex, holistic and dynamic concept and these characteristics revealed to be obstacles to a universally accepted definition. Researchers agree that each organisation needs to define its vision of sustainability, since there is not a model able to fit all (Nawaz and Koç 2018; Porter 2008; Salzmann et al. 2005). Also, the existent definitions of OS rely mostly on the perspective of the most developed countries, leaving behind those in transition or less developed. Finding native definitions of sustainability that fit local realities represents still a challenge in OS research (Goyal et al. 2013; Kourula et al. 2017; Levinsohn 2013).

Another key challenge emerging from literature is the need to find solutions on how to promote the transition of organisations towards sustainability. These would require organisations either to reduce complexity or to find new ways to deal with it. In this context, researchers identified a trend to a more holistic approach to sustainability, instead of the traditional approaches focused in most cases only on one or two dimensions of sustainability (Chang et al. 2017; Engert et al. 2016; Goyal et al. 2013).

Despite the general acknowledgement of the dynamic nature of sustainability, most OS tools are mostly compliance tools and consequently static. Integrating dynamics into the main concepts and tools of OS would be a challenge that organisations transitioning to sustainability must face (Muhammad et al. 2011; Schuler et al. 2017; Searcy 2011, 2012).

The continuous change in stakeholder demand requires organisations to keep adapting to emerging and new situations, reinforcing the need for a dynamic approach in stakeholder engagement (Chang et al. 2017; Engert et al. 2016; Jones et al. 2016b). Though employees are consensual key stakeholders, literature fails to provide accurate guidance on how to involve employees effectively on the transition. Employees are the cornerstone in the change of organisational culture and thus critical to integrate sustainability at an operational level sustainability in organisations (Powell 2011).

Organisational learning is pointed as a key element in the transition towards sustainability, since the capacity to adapt to permanent changeable conditions relies on the organisational capability to keep learning and adapting. In this context, innovation plays a relevant role still lacking clarification and deepen knowledge (Engert et al. 2016).

Figure 3 summarises the key emerging challenges of sustainability in organisations in a conceptual framework. The first challenge in OS transition research is how to support organisations building a shared sustainability vision. The organisational setting allowing the achievement of the envisioned organisational state requires the cascading down of sustainability from management to daily operations. In this process, organisational learning and innovation are key elements, as well as the development of dynamic capabilities (Amui et al. 2017). These are also a key organisational element in addressing stakeholders demand from an innovative approach (Zollo et al. 2016), reinforcing the need of more innovation both at a technological and a managerial level, to find new transitioning solutions (de Almeida and de Melo

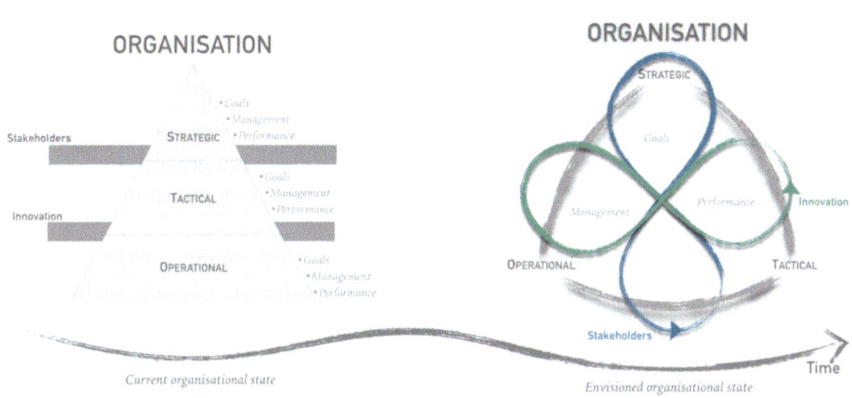

Fig. 3 Conceptual framework on Organisational Sustainability

2017; Amui et al. 2017). After all, the main goal is to support organisations to fill the gap between their intentions and effective sustainability performance (Porter 2008).

6 Conclusion

The main goal of the current study was to identify the key emerging challenges in the transition of organisations towards sustainability. In order to achieve this goal a meta-review of literature reviews was performed focusing on OS and CS, which provided the theoretical background to build a conceptual framework. The conceptual framework constitutes a contribution to literature and future research development, since it allows to identify which are the main research pathways to follow.

The first emergent issue is the very process of defining the concept of sustainability adopted by each organisation and the theoretical background from which it draws to implement its strategy. Sustainability is a value-laden concept and consequently each organisation should be able to define sustainability in relation to its organisational reality. This definition will then affect the pathway selected by the organisation to transition towards sustainability by conditioning stakeholder engagement, strategic management, innovation and performance assessment tools. Developing innovative tools that support organisations in the creation of a shared vision of sustainability and the following change in processes is another emerging topic in this field of research.

Yet another challenge faced by organisations and researchers is how to perform this transition from an integrated perspective, considering both the dimensions of sustainability and a wide concept of organisation, where stakeholders take primacy. This would imply that organisations maintain a holistic definition of sustainability, while integrating and streamlining methods in the change process.

Finally, the long-term perspective underlying sustainability emerges as a key challenge due to the difficulty of guaranteeing a dynamic approach at multiple levels. First of all, is relevant to consider the dynamics of the sustainability concept itself, as well as in the organisation's evolution since constant change is happening both internally and externally. Furthermore, stakeholders' engagement has mostly been studied from a static perspective. However, stakeholder demand is highly dynamic and requires improved and innovative processes to take this characteristic into consideration. From a more internal perspective, innovation in the organisation is a key aspect in its evolution, promoting learning and allowing the organisation to explore its capabilities from a dynamic perspective.

Providing organisations support in the transition to sustainability, while considering all the aforementioned issues represents a great challenge. Also the methods for the transition must be flexible and dynamic, by acknowledging that it is not possible to adopt a unique solution that fits all organisations. Instead, organisations are summoned to develop innovative capabilities to adapt and perform a transition towards a desired sustainability vision.

Acknowledgements The authors would like to acknowledge the support of the Portuguese Science Foundation by providing the Ph.D. fellowship (PD/BD/128446/2017) that supported this work. We also acknowledge CENSE strategic project (UID/AMB/04085/2013) and the NOVA School of Science and Technology of NOVA University Lisbon.

References

Amui LBL, Jabbour CJC, de Sousa Jabbour ABL, Kannan D (2017) Sustainability as a dynamic organizational capability: a systematic review and a future agenda toward a sustainable transition. J Clean Prod 142:308–322

Batista AAS, de Francisco AC (2018) Organizational sustainability practices: a study of the firms listed by the Corporate Sustainability Index. Sustainability (Switzerland) 10(1):226

Brones FA, de Carvalho MM, de Senzi ZE (2017) Reviews, action and learning on change management for ecodesign transition. J Clean Prod 142:8–22

Chang RD et al (2017) Evolving theories of sustainability and firms: history, future directions and implications for renewable energy research. Renew Sustain Energy Rev 72(Nov 2015):48–56

Cho CH, Laine M, Roberts RW, Rodrigue M (2015) Organized hypocrisy, organizational facades, and sustainability reporting. Acc Organ Soc 40:78–94

Clapp J (1998) The privatization of global environmental governance: ISO 14000 and the developing world. Glob Gov 4(3):295–316

Clapp J (2005) Global environmental governance for corporate responsibility and accountability. Glob Environ Polit 5(3):23–34

Daddi T, Magistrelli M, Frey M, Iraldo F (2011) Do environmental management systems improve environmental performance? Empirical evidence from Italian companies. Environ Dev Sustain 13(5):845–862

de Almeida MFL, de Melo MAC (2017) Sociotechnical regimes, technological innovation and corporate sustainability: from principles to action. Technol Anal Strat Manag 29(4):395–413

Engert S, Baumgartner RJ (2016) Corporate sustainability strategy—bridging the gap between formulation and implementation. J Clean Prod 113:822–834

Engert S, Rauter R, Baumgartner RJ (2016) Exploring the integration of corporate sustainability into strategic management: a literature review. J Clean Prod 112:2833–2850

Freeman RE (1984) Strategic management: a stakeholder approach. Pitman Publishing Inc., Boston

Garcia S, Cintra Y, Torres R de CSR, Lima FG (2016) Corporate sustainability management: a proposed multi-criteria model to support balanced decision-making. J Clean Prod 136: 181–196

Geels FW, Schot J (2007) Typology of sociotechnical transition pathways. Res Policy 36(3):399–417

Goyal P, Rahman Z, Kazmi AA (2013) Corporate sustainability performance and firm performance research. Manag Decis 51(2):361–379

Healy H et al (eds) (2013) Ecological economics from the ground up. Routledge, New York

Iraldo F, Testa F, Frey M (2009) Is an environmental management system able to influence environmental and competitive performance? The case of the eco-management and audit scheme (EMAS) in the European Union. J Clean Prod 17(16):1444–1452

Jones P, Comfort D, Hillier D (2016a) Interpretations of the concept of sustainability amongst the UK's leading food and drink wholesalers. Trziste 28(2):213–229

Jones P, Hillier D, Comfort D (2016b) Sustainability in the hospitality industry: some personal reflections on corporate challenges and research agendas. Int J Contemp Hospitality Manag 28(1):36–67

Kang S-W, Lee K-H (2016) Mainstreaming corporate environmental strategy in management research. Benchmarking Int J 23(3):618–650

Kourula A, Pisani N, Kolk A (2017) Corporate sustainability and inclusive development: highlights from international business and management research. Curr Opin Environ Sustain 24:14–18

Kudłak R, Low KYJ (2015) Special issues dedicated to CSR and corporate sustainability: a review and commentary. Long Range Plan 48(3):215–227

Lankoski L (2016) Alternative conceptions of sustainability in a business context. J Clean Prod 139:847–857

Levinsohn D (2013) Disembedded and beheaded? A critical review of the emerging field of sustainability entrepreneurship. Int J Entrep Small Bus 19(2):190–211

Montiel I (2008) Corporate social responsibility and corporate sustainability. Organ Environ 21(3):245–269

Montiel I, Delgado-Ceballos J (2014) Defining and measuring corporate sustainability. Organ Environ 27(2):113–139

Muhammad A, Searcy C, Zutshi A, Ahmad N (2011) An integrated management systems approach to corporate sustainability. Eur Bus Rev 23(4):353–367

Nawaz W, Koç M (2018) Development of a systematic framework for sustainability management of organizations. J Clean Prod 171:1255–1274

Porter TB (2008) Managerial applications of corporate social responsibility and systems thinking for achieving sustainability outcomes. Syst Res Behav Sci 25(3):397–411

Powell SM (2011) The nexus between ethical corporate marketing, ethical corporate identity and corporate social responsibility: an internal organisational perspective. Eur J Mark 45(9/10):1365–1379

Salzmann O, Ionescu-Somers AM, Steger U (2005) The business case for corporate sustainability: literature review and research options. Eur Manag J 23(1):27–36

Schuler D, Rasche A, Etzion D, Newton L (2017) Guest Editors' Introduction: corporate sustainability management and environmental ethics. Bus Ethics Q 27(2):213–237

Searcy C (2011) Updating corporate sustainability performance measurement systems. Meas Bus Excel 15(2):44–56

Searcy C (2012) Corporate sustainability performance measurement systems: a review and research agenda. J Bus Ethics 107(3):239–253

Siew RYJ (2015) A review of corporate sustainability reporting tools (SRTs). J Environ Manage 164:180–195

Zollo M, Bettinazzi ELM, Neumann K, Snoeren P (2016) Toward a comprehensive model of organizational evolution: dynamic capabilities for innovation and adaptation of the enterprise model. Glob Strat J 6(3):225–244

Focusing Sustainable Human Resource Management—Framework for Sustainability Management in Research Organizations

Lisa Kurz, Larissa Jost, Karsten Roth and Peter Ohlhausen

Abstract Research organisations are not only contributing to sustainable development but also contribute to scientific findings. As key influencers of innovation; employers and publicly funded research organisations not only have the social mandate to deal with their responsibilities regarding the environment and society, but also drive to understand their social responsibility for their employees and the impact on research and operational processes. Sponsored by the German Federal Ministry for Education and Research (BMBF), this paper presents the results of the joint research project; LENA—Guidelines for Sustainability Management and describes how 3 of Germany's biggest research organisations (Fraunhofer-Gesellschaft, Leibniz Association and Helmholtz Association) face current challenges in human resource management of research organisations by the integration of a common understanding of sustainability and a broad-based framework. The empirical basis is built by a qualitative organisational-ethnographical study which reflects the expert knowledge, everyday experiences and the subject-oriented interpretation of sustainability in human resource management. The result derives concrete recommendations for the institutional practice and offers structured and methodologically proven options for action addressing the stakeholders in human resource management in research institutions.

Keywords Sustainable human resource management
HR management in research organizations · Social responsibility
Sustainable development

L. Kurz (✉) · P. Ohlhausen
Fraunhofer Institute for Industrial Engineering IAO, Nobelstraße 12, 70569 Stuttgart, Germany
e-mail: lisa.kurz@iao.fraunhofer.de

P. Ohlhausen
e-mail: peter.ohlhausen@iao.fraunhofer.de

L. Jost · K. Roth
German Aerospace Center (DLR), Linder Höhe, 51147 Cologne, Germany
e-mail: larissa.jost@dlr.de

K. Roth
e-mail: karsten.roth@dlr.de

W. Leal Filho (ed.), *Social Responsibility and Sustainability*, World Sustainability Series, https://doi.org/10.1007/978-3-030-03562-4_3

1 Introduction: The Scientific System as Responsible for Sustainability

Humanity is facing major social challenges. Climate change, scarcity of resources, demographic change, global biodiversity loss, equal access to education, lack of access to safe drinking water and healthy eating are just a few. Great efforts have been made in recent years to address them as common tasks, most recently through the adoption of the United Nations Sustainable Development Goals (SDGs): they formulate concrete environmental and development goals for the world community by 2030. Science and research are also explicit to make their contribution to tackling societal challenges, to take over social responsibility and thus to the successful implementation of the UN goals.

In addition to generating knowledge, the science system must also review its values, structures and processes. An important milestone in this reflection process was the discourse on sustainability in science held during the Year of Science "Earth Future Project" (2012). One result of the discussions was that the concepts and guidelines for sustainable action in the context of business, such as the international principles and standards of the Global Reporting Initiative (GRI), can only be applied to science to a very limited extent. This is how the joint project "Guidelines for sustainability management in non-university research institutions" came about. Although the major scientific organisations are linked to the concept of sustainable development and create a large number of individual activities, they do not make them visible internally and externally within the framework of a uniform sustainability strategy with a corresponding training and further education concept. Therefore, despite the already existing diversity of measures, there is a lack of a comprehensive understanding of sustainability and its consistent implementation and design in everyday life at non-university research institutions.

Although non-university research institutions are clearly in favor of their social mission and their importance in sustainability topics, a generally accepted understanding of how non-university research institutions can operationalize the concept of sustainable development and systematically integrate it into their processes is not yet available. Management systems, which are primarily tailored to companies, can only be used to a limited extent, since research institutions as non-profit organizations differ significantly in their mission, business purpose and legal and organizational structure of business enterprises. Another difference is that, in contrast to most of the products and services of business enterprises, the impact of research services on the environment and society is sometimes only noticeable after a considerable time lag. To close this gap, this paper presents the results of the BMBF-funded joint project "Guidelines for Sustainability Management" jointly developed by the Fraunhofer-Gesellschaft, the Helmholtz Association and the Leibniz Association and offers a research-specific interpretation of existing standards in sustainability management and sustainability reporting. Furthermore, it provides an overview of central approaches and implementation options in the fields of action of human resource management, which explicitly take into account the special features of research organizations.

2 The Project LENA—Guidelines for Sustainability Management: Common Understanding and Social Mission

The underlying understanding of sustainable development essentially refers to the definition of the mission statement by the Brundtland Commission as

> […] a development that meets the needs of the present without compromising the ability of future generations to meet their own needs. (Hauff 1987)

Above all, this normative understanding aims at the long-term preservation of natural, social and economic resources in the interests of present and future generations and thus includes the demand for intergenerational and intragenerational justice in the global context. Of central importance is the equal consideration of the three dimensions of environment, economy and social affairs: Environmental protection, economic performance and social responsibility must be brought together in such a way that decisions are sustainable from all three points of view. The preservation of the sustainability of the ecosystem earth is the absolute limit (Hauff 1987). Sustainable development is context-dependent and requires continuous negotiation processes between the various stakeholders due to potential conflicting goals. Against the backdrop of far-reaching environmental, social and economic challenges, politics and civil society increasingly see research as an option for dealing with the urgent issues of social development. These demands can be found in political strategy documents at various levels, for example in the National Sustainability Strategy and the High-Tech Strategy of the German Federal Government, in the EU Sustainability Strategy, in the EU Framework Program Horizon 2020 and in the UN Sustainable Development Goals.

In addition to contributing to solving societal problems and identifying future challenges through research, avoidance of risks and potential harm from research also plays a role, as scientific knowledge and inventions often influence society and the environment over the long term, sometimes disruptively. After all, research organizations, like other organizations and companies, have an immediate responsibility for the impact of their activities on society, employees and the environment. The project “Guidelines for Sustainability Management” is therefore aimed at decision-makers from non-university research organizations as well as at all those who deal with the topic at the level of the umbrella organizations or in the individual institutions, centres and institutes. The guidelines intent to make it easier for organizations that are at the beginning of their sustainability commitment, but also to support those who want to further systematize and expand their already existing sustainability management. Therefore guideline should facilitate the entry into a research-specific sustainability management and present the essential fields of action in a practical way. In this way, it should offer the participants access to already existing approaches and, at the same time, expressly leave room for individual focus. The ability to connect to the guidelines is ensured by linking with existing standards. References and comments clarify the ability of the fields of action to comply with nationally and internationally recognized sustainability reporting standards—in particular the guidelines of the Global Reporting Initiative (GRI) and the German Sustainability Code (DNK). Furthermore, information in this publication is based on internationally recognized standards for sustainability management and social responsibility of organizations, such as the principles of the United Nations Global Compact and principles and areas of action of the international standard ISO 26000.

3 Principles of Sustainability Management in Non-university Research Institutions

In particular, the two basic principles of responsible use of material and non-material resources and good governance are addressed as the basis of sustainability management for non-university research institutions. Research organizations use material and non-material resources to arrive at specific results. The responsible use of resources should be recognized as a natural element of organizational culture. In addition to the employees, infrastructure, equipment, materials, materials and energy, the resources of a research institution also include funds and the knowledge resource.

In order to establish a culture of resource responsibility, it is the task of the organization to create appropriate framework conditions, for example by providing information, measures for competence development and sensitization as well as technical measures. How the responsible use of resources can be implemented in everyday work depends on the area of influence: for example, scientists influence the resource planning of a research project (finances, personnel, infrastructure, material, etc.); human resource managers, in turn, design programs for young researchers, the purchasing team is responsible for ordering effective and efficient equipment and research managers can, for example, specifically promote open access approaches. An intensive networking of research groups for the joint use of research infrastructures via one or several locations increases the utilization and at the same time opens up potential synergies.

Governance or organizational management is the overall system of management and control mechanisms on the basis of which an organization makes and implements decisions. The prerequisite for the establishment of credible sustainability management is compliance with the basic principles of good and responsible organizational governance (good governance) as well as the promotion of an organizational culture in which the principles of good governance are of course integrated. In particular, the following principles should be exemplified by the organizational leadership and anchored by formal mechanisms. These include disclosure and transparency, respect for stakeholder interests, and responsible risk management. Organizational governance is understood to mean the internal and external accountability of a research organization to its stakeholders, for example, in terms of decision-making, compliance and financing. For example, the transparent approach also includes a disclosure of the internal requirements and the underlying values such as an announcement of breaches of the rules and dealing with them. The published information should be timely and factual, as well as clear and objective.

In terms of research activity, transparency goes beyond the requirements of good scientific practice. In addition to a disclosure of the initial situation, the financing and the methodological as well as content-related orientation, it also means the creation of an addressee-appropriate accessibility of the results achieved as well as their effects and consequences.

The consideration of stakeholder interests means the integration of the demands and expectations of internal and external stakeholders and is considered a key element for a successful sustainability management. It is not just about incorporating stakeholder interests in decision-making processes, but also about balancing the different interests of different stakeholders. In order to actively engage stakeholders, an organization should identify and engage with its relevant stakeholders in order to become aware of expectations and to consider them in decision making. The typical stakeholders of research organizations include donors and policy makers, cooperation partners, clients and clients, representatives scientific communities, students and civil society organizations and the media Employees and internal committees are important internal stakeholders. Involvement of stakeholders can take many forms, such as surveys, conferences, workshops, round tables, advisory committees, collective bargaining and joint online platforms.

Organizational decisions that have an impact on the future viability of the organization, staff, environment and society require a balance of opportunity and risk. Risk management means an active, future-oriented control of organizational risks. The tasks of a systematic risk management are initially the identification, analysis, evaluation and monitoring of risks. This information is a prerequisite for sound decision-making at senior management level, so that risks can be avoided or mitigated at an early stage. In this context, risk management is to be understood as a continuous dynamic process in order to continuously improve the risk culture and risk awareness in the organization. Elements of risk management by research organizations include, for example, work aids and learning programs for raising awareness in the field of health and safety at work, corruption prevention or protection of intellectual property as well as confidential, secret and personal data. Similarly, the risk of possible misuse of research results places particular demands on the responsibility and self-control of scientists in all areas of research.

4 Management of Sustainability in Organizational Contexts

The concept of sustainability was interpreted far into the 20th century purely in terms of resource economics. In recent decades, the concept of sustainability has more and more abandoned its original meaning and has become a guiding political concept. Modern sustainability discourse and the current meaning of sustainable development are based on the 1987 definition of the Brundtland Report, published by the United Nations World Commission on Environment and Development. In addition, the Brundtland definition for the first time addresses the sustainability aspects of "limitations of resources" and "needs of people" (Di Giulio 2003, p. 42).

The equal consideration of the three dimensions environmental, economic and social is of central importance. For these so-called three-pillar models, two main arguments crystallize in the discourse. On the one hand, it is the argument of the need to consider all important social sectors and actors, which makes a political pragmatism clear. On the other hand, from a system theoretical point of view, the argument that the environment, economy and society are independent; but at the same time interwoven systems has to be ensured with regard to the lasting sustainability of the system as a whole (Kopfmüller 2007, p. 16). Even if the three-pillar sustainability model receives widespread recognition, it was already criticized early on. This refers to the excessive complexity and the associated arbitrariness of the concept, which should be stopped by a renewed focus on the main objective of environmental protection, taking into account social and economic concerns. In addition, the criticism goes on to say, there would be a risk that further interactions or conflicting goals would be forgotten through the partial optimization within the individual dimensions, and that a purely disciplinary approach would instead be true to nature rather than a desirable interdisciplinary cooperation (ibid., p. 17). In the course of the further development of the Brundlandt definition and the three-pillar models through the scientific and political discourse, instead of an isolated consideration of the three individual dimensions, the demand for a cross-dimensional, integrative sustainability approach becomes more and more loud (Kopfmüller 2011). Environmental protection, economic performance and social responsibility are to be linked in such a way that decisions are sustainable in the light of the dimensions of sustainability. The absolute and natural limit of sustainability is thereby the preservation of the carrying capacity of the earth (Press and Information Office of the Federal Government 2012, p. 24). In addition, the concept of sustainable development implies a participatory and procedural character in that concrete sustainability goals are negotiated continuously and context-specifically.

In terms of the contribution of companies and other organizations to sustainable development, it is usually the integration of social, environmental and ethical concerns into organizational or corporate governance that minimizes negative impacts such as environmental and social externalities internalized and positive effects on society are strengthened (Colsman 2016, p. 75).

5 Focusing Sustainable Human Resource Management: Interpretation of the Concept of Sustainability and Its Necessity in the Scientific System

Since the year 2000, sustainability concepts, sustainability management and the associated sustainability efforts in the area of personnel have become increasingly important, especially in organizations. The sustainability debate sparked beyond the human resources world at the end of the 1980s and attempts to develop the social sustainability model through the work of the Brundlandt Commission and the subsequent UN Conference on Environment and Development on an environmental, economic and social dimension of sustainability define (Hauff 1987). An attempt was made to transfer this three-dimensional approach to the corporate and organizational context, and this is first mentioned in the literature in the form of the "triple bottom line" in 1994 (Elkington 1997). John Elkington embraces the familiar concept of bottom line—the final line of a profit-loss account—and extends it to the dimension of the environment and society. The aim of his Triple Bottom Line is to make clear the added value of organizations and companies, which they produce both economically and ecologically and socially. Born in the Anglo-Saxon financial industry with the aim of contributing to sustainable development, the Triple Bottom Line continued through the executive floors worldwide. These responded positively to the call for a move away from exclusive profit orientation and have been implementing this conceptually as well as practically in their businesses and organizations since the turn of the millennium. At the political level, Elkingtons Triple Bottom Line was also used, with the European Commission requesting listed companies to publish their sustainability efforts in their annual reports (Kuhn 2008, p. 12). Although the three-pillar model already existed, the Triple Bottom Line placed a high priority on the development of today's sustainability reporting formats, such as that of the Global Reporting Initiative or the Dow Jones, particularly as a result of its plea for transparency and orientation towards stakeholder interests Sustainability Indexes.

In order to live up to the vision of leaving to future generations an intact ecological, economic and social fabric, in which the preservation of natural resources, solidarity in society and economic efficiency are ensured, companies and organizations bear the foundation of their social position. their economic and environmental importance. In order to meet the requirements of sustainable development, organizations and companies are adopting different approaches to organization-specific sustainability management. So far, the term sustainability management has mostly been used in connection with companies and municipalities, but it is also occasionally used in relation to other organizations (Schaltegger 2009). The term sustainability management refers to the integration of appropriate concepts, procedures and instruments that can be used to avoid or at least minimize the negative effects of the respective organization on people, society and the environment (Loew and Rohde 2013, p. 13).

Guidelines provide different sustainability criteria and recommendations for action, which are mainly tailored to business enterprises and offer research institutions limited orientation. The relevant recommendations, standards and norms make it clear that humans play a central role in the sustainability management of organizations. Like the Brundtland definition, the concept is based on a anthropocentric approach. Thus, in contrast to biocentrism or ecocentrism, human needs and their satisfaction are at the center of attention. The task of sustainability management in organizations is to imply the rules and guidelines of living together and to target the well-being of society—i.e. the workforce—in which the rights of all individuals are recognized and the interests of the employees safeguarded (Ulrich 2001, p. 237)., The focus here is on human resource management of organizations and becomes an indispensable element of an organization-specific sustainability management. It is obvious that the types and shapes of the range of activities in the field of human resources differ from sector to sector and depend on the situation and dynamics of the labor market; However, they are just as closely tied to the respective organization and its organizational culture and can be operationalized via values, norms, models and leadership. Based on this situation, human resource management has a substantial importance in the communication and integration of sustainability.

The basis of the concept of sustainable human resources management is that of employees and companies on an equal footing. They work together on the development, development and maintenance of personnel potential (Zaugg 2005). The scope of the interpretation of sustainable human resource management is manifold, however, despite different formal design, different depth and granularity, it is possible to pinpoint content-related emphases that are common to a large part of the more detailed definitions and are based on classical human resources management approaches. After extensive studies, Robert Zaugg states that it is the classic human resources management approaches that already lay the foundation for sustainability and sustainability criteria (Zaugg 2009, p. 144). According to this, the Harvard approach, human investment philosophy, people-centered management and corporate human resources management in particular have sustainability references (Beer 1985; Miles and Snow 1995; Pfeffer 1998; Wunderer and Kuhn 1993). Sustainable HR management is less an innovation than an adaptation of existing HRM approaches to current societal and political challenges coupled with employee centricity. One focus of sustainable HR management is therefore the involvement of external stakeholders such as customers, politicians, the media and society; but also internal stakeholders such as committee members or employees. In the basic assumption that sustainability strives for a balance or balance between different interests, the involvement of stakeholders is an important aspect of sustainability management and a prerequisite for success and positive reputation (Zaugg 2009, p. 59). In this context, the already described focus on the needs of the workforce and their individual individuals under the concept of employee orientation as an internal stakeholder group becomes apparent. At the same time, it becomes clear that sustainable human resources management means a professionalization of the existing personnel work and personnel management. Another striking feature is the demand for strategy orientation. In line with this requirement, this means, in the practice of sustainability management, an alignment of personnel management with the strategy of the respective organization and recognition of personnel management as a strategic partner. Sustainable HR management is therefore actively involved in shaping the organizational strategy, makes a contribution to value added and, as a strategic partner, develops competitive advantages at eye level. The different requirements of sustainable human resources management with regard to operational and personnel conditions are closely linked to the strategy and culture of the company or organization (Zaugg 2009, p. 71f). Research institutes contribute to sustainable development not only through their scientific results. As essential elements of the innovation system, as an employer and as publicly (partially) funded organizations, they also have a social responsibility to deal with their responsibility towards the environment, society and employees in their own research and operational processes.

6 About the Method in the Field: Interpretations of Sustainability in Human Resource Management

A reflexive feature of non-university research institutions is their person-centered orientation. For HR management in research organizations, this means supporting the search, development and promotion of outstanding individuals who both have excellent research and an understanding of management (Kurz et al. 2017).

The LeNa sub-project 2, Personnel, consisting of a core team of three scientific staff members and three senior scientists from the participating research organizations, examines the topic of sustainability in human resource management, taking into account the special characteristics of research organizations.

The central research approach in the area of human resources is based on the assumption that personnel management in particular has a substantial importance in integrating and communicating sustainability in the respective organizational culture. From an HR perspective, sustainability is directly related to employee orientation and job satisfaction, which have a positive influence on the organizational culture. Employees are a central source in the investigation of the organizational culture practiced in each case and can provide insights into subjective approaches and practices with regard to sustainable personnel management. The expert survey integrates the following research areas:

- Understanding of sustainability in personnel management for non-university research institutions
- Identification and clustering of action fields
- Challenges for non-university research organizations in terms of sustainable human resource management

As part of the case study, 57 qualitative expert interviews were conducted in various institutions and institutes of the Fraunhofer-Gesellschaft, Helmholtz Association and Leibniz Association. The interviews were conducted in the period from September to December 2014. The length of the interview varied between 45 and 90 min, depending on the reporting needs of the respective interview partner.

The consideration of research institutions as organizations and thus as social systems with basically unpredictable, but specific interactions and interpersonal relationships in the context of sustainable human resource management relativizes both the importance of quantitative methods and their engineering-driven transfer models as well as the approaches of business administration, the organizations predominantly as consider a place of rationality and objectivity (Choi 2010 p. 85). Qualitative methods are important in this case study, as they are able to reconstruct everyday organizational life from the point of view of the subjects—in this case the employees of the examined research institutions—and thereby make human behavior and action within an organization accessible in a procedural approach. Following these considerations, the research organizations have "sites for constructing meaning" (Wright 1994, p. 3) and call for an interpretive approach.

It is not a question of looking at individual causalities in isolation, but of understanding the subjective interpretation and action approaches of the employees and understanding complex interrelationships of research institutions, e.g. the interaction between administration and decentralised research institutions. In the basic assumption that the nature of organisational culture results from the subjective patterns of action and interpretation of the individual employees, the employees are a central source for research into organisational culture (Götz and Moosmüller 1992, p. 2). Employees are, so to speak, the interface of cultural dynamics resulting from the networking of social context and organizational cultural patterns as well as individual courses of action (ibid.). The research institutions house a complex set of meanings whose individual elements—such as the sustainability efforts in personnel management—can only be understood in the overall context of the respective context (Götz and Wittel 2000).

7 Three Pillars of Sustainable Human Resources Management in Non-university Research Institutions

In the context of sustainability management, the Personnel function area particularly addresses the responsibility of the research organization as an employer vis-à-vis its employees and thereby focuses on the social goals of sustainable development. The prerequisite for this is a professional human resources management with sufficient resources to strategically plan and implement HR management activities. Activities in this area relate u. a. the support of staff in career development, the responsible use of fixed-term employees, equal opportunities and the appreciation of diversity, health-sustaining working conditions and competence development for sustainable action. If the results of the interview evaluation are aggregated, in particular three constituent pillars for a sustainable personnel management become clear, which are addressed in the following fields of action (Fraunhofer-Gesellschaft, Helmholtz Association, Leibniz Association 2016, pp. 43ff):

1. service and service-oriented personnel management
2. professional and personal development
3. networking and cooperation.

The service orientation of the personnel management refers to an orientation to the needs of the employees in administrative and vocational training processes. Many requirements in the area of working conditions, such as ensuring adequate and fair social protection or work-related safety and health hazards, are defined by national laws in Germany. Human resources management, which not only complies with these legal framework conditions, but also considers individual and target group-specific needs of employees, contributes significantly to the employees' satisfaction and thus also to employee loyalty. Personnel care will optimally ensure a harmonious framework throughout the entire period of employment of the employees. This includes assistance with employment, contract and collective bargaining issues, including remuneration, occupational safety, social security and occupational pensions. Supporting this can be the establishment of a center of excellence, which will answer collectively agreed, labor and pension law issues. With regard to occupational safety and health, the education and training of specialists and executives, safety experts and regular inspections of research institutions are important.

Another key element is the training of young people. This not only secures the organisation's specialist base, but also assumes social responsibility at the same time: for the trainees, this means an expansion of their technical and social skills, which leads to an increase in their individual employability and their participation in the world of work and employment. The training activities are not only geared to the needs of the own research organisation, but also to the needs of the region. For example, participation in what is known as "collaborative training", which aims at cooperation between different companies and organisations in training, has the advantage that it also secures the specialist staff base and training quality among the regional partners. At the same time, such cooperation can make vocational training more attractive and strengthen regional development.

In practice, the implementation of this field of action "Service and service-oriented personnel management" can be structured as follows:

- Establishment of a center of excellence for answering tariff, labor and pension law issues
- Training of specialists, executives and security experts in matters of health and safety at work, regular inspections of facilities
- Qualification offers for training officers as well as personnel management personnel
- Cooperation in information and training offerings, such as vocational orientation in schools, company training workshops or student laboratories, collaborative education or dual study programs
- Establishment of quality management processes in education ISO 9001
- Development of life-phase-oriented working time models (e.g. part-time training).

The second central pillar is the field of action of "professional and personal development". This field of action addresses lifelong learning, the promotion of diversity and health of employees, the identification of new career paths, the responsible handling of time limits as well as a personnel management based on recognition, respect and fairness. Through the development and design of strategies and concepts for recruitment and competence development, employees are offered long-term career prospects within and outside of the research institution and the framework conditions for excellent research are provided. In particular, the focus is on the social responsibility of the employer for a balanced work-life balance, the responsible use of fixed-term employees, diversity and health promotion.

Personnel management is an important strategic partner of organizational management in order to systematically conceive and design corresponding personnel management activities. The recruitment and retention of highly qualified employees is a central challenge for non-university research institutions in order to continue contributing to the positive development of Germany as a science and business location in the future. One focus is the recruitment of young scientists as a joint task of science organisations and universities. Through a personal culture, the research organization can increase its attractiveness for young scientists.

Personnel development is a central, strategic and long-term task that encompasses all measures for the training, promotion and development of employees in the area of scientific support in a position- and qualification-oriented manner. The focus is on the development of new development and career paths and the prevention of precarious employment relationships, especially for young researchers. Other research institutions and universities also benefit from this. Personnel development enables lifelong learning, contributes to professional and personal development, supports networkable careers beyond research organisations and promotes the permeability of career paths between science and administration.

In the area of strategically oriented development and change processes, personnel development is closely linked to the objectives of organizational development. For example, personnel development takes on corresponding tasks when implementing strategic sustainability management, insofar as this is implemented as a change management process.

Temporary employment contracts are part of a dynamic and flexible science system. However, since fixed-term employment contracts are associated with planning uncertainties for the employees concerned, it is important to ensure responsible employment with fixed-term employees. Personnel and career planning should therefore be transparent, timely and continuous. In addition, temporary employees should be provided with development opportunities that facilitate their successful career in employment.

A cooperation between management and employees based on recognition, respect and fairness is an essential factor for the satisfaction and therefore also for the career opportunities of the employees. The prevailing leadership style is ideally shaped by mutual esteem, trust, self-responsibility and participation. Managers also play an important role in shaping a value-based organizational culture and implementing organizational guidelines and principles. Particularly demanding leadership tasks arise, for example, in interdisciplinary or intercultural research networks, since coordination requires not only strategic and administrative skills, but also a high level of social competence. For the development of a "good leadership culture", the institution should create appropriate framework conditions, for example through qualification offers or guidelines.

By valuing diversity, the organization ensures equal opportunities for work, reconciliation of work, family and private life, equal participation and inclusion. Diversity refers, for example, to gender, age, migration, religion, disability, sexual identity, culture and education of (potential) employees. On the one hand, diversity management means taking into account the different needs and phases of life of employees; but on the other hand also the productive use of the resulting variation of experiences, perspectives and ideas. At the same time, a diverse employee structure promotes innovation potential and is therefore a decisive success factor for research organizations and their research and innovation processes. Occupational health management goes beyond the legal requirements for occupational safety and health. The different groups of employees of a research institution are confronted with different stressors that can burden them in their work environment: Besides physical factors, which can affect the pressure of success of the scientific system, the company culture, the team climate or work organization, these are also physical factors, such as long sitting office workstations. In order to maintain the health and performance of employees, the structural and individual conditions must be tailored to suit the needs of the workplace, task and organization.

In the field of action "professional and personal development" an implementation in research organizations can look like this:

- Transparency in fixed-term policy, for example through internal and external communication or through internal organization guidelines on time limits
- Life-phase-oriented further education and personnel development concepts
- Promoting reconciliation of work and private life, for example through childcare services, eldercare, crisis hotline or flexible working time models
- Leadership awareness for diversity and health responsibilities
- Target group specific qualification offers
- Regular employee survey
- Demand-oriented further development of target group-specific qualification offers, for example for new and/or experienced executives
- Health and safety prevention, such as health days and information on issues such as workplace ergonomics or stress and relaxation.

The third pillar of sustainable personnel management in research institutes is the field of action "networking and cooperation". This involves strategic cooperation, mobility programs, the strengthening of national and international networking in terms of research, teaching, promotion of new talent, infrastructures and transfer.

National and international cooperation and networking are constitutive for the science system and open up the utilization of manifold synergies. Personal and regional cooperations are just as important as project-related, medium-term, long-term or institutionalized forms of cooperation between research institutions. These can relate to research as well as teaching, promotion of young talent, infrastructures and knowledge and technology transfer. The research organization assumes responsibility for providing the necessary framework conditions for international and national exchange both for its scientists and its administrative staff.

The importance of networking and cooperation for personnel management is evident, for example, in the context of personnel marketing in the acquisition of junior scientists, in the formation of integrated training in regional cooperation or in the subject of time limitation by opening up potential employment and career opportunities with cooperation partners. In terms of personnel development, international mobility programs support both the individual expansion of competencies and the increase in knowledge in the organization. By networking the science supporting staff, including the HR managers themselves, important expertise can be further developed.

In addition, the establishment of alumni networks is gaining in importance: as part of expert networks, former employees cooperate with the research organization as project partners or clients, and support personnel development by exchanging their professional experience and mentoring junior researchers. International networking can also serve the exchange of knowledge with developing and emerging countries, for example, by welcoming guest scientists or the consensual secondment of employees to corresponding regions. Among other things, research organizations have a responsibility to support cultural exchange and to integrate foreign scientists. Networking also results in the strengthening of Germany as a location for science. Through cooperation, regional science locations of supraregional importance can be developed and access to global knowledge flows as well as to important research topics and locations secured.

Finally, international research collaborations and networks are urgently needed, in particular, to address societal challenges scientifically, as these are mostly global in nature and involve actors across national and geographical borders.

Collaborations also have advantages in terms of investment, operation and use of research infrastructures (e.g. library networks, scientific collections and databases, laboratory facilities and large scale facilities such as accelerator facilities, telescopes or supercomputers): sharing with national and international partners allows one better utilization of the infrastructures. In its position paper "On the Political Discourse on Major Societal Challenges" (2015), the Science Council emphasizes the importance of cross-border cooperation in the scientific work on global social and environmental challenges.

The implementation in research organization can look like this:

- Offers for the extension of intercultural competences and foreign language skills for domestic and foreign personnel
- Continuing education programs for non-academic staff in order to convey knowledge of international contractual arrangements, social legislation, administrative processes or funding conditions
- Strategic coordination with cooperation partners from science and industry regarding an efficient sharing or utilization of research infrastructures.

8 Conclusion: From Common Vision to Implementation

The article shows how sustainability management can be designed in non-university research institutions and focuses on sustainable personnel management. With regard to sustainability management, the LeNa project is used as an example to illustrate how interorganizational cooperation in research projects can be structured and how the interpretation of terms in an interorganizational comparison with the aim of creating a common understanding has been developed. This approach can serve as a model for future projects that aim to achieve common guidelines or implementations in the context of sustainability through inter-organizational cooperation.

The empirical basis is a qualitative organizational-ethnographic study, which reflects the expert knowledge, the everyday experiences and the topic-oriented interpretation of sustainability in personnel management. The result provides concrete recommendations for institutional practice and offers structured and methodologically proven options for action for personnel management actors in research institutions.

Against the background of sustainable personnel management, the recommendations focus on three areas of action: service and service-oriented personnel management, professional and personal development, and networking and cooperation.

Science itself plays a central role in the realisation of sustainability. The expectations are manifold: science identifies problems and challenges, develops options for identifying alternative ways and applies these principles to itself. The research organizations Fraunhofer-Gesellschaft, Helmholtz-Gemeinschaft and Leibniz-Gemeinschaft have taken the initiative under their own responsibility as free science and have developed this framework. They are certainly role models for other institutions in the field of science. The results are remarkable in many respects: the non-university institutes involved have dealt with the topic, their own structures and many stakeholders beyond the borders of the organizations. Together they worked out what sustainability means for their work and how it can be put into practice. The guidelines and principles were first developed with Lena, but their implementation still needs to be done. The conceptual principles of sustainability management in non-university research institutions have thus been established; the will and degree of implementation are now in the individual areas of responsibility. Following this project, a scientific study on the extent to which the results of the guidelines are communicated and implemented in the research institutions would certainly deliver exciting results and could further reinforce the practical relevance of the recommendations for action.

References

Beer M (1985) Human resource management: a general manager's perspective. Free Press, New York

Choi J (2010) Organisationsethnographie im interkulturellen Kontext. Deutsch-koreanische Projektzusammenarbeit in multinationalen Unternehmen. Waxmann, München
Colsman B (2016) Nachhaltigkeitscontrolling. Strategien, Ziele, Umsetzung, 2nd edn. Springer, Wiesbaden
Di Giulio A (2003) Die Idee der Nachhaltigkeit im Verständnis der Vereinten Nationen. Anspruch, Bedeutung und Schwierigkeiten. Lit Verlag, Münster
Elkington J (1997) Cannibals with forks. The triple bottom line of 21st century business. Capstone, Oxford
Fraunhofer-Gesellschaft, Helmholtz-Gesellschaft, Leibniz-Gesellschaft (ed) (2016) Nachhaltigkeitsmanagement in außeruniversitären Forschungsorganisationen. Handreichung, Munich
Götz I, Moosmüller A (1992) Zur ethnologischen Erforschung von Unternehmenskulturen. Industriebetriebe als Forschungsfeld der Völker- und Volkskunde. Schweizerisches Archiv für Volkskunde 88:1–30
Götz I, Wittel A (2000) Arbeitskulturen im Umbruch – Zur Ethnographie von Arbeit und Organisation. Waxmann, Münster
Hauff V (ed) (1987) Unsere gemeinsame Zukunft. Der Brundtland-Bericht der Weltkommission für Umwelt und Entwicklung. Eggenkamp, Greven
Kopfmüller J (2007) Die Verbindung von Umwelt und Entwicklung im Nachhaltigkeitsdiskurs. Auf dem Weg zu einem integrativen Nachhaltigkeitskonzept. Ökologisches Wirtschaften 1:16–18
Kopfmüller J (2011) Das Integrative Nachhaltigkeitskonzept der Helmholtz-Gemeinschaft. Idee, Anwendung, Perspektiven. Lecture, 15.02.2011 in the course of the workshop "Perspektiven der Nachhaltigkeit am KIT"
Kuhn L (2008) Triple bottom line. Harv Bus Manag 1:12–13
Kurz L, Jost L, Rabelt V, Ohlhausen P, Roth K, Simon D, Fiegel-Kölblin I (2017) Nachhaltigkeit im Personalmanagement. BMBF-Projekt LeNa – Nachhaltigkeitsmanagement in außeruniversitären Forschungsorganisationen. Eigenverlag, Berlin
Loew T, Rohde F (2013) CSR und Nachhaltigkeitsmanagement. Definitionen, Ansätze und organisatorische Umsetzung im Unternehmen, Berlin
Miles RE, Snow CC (1995) The new network firm: a spherical structure built on a human investment philosophy. Org Dyn 23(4):5–18
Pfeffer J (1998) The human equation: building profits by putting people first. Harvard Business School Press, Boston
Press and Information Office of the Federal Government Germany (ed) (2012) Nationale Nachhaltigkeitsstrategie. Fortschrittsbericht 2012, Berlin
Schaltegger S (2009) Nachhaltigkeitsmanagement in der öffentlichen Verwaltung. Herausforderungen, Handlungsfelder und Methoden, Lüneburg
Ulrich P (ed) (2001) Integrative Wirtschaftsethik – Grundlagen einer lebensdienlichen Ökonomie. Haupt Verlag, Bern
Wright S (1994) "Culture" in anthropology and organizational studies. In: Wright S (ed) Anthropology of organizations. Routledge, London
Wunderer R, Kuhn T (1993) Unternehmerisches Personalmanagement. Konzepte, Prognosen und Strategien für das Jahr 2000. Campus, Frankfurt a. M.
Zaugg RJ (2005) Nachhaltiges Personalmanagement Grundlagen, empirische Erkenntnisse und Implementierung in einem KMU. Bern, Lecture
Zaugg RJ (2009) Nachhaltiges Personalmanagement: Eine neue Perspektive und empirische Exploration des Human Resource Management. Springer, Wiesbaden

Lisa Kurz has been working since 2014 at the Fraunhofer Institute for Industrial Engineering IAO as a research fellow in the project LeNa—Guidelines for Sustainability Management in non-university research institutions and in the area of knowledge transfer and teaching at the Institute for Human Factors and Technology Management IAT, University of Stuttgart. Since 2016 she has been a lecturer in work science at the ESB Business School at Reutlingen University and in strategic innovation management and organizational culture at Furtwangen University. Her research focuses on organizational culture, technology assessment and the working world of the future.

Larissa Jost is a change manager and studied business and organizational psychology. Since 2014 she has been working as a Human Resources and organizational developer at the German Aerospace Center e.V. (DLR). As research associate of the project LeNa—Guidelines for sustainability management in non-university research institutions—she worked in cooperation with the Helmholtz Association, the Fraunhofer Association and the Leibniz Association on sustainability in Human Resource Management in research organizations. She is implementation partner for the Helmholtz Association (HGF) in the field of human resources to implement the results of the research project LeNa. Before joining DLR, Larissa Jost worked as a consultant in a consultancy for strategic HR and organizational development near Cologne.

Dr. Karsten Roth has been responsible for the personnel and organizational development of the German Aerospace Center e.V. (DLR) since 2011. He had earlier worked for ten years in German authorities as well as various organizational consultations in the national and international area. Lastly as consultant with responsibility for management development and change management. He studied psychology at the University of Wuppertal, where he received his doctorate degree (Dr.-phil.) in 2007. Dr. Roth lives with his family in Cologne.

Prof. Dr.-Ing. Peter Ohlhausen Head of the department research coordination has been wsorking as a scientific researcher at the Fraunhofer-Institute for Industrial Engineering IAO since 1993. At the Reutlingen University ESB he is responsible for technology and innovation management. The key fields of his research are innovation management, R&D management, knowledge management, diversity and gender in innovation management. He led numerous projects dealing with project management implementation, knowledge management and the reorganisation of R&D departments in industry like Bosch, EADS, BAe, Siemens Medical, Daimler and several SMEs. Furthermore he worked on several projects dealing with diversity in innovation management and diversity and gender in scientific research institutions.

Knowledge, Values and Attitudes Towards Marine Protected Areas in Gozo (Malta)

Mark Mifsud and Silvana Sultana

Abstract Marine Protected Areas (MPAs) serve to protect the marine and coastal biodiversity. Education for sustainable development is of utmost importance for a management plan to be effective. Research on MPAs in Gozo is limited. This study aims to shed light on the knowledge, values and attitudes Gozitan people have towards MPAs and the extent to which education can be a vehicle for these aims. Two Gozitan areas with a MPA in their locality (Xagħra and San Lawrenz) and a locality without a MPA (Victoria) were utilised in this study. A mixed method approach was adopted and entailed the collection of data from qualitative and quantitative aspects. This paper focuses on the qualitative part of the research that consisted of fifteen interviews with stakeholders—five interviewees from each locality. The qualitative part included the analysis of the verbatim that was performed through a thematic approach analysis. The study shows that there is lack of knowledge with regards to MPAs and that the value associated with them is mainly ecological. Moreover, results show that there is a general demand for more education and awareness on MPAs. In response to these findings, a number of recommendations and a list of possible activities are drawn up according to the different stakeholders who could contribute towards incorporating ESD principles in marine education and in fostering social responsibility.

Keywords Marine protected areas · Qualitative research · Gozo
Social responsibility · Education for sustainable development

---

M. Mifsud (✉) · S. Sultana (✉)
CEER, University of Malta, Msida, Malta
e-mail: mark.c.mifsud@um.edu.mt

S. Sultana
e-mail: silvana.sultana.96@um.edu.mt

W. Leal Filho (ed.), *Social Responsibility and Sustainability*, World Sustainability Series, https://doi.org/10.1007/978-3-030-03562-4_4

1 Introduction

The Mediterranean is one of the priority eco-regions in the world. It covers 0.82% of the ocean surface and is home to 17,000 identified marine species. It hosts up to 18% of the global marine biodiversity and boasts of important endemism. It is also known for the reproduction of various pelagic species amongst which the Atlantic Bluefin tuna, the great white shark, sea turtles and other mammal fauna (Gabrié et al. 2012, p. 99).

Moreover, while the Mediterranean's shallow coastal waters protect the key species and sensitive ecosystems like the seagrass beds and corallogeneous assemblages, the deep waters are home to a unique and fragile fauna. In fact, IUCN considers the latter species as threatened or endangered (Gabrié et al. 2012) due to land artificialisation, over exploitation of resources, the proliferation of introduced non-native species, the human impact and even climate change (Mangos et al. 2010; MedPAN, UNEP/MAP & RAC/SPA 2016).

However, the risks linked to biodiversity loss are not only ecological but also moral, sociocultural and economic. In fact, the Millennium Ecosystem Assessment (2005) associates biodiversity with 'the ongoing provider of ecosystem services and the well-being of the individuals who enjoy them'. From the economic side, it was calculated that Mediterranean riparian countries gain around 26,128 million Euro per year from the marine ecosystems (Mangos et al. 2010, p. 9).

2 MPAs in Malta

The first Marine Protected Area declared under the Environment Protection Act in Malta dates back to 2005 and covers the area between *Rdum Mejjiesa* and *Ras ir-Raħeb* in Malta. This area forms part of the EU Natura 2000 Network of Protected Areas. Two years later, the area in the limits of *Dwejra*, Gozo was recognised too for its richness in biodiversity (MEPA (a)).

In the year 2010, four areas were officially given protection and declared as MPAs. These include: the stretch area in the North East of the islands, *Mġarr ix-Xini*, *Dwejra* and the area from *Għar Lapsi* to the island of *Filfla*. MEPA (2010) says that the designation of such areas is of utmost importance also to ensure that future generations can enjoy the benefits of such marine environments. Collectively Maltese MPAs cover 191 km^2 (Gabrié et al. 2012) that is 5% of Malta's territorial waters (Mifsud and Verret 2015).

All five MPAs carry the label of 'Special area of conservation of international importance'. The area around *Filfla* is considered as a National Park (Category II) by IUCN with the objective of protecting the natural biodiversity along with its underlying structure, supporting environmental processes and promoting education and recreation. All other sites, including the area from *Għar Lapsi* till *Filfla* are listed with the Category IV—habitat/species management areas. The objective of these areas focuses on the maintenance, conservation and restoration of habitats and species (Gabrié et al. 2012, p. 194).

Malta's MPAs are primarily protected due to the *Posidonia oceanica* beds which are endemic for the Mediterranean and constitute an important marine autotrophic ecosystem (Borg et al. 2005). According to Nature Trust (2015), this Neptune's grass in the Maltese waters is in a rather good state though some of the species inhibiting it, including the *Pina nobilis*, are currently under threat.

Malta benefits from the Mediterranean Sea's marine ecosystem services and gains approximately 83 million Euro per year. Such benefits derive from fisheries, renting sources for recreational support, benefits related to climate regulations like CO_2 absorption, protection against erosion and waste treatments (Mangos et al. 2010, 21).

In the MPAs concerning this study, mainly the sites of Dwejra and Ramla, there are not any people employed specifically for the MPAs. There are the cleaners who are in charge of cleaning the area and the public convenience in the vicinity. These are workers employed by either the government through the Public Cleansing Department or else employed by the respective Local Council (N. Formosa, personal communication, March 7, 2018). However, one finds hawkers in the areas that sell food and beverages or rent deck-chairs and umbrellas. These self-employed citizens get the necessary permits from the Health Authorities and the Trade Licencing Unit to operate (Commerce Department 2018).

Both sites are popular with locals and tourists alike and are among the most highly frequented areas in the Maltese Islands. Unfortunately, statistics show only the number of tourists visiting the areas and data concerning locals is non-existent. In 2015, the number of inbound visitors in Malta amounted to 1,807,269 with an expenditure of €1,643,944. The main expenses incurred were related to food and beverages (57.1%), followed by recreation including excursions and site visits (10.8%). Other statistics show that the majority of tourists visit the islands for the sun (26.8%), sun and culture (18.9%) and diving (6.3%). With regards to the most popular cultural activities, there is sightseeing (83.4%) and the most favoured outdoor sport is swimming (42.2%) (Malta Tourism Authority 2016a).

Another study indicates that in 2016, out of 1,196,630 tourists that visited Gozo, only 190,600 resided on the island for one night or more. However, it also proved that 50.8% of the tourists were returning after a day tour or a cruise trip visit (Malta Tourism Authority 2016b). This same report shows that the majority visit Victoria (94%) while 69.4% visited San Lawrenz and the Azure Window and 66.1% visited Ramla.

Nevertheless, designation on its own or retrieving the economic benefits are not sufficient. Management of such MPAs is of utmost importance and it is the duty of individuals to safeguard such unique natural environment (MEPA (b)). But in order to safeguard such heritage, the locals need to know about its value, importance and necessity and it is also the duty of the authorities, to educate (Fig. 1).

When the last MPAs were launched in 2010, Environment Protection Officer Cousin explained that the areas will 'enjoy high degree of environment protection'. However, he continued that activities happening or being planned in such areas will

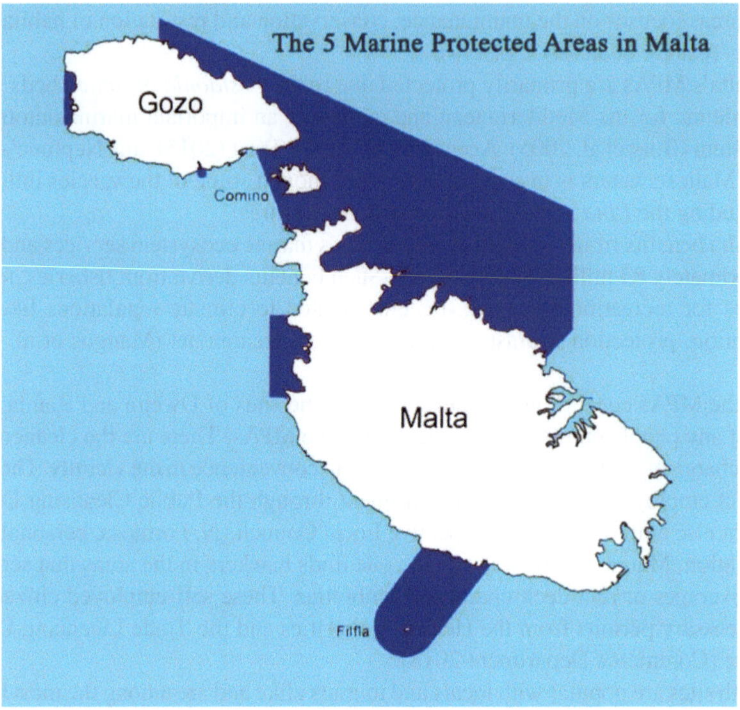

Fig. 1 The five marine protected areas in Malta

not be prohibited though permissions need to be granted. Assessments need to be carried out prior to certain events due to possible negative impacts on the biodiversity of the sites (MEPA 2010; MEPA (b)).

3 Management of MPAs in Malta

The first Action Plan for Dwejra dates back to 2005 and was prepared by MEPA, NTM and WWF Italia. This plan covers both the terrestrial and marine areas (Dwejra Action Plan 2005). In 2006, a Draft Management and Monitoring Report was issued for the island of Filfla and was based on scientific data (MEPA 2006).

Moreover, with a co-financing of 1.3 million euro from EU Agricultural Fund for Rural Development, ERA launched the 'Natura 2000 Management Planning for Malta and Gozo' in December 2016. This established management plans and provides legal provisions for almost all the Natura 2000 sites across the islands. Conservation orders were considered for smaller sites. With regards to MPAs Management Plans, there is an ongoing consultation process that terminated in May 2017 (ERA 2017).

However, as evident in modern literature on MPAs management, the new direction leads towards ecosystem-based Marine Spatial Planning. This 'term' appeared in 2006 by the EU Commission Green Paper that acknowledged MSP as a 'key instrument in managing the growing and increasingly competitive marine economy while at the same time safeguarding biodiversity' (Deidun et al. 2011). Indirectly, the concept of MSP was introduced in Agenda 21, Chap. 17, that talks specifically on integrated management and sustainable development of coastal zones including economic zones and involvement of local stakeholders in the management (UN Sustainable Development 1992).

4 Education and MPA's

Undoubtedly, education is an essential component in managing MPAs. The main conservation organisations like the WWF, CI and TNC invest in education and outreach activities in and around MPAs with the aim to provide knowledge and consequently change attitudes and behaviour (Gabrié et al. 2012).

Leisher et al. (2012) carried out one of the first studies to assess the efficacy of using education and outreach activities to improve community knowledge and attitudes about MPAs. Their results showed that people who lacked knowledge about conservations were the primary beneficiaries. Moreover, the youths were more receptive to conservation and absorbed provided information more quickly. Youths are an important stakeholder in acquiring knowledge (Jonsson 2005). Driskell (2002) considers education for youths as a 'wise long-term investment to sustain conservation efforts'.

5 Methodology

The following research questions were constructed to provide answers to lacunae:

- What knowledge do Gozitans presently have on Marine Protected areas?
- Which values are attributed to the specific MPAs in Gozo?
- How can education be the means to enact change and transform the present behaviour into a sound environmental knowledge and awareness?

For this study, a mixed-method approach was taken however, this paper will focus on the qualitative data of this research. In fact interviews were administered. Based on the traditional view of mixed methods, interviews may be combined to questionnaires to triangulate or validate findings in a way that they may mutually corroborate (Bryman 2008, p. 608). However, Fielding's (2012) concept of further analysing the data for a 'wider' image was taken into consideration.

The research design of this study originated in March 2016 and included a pilot study for both the quantitative and qualitative research. This served to what Van

Table 1 Interviews' thematic categories

Sustainable Development Pillars	Thematic Sections
Social	Awareness
	Education
Environment	Environment Consciousness
Economy	Economic Value
Culture	Cultural Activity

Teijlingen and Hundley (2001) call a 'trial run'. It was done in preparation for the major study and was considered instrumental for checking about failures, ambiguities or complications. The researcher approached one interviewee for the pilot study. The questions were easily understood and consequently the original format was retained. Even the replies received during the piloting phase were retained and were included with the others received later. In fact the other interviews were carried out in November and December 2016. The analysis was done in March 2017 and the study was completed by May 2017.

The interview contained 7 core questions that were asked to all participants and other more specific questions (between 2 and 5) were asked according to the designation of the participant. In addition, three more questions were prepared in case the participants did not bring forth aspects concerning the well-being, economy and sustainable development in relation to Marine Protected Areas.

Each face-to-face interview took approximately 20 min in length and was held at the venue, date and time suggested by the participant. Thirteen interviews were recorded on the researcher's mobile phone and consequently, transcribed by the researcher. Two interviewees preferred to reply to the questions in writing and in these cases, emails were sent. Since some interviews were done in the Maltese language, a translation of the transcript followed.

The interviews were organised in five different thematic sections based on the four pillars of sustainable development (Table 1).

The researcher's theme classification is rather straightforward except for the social aspect that is divided in two themes: *awareness* and *education*. UNESCO (2017) considers 'awareness' as knowledge and information that may have significant impact on people's lives, particularly where Information and Communication Technologies (ICT) are involved. It even states that 'awareness' can transform economies and societies. On the other hand, *education* holds a broader dimension and has the ability to transform lives. Helen Clark, UNDP Administrator (UNESCO, Incheon Declaration 2015) states:

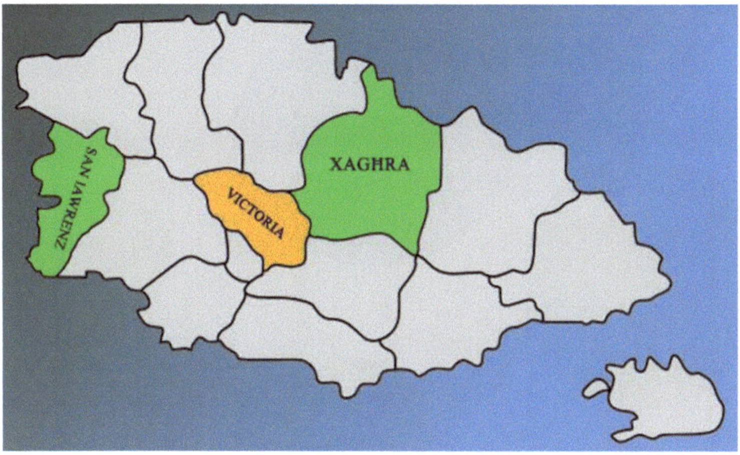

Fig. 2 Map of Gozo showing the three localities involved in this study

> In our world, knowledge is power and education empowers. It is an indispensable part of the development equation. It has intrinsic value – extending far beyond the economic – to empower people to determine their own destiny. That is why the opportunity to be educated is central to advancing human development.

6 Sampling Strategy

For the data collection, Gozitans from two MPA zones and from a non-MPA zone were approached. This was done with the intention of perceiving the awareness, values and even opinions of participants' with the possibility of comparing the results. The participants chosen were from the localities of Xagħra, San Lawrenz and Victoria. Xagħra lies on top of the largest north-eastern stretch of MPA in Malta. San Lawrenz, that incorporates Dwejra, is another chosen locality in the western coast of Gozo and Victoria is the main town of the island that confines with other villages and not with beaches (Fig. 2).

The fifteen semi-structured interviews were carried out in the concerned localities. These included a member from the respective Local Council, a member from the School Management Team, a youth, a tourist operator and a senior citizen. Figure 3 gives further details about the interviewees.

The interview was divided into four sections. The first part included some ice-breaking questions mainly related to the interviewee's job or relationship to the Marine Protected Area. The second part focused on awareness while the third concerned the values and opinions of the interviewee. The final part was specifically concerned with the interviewee's designation or status.

Fig. 3 The interviewees

The researcher opted for the semi-structured interviews since these provide 'sufficient flexibility to approach different respondents differently while still covering the same area of data collection' (Noor 2008). This approach gives the possibility to interviewees to express themselves freely and give insights of how they view the world (Markham as cited by Bryman 2008, p. 633). Moreover, they could freely express their opinions and ideas in their own words (Johl et al. 2012).

7 Limitations of the Study

Although every possible avenue of reducing limitations was explored and the research design tried to limit the weaknesses, limitations can still be identified. A limitation could be the choice of interviewees. There are sections within society that could have given other interesting comments but, due to time constraints and the word count limit of the study, it was not possible to carry out other interviews. Another issue is that this study assesses the participants' values that are attributed to MPAs. The term 'value' here is intertwined with 'behaviour' and 'attitudes'. In fact, in his studies on e-shopping, Jayawardhena (2004), found that personal values were significantly related to positive attitudes that consequently served as a predictor for behaviour. Even Sharma and Jha (2017) concluded that personal values impact behaviour.

8 Results

This study tried to compare participants residing in localities with MPAs sites (Xagħra and San Lawrenz) with those living in a non MPA site (Victoria). On the whole, replies were quite similar and the interest in the topic was generated among all participants. This may be due to the fact that they associate the sites with the island's heritage and identity.

All qualitative data gathered was analysed using the thematic approach which, according to Braun and Clarke (2006), serves to identify, analyse and report patterns or themes within data. In reality, this kind of research is still 'hotly debated' because it is not 'generalisable' and is rather 'specific to a particular context, participants and time' (Thomas and Harden 2008).

The following illustrate the results that emerged from the five topics of the thematic approach:

- Awareness
- Education
- Environment Consciousness
- Economic Value
- Cultural Activity.

9 Awareness

'Awareness' is the theme that was brought forth by every interviewee. As already mentioned, 'awareness' is intertwined with education. However it is closely linked with information and knowledge that are the first mediums towards protection.

> Iva l-għarfien huwa importanti ħafna. Jekk int ma tkunx 'aware' ma tistax tipproteġi dawn is-sites.
>
> Yes, information is important. If you lack 'awareness' you cannot protect these sites.
>
> [Interview with SMT member – Victoria]

Various proposals of information dissemination and awareness campaigns were mentioned. The researcher however thinks that the SMT members in the three schools have a wider vision. They understand the importance of including the matter as a cross curricular theme. At the same time they require experts in the field because they are conscious of their limits and that of their staff. Moreover, they are also in favour of outdoor learning activities.

Additionally, the senior citizens are still keen on increasing their knowledge. They said that the knowledge they possess was obtained either through tradition or else through personal interest. They admit that they have heard no one talking about these areas and their values. However, the major surprise came from the youths. This younger group was the least confident during the interview and confessed clearly that the subject was rather new to them. The researcher felt also that lack of awareness was not the only issue. Youths seemed to find it hard to express an opinion and, in fact, the researcher helped with prompting to keep the conversation flowing.

10 Education

Education was also mentioned several times during the interviews. Education is more than just awareness. It has the power to transform lives and as Annan (1999) said:

> Education is a human right with immense power to transform. On its foundation rest the cornerstones of freedom, democracy and sustainable human development (UNICEF 1999, p. 4).

The term 'education' was mostly highlighted by participants from the SMT of schools and from members from the Local Council who may perhaps have a wider and long-term vision.

11 Environment Consciousness

Environment was the theme mentioned most frequently and justly so since MPAs are part of the nation's environment. All participants agree with the protection and conservation of the marine environment which was highly associated with fishing. The researcher noticed that the participants gave a lot of weight to fishing, most probably due to the fact that they ignore other important factors like marine habitats and flora.

12 Economic Value

The economic value was not on top of the discussions but emerged every now and then. There were two different points of view: those who see economy as a benefit and others who see economy as an abuse.

As expected, the tourist operator (Xagħra) sees the potential of a marketing campaign which will attract more tourists. Even the member from San Lawrenz Local Council agrees that MPAs can attract more tourists but insists on sustainable tourism to avoid additional pressures on the environment.

Another suggestion came from the SMT member (San Lawrenz) who showed interest in a sunken wreck close to the MPA that will attract more divers and mitigate local habitat and species.

Another interviewee has a perception that links economy with harm to the environment. He reflects on the current situation where egoism and consumerism dominate over reason and righteousness.

> Għax once li ma tkunx protected tinbelgħa mis-settur ekonomiku. Issa issirx lukanda jew issirx tourist resort once li ma tkunx protected, l-egojiżmu tal-bniedem jew il-konsumiżmu tallum naħseb jibla' kollox.
>
> Since a site is not protected, the economic sector gets hold of the area either by building a hotel or a tourist resort. Once a site is not protected, man's egoism or modern consumerism takes over.
>
> [Interview with SMT member – Xagħra]

13 Cultural Activity

The cultural aspect was the least mentioned. The aspects that were most mentioned were fishing and the recreational aspect that is often linked with swimming. One particular comment from the Tourist Operator (Xagħra) mentioned the banning of certain cultural activities in a MPA. Another interesting comment mentions the legends related with the areas forming part of the north-eastern stretch of Gozo.

Tabel 2 presents a synopsis of the thematic results obtained from the qualitative data.

The social and environmental themes were linked in a way that they depend on one another. However links with the other themes—economy and culture—were not brought forth. This shows again that participants are still not connecting the four pillars of sustainability.

The responses received during the interviews were very interesting and each interviewee gave useful insights. All the information gathered serve to back up and strengthen the scientific results obtained from the quantitative data.

Nevertheless, the qualitative results can be grouped into six main points that will lead the researcher in establishing recommendations for future research.

1. The results show that there is a general need and eagerness for awareness of MPAs. All participants in this research felt that they lack knowledge on MPAs, though they agree on their protection.
2. Moreover, since the MPAs are related to the environment, most participants think of this aspect only without relating it with economy and culture. This means that the Sustainable Development concept is not yet ingrained in the mentality and most of the participants are still unable to connect the sustainable pillars.
3. Members from the SMT of schools did mention that marine education should be a cross-curricular theme but no one referred specifically to the term 'Education for Sustainable Development'. The researcher assumes that the concept of ESD is understood up to a certain extent and it still remains a 'buzz-word'.
4. Youths interviewed were the less knowledgeable on the topic but what is worse, according to the researcher, is the fact that they found it difficult to give an opinion. The problem may be more deeply rooted and may be linked with their thinking process. They seemed unable to express themselves and this is undoubtedly

Table 2 Synopsis of qualitative results

	Sustainable Development Pillars			
	Social	**Environment**	**Economy**	**Culture**
Thematic Sections	• Awareness and education were repeatedly mentioned. Participants are aware of their lack of knowledge. • SMTs did not mention the term 'ESD' though they mentioned marine education as a cross-curricular theme. • Youths were the less knowledgeable and were not confident in holding the conversation.	• All participants are environmentally conscious. • All participants agree on protecting the marine sites. • Fishing was frequently mentioned. • Only two participants mentioned the need of man to live in harmony with nature.	• Two different points of view emerged: economy as a benefit and economy as harm to the environment. • The ecosystem services' value was never brought forth.	• Fishing and the recreational aspect were mentioned. • Legends linked with *Ghajn Barrani* and *Ramla Bay* which are part of the MPA at the north-eastern stretch of the Maltese Islands.

Table 3 The recommendations

	Recommendations
Knowledge	1. More information provided by authorities that currently are in charge of MPAs
Value	2. Effective management that puts at the forefront the ecosystem services 3. A green economy that safeguards, exalts the natural heritage and provides job opportunities 4. Appreciation of the cultural value
Education	5. Strengthening of ESD principles and practices 6. Education and awareness campaigns on MPAs

against ESD principles that bring to the forefront communication, discussions and fora.

5. The value of the ecosystem services was never mentioned during the interviews. Though this is still a much debated topic, it is important that citizens are aware of what constitutes a MPA, of its value for the environment and also of its importance for the well-being of the locals.
6. The cultural aspect was also left apart. Certain legends were mentioned during the interviews and this could be one cultural aspect that can serve as a good introduction for MPAs to the younger generation. Even certain social gatherings or festivals in MPAs could serve to educate and inform the public about the value that lies in the national waters.

14 Recommendations

Table 3 presents a set of six recommendations brought forward from this study. Though the recommendations are classified in three sections—knowledge, value and education—they are all interconnected. Consequently, Fig. 4 provides a list of possible initiatives and 'expert bodies' that should be involved in order to create consciousness and responsibility (Fig. 5).

Recommendations:

- Communication Strategy (ERA, UoM, CEER)
- Marine education in schools (MEDE, CEER)
- Marine Biology Courses in Gozo (UoM, Gozo Campus)
- Outreach and on site activities (ERA, Local Councils)
- Underwater documentaries in prime-time hours on TV (UoM, diving companies)
- Religion lessons linked to the marine environment (Church Authorities, MEDE)
- Courses for people in authority like politicians, policy-makers and journalists (ERA, UoM)
- Investment in Interpretation Centres and proper signage (Local Councils, ERA)
- Nutrition and cooking sessions linked to the MPAs (Local Councils, NGOs, Institute of Tourism Studies, Health Promotion Unit)
- More wrecks or underwater museums to attract divers (Diving Schools, Local Councils, Art Funds)
- More resources or guidelines for teachers, youth leaders, Private Sector on including ESD in teaching, at work or at leisure (UoM, CEER, MEDE)

Fig. 4 Possible initiatives

15 Further Studies

On considering the results obtained from this research, the following ideas for further research are being proposed:

1. A study at *Dwejra* MPA to consider whether its popularity has increased or decreased after the Azure Window collapsed in March 2017.
2. A study on whether there is a real need for more MPAs in Gozo, as proposed in the electoral programme for the general elections in Malta (Forza Nazzjonali 2017, p. 22), which areas should be proposed and how can the locals be involved in the selection of the new MPAs.
3. Research on how ESD practices can bring forth, among the locals, a sense of social responsibility that contributes towards protection, appreciation and ownership of the MPAs.

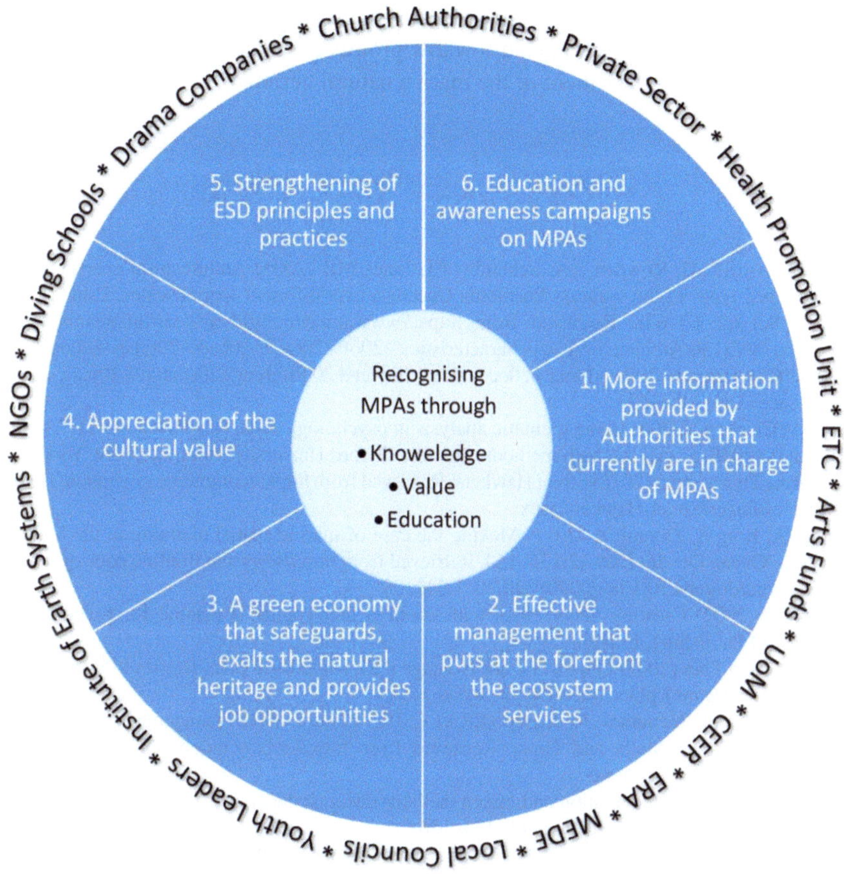

Fig. 5 Summary of the recommendations and the 'expert bodies'

16 Conclusion

This research gives a clear picture of the present awareness and perceptions of MPAs and marine education in Gozo. To safeguard MPAs and their ecosystem services, action needs to be taken before it might become too late. The study also shows that the global vision of Gozitan people towards Education for Sustainable Development needs to improve.

From a personal point of view, this study provided an enriching experience particularly during meetings with different individuals during the interviews. It helped also the researcher to appreciate more the marine environment and encouraged her to work more in the field of ESD.

It is hoped also that this study serves as a starting point for appropriate and necessary actions to enact change towards protection, appreciation and above all, sustainable use and enjoyment of the marine natural heritage.

References

Borg JA, Attrill MJ, Rowden AA, Schembri PJ, Jones MB (2005) Architectural characteristics of two bed types of the seagrass Posidonia Oceanica over different spatial scales. Estuar Coast Shelf Sci 62:667–678. Retrieved from https://www.um.edu.mt/library/oar/bitstream/handle/123456789/5130/Architectural%20characteristics%20of%20two%20bed%20types%20of%20the%20seagrass%20Posidonia%20oceanica%20over%20different%20spatial%20scales.pdf?sequence=1&isAllowed=y

Braun V, Clarke V (2006) Using thematic analysis in psychology. Qual Res Psychol 3(2):77–101

Bryman A (2008) Social research methods, 3rd edn. Oxford University Press Inc., New York

Commerce Department (2018) Street Hawkers. Retrieved from https://commerce.gov.mt/en/Trade_Services/Pages/Street-Hawkers.aspx

Deidun A, Borg S, Zammit A (2011) Making the case of marine spatial planning in the Maltese Islands. Ocean Dev Int Law 42:136–154. Retrieved from http://www.tandfonline.com.ejournals.um.edu.mt/doi/abs/10.1080/00908320.2011.542108

Driskell D (2002) Creating better cities with children and youths: a manual for participation. UNESCO Publishing, London

Dwejra Action Plan (2005) Qawra/Dwejra Heritage Park—approved plan. Retrieved from https://www.mepa.org.mt/LpDocumentDetails?syskey=541

Environment and Recourses Authority (ERA) (2017) Natura 2000 management planning for terrestrial sites in Malta and Gozo. Retrieved from http://era.org.mt/en/Pages/Natura-2000-Management-Planning.aspx

Fielding NG (2012) Triangulation and mixed methods designs: data integration with new research technologies. J Mix Methods Res 6(2):124–136

Gabrié C, Lagabrielle E, Bissery C, Crochelet E, Meola B, Webster C, Claudet J, Chassanite A, Marinesque S, Robert P, Goutx M, Quod C (2012) The status of marine protected areas in the Mediterranean Sea. *MedPAN & RAC/SPA*. Ed: MedPAN Collection. Retrieved from http://www.medpan.org/documents/10180/0/The+Status+of+Marine+Protected+Areas+in+the+Mediterranean+Sea+2012/069bb5c4-ce3f-4046-82cff72dbae29328

Jayawardhena C (2004) 'Personal values' influence on e-shopping attitude and behaviour. Internet Res 14(2):127–138

Johl SK, Bruce A, Binks M (2012) A study on the use of mixed method approach via sequential procedure to investigate corporate governance in corporate entrepreneurship among the 100 U.K. Financial Times Stock Exchange (FTSE) companies. Afr J Bus Manag 6(21):6369–6377. Retrieved from http://www.academicjournals.org/journal/AJBM/article-full-textpdf/31405A729890

Jonsson A (2005) Public participation in water resources management: stakeholder voice on degree, scale, potential and methods in future water management. Ambio 34(7):495–500

Leisher C, Mangubhai S, Hess S, Widodo H, Soekirman T, Tjoe S, Wawiyal S, Neil Larsen S, Rumetna L, Hamlin A, Sanjayan M (2012) Measuring the benefits and costs of community education and outreach in marine protected areas. Mar Policy 36(5):1005–1011. Retrieved from http://www.sciencedirect.com/science/article/pii/S0308597X12000395

Malta Environment & Planning Authority (a) Which are the current marine protected areas of Malta? Retrieved from https://www.mepa.org.mt/currentareas

Malta Environment & Planning Authority (b) What can I do for marine protected areas? Retrieved from https://www.mepa.org.mt/canido

Malta Environment & Planning Authority (2006) Marine scientific surveys around Filfla for its conservation—draft management and monitoring report. Retrieved from http://www.mepa.org.mt/erdf_filfla

Malta Environment & Planning Authority (2010) Four new marine protected areas. Retrieved from https://www.mepa.org.mt/news-details?id=617

Malta Tourism Authority (2016a) Tourism in Malta 2015. Retrieved from https://www.google.com.mt/search?rlz=1C1GCEA_enMT763MT763&q=MTA+statistical+report(OF)2016+Edition&spell=1&sa=X&ved=0ahUKEwiAk5_yzc3ZAhVQ2aQKHWmrDiUQBQgjKAA&biw=1600&bih=794

Malta Tourism Authority (2016b) Tourism in Gozo 2016. Retrieved from http://www.mta.com.mt/page.aspx?id=134

Mangos A, Bassino J-P, Sauzade D (2010) The economic value of sustainable benefits rendered by the Mediterranean marine ecosystem. Plain Beau, Valbonne. Retrieved from https://planbleu.org/sites/default/files/upload/files/Cahier8_marin_EN.pdf

MedPAN, UNEP/MAP & RAC/SPA(2016) The 2016 status of marine protected areas in the Mediterranean 2017 (in prep.). *Brochure*. Retrieved from http://www.medpan.org/documents/10180/0/Brochure+The+2016+status+of+MPAs+in+the+Mediterranean/2c19ef9e-fe45-40ed-9756-391aff46cc9a

Mifsud M, Verret M (2015) Perceptions of the Maltese public towards local Marine protected areas. J Teach Educ Sustain 17(1):48–57

Millennium Ecosystem Assessment (2005) Retrieved from http://www.millenniumassessment.org/en/index.html

Nature Trust Malta (2015) Current and future impacts on the marine environment: The challenge to achieve good environmental status. *A Summary Report*. MedTrends Project, Malta

Noor KBM (2008) Case study: a strategic research methodology. Am J Appl Sci 5(11):1602–1604. Retrieved from http://docsdrive.com/pdfs/sciencepublications/ajassp/2008/1602-1604.pdf

Sharma R, Jha M (2017) Values influencing sustainable consumption behaviour: exploring the contextual relationship. J Bus Res 76:77–88

Thomas J, Harden A (2008) Methods for the thematic synthesis of qualitative research in systematic reviews. BMC Med Res Methodol. Retrieved from https://bmcmedresmethodol.biomedcentral.com/articles/10.1186/1471-2288-8-45

UNESCO (2015) Incheon declaration—towards inclusive and equitable quality education and life-long learning for all. Retrieved from http://en.unesco.org/world-education-forum-2015/incheon-declaration

UNESCO (2017) Building knowledge societies. Retrieved from http://en.unesco.org/themes/building-knowledge-societies

UNICEF (1999) The state of the world's children—education. Retrieved from https://www.unicef.org/sowc99/sowc99a.pdf

United Nations Sustainable Development (1992) Agenda 21. Retrieved from https://sustainabledevelopment.un.org/content/documents/Agenda21.pdf

Van Teijlingen ER, Hundley V (2001) The importance of pilot studies. Soc Res Update 35. Retrieved from http://sru.soc.surrey.ac.uk/SRU35

Social Projects and the Internalization of Sustainability and Social Responsibility: Concepts for the Improvement of Quality of Life

Sônia Regina da Cal Seixas, Michelle Renk, João Luis de Moraes Hoefel and Gabriela Farias Asmus

Abstract Based on studies carried out in São Paulo State, Brazil, it was observed that several municipalities have faced huge environmental changes that can be noticed in places where sustainability of natural resources was not prioritized. It was observed the prevalence of an economic development model supported by technological megaprojects in areas of environmental conservation, which also represent areas prone to social and technological risks. The municipality of Bom Jesus dos Perdões, located in São Paulo State, showed very different average values for some social indicators in a set of municipalities of the region. The analysis of data on violence and crime, between 1996 and 2015, allows us to present a general local diagnosis that shows decreasing social security. Simultaneously, it allowed to identify how it is possible to reflect on social reality through two projects: (1) voluntary social work with children and adolescents at risk, through sports practices and (2) Integrated Management Office [Gabinete de Gestão Integrada]—GGI—a project created by the Brazilian Federal Government, of optional municipal adhesion, where it is proposed the integration of several local institutions aiming a joint management of public security. Both projects seek to improve social reality with sustainability and social responsibility of the population.

S. R. da C. Seixas (✉) · M. Renk
Center for Environmental Studies and Research, NEPAM, State University of Campinas, UNICAMP, Rua dos Flamboyants, 155—Cidade Universitária, Campinas, SP 13083-867, Brazil
e-mail: srcal@g.unicamp.br

M. Renk
e-mail: michellerenk@yahoo.com.br

J. L. de M. Hoefel
Atibaia College (FAAT)/Center for Sustainability and Cultural Studies (NESC/FAAT), Estrada Municipal Juca Sanches 1050, Atibaia, SP 12954-070, Brazil
e-mail: jlhoeffel@gmail.com

G. F. Asmus
Federal University of ABC, Avenida dos Estados n. 5001, Santo André, SP 09210-580, Brazil
e-mail: a.gabriela@ufabc.edu.br

W. Leal Filho (ed.), *Social Responsibility and Sustainability*, World Sustainability Series, https://doi.org/10.1007/978-3-030-03562-4_5

Keywords Social projects · Sustainability · Social responsibility · Quality of life

1 Introduction: Socioenvironmental Issues and Sustainable Development

The Sustainable development concept has been part of the public arena since the establishment of the *World Conservation Strategy* (IUCN 1980; Baker 2006). During the 1980' the concept incorporated social, economic and ecological approaches toward development, as addressed by the *Our Common Future* document in 1987 (WCED 1987; UN 2012-2015 2015a).

This approach revolutionized the concept, broadening it strongly and allowing it to lose its exclusive disciplinary character, be it biological or economic, implying the researchers of the area to review their own approaches (Baker 2006, pp. 18, 19). It also allowed extending our vision of the world and beginning to treat the concept in several lines of reasoning (Bawden and Allenby 2017; D'Harcourt et al. 2017; Gupta 2017; Lang et al. 2012; Lindahl et al. 2016; Saito 2017).

In this perspective, development with sustainability impute an approach that takes into account the participation of society as a whole in the search for and consolidation of development that respects the nature and the different social demands. This aspect is highlighted by Lindahl et al. (2016) and van der Leeuw (2014), which evidenced the similarity between sustainability and democracy, freedom and social justice concepts (Lindahl et al. 2016). Considering that such concepts have a political character and are strongly disseminated between societies, it is not easy to incorporate them in the debate arena of a development model which highlights the importance of natural resources and social political participation in building a sustainable society. Moreover, van der Leeuw (2014) conceptual approach highlights culture and individual commitment to the construction of a more just, egalitarian and sustainable society. In order to complement this contribution, two aspects that need to be rescued and present in the global sustainability agenda must be highlighted: the role of education and social responsibility, to stimulate and achieve a better quality of social life with sustainability. These aspects are important considering the difficult moment the world is going through, from an economic, cultural and environmental point of view, with special emphasis on serious situations of social, ethnic, racial intolerance, conflicts, as well as the strong presence of extreme events related to climate change.

Robert et al. (2005), in reviewing the concept, its goals and values from the big global reports, draw attention to the global sustainable development project that is being pursued, highlighting the values of the 2000 Millennium Declaration. They point out that six crucial points should be considered: Freedom, equality, solidarity, tolerance, respect for nature and shared responsibility. Moreover, according to Leal Filho et al. (2015), environmental education plays an essential role in addressing such values. The authors point out that sustainable development is and will continue to be a matter of great international concern, and that despite advances in recent decades much still needs to be done to improve future prospects for sustainable development.

1.1 Sustainability and Social Responsibility

It is fundamental to understand the role of social responsibility to improve the quality of life and solidify a local sustainability project. The concept of social responsibility has been widely used in the corporate world. Indicating the profile of responsibility, strategy and actions related to a company's social and environmental commitment is its corporate responsibility (CSR), expressed in its sustainability reports, which are vehicles used to demonstrate how a particular company behaved in relation to environmental issues during a given financial period, and what are their future strategies for this type of action (Gatti and Seele 2014; Camilleri 2017). In the context of this paper, the concept of social responsibility will be used in a more comprehensive way, seeking to link the commitment of a certain group or institution with strategic actions to improve the quality of life of a given population from an environmental and social point of view.

We share with Baker and Mehmood (2015) the view that the recognition of crisis situations can help in the constitution of socially innovative actions that can become a tool for both social promotion and improvement of the quality of life of the population, thus contributing to social transformation and the creation of sustainable places. In the following items we will deal specifically with this premise. First, we will show how the identification of significant changes in a given region and its consequent social problems was the start point to the constitution of two social projects in the municipality of Bom Jesus dos Perdões, which were considered an opportunity toward sustainability with social responsibility.

As Baker and Mehmood (2015, pp. 321, 322) pointed out, social innovation initiatives from a community perspective can positively change state participation and commitment, while at the same time putting strong pressure on more creative and participatory forms of governance. The authors believe that governance processes that strengthen the role of economic and social actors in the direction of social change can contribute to instill democratic practices in a given society. Considering the social, economic and state actors that mix as agents of social change, social innovation can play a key role in enhancing human-environment-sustainable interactions.

In this way, from experience accumulate trough studies carried out in São Paulo State, Brazil, it was observed that several municipalities have faced enormous environmental changes that can be noticed in places where sustainability of natural resources was not prioritized. It was perceived the prevalence of an economic development model supported by technological megaprojects in areas of environmental conservation, which also represent areas prone to social and technological risks.

In this article, our main objective is to analyze how social responsibility projects can contribute to sustainability and improve the quality of life of population in a municipality. The municipality chosen is Bom Jesus dos Perdões, located in São

Paulo State, which showed very different average values for some social indicators in a set of municipalities of the region studied in a previous research (Hoeffel et al. 2015, 2015a; Seixas et al. 2015). In this paper we will present and analyze only the data from Atibaia as it is the largest municipality close to Bom Jesus dos Perdões. Selected secondary data such as homicides, transport accidents and suicides were extracted from the National Health Department Database (DATASUS) for the period 1996 to 2015, in accordance to the International Disease Classification (ICD-10). Others indicators such as knavish homicide, rape, robbery and theft were extracted from the Statewise System for Data Analysis Foundation (SEADE) for the period 2001 to 2015. All indicators were presented as rates per 100 thousand inhabitants, as standardized by the World Health Organization (WHO).

We also proceeded semi-structured interviews with political and community leaderships in the municipalities of Bom Jesus dos Perdões and Atibaia, during the period from March to December 2017.

The analysis of data on violence and crime, allows us to present a general local diagnosis that shows decreasing social security. Simultaneously, it allowed to identify how it is possible to reflect on social reality through two projects: (1) voluntary social work with children and adolescents at risk, through sports practices and (2) Integrated Management Office—GGI—a project created by the Brazilian Federal Government, of optional municipal adhesion, where it is proposed the integration of several local institutions aiming a joint management of public security. Both projects seek to improve social reality with sustainability and social responsibility of the population.

1.2 Bom Jesus dos Perdões, São Paulo/Brazil: Local Problems, Violence and Crimes

Bom Jesus dos Perdões is a municipality in the state of São Paulo, with an intense religious tradition and natural attractions that give it recognition as both tourist and religious circuit (Prefeitura de Bom Jesus dos Perdões 2018). It covers an area of 109 km^2, with estimated population reaching 24,023 inhabitants in 2017 and population density of 181.87 hab/km^2 in 2010 (IBGE 2018), with significant urban and especially industrial growth occurring alongside Dom Pedro I Highway (SP-65) (Prefeitura de Bom Jesus dos Perdões 2018).

According to Hoeffel et al. (2009, 2010), this municipality is peculiar for its technical- scientific and political-economic not totally justified exclusion from a very big regional conservation unit, the Environmental Protection Area (EPA) of the Cantareira System. This EPA was created in 1998 with the objective of maintaining and improving water quality, mainly in the municipalities around Cantareira System reservoirs, which supply the São Paulo Metropolitan Region (São Paulo 2000) and regulate the flow of water to the Metropolitan Region of Campinas. Bom Jesus dos Perdões is completely surrounded by this EPA. It is also important to note that the

D. Pedro I highway cuts a significant part of this EPA and the Atibainha River Reservoir, part of the Cantareira Water Reservoirs System.

Thus we can observe in Bom Jesus dos Perdões the existence of an Industrial Pole located alongside the D. Pedro I highway, in the Atibainha River floodplain, which has determined, besides a great urban and industrial expansion, several environmental problems. Another relevant aspect is the fact that this municipality is classified as a priority area (priority 7—on a scale of 1 to 8) in the Connectivity Map of the state of São Paulo, which led to the creation of Itapetinga State Park in 2010, which comprises part of the territory of the municipality, in a portion not cut by the D. Pedro I (São Paulo 2010). These data demonstrate that the municipality presents an ecological richness and biodiversity that must be preserved, a fact that reinforces its inclusion in the EPA of the Cantareira System and the necessity of the implantation of suitable conservationist measures.

The D. Pedro I highway is 145.5 km long and interconnects two important regions of the state—Vale do Paraíba and Metropolitan Region of Campinas—(Rota das Bandeiras 2018). In 2006, the state of São Paulo invested in the project "Export Corridor Campinas—São Sebastião", which includes the route from Viracopos International Airport (in Campinas) to the Port of São Sebastião on the north coast of the state. Investments involving the duplication of the Tamoios highway, the recovery of the D. Pedro I highway, the construction of the road layout of São Sebastião and Caraguatatuba, and the expansion of the Port of São Sebastião were carried out with the objective of creating a flow route for the industrialized products of São Paulo (Braga 2008).

As an expected result of these investments, there was a significant increase in the number of companies, industries and especially logistic sheds installed along the D. Pedro I highway. From the year 2006, there was an increase in the number of employers' establishments in the municipalities along this road axis. In Atibaia, the largest municipality that borders Bom Jesus dos Perdões, growth was approximately 48%. Bom Jesus dos Perdões almost doubled the number of establishments in 10 years, presenting growth of just over 70%, according to Fig. 1.

The increase in the number of employers in the region, especially in Atibaia from 2006 to 2011, certainly heightened the population pressure in the neighboring municipality, Bom Jesus dos Perdões. In the 1980s, Atibaia had a geometric population growth rate almost double of that observed in the state of São Paulo; however, it began to decline in the following decade. Meanwhile, the population of Bom Jesus dos Perdões began to grow significantly in the 2000s, reaching its peak in the following decade, with a geometric growth rate 266% higher than the observed in São Paulo and 205% higher than that of Atibaia (Fig. 2).

The pressure of the population growth ends up generating the disordered occupation of the urban and peripheral areas accentuating problems of deficit infrastructure. For example, the contingent in the areas of safety, health and transportation do not follow the new demands of the municipality, which can also generate new impacts such as increased urban violence.

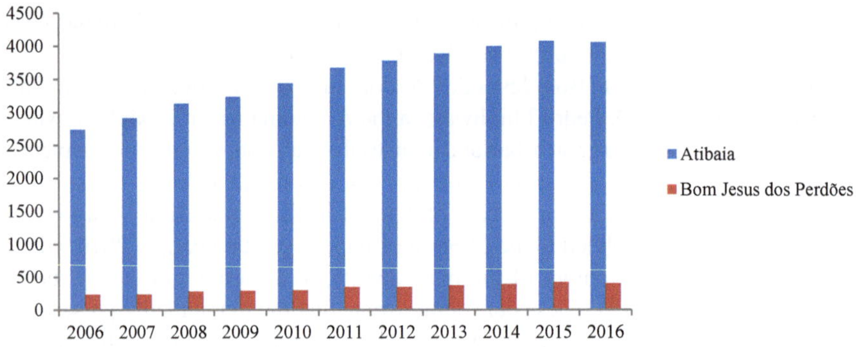

Fig. 1 Evolution of the number of employers' establishments in the municipalities of Atibaia and Bom Jesus dos Perdões from 2006 to 2016. *Source* SEADE (2018)

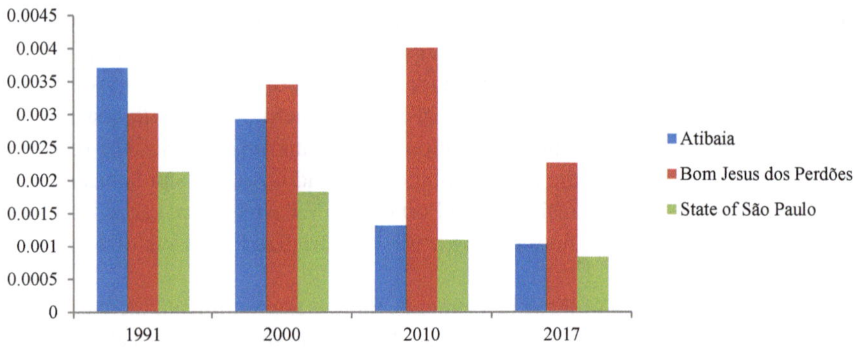

Fig. 2 Geometric rate of annual growth of the population of Atibaia, Bom Jesus dos Perdões and state of São Paulo in the 1980s, 1990, 2000, 2010 (in % per year) *Source* SEADE (2018)

According to an interviewee:

A public opinion survey was conducted in the city (Pesquisa Perdões 2020) and the 3 areas pointed out as priority by the community were safety in the first place, followed by education and health. (Interview with a city councilman 2017).

1.2.1 Urban Violence and Crimes

Homicides

Considering the period of analysis (1996–2015), the homicide rate decreased 70% over the period in the state of São Paulo. Atibaia surpassed the state average during the period 2007–2011, when it then became similar to state averages. Bom Jesus dos Perdões equates to the state media from 2003, with greater peaks in 2005 and 2007 and a tendency of decrease in homicide rates (Fig. 3).

Fig. 3 Homicide rate per 100 thousand inhabitants from 1996 to 2015. *Source* DATASUS (2018)

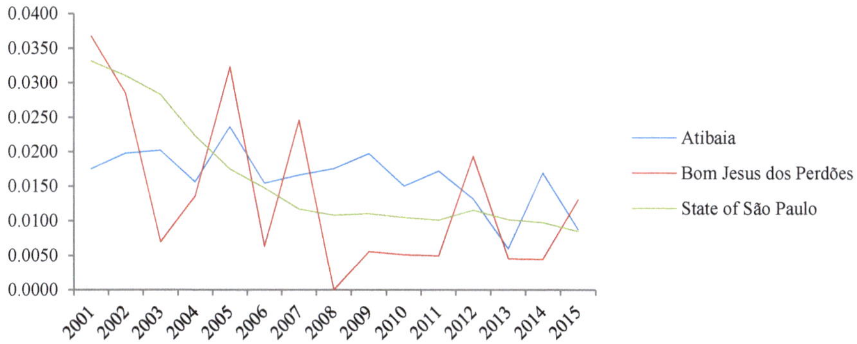

Fig. 4 Knavish homicides rates per 100,000 inhabitants from 2001 to 2015. *Source* SEADE (2018)

Knavish homicides

The rate of occurrence of intentional homicide has progressively decreased in the state of São Paulo. In the evaluated municipalities, the rates exceeded the state average in 2004, fact that can be observed in the following years except from the period between 2008 and 2011, when Bom Jesus dos Perdões drops its rates (Fig. 4).

Transport accidents

The number of deaths due to traffic accidents decreased over the period, however, the municipalities studied have rates higher than the state of São Paulo, with few exceptions, during the whole period. This suggests a possible influence of Dom Pedro railway over the municipalities' everyday (Fig. 5).

Suicide

Suicides have presented constant occurrence rates in the State of São Paulo. However, Bom Jesus dos Perdões highlights for presenting rates higher than the state average in some years. Similarly, Atibaia presents rates equivalent or higher than the state averages during the period evaluated (Fig. 6).

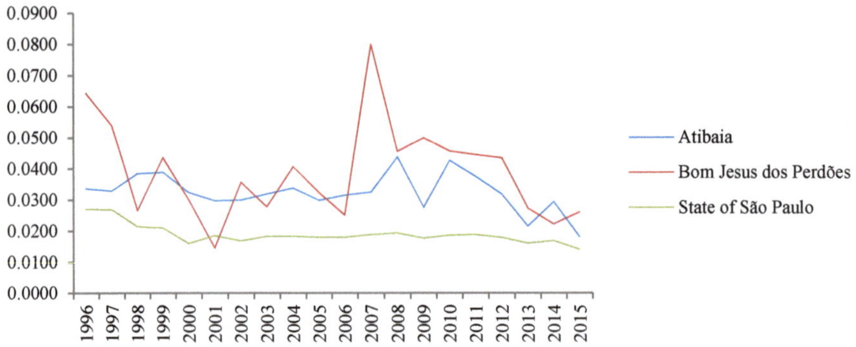

Fig. 5 Death rate from transport accident per 100 thousand inhabitants from 1996 to 2015. *Source* DATASUS (2018)

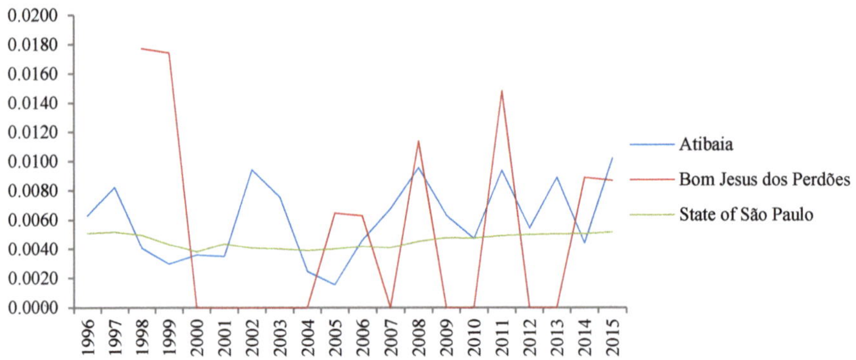

Fig. 6 Suicide rate per 100 thousand inhabitants from 1996 to 2015. *Source* DATASUS (2007)

Rape

The rape rate in the analyzed region remained below the São Paulo state average until 2009, when rates increased. In 2012, Bom Jesus dos Perdões presented a rate 104% higher than the state of São Paulo. In 2014, Atibaia presented a rate 117% higher than the state (Fig. 7) .

Theft

It is observed that both the evaluated municipalities and the State of São Paulo presented a tendency of reduction in the theft rates during the period 2001–2015. However, the theft rate in the municipality of Atibaia is high when compared to Bom Jesus dos Perdões and also to the state average (Fig. 8).

Robbery

With regard to robberies, the studied municipalities presented rates lower than the state during the whole analyzed period. It is noted that, at the beginning of the period, the studied region showed declining rates. Then, rates again increased from 2005 to 2008, accompanying the tendency of the state (Fig. 9).

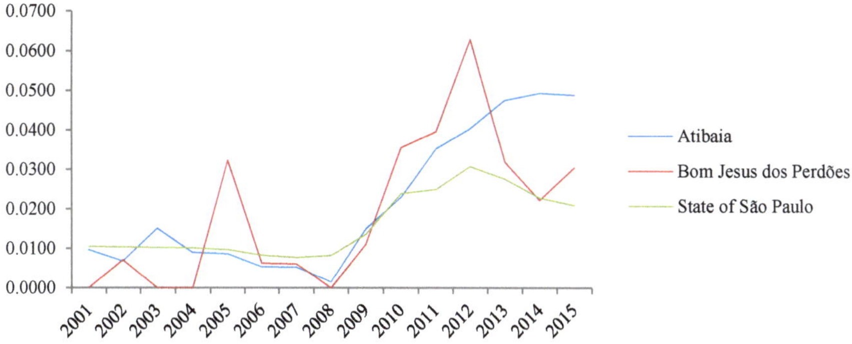

Fig. 7 Rape occurrence rate per 100 thousand inhabitants from 2001 to 2015. *Source* SEADE (2017)

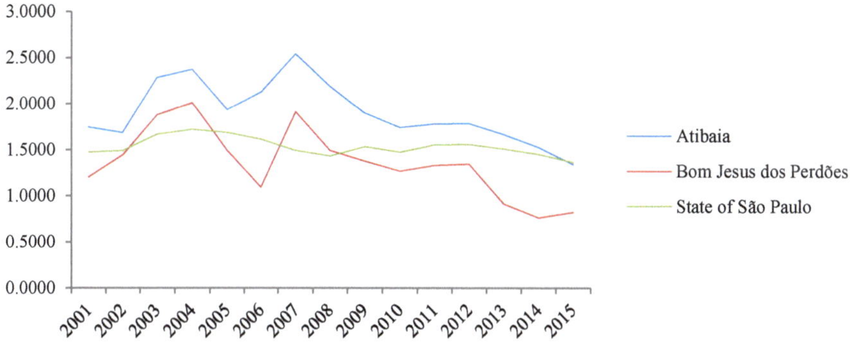

Fig. 8 Theft rate per 100 thousand inhabitants from 2001 to 2015. *Source* SEADE (2017)

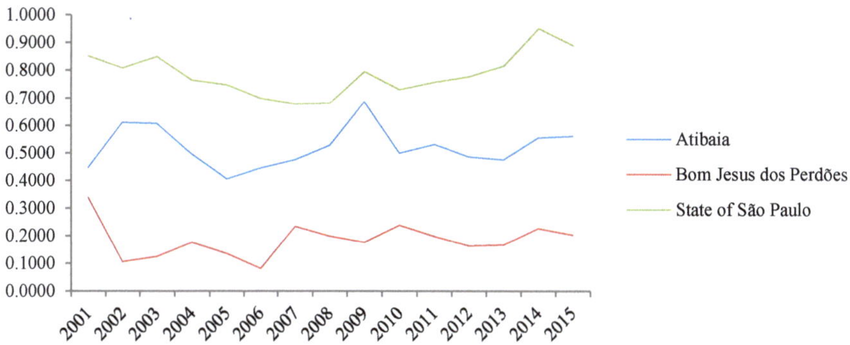

Fig. 9 Robbery rate per 100 thousand inhabitants from 2001 to 2015. *Source* SEADE (2017)

According to the selected indicators, it is observed that from 2006 the rates of homicides, transportation accidents and rape occurrence increased in both municipalities evaluated. This period coincides with the opening of the export corridor. With regard to robberies and thefts, we observe rates with fluctuations similar to those of the state. Although this scenario creates greater difficulties for a regional sustainability project, it is possible to highlight some measures observed in the municipality of Bom Jesus dos Perdões that can be considered proactive in the construction of social projects that seek sustainability with social responsibility.

2 Social Projects to Improve Quality of Life

The identification of social problems in a given region can contribute to the construction of opportunities and challenges to improve the quality of life of the population based on sustainability principles (Seixas et al. 2018; Rodrigues 2017; Wittmayer and Schäpke 2014). As highlighted in the previous topic, Bom Jesus dos Perdões has a number of problems associated with rising rates of violence and crime indicators. At the same time, we were able to identify two social alternatives that have contributed to leverage sustainability processes involving the community in a perspective of social responsibility.

2.1 Voluntary Social Work with Children and Adolescents at Risk

This project, informally called the "Community Association of Bom Jesus dos Perdões", has been in the municipality since 2010. It is the result of collective action initiated by a pastor of a Christian church, without any official institutional support from municipal administration or its social secretaries. The activities are connected to sports, mainly *futsal*, which is the modality most appreciated by the children and adolescents enrolled in the project.

The activities take place through a weekly meeting, coordinated by 6 volunteers, one of whom is a physical education teacher. Teams are divided according to the age group of the participants.

In the year 2017 70 children and young people were attended. The meetings seek to link children and youth in moral (Christian) and disciplinary values, seeking to improve personal conduct, social responsibility with the project, with society, and with the school of those involved.

In addition to sports, the project also includes English classes, through immersion among American volunteers who are visiting the municipality, due to the Church's connection with the United States. The basis of the project is:

> [...] achieve the individual in its entirety, to be effective in the results which are the result of persistence, presuming an internal change, which will result in external change regarding the character and conduct in social life (Interview with the project coordinator 2017).

In 2018 they intend to expand the activities developed, starting new projects such as: Judo, Musicalization, English, basketball, volleyball, technical courses, among others.

Due to the positive evaluation of the project by the community, they were invited by the Public Prosecutor of the city to establish a partnership with the Foundation Center for Socio-educational Assistance to the Adolescent—*CASA*[1] - aiming to broaden the sense of social responsibility of young people who are cared for by this institution. The CASA Foundation, an institution linked to the State Department of Justice and the Defense of Citizenship, provides assistance to young people aged 12–20 in the whole state of São Paulo.

The young people are inserted in socio-educational measures of deprivation of liberty (inmates) and semi-freedom, in accordance with the directives and norms foreseen in the Statute of the Child and the Adolescent (ECA) and in the National System of Socio-educational Assistance (SINASE). The measures, determined by the Judiciary, are applied according to the infraction act and the age of the adolescents. The service is decentralized throughout the state, so that young people are cared for close to their family and within their community, which facilitates social reintegration.

The partnership established between the project and CASA Atibaia—with the mediation of the public prosecutor—proposes to develop recreational and sports activities between two groups of young people: those at risk, but free, and those with deprivation of freedom, residents of CASA. Thus, about eight young people from the project are taken to play with the internal adolescents of the CASA Foundation. The activities encourage the young inmates to talk about life in the internal regime, deprivation, obedience to regulations, etc. Volunteers believe that this technique of social involvement encourages both groups to analyze the objective and restrictive conditions of each group, increasing the possibility of improving their behavior and social responsibility towards the community and society as a whole.

The success of the strategies used in the seven-year project is assessed by observing positive changes in the behavior of young people, both at home and at school. Many of the participants at the beginning of the project are already returning adults to interact with the young people they are currently attending, many of whom are integrated into social life, and others are studying to pursue a more promising career. Thus, it is considered that the project has favored the improvement of the quality of life of young people and their respective families in the municipality.

The volunteers interviewed consider the project a good example of social responsibility, since it meets the greatest social challenge they have faced:

> [...]making the in need of social assistance independent of social assistance (Interview with project coordinator 2017).

Regarding sustainability, they believe that this will happen when people increase the investment in education:

---

[1]http://www.fundacaocasa.sp.gov.br/View.aspx?title=a-funda%C3%A7%C3%A3o&d=10.

> Everything we do must be done with excellence, seeking excellence. Then will come the credibility, support, investment and human and economic sustainability. Lives are the biggest investment that overrides things … worth it. (Interview with project coordinator 2017).

2.2 *Integrated Management Office [Gabinete de Gestão Integrada]—GGI: The Integration of Several Local Institutions Aiming a Joint Management of Public Security*

This other project has proved to be efficient in a vision of social responsibility, since it involves several social secretariats of the municipality, as well as civil society organizations. It is important to note that GGI is an institutional arrangement, created by the Brazilian Federal Government, within the scope of the II National Public Security Plan—PNSP (2003–2006). The GGI was implemented in the Brazilian states during the process of institutionalization of the Public Security System (SUSP), since this organizational arrangement would be one of the institutional engineers that would enable greater articulation between the different institutions of the public security and the criminal justice system which operates at the municipal level.

Therefore, the GGI would make possible the monthly meeting of the following institutions: Municipal Guard and Community Council of Public Security (CONSEG), police (Military and Civil); Public defense; Public ministry; Judiciary; Prison System and System for the execution of Socio-educational Measures. The proposal was that these bodies meet once a month to discuss the main problems of security and public order of the locality, delineation of actions of prevention and repression to crime, monitoring and evaluation of the results achieved with the actions developed.

According to Balestreri (2009, p. 18), the GGI appear in the context of the Unified Public Security System (SUSP), articulated by the National Secretariat of Public Security, which is the body responsible for the induction of public security policies. The GGI is an executive and deliberative forum, whose mission is to systematically integrate federal, state and municipal bodies and institutions, prioritizing the planning and execution of integrated actions to prevent and combat violence and crime. It also seeks to consolidate or move towards a new paradigm in public security with a focus on good results management practices.

Currently, GGIs constitute a national network at the regional, state and municipal levels. They have a heterogeneous functional dynamics, as some have advanced in terms of results, while others follow an ambition process with a participatory management model oriented to solve problems.

This new paradigm seeks to organize a new dialogue and social and institutional participation aiming at a new form of management that will certainly reorganize the participative model of governance and social responsibility, focused on the right of citizenship. It is also necessary to instigate the integration of basic services performed by public security operators at the various levels of management, especially those perceived by the community. It is vital to develop mechanisms capable of disseminating the GGI philosophy and its practices in routine work, through courses, curricular structures and other intra-institutional and inter-institutional processes (Balestreri 2009, p. 18; Brasil 2009–2010).

In Bom Jesus dos Perdões, the project began in October 2017, basically in response to a significant increase in the municipality's violence and crime indicators, as we demonstrated in the previous item. The premise of the Program, highlighted during its opening, is that through the institutions, it is intended to make a deep analysis of the difficulties of security in the city. The GGI will involve the whole municipality in order to be able to combat crime intensely and effectively, providing greater security to the population and is expected to foster greater responsibility and social sustainability.

3 Conclusion: A New Line of Challenges and Opportunities for Achieving Sustainability

The population and urban growth experienced in the last decades in the municipality of Bom Jesus dos Perdões does not represents just local economic development. It also represents a more comprehensive development project, since the D. Pedro I Railway region is considered strategic for the development of the State of São Paulo and Brazil. However, the lack of planning and infrastructure needed to meet the new needs of the municipality is evidenced by the overload in existing structural services, by anthropogenic pressure on natural resources, and by the configuration of new social problems at a rapid rate for which municipal management demonstrates lethargy in solutions.

In this sense, identifying the various problems that the community has faced, from the economic, social and environmental points of view, is one of the first steps to seek sustainability and improve the quality of life of the population. Thus, the social responsibility projects identified seem to us concrete challenges and opportunities, insofar as they show social and institutional actions, together with strategic actions, to improve the quality of life of the population from an environmental and social point of view.

The identification by privileged actors, including community consultations, of the significant changes in Bom Jesus dos Perdões, such as the disorderly increase of population and consequent social problems as expressive crime rates (homicides, deaths related to traffic accidents, rapes), as we presented in this paper, was the starting point for the construction/ adhesion of the social projects in the municipality that are constituted in a privileged opportunity of sustainability with social responsibility for the improvement of the local quality of life.

The Project *Community Association of Bom Jesus dos Perdões* essentially captures the importance of culture and encourages education and individual commitment to building a more participatory, just and sustainable society seeking a development model that considers environmental and social demands as well as economic ones. At the same time, the GGI responds to the immediate need to stop the course of local violence in an integrated way, highlighting the role of the local population in identifying and solving problems, reorganizing the participative model of management and social responsibility, focused on a right of citizenship.

Acknowledgements We thank São Paulo State Foundation for Research Support (FAPESP, Brazil) for financial support to the research that led to this article (FAPESP n. 2016/18585-3), and for the National Council for Scientific and Technological Development (CNPq/Brazil) for the Fellowship granted to the first author.

References

AASHE (2014) Association for the advancement of sustainability in higher education. Sustainability tracking, assessment and rating system. https://stars.aashe.org/. Last Accessed 27 Mar 2014

Baker S (2006) Sustainable development. Routledge, London, p 245

Baker S, Mehmood A (2015) Social innovation and the governance of sustainable places. Local Environ 20(3):321–334

Balestreri RB (2009) Prefacio in Brasil. Gabinetes de gestão integrada em segurança pública: coletânea 2003–2009. Secretaria Nacional de Segurança Pública, Brasília, p 343

Bawden R, Allenby B (2017) Sustainability science and the epistemic challenge: some matters philosophical and why we ought to come to know them better. Sustain Sci 12:901–905

Braga V (2008) Logística, planejamento territorial dos transportes e o projeto dos Centros Logísticos Integrados no Estado de São Paulo. e-premissas, Revista de estudos estratégicos 3: 68–92

Brasil. Secretaria Nacional de Segurança Pública (2009) Gestão e disseminação de dados na Política Nacional de Segurança Pública. Gabinetes de Gestão Integrada no Brasil: à guisa de institucionalização? Fórum Brasileiro de Segurança pública. http://www.justica.gov.br/sua-seguranca/seguranca-publica/analise-e-pesquisa/download/fbsp_termoparceria_2/5-2-e-5-3-gabinetes-de-gestao-integrada.pdf. Last Accessed 23 Feb 2018

Brasil. Secretaria Nacional de Segurança Pública (2010) Gabinetes de gestão integrada em segurança pública: coletânea 2003–2009. http://www.justica.gov.br/sua-seguranca/seguranca-publica/analise-e-pesquisa/download/outras_publicacoes/pagina-1/15livro-ggi-capa-verde.pdf. Last Accessed 23 Jan 2018

Camilleri MA (2017) Corporate sustainability and responsibility: creating value for business, society and the environment. Asian J Sustain Soc Responsib 2:59–74

Departamento de Informática do SUS [DATASUS] (2007) Ministério da Saúde, Informações de Saúde. http://datasus.saude.gov.br. Accessed 14 Oct 2007.

Departamento de Informática do SUS [DATASUS] (2018) Ministério da Saúde, Informações de Saúde. http://datasus.saude.gov.br. Accessed 14 Jan 2018.

D'Harcourt E, Ratnayake R, Anna Kim A (2017) How can the sustainable development goals improve the lives of people affected by conflict? Bull World Health Organ 95:157–158

Fundação Sistema Estadual de Análise de Dados [SEADE] (2017) Perfil municipal, regional e do estado de São Paulo [Online]. http://www.seade.gov.br/produtos/perfil/perfilMunEstado.php. Accessed 5 Sep 2017.

Fundação Sistema Estadual de Análise de Dados [SEADE] (2018) Perfil municipal, regional e do estado de São Paulo [Online]. http://www.seade.gov.br/produtos/perfil/perfilMunEstado.php. Accessed 3 Feb 2018.

Gatti L, Seele P (2014) Evidence for the prevalence of the sustainability concept in European corporate responsibility reporting. Sustain Sci 9:89–102

Gupta GS (2017) The paradox of sustainable development: a critical overview of the term and the institutionalization process. Periodica Polytechnica Soc Manage Sci 25(1):1–7

Hoeffel JL, Fadini AAB, Reis JC, Jesus CR (2009) Boa vista—sustentabilidade e desenvolvimento local: percepções ambientais e uso do solo na bacia hidrográfica do Rio Atibaia. OLAM 9(1):107–140

Hoeffel JL, Fadini AAB, Reis JC, Jesus CR (2010) Alterações ambientais na APA Cantareira: um estudo na Bacia Hidrográfica do Rio Atibaia. OLAM 10(1):61–90

Hoeffel JL, Seixas SRC, Oliveira L, Morais ABS, Gomes AK (2015) Transformações socioeconômicas e ambientais no corredor de exportação D. Pedro? Um estudo em Nazaré Paulista e Bom Jesus dos Perdões/SP. In: I Congresso Novos Direitos - Cidades em Crise?. http://www.novosdireitos.ufscar.br/congresso/copy_of_ANAISDOICONGRESSONOVOSDIREITOS.pdf. Last Accessed 3 Feb 2018

Hoeffel JLM, Seixas SRC, Oliveira KES, Rocha JV, Lima FB (2015a) Urbanização e Mudanças no Uso do Solo no Eixo do Corredor de Exportação – Rodovias D. Pedro I/Tamoios – SP. XVI ENANPUR: Espaço, Planejamento e Insurgências, Belo Horizonte, Anais. http://anpur.org.br/xvienanpur/geral/arquivos/Caderno-de-Resumos-XVI-Enanpur.pdf. Last Accessed 6 Jan 2018

Instituto Brasileiro de Geografia e Estatística [IBGE] (2018) Produto Interno Bruto dos Municípios [Online]. Brasília, IBGE. http://www.cidades.ibge.gov.br. Accessed 4 Feb 2018.

International Union for Conservation of Nature and Natural Resources [IUCN] (1980) World conservation strategy. Living resource conservation for sustainable development

Lang DJ, Wiek A, Bergmann M, Stauffacher M, Martens P, Moll P, Swilling M, Thomas CJ (2012) Transdisciplinary research in sustainability science: practice, principles, and challenges. Sustain Sci 7(1):25–43

Leal Filho W, Manolas E, Pace P (2015) The future we want: key issues on sustainable development in higher education after Rio and the UN decade of education for sustainable development. Int J Sustain High Educ 16(1):112–129. https://doi.org/10.1108/IJSHE-03-2014-0036. Last Accessed 3 Feb 2018

Lindahl KB, Baker S, Rist L, Zachrisson A (2016) Theorising pathways to sustainability. Int J Sustain Dev World Ecol 23(5):399–411

Prefeitura Municipal de Bom Jesus dos Perdões. Gabinete de Gestão Integrada é implantado em Bom Jesus dos Perdões. O programa federal começa sua fase de implantação na cidade. https://www.bjperdoes.sp.gov.br/2017/10/18/gabinete-de-gestao-integrada-e-implantado-em-bom-jesus-dos-perdoes/. Last Accessed 3 Aug 2017

Robert KW, Parris TM, Leiserowitz AA (2005) What is sustainable development? Goals, indicators, values, and practice. Environ Sci Policy Sustain Dev 47(3):8–21

Rodrigues TNHC (2017) Entre faltas e oportunidades: ONGs e prevenção da violência. Rev Bras Segur Pública 11(1):130–146

Saito O (2017) Future science-policy agendas and partnerships for building a sustainable society in harmony with nature. Sustain Sci 12:895–899

São Paulo (2000) Secretaria de Estado do Meio Ambiente. Atlas das Unidades de Conservação Ambiental do Estado de São Paulo. SMA, São Paulo

São Paulo (2010) Decreto Estadual nº 55.662, de 30 de março de 2010, que cria o Parque Estadual Itapetinga. São Paulo, Assembléia Legislativa. http://www.al.sp.gov.br/repositorio/legislacao/decreto/2010/decreto-55662-30.03.2010.html. Last Accessed 21 Mar 2018

Seixas SRC, Hoeffel JLM, Rocha JV, Vernalha EB, Lima FB, Nunes R (2015) Urbanization and socioenvironmental dynamics in the D. Pedro I - Tamoios Road Axis: some environmental sociology contributions. 12nd Conference of the European Sociological Association, Prague

Seixas SRC, Hoeffel JLM, Barrett, P (2018) Water resources in the context of global environmental change: Some perspectives for sustainability. In: Leal Fo W (ed) Handbook of Sustainability Science and Research. Serie: Climate Change Management, Springer, pp 87–102

United Nations [UN] (2012) Back to our common future. Sustainable development in the 21st century (SD21) project. Summary for policymakers. New York, USA

United Nations [UN] (2015) Transforming our world. The 2030 agenda for sustainable development. A/RES/70/1. UN General Assembly, New York, USA
United Nations [UN] (2015a) The millennium development goals report. New York, USA. http://www.un.org/millenniumgoals/2015_MDG_Report/pdf/MDG%202015%20rev%20(July%201).pdf. Last Accessed 5 Feb 2018
van der Leeuw SE (2014) Sustainability, culture and personal responsibility. Sustain Sci 9:115–117
Wittmayer JM, Schäpke N (2014) Action, research and participation: roles of researchers in sustainability transitions. Sustain Sci 9:483–496
World Commission on Environment and Development [WCED] (1987) Our common future. Oxford University Press, Oxford

Professor Sônia Regina da Cal Seixas has a Ph.D. in Social Sciences, from Campinas State University, UNICAMP, Brazil and Post-doctoral in University of Reading/UK. Nowadays she is researcher at the Center for Environmental Studies and Research, NEPAM, State University of Campinas, UNICAMP. She is professor of Ph.D. Program Environment & Society (NEPAM-IFCHUNICAMP) and of Postgraduate Program Energy Systems Planning (FEM-UNICAM). She has a productivity research fellowship CNPq—Brazil's National Council for Scientific and Technological Development and Group Leader-Directory CNPq: LEMAS—Lab of Environmental changes studies, Quality of life and subjectivity.

Dr. Michelle Renk has a Ph.D. in Environment and Society by Center for Environmental Studies and Research, NEPAM, State University of Campinas, UNICAMP (2016), Master in Planning of Energy Systems by the Faculty of Mechanical Engineering FEM/UNICAMP (2010). Specialization in Environmental Risk Analysis by the Faculty of Mechanical Engineering FEM/UNICAMP (2008). Graduation in Biological Sciences Degree and Bachelor's degree from the University of Grande ABC (2005). I completed the BEPE/Fapesp Research Internship at the Horton Horton Social Science Center (HSSC) and The Carsey Institute of the University of New Hampshire (NHS) in New Hampshire (USA). I am a researcher at the Laboratory for Environmental Change, Quality life and Subjectivity (CNPq -EPAM-UNICAMP).

Professor João Luiz de Moraes Hoefel has a Ph.D. in Social Sciences at Campinas State University (IFCH/UNICAMP), São Paulo, Brazil, with a focus on environmental issues and developed a Post-Doctoral research at the Center for Environmental Education and Public Policies at São Paulo University (ESALQ/USP). He teaches at Atibaia College (FAAT), Atibaia, São Paulo, where he also coordinates the Center for Sustainability and Cultural Studies (NESC/FAAT) developing research projects on Natural Resources Conservation, Environmental Education and Environmental Planning, and at University of São Francisco, Bragança Paulista. He also develops research activities at NEPAM/UNICAMP on Global Environmental Change and Quality of Life as a Collaborator Researcher.

Professor Gabriela Farias Asmus Biologist, Master in Tropical Biology and Natural Resources—Ecology (INPA-2008) and Ph.D. in Environment and Society (NEPAM/UNICAMP-2014). Post-doctorate in NEPAM (FAPESP 2014/26007-4). Assistant Professor at UFABC—Santo André. Research theme addresses environmental health in the context of environmental changes and global climate change.

Enhancing Organizations' Social Responsibility by Workplace Health Promotion?

Eva Bamberg, Grit Tanner, Carolin Baur and Marlies Gude

Abstract The idea behind the social responsibility of organizations is that firms have responsibilities beyond maximizing profit. They are accountable not only to their owners and shareholders but also to stakeholders, such as customers. In recent years, the social responsibility of organizations has become increasingly important. Models and management systems using the term corporate social responsibility (CSR) have been developed. However, the concept of CSR is not uncontroversial. CSR is criticized as a new trend with no content. CSR programmes are often decoupled from the core business, and they are often only marketing and symbolic gestures. In this paper, we argue that an integration of CSR into other management systems such as human resource management or health promotion may have various positive effects. These effects concern both CSR and the other management systems. Using the example of workplace health promotion (WHP), positive reciprocal effects can be explained as follows. On the one hand, by conducting WHP, organizations show responsibility towards their employees, therefore enhancing CSR. CSR can be improved by considering the WHP structure and topic, such as a focus on interventions. On the other hand, the concepts of CSR can contribute to the spread of WHP. Organizations' influences on WHP can be extended by the inclusion of other organizations in the supply chain and by networking. In this context, the quality of interactions between the organizations must be considered. The development and use of criteria are needed to guarantee positive effects on both CSR and WHP.

Keywords Workplace health promotion · (Corporate) social responsibility
Management systems · Supply chain · Networks

E. Bamberg (✉) · G. Tanner · C. Baur · M. Gude
Faculty of Psychology and Human Movement, Institute of Psychology, Work and Organizational Psychology, Universität Hamburg, Von Melle Park 11, 20146 Hamburg, Germany
e-mail: bamberg@uni-hamburg.de

W. Leal Filho (ed.), *Social Responsibility and Sustainability*, World Sustainability Series, https://doi.org/10.1007/978-3-030-03562-4_6

1 Corporate Social Responsibility

Sustainability in the context of organizations and enterprises includes a tripartite focus on economic, environmental, and social phenomena. Therefore, sustainability also includes social responsibility (Alvarado-Herrera et al. 2017; Eriksson and Svensson 2016). The idea that companies have responsibilities to society that extend beyond economic aims is taken up in concepts and programmes termed Corporate Social Responsibility (CSR). In recent years, CSR has become a topic of growing academic and practical interest (Tanner et al. 2018a). Governments have developed CSR strategies, and international and national organizations have begun to put CSR into practice. As Carroll (2015, p. 95) stated in an overview, "now CSR/Business Ethics/Corporate Citizenship/Sustainability is part of virtually every company's agenda". Alvarado-Herrera et al. (2017) define CSR as the commitment of firms to maximize long-term economic, societal and environmental well-being. These effects are realized by business practices, policies, and the allocation of resources (Du et al. 2011).

According to the European Commission, "Companies can become socially responsible by following the law; integrating social, environmental, ethical, consumer, and human rights concerns into their business strategy and operations" (Commission of the European Communities o.J.). Analysing the definitions of CSR, Dahlsrud (2008) notes five dimensions that are considered core features of the concept of CSR. The voluntariness dimension (which is controversial in the literature) refers to actions that are not prescribed by law. The other four dimensions can be seen as different fields of action: the environmental dimension refers to care for the natural environment; the social dimension concerns the relationship between business and society; the economic dimension includes socio-economic or financial aspects; and the stakeholder dimension considers the interactions with stakeholders such as employees, suppliers, customers and communities. The Commission of the European Communities (2001) distinguishes between external and internal CSR. External CSR concerns local communities, business partners, suppliers and consumers, and human rights issues. Internal CSR concerns human resources management, health and safety at work, adaption to change, and the management of environmental impacts and natural resources.

Social responsibility does not stop at a single company. Every company is connected to other enterprises in various ways. With the decision to cooperate with specific enterprises, a company accepts (and thus might also promote) the processes and conditions under which the services and products of these enterprises are made. With their price policy and their given time schedule, companies influence the working conditions of other organizations. For example, a municipal authority that awards a contract to the most inexpensive cleaning company accepts that the wages of the cleaning staff who will work for this municipal authority are low and that the work pressures are high. Thus, CSR efforts cannot be reduced to one company's perspective but must include the supply chain (Eriksson and Svensson 2016).

Even though the relevance of CSR is emphasized worldwide by governments, international and local enterprises, and non-governmental organizations, there is a critical view of CSR (Gröneweg and Matiaske 2012). Simulated commitment to CSR to improve the competitive standard is not unusual (CSR-washing; Pope and Wæraas 2016). Companies attempt to position themselves as socially responsible brands by being linked to social causes such as sustainable development or the fight against climate change (Alvarado-Herrera et al. 2017). CSR initiatives may also help control the interests of external stakeholders (Banerjee 2012) so that the company can be stabilized in the long term against external disturbances. Companies can turn CSR into business and profit generation (Bondy et al. 2012). In most cases, so says the critique, CSR is more the result of the financial interests of organizations than of true concern for people, animals, or the environment. Alternative economic models that include profit *and* responsible conduct as equally necessary have not been developed.

This critique is taken into account by the differentiation between (easily) decoupled and integrated practices, which was formulated by Weaver et al. (1999) with respect to ethics. Easily decoupled practices are those that provide the image of conforming to expectations; however, they insulate the organization from many of those expectations. Integrated practices and programmes affect ethics-oriented decisions and actions in different fields of an organization. These programmes are connected to other structures and processes within organizations (Weaver et al. 1999). With respect to compliance programmes, MacLean and Behnam (2010) stated that enterprises create formal structures and programmes but avoid integrating the programmes into everyday business: "This decoupling produces formal compliance programmes that are a form of window dressing, symbolic gestures designed to give the appearance of satisfying regulatory requirements and enhance external audiences' perceptions of organizational legitimacy while still allowing for 'business as usual'" (McLean and Behnam 2010, p. 1499; see also Weaver et al. 1999; Eriksson and Svensson 2016).

Integrated programmes represent a promising possibility for improving the implementation of CSR in practice. Integrated programmes strengthen the connection between CSR and different fields of action. One possible field is given by workplace health promotion (WHP). A connection between CSR and WHP makes it possible to develop integrated programmes. In the following passage, we demonstrate how the social responsibility of organizations can be enhanced by WHP. We show that the establishment of criteria for WHP that extend beyond the borders of organizations is an important step for improving social responsibility in organizations.

The following deliberations were worked up within the project *Occupational Health Promotion—An Inter-Organizational Perspective (**Ge**sundheitsmanagement aus **i**nter-**o**rganisationaler **P**erspektive—GESIOP).*[1] The project aims to bring together CSR and WHP. Within the project, two universities, three enterprises and an NGO have participated for three years. In the first part of the project, we conducted interviews, group discussions, and questionnaire studies. Longitudinal studies, diaries and experiments are works in progress. In addition to the review of scientific practical literature, the interviews and the group discussions are the main bases of these deliberations. We conducted approximately 16 interviews, each last-

---

[1]The project was funded by the German Federal Ministry of Education and Science (Funding No. FKZ 02L14A040).

ing approximately one hour, with CSR- and health-managers of different enterprises. The interviews were then analysed by content-analysis (Tanner et al. 2018b).

2 Improving CSR by Workplace Health Promotion

In European countries, health spending accounts for nearly 10% of the GDP (OECD 2016). There are considerable costs for enterprises and for society in general due to sickness-related absences of employees. In 2015, employees in Europe had an average of 12 absenteeism days from work due to illness (WHO 2018a). There are several reasons for sickness-related absences, one of which lies in working conditions. According to a study from the German Social Accident Insurance, natural scientists had an average of 4.5 sick days, whereas forestry workers had a sevenfold higher average of 28.5 days (Dieterich et al. 2000). Therefore, WHP, which aims to improve healthy working conditions and reduce work-related diseases and accidents, is a central issue in working life.

Within WHP, the physical and psychosocial work environment is included (WHO 2010). The Luxembourg Declaration of the European Network for Workplace Health Promotion (ENWHP) defines WHP as the "combined efforts of employers, employees, and society to improve the health and well-being of people at work". These improvements can be achieved through a combination of the following:

- improving the work organization and the working environment,
- promoting active participation, and
- encouraging personal development (ENWHP 2007).

As the figures cited above show, the protection and promotion of health are an economic necessity. However, they are primarily humanitarian and social obligations.

The concept of WHP is rather ambitious. Indeed, studies on this subject suggest that it is only partially implemented in reality. Most interventions focus on the individual level, some on the organizational level, and few on the community level (Harden et al. 1999). Furthermore, only some enterprises actually carry out WHP. There has been a moderately increased prevalence of WHP over the last years. In 2012, approximately 45% of employees reported WHP activities in their companies, versus 38% in 2006 (Beck et al 2015). There is a need for WHP-related advice and support, especially in small enterprises. The company size seems to be crucial for how health promotion is practised. Large companies typically institutionalize health protection within the company. They employ special experts for health protection and social services with distinct budgets for health promotion. Some large companies integrate human resources, job design, and health promotion. Activities are diverse, addressing both physical and psychosocial aspects of health, and are based on problem analyses and data obtained by a health surveillance and reporting system (Bamberg et al. 2011).

Within small- and medium-sized companies, health protection usually exists, at best, informally. Often, there is no person in charge of WHP. If and to what degree health protection is realized in these companies often depends on the knowledge and attitude of single persons (like the owner), who might be overloaded with the numerous functions and responsibilities assigned to one person. Worldwide and in Western industrialized countries, the majority of workers work in small- and medium-sized companies (McCoy et al. 2014). The risk for accidents in these companies is higher than in large companies (Zwetsloot 2017). Thus, initiatives for health protection are crucial in small- and medium-sized companies.

Furthermore, WHP is widely limited to industrialized countries. It is not uncommon for an enterprise within an industrialized country to practice well-developed/sophisticated WHP while the suppliers of this enterprise ignore standards of safety and health promotion at work. If companies outsource specific business branches, it may be that these business units had well-developed WHP in the past but lost it during the outsourcing process.

Therefore, on the one hand, there is the necessity, political willingness, and knowledge of WHP; on the other hand, there are several limitations. As shown below, networking and considering the supply chain can overcome some of the limitations of WHP. Furthermore, a socially responsible perspective that incorporates strategic human resource management can contribute to the achievement of employee well-being (Dezmar-Krainz 2015). WHP can be seen as an essential part of CSR (Zwetsloot and Starren 2004; Tanner et al. 2018b). Accordingly, the WHO views ethics and sustainability as key factors of healthy workplaces (WHO 2018b). The promotion of well-being at work seems to be a fundamental part of CSR, and work design is an important part of health promotion. However, our own studies showed that there is a strong predominance of decoupled CSR programmes. Persons in charge of CSR often work in the marketing department. They have marginal knowledge and experience of the structures, programmes, and processes of human relations or WHP, which are actually important areas of CSR. In contrast, persons in charge of fields such as health promotion are rarely informed about CSR programmes in general. Therefore, the link between CSR and work design is still unfamiliar to stakeholders (Jain et al. 2011). Nevertheless, Jain et al. (2014) propose a "CSR-inspired" approach to health promotion at work. Such a CSR-inspired approach to WHP is useful for not only CSR but also WHP programmes (Tanner et al. 2018b; Bamberg et al. 2016; Tanner and Schmidt 2017).

One way to integrate WHP and CSR is to formulate clear goals. Goals are a prerequisite to building an appropriate structure that enables companies to establish WHP-oriented processes (Ha-Brookshire 2017). The goals of WHP can be deduced from programmatic concepts, such as the abovementioned Luxembourg Declaration, and from studies on work and health. Studies on the effects of work on health have shown that working conditions and personal factors influence health and well-being (Cox 1997; Hammer et al. 2015). Resources (such as social support) and stressors (such as time pressures) at work and personal resources and risk factors influence both the appraisal of the working conditions processes (for example, as a challenge or as a threat) and the coping processes (such as problem- or emotion-oriented coping). Accordingly, the Luxembourg Declaration of the ENWHP proposes the consideration of the following aspects:

- the combination of individual- and environment-directed measures;
- a resource-oriented approach ("enhancing health-promoting potentials and well-being in the workforce"), thus expanding risk reduction;
- a continuous improvement process comprising needs analysis, priority setting, planning, implementation, continuous control, and evaluation; and
- employees' participation in this process (Beck et al. 2015).

The guidelines for WHP suggest that WHP programmes need to apply at different levels to facilitate sustained changes. At the individual level, a programme should incorporate a range of educational strategies; at the organizational level, support mechanisms throughout the organization should reinforce and encourage positive health actions. At the community level, WHP programmes should have the potential to be actively disseminated by employees to their families and social networks (Harden et al. 1999).

Based on studies of work and health, the Luxembourg Declaration, and the above-mentioned guidelines, the criteria for WHP should include both enabling factors and qualified health promotion content.

Criteria of WHP I: ENABLING FACTORS
Enabling structures and processes
Person in charge of WHP
Structure and budget
Steering group
Periodic risk management (assessment, intervention, evaluation)
Participation of employees
Informing employees
Participation in generating health-oriented interventions
Criteria of WHP II: CONTENT
Condition-oriented: Promotion of resources and reduction of stressors at work and in the organization
Safety and health at work (work organization, working conditions, and work tasks)
Motivating work and opportunities to learn within the work process
Relevance of health to the decisions in organizations
Relevance of health to leadership
Person-oriented: Promotion of resources and reduction of risk factors
Enabling personal growth and qualifications
Enabling work-life balance
Enabling healthy behaviours
Employment contract
Job security
Adequate payment
Consideration of special groups of employees
Considering employees with high strain
Health-oriented interventions for special groups

At first glance, criteria such as these seem to be obvious. Indeed, they are broadly shared in the science and practice of WHP. However, the realization of these criteria is not simple. In many companies, it is unclear how to ensure healthy workplaces or how to involve employees. As we mentioned above, in many organizations, there is a lack of experience in implementing WHP. In these cases, networks between companies may be useful.

3 Networks and the Supply Chain

Building networks enables members of organizations to share their experiences and gain knowledge of WHP. Furthermore, resources can be used together, or joint projects are possible (Tanner et al. 2018b). Networks may include different organizations that cooperate in one supply chain. In this case, WHP could be connected with supply chain management. In terms of work design, as a field of WHP, Zink (2014, p. 128) proposes "a systemic and holistic approach regarding whole value creation chains including all working conditions along the value creation". Ha-Brookshire (2017) stated that for truly sustainable corporations, the entire supply chain must be sustainable.

Networks are useful for spreading the ideas, instruments, and strategies of WHP across organizations. In terms of the supply chain, networks help prevent WHP from stopping at one enterprise. Therefore, building up networks and including the supply chain are appropriate possibilities to realize the abovementioned CSR-inspired approach to WHP. Cooperation with other organizations and the inclusion of the supply chain are central elements of CSR. Concerning WHP, communication in networks is already realized in some cases, but supply chain management is new (Tanner et al. 2018b).

WHP-oriented supply chain management presupposes that the organizations that are members of a supply chain are interested in the health-relevant conditions of each other. This interest may be realized in different ways: health-related conditions and processes may be an issue in organizations' communications, these conditions and processes may be criteria for cooperation, or organizations may consider how their contract conditions influence the health-relevant conditions of others.

However, networks and supply chain management per se are not enough to realize the link between CSR and WHP. The communication and cooperation within networks and within the supply chain may be quite different, which leads to the question of how to communicate or cooperate within such networks and within the supply chain. Presently, there are no known studies on WHP-oriented communication and cooperation within networks. However, there are several studies and programmatic papers on interactions within the supply chain.

Organizations may have different reasons and motivations for interactions within a WHP-oriented network. Sometimes the partners want to gain knowledge, for example, on the connection between work and health. In these cases, the exchange of information is relevant. Sometimes partners want support for a project, for example, a network of enterprises from the same branch and the responsible health insurance organization intend to develop health interventions together. In these cases, a joint planning process is necessary. Therefore, the purpose of the network interaction affects the intensity of the contact and may also affect the quality of these networks.

With respect to the quality of interactions within networks and the supply chain, Ming et al. (2014) distinguish between coordination on a basic level, which means the synchronization of activities. High-quality coordination is associated with cooperation, which includes the sharing of information. An even higher quality is achieved by collaborating with other organizations. Collaboration includes the possibility that a company modifies its behaviour based on the partner. There are several factors that influence the interaction and therefore sustainability in a supply chain. Commitment is a crucial factor (Eriksson and Svensson 2016; Ming et al. 2014; Pagell and Wu 2009). In cases of interventions for work and health, management commitment seems to be the most relevant factor (Janetzke and Ertel 2017).

Within organizations, participation is a prerequisite for the commitment of the members. Dachler and Wilpert (1978) distinguish different forms of participation within organizations, which vary with respect to the extent of the decision that can be made: (1) receiving information about the decision; (2) having the possibility of giving an opinion; (3) taking into account the opinion of organizations' members; (4) having the right to veto; (5) making a decision by organization members. These forms of participation can be generalized to supply chains. Locke and colleagues stated that global brands could help improve the health of the employees in supply chain factories by working with suppliers to help them introduce WHP (e.g., Locke and Romis 2007). This WHP occurs primarily with a compliance approach; there are demands on the members of the supply chain, which are given and controlled by codes of conduct and monitoring systems. In addition, Locke et al. (2009) proposed a commitment approach. The central elements of a commitment approach are joint problem solving, information exchange, and the diffusion of best practices. The commitment approach leads to improvements in working conditions and labour rights at the workplace (Locke et al. 2009; Tanner et al. 2018b). The commitment approach includes more possibilities for participating than the compliance approach.

Following the ideas of participation and the commitment approach within networks and supply chains, several further WHP criteria are relevant.

Criteria of WHP II: WHP-oriented interactions within networks and the supply chain
Building up WHP-oriented networks
Building up WHP-oriented supply chains
(Mutual) Information on health-related conditions and activities
Interactions or exchanges of experiences
Joint projects
Possibilities of others (such as the staff of suppliers, temporary agency workers) joining health-oriented interventions
Adaption of processes, structures, and demands in companies to the conditions of other persons or organizations
Health-oriented conditions as criteria for the selection of partners to cooperate with (e.g., suppliers)

There are different ways in which criteria can contribute to a CSR-inspired health promotion. In general, criteria are helpful in discussing the standards of CSR and WHP. They offer an opportunity to either describe a possible field of action or define key points that should be emphasized. For organizations, criteria can be used as a guide to describe their own WHP. For the business partners, criteria can be used to appraise the WHP of others. Within cooperation processes, criteria are helpful in gaining a mutual understanding of the WHP, to set minimum standards and to develop standards. Using the criteria as a guide for self-descriptions may help to reduce the abovementioned risk of CSR-washing. To fulfil these functions and work with these criteria, it is useful to operationalize them by developing questions or items. Accordingly, we established a list of items concerning the criteria (see example).

Example: Criteria and items

(Mutual) Information on health-related conditions and activities, such as audits, codes of conduct	Is there a mutual exchange of information and experiences? Are other organizations informed on the WHP and quality of work? Has information on other organizations been obtained? Is the public informed of the WHP?

4 Summary and Discussion

Despite the importance and dissemination, CSR is often criticized for being more hype than substance. Instead of realizing the social responsibility for humans, environment, and society, several companies use CSR for their economic interests. Instead of integrating CSR programmes with the organization's core or day-to-day business, decoupled programmes are institutionalized. Decoupled programmes immunize organizations against demands for responsible behaviour. We propose several steps to address this critique.

1. One way to improve the implementation of CSR is to integrate CSR into specific organizational fields of action, such as WHP. WHP influences the health of working people and their families, the well-being of our society and the monetary costs associated with illness and health-related retirement. Therefore, WHP is a relevant element of sustainability and social responsibility in organizations. WHP is a part of the practice of CSR that lies between rules and autonomy. In industrialized countries, there is a system of elaborate rules and regulations concerning health and safety at work; however, organizations have broad autonomy to work with these rules and regulations. Institutionalizing WHP in organizations and building up integrated programmes for WHP require criteria and goals. Criteria can be deduced from studies on work and health and from programmatic

papers, such as from the WHO or the EU. These well-known criteria and goals concern, in addition to other aspects, the structures and responsibilities for WHP, the inclusion of person-and environment-oriented methods and the participation of employees.

2. Programmes that integrate CSR and WHP cannot be restricted to single organizations. Therefore, the abovementioned criteria are not sufficient; they must be expanded. A CSR-inspired WHP, which reaches beyond the borders of organizations includes interactions with other organizations, in networks and in the supply chain. Interactions make it possible for enterprises to share experiences on WHP, achieving the joint use of resources.
3. The type of interaction between organizations may differ, ranging from an exchange of information to joint decision making or to the modification of processes independent of the conditions of the partner (Tanner et al. 2018b).
4. A central aim of the interaction between organizations is to build up commitment towards a CSR-inspired WHP. As studies on participation and cooperation within the supply chain show, the participation of members with equal rights influences their commitment.

Thus far, there are single studies on cooperation within the supply chain; however, the development of theory and praxis on this issue is rare. Therefore, the above proposed steps concern a new scientific field and a new field of application. The conditions under which organizations are willing to interact in this manner remain an open question.

There are several prerequisites for WHP-oriented interaction within the supply chain or within networks. To practise WHP requires resources. To interact in networks and the supply chain, additional resources are necessary. Small organizations in particular lack these resources. Furthermore, to interact in networks, enterprises must provide information that concerns not only their benefits but also their health-related weak spots. Communicating an organization's own weak points between enterprises is not self-evident in our society. In addition, enterprises must spend time collaborating with others. This is also rather unusual in a system based on competition. Our research showed several possibilities that can encourage communication within networks, such as financial support from the government or advice from health insurance. However, more research on this issue is necessary.

In general, in some branches and countries, the conditions for WHP and especially for WHP in the supply chain are insufficient. In countries with high unemployment or in branches with low-skill work, employees are easily replaceable. In these cases, companies rarely care about the health of employees. However, in other branches and countries, such as industrialized countries and branches with highly qualified workforces and a shortage of labour, the prerequisites for WHP and CSR are rather good. In these cases, companies compete for customers and employees. These companies have the potential for improving their CSR and WHP programmes. By doing so, these companies gain a competitive advantage over other enterprises. In recent years, some organizations have gained knowledge and experience related to WHP, while others have not had this opportunity. In particular, small-sized enterprises often lack the knowledge of and resources required for CSR and WHP. Therefore, they may profit crucially from support within a network. Nevertheless, even experienced organizations can profit from a WHP-oriented exchange with others. For organizations that have practised WHP for a period of time, it may be useful to verify their knowledge and perhaps modify their praxis. Moreover, networks offer more than an enhancement of knowledge; resources may be used together, and joint projects are possible. Specifically, for the supply chain, the joint use of resources and collective projects are useful. Particularly when enterprises influence the working conditions of other companies, WHP-oriented cooperation within the supply chain is appropriate (Tanner et al. 2018b).

The inclusion of networks and the supply chain in WHP makes it necessary to work on several issues that have rarely been considered in the past. Our remarks on WHP-oriented interactions within networks and the supply chain require further elaboration. A systematic review of experiences within this field is necessary, as is additional research on this topic. Similarly, research exists regarding psychological barriers to socially responsible behaviour in the supply-chain context. Baur (2018) discusses the reasons that actors along the supply chain would behave in ways that are not in line with their own moral standards and sometimes in ways that violate ethical guidelines. However, there is a lack of empirical knowledge on the factors that influence the assumption of responsibility in supply chains. Supply chains are increasingly complex and difficult to manage. Hence, the establishment of WHP is a true challenge, and the experience and knowledge required to meet this challenge are scarce.

Another issue concerns customers. Customers affect companies by using services and buying products (Gude et al. 2017; Gude 2018). They are part of the supply chain. The significance of customers has been considered in studies on CSR and in studies on sustainability, but rarely in research on WHP. One exception is fair trade processes. Some campaigns and studies on fair trade have considered work security. The inclusion of customers in WHP in general is a promising task for science and in practice.

References

Alvarado-Herrera A, Bigne E, Aldas-Manzano J, Curras-Perez R (2017) A scale for measuring consumer perceptions of corporate social responsibility following the sustainable development paradigm. J Bus Ethics 140(2):243–262

Bamberg E, Ducki A, Metz A-M (2011) Gesundheitsförderung und Gesundheitsmanagement in der Arbeitswelt. Ein Handbuch. Hogrefe, Innovatives Management, Göttingen

Bamberg E, Dettmers J, Tanner G (2016) Diffundierende Grenzen von Organisationskulturen – die Rolle von Kundinnen und Kunden. In: Badura B, Ducki A, Schröder H, Klose J, Meyer M (eds) Fehlzeiten-Report 2016. Unternehmenskultur und Gesundheit - Herausforderungen und Chancen: Zahlen, Daten, Analysen aus allen Branchen der Wirtschaft: mit 130 Abbildungen und 253 Tabellen. Springer, Berlin, pp 193–200

Banerjee SB (2012) Corporate social responsibility. The good, the bad and the ugly. Critical Sociol 34(1):51–79

Baur C (2018) Nachhaltigkeit in der Wertschöpfungskette: Das Problem des eingeschränkten moralischen Bewusstseins. In: Schmitt C, Bamberg E (eds) Psychologie und Nachhaltigkeit. Theoretische Grundlagen, Anwendungsbeispiele und Zukunftsperspektiven. Springer, Heidelberg pp 149–163

Beck D, Lenhardt U, Schmitt B, Sommer S (2015) Patterns and predictors of workplace health promotion: cross-sectional findings from a company survey in Germany. BMC Public Health 15:343

Bondy K, Moon J, Matten D (2012) An institution of corporate social responsibility (CSR) in multi-national corporations (MNCs) form and implications. J Bus Ethics 111(2):281–299

Carroll AB (2015) Corporate social responsibility. Org Dyn 44(2):95

Commission of the European Communities (2001) Green paper. Promoting a European framework for corporate social responsibility. European Commission, Brussels 18 July, 2001. http://europa.eu/rapid/press-release_DOC-01-9_en.pdf. Last accessed 05 Feb 2018

Commission of the European Communities (o.J.) Corporate social responsibility (CSR). European Commission. http://ec.europa.eu/growth/industry/corporate-social-responsibility_en. Last access 20 Feb 2018

Cox T (1997) Workplace health promotion. Work & Stress 11(1):1–5

Dachler HP, Wilpert B (1978) Conceptual dimensions and boundaries of participation in organizations: a critical evaluation. Adm Sci Q 23(1):1–39

Dahlsrud A (2008) How corporate social responsibility is defined. An analysis of 37 definitions. Corp Soc Responsib Environ Manag 15(1):1–13

Dezmar-Krainz K (2015) Enhancing wellbeing of employees through corporate social responsibility context. Megatrend Revija 12(2):137–153

Dieterich C, Vetter C, Naji N (2000) Krankheitsbedingte Fehlzeiten in der deutschen Wirtschaft. In Badura B, Litsch M, Vetter C (eds) Fehlzeiten-Report 1999. Psychische Belastung am Arbeitsplatz Zahlen, Daten, Fakten aus allen Branchen der Wirtschaft. Springer, Heidelberg, pp 363–622

Du S, Bhattacharya CB, Sen S (2011) Corporate social responsibility and competitive advantage. Overcoming the trust barrier. Manag Sci 57(9):1528–1545

ENWHP (2007) Luxembourg declaration on workplace health promotion. European Network for Workplace Health Promotion. http://www.enwhp.org/fileadmin/rs-dokumente/dateien/Luxembourg_Declaration.pdf. Last accessed 12 Feb 2018

Eriksson D, Svensson G (2016) The process of responsibility, decoupling point, and disengagement of moral and social responsibility in supply chains. Empirical findings and prescriptive thoughts. J Bus Ethics 134(2):281–298

Gröneweg C, Matiaske W (2012) Gullivers Fesseln—corporate social responsibility. Werkstatt für Organisations- und Personalforschung e.V., 25. https://www.econstor.eu/obitstream/10419/70813/1/73372115X.pdf. Last accessed 23 Mar 2018

Gude M, Bamberg E, Etzold M, Wolf KJ (2017) Macht sich Gesundheitsmanagement bezahlt? Eine empirische Untersuchung des Einflusses von Informationen zu betrieblichem Gesundheitsmanagement auf konsumbezogene Reaktionen von VerbraucherInnen. Zeitschrift für Wirtschafts- und Unternehmensethik 18(3):328–346

Gude M (2018) Konsumhandlungen und Nachhaltigkeit: (Wie) Passt das zusammen? In: Schmitt C, Bamberg E (eds) Psychologie und Nachhaltigkeit. Theoretische Grundlagen, Anwendungsbeispiele und Zukunftsperspektiven. Springer, Heidelberg pp 135–148

Ha-Brookshire J (2017) Toward moral responsibility theories of corporate sustainability and sustainable supply chain. J Bus Ethics 145(2):227–237

Hammer SS, Liebherr M, Kersten S, Haas CT (2015) Adherence to worksite health interventions. Practical recommendations based on a theoretical framework. J Work Behav Health 30(4):325–343

Harden A, Peersman G, Oliver S, Mauthner M, Oakley A (1999) A systematic review of the effectiveness of health promotion interventions in the workplace. Occup Med 49(8):540–548

Jain A, Leka S, Zwetsloot G (2011) Corporate social responsibility and psychosocial risk management in Europe. J Bus Ethics 101(4):619–633

Jain A, Ripa D, Herrero J (2014) An analysis of the coverage of psychosocial factors at work in corporate social responsibility instruments and standards. In: Leka S, Sinclair RR (eds) Contemporary occupational health psychology. Global perspectives on research and practice, vol 3. Wiley-Blackwell, Chichester, pp 218–240

Janetzke H, Ertel M (2017) Psychosocial risk management in more and less favorable workplace conditions. Int J Work Health Manag 10(4):300–317

Locke R, Romis M (2007) Improving work conditions in a global supply chain. MIT Sloan Manag Rev 48(2):54–62

Locke R, Amengual M, Mangla A (2009) Virtue out of necessity? Compliance, commitment, and the improvement of labor conditions in global supply chains. Polit & Soc 37(3):319–351

MacLean TI, Behnam M (2010) The dangers of decoupling. The relationship between compliance programs, legitimacy, perceptions, and institutionalized misconduct. Acad Manag J 53(6):1499–1520

McCoy K, Stinson K, Scott K, Tenney L, Newman LS (2014) Health promotion in small business: a systematic review of factors influencing adoption and effectiveness of worksite wellness programs. J Occup Environ Med 56(6):579–587

Ming Y, Grabot B, Houé R (2014) A typology of the situations of cooperation in supply chains. Comput Ind Eng 67:56–71

OECD (2016) Health at a Glance: Europe 2016. State of Health in the EU Cycle. Organisation for Economic Co-operation and Development. https://www.oecd.org/els/health-systems/Health-at-a-Glance-Europe-2016-CHARTSET.pdf. Last accessed 16 Mar 2018

Pagell M, Wu Z (2009) Building a more complete theory of sustainable supply chain management using case studies of 10 exemplars. J Supply Chain Manag 45(2):37–56

Pope S, Wæraas A (2016) CSR-washing is rare. A conceptual framework, literature review, and critique. J Bus Ethics 137(1):173–193

Tanner G, Schmidt T (2017) Do CSR activities always promote organizational identification? Organizational identification as pathway between CSR activities and commitment. In: Oral presentation at the 18th conference of the European Association of Work and Organizational Psychology, Dublin

Tanner G, Bamberg E, Gude M (2018a) Corporate social responsibility: Verständnis, Bedingungen und Wirkungen. In Schmitt C, Bamberg E (eds) Psychologie und Nachhaltigkeit. Theoretische Grundlagen, Anwendungsbeispiele und Zukunftsperspektiven. Springer, Heidelberg pp 113–126

Tanner G, Bamberg E, Gude M (2018b) Workplace health promotion inspired by corporate social responsibility—interactions within supply chains and networks. Manag Rev (submitted)

Weaver GR, Trevino LK, Cochran PL (1999) Integrated and decoupled corporate social performance. management commitments, external pressures, and corporate ethics practices. Acad Manag J 42(5):539–552

WHO (2010) WHO healthy workplace framework and model. Background and supporting literature and practices. World Health Organization, Geneva, Switzerland. http://www.who.int/occupational_health/healthy_workplace_framework.pdf. Last accessed 20 Feb 2018

WHO (2018a) Absenteeism from work due to illness, days per employee per year. World Health Organization. https://gateway.euro.who.int/en/indicators/hfa_411-2700-absenteeism-from-work-due-to-illness-days-per-employee-per-year/visualizations/#id=19398&tab=graph. Last accessed 16 Mar 2018

WHO (2018b) Five keys to healthy workplaces. World Health Organization. http://www.who.int/occupational_health/5_keys_EN_web.pdf?ua=1. Last accessed 19 Feb 2018

Zink KJ (2014) Designing sustainable work systems. The need for a systems approach. Appl Ergon 45(1):128

Zwetsloot G, Starren A (2004) Corporate social responsibility and safety and health at work. Office for Official Publication of the European Communities. European Agency for Safety and Health at Work, 60

Zwetsloot GIJM (2017) OSH management in small and micro enterprises. OSHwiki. https://oshwiki.eu/wiki/OSH_management_in_small_and_micro_enterprises. Last accessed 16 Mar 2018

Solidarity and Subsidiarity—How to Widen Access to Higher Education?

Neil M. Speirs

Abstract Bourdieu (The logic of practice. Polity Press, Cambridge, 1989) notes that 'agents shape their aspirations according to concrete indices of the accessible and the inaccessible, of what is and is not 'for us'.' In order to affect positive change on the reproduction of what young learners see around them we require the employment of a critical pedagogy. Aspiration has been classified by some authors (Brown in ACME: Int E-J Critical Geogr 12(3):419–430, 2013) as a neoliberal social hope that detracts from the underlying class anchorage in the educational system. Despite this, a pedagogy that promotes belief and hope followed by informed action is vital in the chain of events and circumstances that can change the life trajectory of individuals. This paper will build on early notions of solidarity and subsidiarity as outlined by E.U. architects Jean Monnet and Robert Schuman. This will be used as a lens to consider the social responsibility that a university has to its local community, in particular with regards to educational achievement and positive destinations of widening participation students. It will look at the transitions & life events that such students from lower socio-economic status backgrounds may experience and how these sit as barriers to engagement, attainment and achievement. It will then consider how we can work with community to address and overcome some of these barriers.

Keywords Widening participation · Social justice · Solidarity · Subsidiarity
Critical pedagogy

1 Introduction

Internationally, there have been targets, policies and recommendations put into place to widen participation in higher education (Lewis 2002; Gale and Parker 2013; Osborne 2003; Pugh et al. 2005; Jones and Thomas 2005; O'Neil 1971). This is centred on the need to provide and indeed fulfil the promise of access, participation

N. M. Speirs (✉)
Widening Participation, University of Edinburgh, 33 Buccleuch Place, Edinburgh EH8 9JS, UK
e-mail: neil.speirs@ed.ac.uk

W. Leal Filho (ed.), *Social Responsibility and Sustainability*, World Sustainability Series, https://doi.org/10.1007/978-3-030-03562-4_7

and inclusion. So that, those students from lower socio-economic status backgrounds can access and fully take part in their education, from primary through secondary and onto higher education. However every schooling system within the OECD suffers from school failure (OECD 2012). The same report notes that failure in school will affect a child for life, ultimately leading them to be 'less able to participate fully in the civic and democratic aspects of modern societies' (ibid, p. 3). In a 2009 (p. 2) communiqué the Conference of European Ministers Responsible for Higher Education called for member states to 'set measurable targets for widening overall participation and increasing participation of underrepresented groups in Higher Education'. While a few years prior to this, The European Commission (2006, p. 2) stated that; 'equitable systems ensure that the outcomes of education and training are independent of socio-economic background and other factors that lead to educational disadvantage and that treatment reflects individuals' specific learning need'. The recently published strategic plan of The Office for Fair Access in England (OFFA 2015) reminds universities that they are the access regulator and will set expectations for the sector. They outline the need for quicker and sustained change, while acknowledging that there has been a 40% increase in students from disadvantaged backgrounds entering universities with high entry tariffs since 2011. However, this is met with the fact that only 3.2% of students from the most disadvantaged backgrounds matriculate at highly selective universities. OFFA then, similarly to other such organisations and political bodies, outline their key aims linked to widening participation. Namely to see an increase in the proportion of students from disadvantaged backgrounds in higher education and to specifically address the gap in participation between the most advantaged and disadvantaged at the most selective of institutions. However, recent PISA data (OECD 2016, p. 39) shows that still; 'Socio-economic status is associated with significant differences in performance in most countries and economies that participate in PISA' (p. 39), and again the report outlines that 'Advantaged students tend to outscore their disadvantaged peers by large margins' (ibid, p. 40). The particular emphasis within the report is on science based learning and it rather worryingly comments on the numbers of young learners that 'fail to reach even the most essential learning outcomes' (ibid, p. 3). This lack of equity across OECD countries in science based learning is summarised (ibid) perhaps by the fact that those students from lower socio-economic status backgrounds are almost three times more likely than advantaged students not to attain the baseline level of proficiency in science. Government bodies, politicians and regulatory bodies all seem to have something to say about widening participation. Just last year, the UK Prime Minister Theresa May used her party conference speech to claim that 'for more than a century it's been Conservative education secretaries who've driven the reforms that have widened access and raised standards' (THE 2017). While French President Emmanuel Macron's education reforms are tackling inequality 'at the root, which means that it requires an *ex ante* solution such as education, not strictly *ex post* measures such as redistributive taxation' (Aghion and Berner 2018). Former President of the United States Barack Obama stated in his second term inaugural address that 'We are true to our creed when a little girl born into the bleakest poverty knows that she has the same chance to succeed as anybody else.' (White House 2013).

Despite discussions around the socially just reasons for widening participation, there appears to be a consistent parallel narrative that accompanies the inspiring words about seeing the fulfilment of equitable access and participation in all forms of education. We can begin by recalling that education has indeed become a central element of OECD countries' growth strategies (OECD 2012, p. 9). Perhaps any early notion that this was simply a drive for social justice might be met now with scepticism, the same OECD report comments on the results of educational failure ultimately imposing high costs on society, in particular imposing 'additional costs on public budgets to deal with the consequences—higher spending on public health and social support and greater criminality, among others' (ibid, p. 3). Other authors have presented a similar case, for example Heckman (2011) warns that underdeveloped human capital hampers productivity growth and limits the effective and full use of resources. While in the USA data collected by Rouse (2005) reveals that secondary school dropouts could be costing the United States 1.6% of the country's GDP. In California the value of high school graduates is clear, Belfield and Levin (2007) estimate that each such graduate produces a net fiscal gain for the total public sector during the lifetime of the graduate of about USD 169,000. Belfield and Levin (2007, p. 2) summarise that 'efforts to improve educational outcomes for at-risk populations' must be viewed 'as public investment that may yield benefits considerably in excess of investment costs'. As Camilleri and Mühleck (2010) discuss, the knowledge economy is perhaps currently one of the most popular political slogans. This transition to the so-called knowledge economy has placed education and knowledge at the centre of economic policies. Consequently we see a commercial value being ascribed to education, this is the neo-liberal commodification of education.

This is evident in a number of the documents and publications already mentioned. The Office for Fair Access in England talk of cost efficiencies, corporate planning objectives and working to improve processes (OFFA 2015). Very quickly their strategy goes from a desire to see an 'increase in the proportion of students from disadvantaged backgrounds in higher education' (ibid) to a neoliberal culture of metrics, performativity and education as human capital. Any hint of the potential for radical reform has taken a back seat. The Commission on Widening Access in Scotland begin their final report (2016) in a similar manner; 'Fundamentally, access is an issue of fairness' before within the same sentence pointing out that Scotland has an 'economic duty to tackle this inequality' (ibid, p. 3). The report shortly re-emphasises that there is an 'economic imperative' to this work, while again the shift to the knowledge based economy is emphasised and human capital theory employed in noting 'the key economic asset of any nation is the talent and skills of its people' (ibid, p. 3) and then warns of the dangers of 'missing out on the economic potential of some of our finest talents' (ibid, p. 3). The Irish government's National Plan for Equity of Access to Higher Education (HEA 2015), follow others in presenting the duality of widening participation—that it makes sense both socially and economically. The HEA warns that in order to continue with a stable economy 'more graduates will be needed to supply the expertise and skills required in vital areas'. The report states it is possible to observe 'skills shortages in vital and dynamic areas of our economy' and then re-emphasises the need for the economy 'to tap into the skills and talents of all our people so that the economy can continue to grow and prosper'. They do however make clear reference to the 'non-financial benefits to having a higher education' and are one of the few to make direct reference to the need for consideration of the wider social policy context of widening participation with their call to build 'social capital in communities with low levels of participation'. In their overview of access to higher education in the EU, Camilleri and Mühleck (2010) similarly acknowledge the need for public policy to 'Empower civil society in the enrichment of cultural capital'. It is vital that we return to the deeper healing of finding solutions to Bourdieu's (1989, p. 64) warning that 'agents shape their aspirations according to concrete indices of the accessible and the inaccessible, of what is and is not for us' rather than simply developing human capital to satisfy the economic needs of the state.

This drift from widening participation and diversity being seen through a social justice rational to one of economics, leads us to the point of diversity as 'a barometer of societal inequalities' versus diversity as 'good economic/business sense' (Archer 2007). This tension arises from an imbalance of legitimacy, 'resulting in the privileging of the economic.' (ibid). Archer rightly concludes that the economic framing of diversity and widening participation involves a 'reification of diversity that is untenable within a social justice agenda.' Perhaps it is that social justice invites the uncomfortable notion of class and the inequalities therein to be seriously addressed. Beck (1992) questioned whether it was difficult to investigate and observe class inequalities because of a 'social surge of individualisation'. If class identities have been submerged (Bradley 1996), then truly 'class has been the loser in the identity politics of the last decade' (ibid). This apparent reluctance to talk in terms of class, may indeed be the result of an uncomfortable truth, as Bradley suggests, if society desires to be classless, it must be aware that this has not be attained. Smith (2000) recalls the 'flood of erstwhile social democratic political parties' and their notion of a 'new middle' or 'third way' in which 'class consciousness is sublimated beneath the familiarly American self-deception that we are all middle class now'. In a recent article, Michaels (2016) argues that in fact identity politics is 'the politics of an upper class that has no problem with seeing people being left behind as long as they haven't been left behind because of their race or sex'. Of course, no person should be left behind or excluded on matters of race or sex, but Michaels argues that identity politics has no worries in leaving behind or excluding those from the working classes, he continues 'That's why elite institutions like universities make an effort to recruit black people as well as white into the ruling class. They're seeking to legitimate the class structure, not abolish it.' He concludes that the left should be 'more committed to doing something for the vast majority of people of all races, genders, and sexual orientations who will never belong to that class'. However, Scott (2000) makes it clear that class meanings have not vanished; 'Everybody does not have to believe in the existence of class, nor constantly think of themselves in terms of class identity, for class to be a social division. The system of class situations is not dependant in people's awareness of it.' So class meanings may not have disappeared, they are just 'more easily judged to be so' (Bradley 1996) by the political classes and other elites. However, widening participation must be viewed as a matter of social justice, addressing class inequalities—this in itself is truly enough of a justification for a fair and right minded society. But as Jacques Attali (2004, pp. 9–10) noted; 'No one, or almost no one, believes any longer that changing lives of others has importance for him or her'. Paulo Freire (2004, p. 55) provides a notion of hope here by saying that the 'most fundamental lesson is the one of non-conformity before injustice, the teaching that we are capable of deciding, of changing the world, of improving it.' However, for students from lower socio-economic status backgrounds there indeed may be a sizeable risk attached to engaging with higher education. As Beck (1992) outlines; 'The history of risk distribution shows that, like wealth, risks adhere to the class pattern, only inversely: wealth accumulates at the top, risks at the bottom'. So the very act of entering the field of higher education, perhaps with a lower volume social and cultural capital, could be seen as a risk that might not be worth taking. Perhaps this is another example of the fact that 'Poverty attracts an unfortunate abundance of risks' (ibid). However, Bauman (2005) states that 'Every choice involves risks that may be produced by forces transcending the comprehension and active capacity of the individual', how then are these risks to be taken if they are beyond the capacity of individuals, particularly for those potential students with low volume social and cultural capital. As Beck concludes (1992), it is possible that the income, education and power of the wealthy 'can purchase safety and freedom from risk.' While Bauman's notion of liquid modernity (2005) certainly warms us that 'Not everything works in life as one would like it to work', it is also the case that habitus (Bourdieu and Passeron 1977) and capital can go a long way to assisting in the navigation of uncertainty and risk—while the structures and various forms of institutional habitus that create and frame unequal participation in higher education require radical social policies to be dismantled.

As a result, we can see the social responsibility that a university, community or indeed a state or country has, to act on matters of social justice in order to widen participation. Wherein the legitimacy of this social responsibility is sufficient so as not to require neoliberal economic arguments—which only serve as a way to legitimatise class inequities rather than dismantle them.

2 The Power of Pedagogy

Having established a social justice rationale for why we should widen participation, we must think about how this can be realised. Firstly we must consider the schools that pupils attend day in and day out. The OECD (2012, p. 11) commented that; 'disadvantaged schools often lack the internal capacity or support to improve, as school leaders and teachers and the environments of schools, classrooms and neighbourhoods frequently fail to offer a quality learning experience for the most disadvantaged'. Authors such as Hanushek et al. (2014, p. 26) have suggested that 'attracting the best' to the teaching profession would increase equity, they note the 'effect of teacher cognitive skills on student performance is substantially larger for students with low socioeconomic background' (ibid). While the European Commission (2006, p. 5) note that 'the quality of teachers plays a prominent role in achieving high and equitable educational results.' According to Dunne and Gazeley (2008) the identification of underachieving pupils by teachers 'overlapped with, and were informed by, their tacit understanding of pupils' social class position.' The authors acknowledge the continuing 'conditions in which middle-class pupils were encouraged to achieve while the underachievement of many working-class pupils was normalised.' which as Diamond et al. (2004) notes, is rooted 'in school-based organizational habitus through which expectations of students become embedded in schools.' Rist's study from (1970) asked the question 'Given the treatment of low-income children from the beginning of their kindergarten experience, for what class strata are they being prepared other than that of the lower class?' There does, however, need to be a complete rejection of this social fatalism (Merton 1948), despite the self-fulfilling prophecy of the practice of some schools and teachers. Universities, too, must be careful that their pedagogical practice is socially inclusive and that staff do not become the 'Old, learned, respectable bald heads' of Yeats (2016). We must ensure that academic staff avoid the drive to become gated intellectuals, as Giroux (2014, p. 89) notes, they become 'walled off from growing impoverished populations….cut loose from any ethical mooring or sense of social responsibility.' The risk is the formation of a gated pedagogy, one that 'establishes boundaries to protect the rich, isolates citizens from each other, excludes those populations considered disposable and renders invisible young people, especially poor youth of colour, along with others marginalised by class and race.' (ibid). Correctly, Giroux's (2014, p. 149) preference is the public intellectual, those that 'have a responsibility to unsettle power, trouble consensus, and challenge common sense. The very notion of being an engaged public intellectual is neither foreign to nor a violation of what it means to be an academic scholar but central to its very definition.' The self-fulfilling prophecies must be met with Freire's (2014, p. 89) reminder that 'though we are programmed we are nevertheless not determined.', and that of Bourdieu (1984) when he suggested that fractions of a class will deviate from the most common trajectories of their class as a whole, and follow the most probably trajectory for members of another class. Giroux (1988, p. 2) posed the question 'How can we make schooling meaningful so as to make it critical and how can we make it critical so as to make it emancipatory?' This forces us to re-emphasize the fact that education can be emancipatory and address the need for a critical pedagogy. Building a critical pedagogy will allow us (ibid, p. 7) to 'help students, particularly from the oppressed classes, recognize that the dominant school culture is not neutral and does not generally serve their needs.' Giroux (ibid, p. 7) continues; 'Such an activity calls for a mode of dialogue and critique that unmasks the dominant school culture's attempt to escape from history and that interrogates the assumptions and practices that inform the lived experiences of day to day schooling.'

Paulo Freire's Pedagogy of the Oppressed (1970a, b, p. 71) considered the teacher-student relationship. Freire notes its essentially narrative character; 'This relationship involves a nar-rating Subject (the teacher) and patient, listening objects (the students). The contents, whether values or empirical dimensions of reality, tend in the process of being narrated to become lifeless and petrified. Education is suffering from narration sickness.' This is far from the desired critical pedagogy, but rather leads to his notion of the banking concept of education; 'Narration (with the teacher as narrator) leads the students to memorize mechanically the narrated content. Worse yet, it turns them into "containers," into "receptacles" to be "filled" by the teacher. 'The more completely she fills the receptacles, the better a teacher she is. The more meekly the receptacles permit themselves to be filled, the better students they are' (ibid). The preferred pupil or student centred approach offered by critical pedagogy is to be found in the writings of Pádraig Mac Piarais (1916, p. 28) who wrote that 'What the teacher should bring to his pupil is not a set of ready-made opinions, or a stock of cut and dry information, but an inspiration and an example… so infectious an enthusiasm as shall kindle new enthusiasm.' Akin to Freire's disapproval of the mechanistic banking approach to education, Pearse (1924, p. 12) had criticised the 'conception of education as some sort of manufacturing process.' whereby children are seen as the 'raw material', he continues; 'we desiderate for their education 'modern methods' which must be efficient but cheap; we send them to Clongowes to be 'finished'; when finished they are 'turned out'; specialists 'grind' them for the …so called liberal professions.' Freire (1970a, b, pp. 76–77) describes the banking concept of education as serving 'the interests of oppression. Based on a mechanistic, static, naturalistic, spatialized view of conscious-ness, it transforms students into receiving objects. It attempts to control thinking and action, leads women and men to adjust to the world, and inhibits their creative power.' Perhaps Mac Piarais (2013, p. 11) was correct when he said that 'What is needed here, is not even a revolution, but a vastly bigger thing—a creation', after all, Bourdieu and Passeron (1977, p. 18) reminds us that 'pedagogic action is objectively a symbolic violence to the extent to which it is an imposition of a cultural arbitrary by an arbitrary power.' This notion of symbolic violence or symbolic control is discussed further by Bernstein and Solomon (1999) 'Symbolic control, through its pedagogic modalities, attempts to shape and distribute forms of consciousness, identity and desire'. We can see that the set of pedagogical practices that teachers and lecturers employ have the huge potential to reveal education as truly emancipatory, and yet teachers and lectures are also at risk of employing forms of pedagogy that will work against the drive to widen participation. For too long, teachers and lectures have held class driven expectations of what students can achieve in the classroom. Coupled with the institutional habitus that frames the structures in which our pupils and students learn, surely the case for a critical pedagogy has been made?

3 Solidarity and Subsidiarity

On the 9th of May 1950, the Schuman declaration proposed a European project of unification that would bring together a continent divided by war and transform it into a place of peace and security. Within this plan Jean Monnet and Robert Schuman, two of the early architects of the European Union, believed that both solidarity and subsidiarity were key principles. In a world that has since seen the drive for the individual's self-interests and rights, it is clear that the notion of solidarity (and subsidiarity) has been pushed to the margins. As Kelly (1998) notes 'Appeals for solidarity help us to become aware of the world of the other, especially the suffering and oppressed.' He continues 'solidarity is the name given to the broad movement to side with the marginalized and the oppressed against systems that deprive them of basic human dignity'. Kelly (ibid) makes reference to Paulo Freire's (1970a, b) call to 'problematize' all that is currently in order, so that we may be 'able to reveal our historical conditioning and as a result of this open our world to the presence of self-understanding and ultimately structural transformation'. According to Schweigert (2002) solidarity involves a form of shared membership that is characterised by a mutual care and respect, 'revealing a sense of belonging enriched by a commitment to human dignity'. Putnam et al. (1994) takes this further and speaks of the desire to give to others but without the expectation of direct personal reciprocation. In their article from 2007 Lynch et al. outline that the focus of education is clearly on the development of the rational public actor in the liberal educational tradition—however neoliberal agendas have pushed this further to see the production of the 'self-sufficient, rational economic man'. The title of their paper, 'Breaking silence—educating citizens for love, care and solidarity', hints at Freire's (1970b, p. 457) idea of a culture of silence and the need to challenge this. Lynch et al. (ibid) rightly contest an unsatisfactory status quo, where learning about and taking part in solidarity 'is not on the education agenda'. The authors are in agreement with the expression of care for the planet through environmentalism, but they are confused as to why the care for 'all of humanity' appears to be excluded. They authors propose that this may be because 'Neo-liberal thinking in education has succeeded in doing what classical liberalism did not do; it subordinates and trivialises education that has no market value.' If the present is 'fed on a culture of insecurity that induces anxiety, competition and indifference to those more vulnerable than oneself' (ibid) and as Bauman (2005) notes is 'largely out of control and unpredictable, is increasingly turning from a land of hope into a major source of apprehension', then clearly we are struggling to find a place to value solidarity and indeed care and love as Lynch et al. (ibid) refer to. Interestingly, there may be a narrative connection between what Lynch et al. (2007) refer to as 'nurturing capital' and its necessity in order to relate to and support one another in a manner of solidarity—and the pupil focussed pedagogy and commitment to the idea of fostering and developing pupils, that Pearse (1916) had implemented.

Durkheim (1984, p. 331) spoke of 'The totality of bonds that bind us to one another and to society, which shape the mass of individuals into a cohesive aggregate' which is, for example, illustrated by the Solidarity movement in Poland from the 1980s which 'was predominantly a working-class creation whose formation was possible only through the collective efforts of hundreds of thousands of workers on Poland's Baltic Coast.' (Karabel 1993) and employed decision making processes at a local level in the manner of subsidiarity. There are of course many current examples in education, of individuals and institutions working to widen participation through actions of solidarity (Speirs 2017; Speirs et al. 2017), there are many that still believe in and want to act on the values of solidarity, aiming to break the silence. A particular example of this are the 'Social frontistiria' (Zambeta 2014) currently active in Greece. Framed by the recent economic crisis in the country, they provide support for students that wish to prepare for the entrance examinations to university. Each social frontistiria is led by a team of volunteer teachers, some of whom may no longer be employed in the teaching profession due to the local economic situation. Zambeta (ibid) notes that the classes prioritise those students from low income families and those with no household employment—'the most vulnerable social groups' (ibid). Local activism or the local political Municipalities are leading the initiation of these social frontistiria—truly in the notion of subsidiarity. As Schweigert (2002) notes, subsidiarity directs 'decision making to the social level that is most effective, with particular respect for the power of local communal levels of society.' The author continues to remind us that while subsidiarity can certainly limit state interference with local community matters, it does not suggest that the local level is 'intrinsically more just and righteous than the higher level.' However within education, we can see how subsidiarity promotes local level decisions and indeed strikes against the managerialism and performativity of the neoliberal ideology. As a result, we can understand the need for solidarity and subsidiarity as a collective lens that might frame our rationale for widening participation and how to actively achieve it. Through solidarity and subsidiarity, social justice can be achieved. This may indeed be a struggle (Freire 2016, p. 14), but 'a struggle that must incorporate love, conciliation, freedom, and hope.' Robert Schuman (1964, p. 49) cried out for 'ambitions of hegemony, narrow-minded political nationalism, self-centred protectionism and cultural isolationism' to be thrown aside and the 'notion of solidarity, that is, the conviction that everyone's true self-interest lies in recognizing and accepting everyone's interdependence in practice' to be strived for. While the present day E.U. may not live up to the standards set by Monnet and Schuman, as European citizens or indeed citizens of the world we are challenged by these laudable aims. We must continue to work in our local communities with pupils, parents, carers, teachers and community to achieve this vision of social justice. So that, those students from lower socio-economic status backgrounds can access and fully take part in their education, from primary through secondary and onto higher education, for 'Without a vision for tomorrow, hope is impossible' (Freire 1998, p. 45).

4 Exclusion Versus Hope

We have seen that despite the calls from Government and other political institutions to widen participation to higher education, quite often the rationale employed is not that of a simple call to social justice. Too often the case is made in terms of the knowledge economy, human capital and the economic imperative. This favour carried by the economic is synonymous with the neoliberal agenda whereby education is bought and sold, a commodified object traded on the market floor. Social justice, while mentioned in white papers and policy documents, is simply not the priority. Davies (2003) notes that, within the EU, it was in the late 1980s 'when education and training were part of the major push towards the achievement of the single market and a growing concern with the globalisation of markets and the restructuring of employment and the labour market.' Davis goes on to note how the 'period leading up to the Maastricht Treaty was significant for the emergence of another key narrative in the EU policy discourse: social exclusion.' The term social exclusion is attributed to Jacques Delors, used during negotiations leading up to the Treaty. Allen et al. (1998, p. 13) suggest that eliminating social exclusion is therefore about seeking 'the social, political and moral insertion of subjects within this wider French social order'. Quite clearly, Davis (ibid) states that this insertion of subjects, or citizens, back into social order cannot rely purely on employment as the only way to achieve this. Indeed a wider form of social justice is required in order to combat social exclusion. However, Freire (1970a, b, p. 74) had already argued that individuals are not excluded or outside society, 'They have always been "inside"—inside the structure which made them "beings for others." The solution is not to "integrate" them into the structure of oppression, but to transform that structure so that they can become "beings for themselves"'. Freire's vision can be promoted through the employment of a critical pedagogy, one that does not relate the academic expectations of learners to their socio-economic status by teachers or lecturers. It is therefore necessary that we take on the role of the public intellectual that challenges 'what is and is not for us.' (Bourdieu 1989). Within the field of widening participation, as Brown (2013) notes, aspiration has been the key word, but it is politicised. Brown argues that aspiration in this context is merely 'a neoliberal social hope'. Beuret (2011) asserts that this neoliberal aspiration is all about 'a better job, more money, more things and a higher rung on the career ladder'. Beuret concludes that 'Hope is individual in our world, never collective'—this is of course not the hope of solidarity. We must insist on Freire's notion of hope and care, and a critical pedagogy that can pierce this neoliberal ideology, through learning that promotes belief and critical curiosity that is vital in the chain of events and circumstances that can change the life trajectory of individuals. Yet, generation after generation lives with the fact that poverty and in many cases extreme poverty has not been eradicated. Students from lower socio-economic status backgrounds can experience many challenging situations, life events and obstacles to participation long before they have considered degree level study or successfully matriculated as an undergraduate. These act as barriers to engagement, attainment and achievement. For example as Welsh et al. (2010) state, there are 'substantial achievement gaps between middle-income children and low-income children at school entry that widen over time and contribute to serious disparities in learning difficulties, educational attainment, and long-term employment potential.' While the development of various mental process are reported to be 'delayed in children growing up in poverty and appear to play a central role in predicting school adjustment and academic attainment.' (Noble et al. 2007; Blair and Razza 2007). With regards to diet, Patrick et al. (2005) note that 'Adolescents whose parents were relatively more educated had higher intakes of carbohydrates, protein, fibre, folate, vitamin A, and calcium; higher consumption of vegetables; and greater likelihood of consuming the recommended servings of dairy products' Reidpath (2002) places this in context; 'those living in areas with the lowest individual median weekly incomes have 2.5 times the exposure to fast-food outlets compared to those living in areas with the highest individual median weekly incomes'. While O'Neill et al. (2003) outlines how 'poor and working-class communities are often more exposed to air pollution' and that there is a 'differential distribution of the health impacts from air pollution', all of which is related to class. The Chief Medical Officer of the UK (2018, p. 35) recently reported again on the 'evidence around pollution and health inequalities' and the 'interaction between the pollution and socioeconomic position—through an underlying susceptibility'. These few examples are presented to begin to illustrate the challenges that pupils and students from lower socio-economic status background face even before we consider social and cultural capital or indeed the class anchorage within current curriculum and pedagogy. These few examples show us what Freire (1970a, b) meant—the idea of people restricted by social structures and so not able to become 'beings for themselves'.

5 Conclusions

Pauly (1995) referred to the notion of shifting baselines within a context of conservation work with fisheries management. He discussed how each generation sees a lower baseline of fisheries stock and that is then normalised, an amnesia sets in, failing to acknowledge the drop in stocks over generations—leading to the idea of shifting baselines as a result of no historical memory. While not directly applicable to the situation regarding social justice and widening participation, I refer to the work of Pauly because I believe there is indeed an overlap in terms of the somewhat laissez-faire approach by the political classes and indeed some elements of the electorate when it comes to issues surrounding class and the associated inequalities. Each generation is faced with the baseline figures of poverty and associated inequalities, there is then a normalisation of this inequity whereupon it no longer interests the public sphere for meaningful debate, and a collective amnesia has set in. Social and environmental issues can be often the victims of lip service, each generation accepting the fact that there will be social inequalities—a normalised baseline that is accepted. Society becomes ever more populated by the 'cheerful robots' that Giroux (2012, p. 3) refers to. This is where the cultural silence must be challenged and questioned through the values of solidarity and subsidiarity. So as Nicholson (1989) notes, the educated individual is no longer somebody that 'has no sense of solidarity with the human community or sensitivity to the natural environment.' indeed they would be socially responsible, a critically thinking citizen.

In the same way that John Keats (Forman 1953, p. 140) felt philosophy would 'Conquer all mysteries by rule and line' so too has neoliberalism removed the care, the solidarity and the hope from education. Just as the romantic poets would write about the 'disparity between appearance and reality' (Pedrini and Pedrini 1966, p. 75) we are very aware of the political sheen placed on commitments to widen participation, be it by governments or regulatory bodies. We have seen the legitimacy of the economic argument dominate that of social justice and social responsibility. As Pedrini et al. (ibid) remind us 'The romantic poets held that appearances are not always what they seem, and it is only by exercise of imagination that man can strip away illusions and get to the core of life'. Might this be the sociological imagination of Mills (2000)? Perhaps a wide spread understanding of such an imagination could well serve us as an analytical tool in order to strip bare political spin on matters concerning inequity. Freire certainly (2016, p. 23) outlines, it is the role of a university to 'stimulate different ways of thinking, of dreaming' the responsibility of a university to produce graduates that are literate in critical thinking.

What we have seen is the drive to widen participation is anchored in social justice but has increasingly been hi-jacked by the neoliberal political movement through the commodification of education. The power of critical pedagogy as a tool to widen participation is clear, it allows us to rediscover education as truly emancipatory—removing the class anchorage in curriculum and pedagogical practices. Adjoining this critical pedagogy with true notions of solidarity and subsidiarity has the potential to be truly radical and allow major steps forward in widening participation, through showing a genuine sense of care and hope to those around us. Solidarity and subsidiarity are not only reasons for widening participation but also ways to widen participation successfully. This care and hope is an antidote to the normalised feelings and experiences of exclusion that many go through. Indeed, there are a myriad of environmental and personal circumstances and barriers preventing learners from engaging with their education as we have seen. Through local community led initiatives—employing the notion of subsidiarity—we can achieve further successes in widening participation. This will see the acknowledgment of the social structures that frame inequality in the first place and govern social reproduction, but will also be filled with Freire's (2016, p. 10) notion of hope, which is based on 'human agency and a full scale attack on the fear of freedom'. It is vital that those of us in universities involved in creating and delivering such widening participation initiatives engage with social justice through the lens of solidarity and subsidiarity—employing a critical pedagogy that is full of hope and reveals the emancipatory nature of education, thus removing its class anchorage. This is our social and moral responsibility.

References

Aghion P, Berner B (2018) Macron's education revolution. Project syndicate. https://www.project-syndicate.org/commentary/macron-education-reforms-by-philippe-aghion-and-benedicte-berner-2018-03. Last accessed 13 Mar 18

Allen J, Cars G, Madanipour A (1998) Introduction. In: Madanipour A, Cars G, Allen J (eds) Social exclusion in European cities. Processes, experiences and responses. Jessica Kingsley Publishers, London

Archer L (2007) Diversity, equality and higher education: a critical reflection on the ab/uses of equity discourse within widening participation. Teach High Educ 12(5–6):635–653

Attali J (2004) La Voie Humaine. Fayard, Paris

Bauman Z (2005) Education in Liquid Modernity. Rev Educ Pedagog Cult Stud 27(4):303–317

Beck U (1992) Risk society: towards a new modernity. Sage, London

Belfield CR, Levin HM (2007) The price we pay: economic and social consequences of inadequate education. Brookings Institution Press, Washington, D.C.

Bernstein B, Solomon J (1999) Pedagogy, identity and the construction of a theory of symbolic control. Br J Sociol Educ 20(2): 265–279
Beuret N (2011) Hope against hope: a necessary betrayal. The Paper. https://wearethepaperdotorg.wordpress.com/2011/02/04/22/. Last accessed 15 Mar 18
Blair C, Razza RP (2007) Relating effortful control, executive function, and false-belief understanding to emerging math and literacy ability in kindergarten. Child Dev 78(2):647–663
Bourdieu P (1984) Distinction: a social critique of the judgement of taste (trans: Nice R). Polity Press, Cambridge
Bourdieu P (1989) The logic of practice (trans: Nice R). Polity Press, Cambridge
Bourdieu P, Passeron J-P (1977) Reproduction in education, society and culture (trans: Nice R). Sage, London
Bradley H (1996) Fractured identities: changing patterns of inequality. Polity Press, Cambridge
Brown G (2013) The revolt of aspirations: contesting neoliberal social hope. ACME: Int E-J Critical Geogr 12(3):419–430
Camilleri AF, Mühleck K (2010) Evolving diversity—an overview of equitable access to higher education in europe. MENON Network, Brussels
Chief Medical Officer (2018) Annual report of the Chief Medical Officer 2017, health impacts of all pollution—what do we know?. Department of Health and Social Care, UK
Commission on Widening Access (2016) A blueprint for fairness. Scottish Government, Edinburgh
Davies P (2003) Widening participation and the European Union: direct action—indirect policy? Eur J Educ 38(1):99–116
Diamond JB, Randolph A, Spillane JP (2004) Teachers' expectations and sense of responsibility for student learning: the importance of race, class, and organizational habitus. Anthropol & Educ Q 35(1):75–98
Dunne M, Gazeley L (2008) Teachers, social class and underachievement. Br J Sociol Educ 29(5):451–463
Durkheim Emile (1984) The division of labour in society. Macmillan, Basingstoke
European Commission (2006) Communication from the Commission to the Council and to the European Parliament. Efficiency and equity in European education and training systems, COM (2006) 481 final. Brussels
Forman HB (ed) (1953) The poetical works of John Keats. Oxford University Press, Oxford
Freire P (1970a) Pedagogy of the oppressed. Continuum, New York
Freire P (1970b) Cultural action and conscientization. Harv Educ Rev 40:452–457
Freire P (1998) Pedagogy of the heart. Continuum, New York
Freire P (2004) Pedagogy of indignation. Paradigm, Boulder, CO
Freire P (2014) Pedagogy of hope. Bloomsbury Academic, London
Freire P (2016) Pedagogy of solidarity. Routledge, New York
Gale T, Parker S (2013) Widening participation in Australian Higher Education: report submitted to HEFCE and OFFA. Project Report. CFE (Research and Consulting) Ltd., Leicester
Giroux HA (1988) Teachers as intellectuals: toward a critical pedagogy of learning. Bergin & Garvey, New York
Giroux H (2012) On critical pedagogy. Continuum, London
Giroux HA (2014) Neoliberalism's war on higher education. Haymarket, Chicago
Hanushek EA, Piopiunik M, Wiederhold S (2014) The value of smarter teachers: international evidence on teacher cognitive skills and student performance (Munich Discussion Paper No. 2014-51). Department of Economics, University of Munich, Germany
HEA (2015) National plan for equity of access to higher education 2015–2019. Department of Education and Skills, Dublin
Heckman J (2011) The economics of inequality: the value of early childhood education. Am Educ 35(1):31–35
Jones R, Thomas L (2005) The 2003 UK Government Higher Education White Paper: a critical assessment of its implications for the access and widening participation agenda. J Educ Policy 20(5):615–630

Karabel J (1993) Polish Intellectuals and the origins of solidarity: the making of an oppositional alliance. Communist Poet-Communist Stud 26(1):25–46
Kelly MA (1998) Solidarity: a foundational educational concern. Relig Educ 93(1):44–64
Lewis B (2002) Widening participation in higher education: the HEFCE perspective on policy and progress. High Educ Q 56(2):204–219
Leuven/Louvain-La-Neuve, Communiqué (2009) European Commission, Brussels
Lynch K, Lyons M, Cantillon S (2007) Breaking silence: educating citizens for love, care and solidarity. Int Stud Sociol Educ 17(1–2):1–19
Mac Piarais P (1916) The murder machine. Whelan & Son, Dublin
Mac Piarais P (1924) Political writings and speeches. Phoenix Publishing Co. Ltd, Dublin
Mac Piarais P (2013) Political writings and speeches. Éire-Gael Society, Dublin
Merton RK (1948) The self-fulfilling prophecy. Antioch Rev 8:193–210
Michaels WB (2016) A universe of exploitation. The Nation. https://www.thenation.com/article/what-is-the-left-without-identity-politics/. Last accessed 14 Mar 18
Mills CW (2000) The sociological imagination. Oxford University Press, Oxford
Nicholson C (1989) Postmodernism, feminism and education: the need for solidarity. Educ Theory 39(3):197–205
Noble KG, McCandliss BD, Farah MJ (2007) Socioeconomic gradients predict individual differences in neurocognitive abilities. Dev Sci 10(4):464–480
OECD (2012) Equity and quality in education: supporting disadvantaged students and schools. OECD Publishing, Paris
OECD (2016) PISA 2015 results (vol I): excellence and equity in education. PISA, OECD Publishing, Paris
OFFA (2015) Strategic plan 2015–2020. Office for Fair Access, Bristol
O'Neil RM (1971) Preferential admissions: equalizing the access of minority groups to higher education. Yale Law J 80(4):699–767
O'Neill MS, Jerrett M, Kawachi I, Levy JI, Cohen AJ, Gouveia N, Wilkinson P, Fletcher T, Cifuentes L, Schwartz J (2003) Health, wealth, and air pollution: Advancing theory and methods. Environ Health Perspect 111(16):1861–1870
Osborne M (2003) Increasing or widening participation in higher education?—a European overview. Eur J Educ 38(1):5–24
Pauly D (1995) Anecdotes and the shifting baseline syndrome of fisheries. Trends Ecol Evol 10(10):430
Patrick H, Nicklas TA (2005) A review of family and social determinants of children's eating patterns and diet quality. J Am Coll Nutr 24(2):83–92
Pedrini LN, Pedrini DT (1966) serpent imagery and symbolism—a study of the major English romantic poets. College and University Press, New Haven
Pugh G, Coates G, Adnett N (2005) Performance indicators and widening participation in UK higher education. High Educ Q 59(1):19–39
Putnam RD, Leonardi R, Nanetti RY (1994) Making democracy work: civic traditions in modern Italy. Princeton University Press, Princeton
Reidpath DD, Burns C, Garrard J, Mahoney M, Townsend M (2002) An ecological study of the relationship between social and environmental determinants of obesity. Health & Place 8(2): 141–145
Rist RC (1970) Student social class and teacher expectations. Harv Educ Rev 40(3):411–451
Rouse C (2005) The labour market consequences of an inadequate education, prepared for teachers' college symposium on educational equity. Columbia University Press, New York
Schuman R (1964) Pour l'Europe. Nagel, Paris
Schweigert FJ (2002) Solidarity and subsidiarity: complementary principles of community development. J Soc Philos 33(1):33–44
Scott J (2000) Class and stratification. In: Payne G (ed) Social divisions. Macmillan, Basingstoke
Smith N (2000) What happened to class? Environ Plan A 32:1011–1032

Speirs NM (2017) In: Judy Robertson J, Williams S, Jones D, Isbel L, Loads D (eds) EqualBITE: gender equality in higher education. Sense Publishers, Rotterdam

Speirs NM, Riley SC, McCabe G (2017) Student-led, individually-created courses: using structured reflection within experiential learning to enable widening. J Perspect Appl Acad Pract 5(2):51–57

Times Higher Education (2017) Text from the prime minister's 2018 party conference address. October 4th 2017. THE, London

Welsh JA, Nix RL, Blair C, Bierman KL, Nelson KE (2010) The development of cognitive skills and gains in academic school readiness for children from low-income families. J Educ Psychol 102(1):43–53

White House (2013) Inaugural address by President Barack Obama, Office of the Press Secretary. https://obamawhitehouse.archives.gov/the-press-office/2013/01/21/inaugural-address-president-barack-obama. Last accessed 13 Mar 18

Yeats WB (2016) Collected poems. Macmillan Collector's Library, London

Zambeta E (2014) Education and social solidarity in times of crisis: the case of voluntary shadow education in Greece. Educ Inq 5(1):69–88

An Indigenous 'Right Way' Environmental, Social and Cultural Core-Benefits Verification Standard

Lisa McMurray, Rowan Foley and Carl O'Sullivan

Abstract There is a lot of pressure on Indigenous people to conform, to think and act like Europeans. There is an unspoken belief that it would be much easier for all concerned to simply use European models and accepted western ways of doing things. Developing a new standard for the cultural, social and environmental core benefits of carbon projects that uses the *Indigenous-to-Indigenous* way of working has not been easy. It has involved a lot of thinking, reflection and discussion with many wonderful Aboriginal and non-Aboriginal people. It also involved more formal consultation and peer review processes as well as presentations at industry forums and community meetings. Ultimately, we do not see a role for non-Aboriginal people in leading this process nor are we 'massaging' a western monitoring and evaluation (M&E) model into an Aboriginal context. We can't tweak existing standards whose soul is fundamentally different. There is, however, a support role for non-Aboriginal people which affords for a generosity of spirit to not dominate or dismiss Indigenous ways of working. The concept of Indigenous people working with Indigenous people using Indigenous expertise to verify core-benefits has been seen by some as an inferior process, a bit soft perhaps, lacking rigour even. The idea that Indigenous people have something to offer challenges the dominate neo-colonial concepts of dependence and is a form of environmental racism that needs to be called out. We are holding fast in our belief that in order to be part of the solution and to recognise Indigenous people's expertise we cannot build the capacity of NGOs, government agencies and M&E specialists. We are fully aware this approach can be disarming because it requires 'white experts' to move aside. There are many intelligent Indigenous people with expertise that did not receive a formal education. On a personal level I have drawn

L. McMurray (✉)
Aboriginal Carbon, Fund 4 Porter Ave, Marrickville, NSW 2204, Australia
e-mail: lisa@abcfoundation.com.au

R. Foley
Aboriginal Carbon Fund, PO Box 2429, Alice Springs, NT 0871, Australia
e-mail: rowan@abcfoundation.com.au

C. O'Sullivan
Carl O'Sullivan, 7/52-54 Holden St, Gosford, NSW 2250, Australia
e-mail: carl.osullivan85@gmail.com

W. Leal Filho (ed.), *Social Responsibility and Sustainability*, World Sustainability Series, https://doi.org/10.1007/978-3-030-03562-4_8

inspiration for this work through my mother, a strong spiritual Badtjala (Butchulla) woman. As a child she was one of the 'dump house mob' spending four years of her childhood growing up on a rubbish dump, left school by grade 7, lost all her teeth by 16 years and started work as a domestic servant. Despite all these difficulties she raised four children all going to University, negotiated the first hand back of land on K'gari (Fraser Island) from the infamously conservative Joh Bjelke-Petersen Queensland Government and was instrumental in launching our successful Native Title claim. I have also drawn inspiration from Oodgeroo Noonuccal, Vincent Lingiari, Tony Tjamawa and Eddie Mabo. All of whom challenged the status quo for the benefit of Indigenous people. I would like to thank the Aboriginal Carbon Foundation team and Caritas Australia for their ideas, dedication and patience. I would also like to thank the Queensland Government for funding this innovative work.

Keywords Indigenous · Carbon farming · Core-benefits · Verification Self-determination

1 Introduction

There is a lot of pressure on Indigenous people to conform, to think and to act like Europeans. There is an unspoken belief it would be much easier for all concerned to simply use European models and accepted western ways of doing things.

Developing a new standard for the environmental, social and cultural core-benefits of carbon projects that uses the *Indigenous-to-Indigenous* way of working has not been easy. It has involved a lot of thinking, reflection and discussion with many knowledgeable and wise Aboriginal[1] and non-Aboriginal people. It also involved more formal consultation and peer review processes as well as presentations at industry forums and community meetings.

Ultimately, we do not see a role for non-Aboriginal people in leading this process nor are we 'massaging' a western monitoring & evaluation (M&E) model into an Aboriginal context. We can't tweak existing standards whose soul is fundamentally different. There is, however, a support role for non-Aboriginal people which affords for a generosity of spirit to not dominate Aboriginal ways of working.

The concept of Aboriginal people working with other Aboriginal people using Aboriginal expertise to verify core-benefits has caused some non-Aboriginal experts to raise concerns that the process is inferior, a bit soft perhaps, lacking rigour even. The idea that Aboriginal people have something to offer challenges the dominant neo-colonial concepts of dependence and is a form of environmental racism that needs to be called out.

There are many intelligent Aboriginal people with expertise that did not receive a formal education. We are holding fast in our belief that by being part of the solution

[1]Throughout the paper Australian Aboriginal and Torres Strait Islanders have been referred to as Aboriginal Australians. The term Indigenous refers to First Nations people globally.

and thereby recognising Aboriginal people's expertise, we cannot continue to build the capacity of NGOs, government agencies and M&E specialists, at the expense of Aboriginal people's knowledge gain. We are fully aware this approach can be disarming because it requires 'non-Aboriginal experts' to move aside.

This work is personal for AbCF Chief Executive Officer, Rowan Foley:

> On a personal level I have drawn inspiration for this work through my mother, a strong spiritual Butchulla[2] woman. As a child she was one of the 'dump house mob' spending four years of her childhood growing up on a rubbish dump, left school by grade 7, lost all her teeth by 16 years and started work as a domestic servant. Despite all these difficulties she raised four children all going to University, negotiated the first hand back of land[3] on K'gari (Fraser Island) from the infamously conservative Joh Bjelke-Petersen Queensland Government[4] and was instrumental in launching our successful Native Title claim.[5] I have also drawn inspiration from Oodgeroo Noonuccal,[6] Vincent Lingiari,[7] Tony Tjamawa[8] and Eddie Mabo.[9] All of whom challenged the status quo for the benefit of Aboriginal people.

---

[2]The Butchulla people are the Traditional Owners of K'gari (Fraser Island, South-East coast of Queensland, Australia.). The Butchulla people are formally Native Title Holders after an 18-year claim process with the Australian and Queensland Government.

[3]The Queensland Government gazetted a 20-hectare lease to Thoorgine Aboriginal Culture Centre on K'gari Fraser island.

[4]Sir Johannes Bjelke-Petersen (13 January 1911–23 April 2005) was the Premier of Queensland from 1968 to 1987. Renowned for extreme conservatism leadership of a government that was found to be corrupt in later years, made him one of the most controversial political figures of Australian politics.https://en.wikipedia.org/wiki/Joh_Bjelke-Petersen (viewed March 2018).

[5]The 1992 High Court Mabo case accepted that traditional law and custom could be a basis for claiming land ownership by Aboriginal and Torres Strait Islander Australians. The subsequent Native Title Act 1993 (NTA) established the procedure for making native title claims. 'Native Title' recognises a set of rights and interests over land or waters where Aboriginal and Torres Strait Islander groups have practiced and continue to practice, traditional laws and customs prior to colonisation. https://auroraproject.com.au/what-native-title (viewed March 2018).

[6]Oodgeroo Noonuccal (3 November 1920–16 September 1993) was a Noonuccal woman, the traditional inhabitants of Minjerribah (North Stradbroke Island, Queensland). Oodgeroo Noonuccal was an Australian poet, political activist, artist and educator. She was also a campaigner for Aboriginal rights. https://www.poetrylibrary.edu.au/poets/noonuccal-oodgeroo (viewed March 2018).

[7]Vincent Lingiari (1919?–1988), a Gurindji Aboriginal stockman and land rights leader from Victoria River Gorge, Northern Territory. On 23 August 1966, fed up with Aboriginal stockmen and women being 'treated like dogs' in their own country, Lingiari led two hundred people, employees and their families of Wave Hill station (owned by Vesteys British pastoralist company) in a 'walk-off'. The Gurindji strike was to last nine years, the longest in Australian history. Although initially an employee-rights action (for better conditions and wages), it soon became a major federal issue when the Gurindji people demanded the return of their traditional lands. The strike culminated in Prime Minister Gough Whitlam pouring sand into Vincent Lingiari's hands in 1975 to symbolise the handing back of Wave Hill station to the Gurindji people. http://indigenousrights.net.au/land_rights/wave_hill_walk_off,_1966-75 (viewed March 2018).

https://en.wikipedia.org/wiki/Vincent_Lingiari (viewed March 2018).

[8]Tony Tjamawa was a senior Traditional Owner and lead Aboriginal negotiator for the hand back of Uluru-Kata Tjuta National Park (Northern Territory) to the Indigenous custodians.

[9]Eddie Koiki Mabo (1936–1992) was born in 1936 on Mer Island one of the Torres Strait Islands (Queensland) and was famous for his campaigning for indigenous land rights and the 1992 High Court decision that overturned the 'terra nullius' legal notion that meant land was unowned which

Within our sector of carbon farming, environmental, social, cultural and economic sustainable development and social responsibility intersect through the verification of carbon farming[10] core-benefits. This paper outlines the 'Indigenous to Indigenous' philosophy behind the core-benefits verification and articulates how this approach enables Aboriginal people to genuinely lead this process.

2 Aboriginal Carbon Foundation (AbCF)

The vision of the AbCF is to nurture and build a sustainable Aboriginal carbon industry. In doing this, our aim is to build wealth for Traditional Owners with environmental, social and cultural core-benefits through the ethical trade of carbon credits with Corporate Australia, government agencies and international bodies.

AbCF was established in 2010 as a *not-for-profit* company limited by guarantee. The founding Directors were Tracker Tilmouth,[11] David Ross[12] and Allan Cooney.[13] The Aboriginal staff and associates are passionate about working to build viable economies on Aboriginal lands. We are supported by a range of project partners who are equally committed to supporting Aboriginal people.

The AbCF has a strong culture of innovation and collaboration. It takes risks and spends a lot of time on R&D, creating carbon products and services that benefit Aboriginal people and address climate change locally, nationally and internationally.

The AbCF operates at the cutting edge of ideas and community-based approaches. It brings together people with fresh concepts, professional experience and a desire to achieve outcomes that tackle Aboriginal poverty and climate change through a strengths-based approach.

---

characterised Australian law with regards to Indigenous land title. http://www.biographybase.com/biography/Mabo_Eddie.html (viewed March 2018).

[10]A formal system for either carbon abatement or sequestration of greenhouse gasses through approved methodologies in Australia which is regulated by the government Clean Energy Regulator.

[11]'Tracker' who passed away in 2015 was an Arrente man and a Northern Territory Aboriginal activist. Tracker helped establish the Central Australian Aboriginal Legal Aid Service and the Aboriginal health service in the southern region of the Northern Territory.).

[12]David Ross is of Arrernte and Kaytetye descent and is the CEO of the Central Land Council based in Alice Springs in the Northern Territory. https://www.clc.org.au/ (viewed March 2018).

[13]Allan Cooney was the 2013 Casey Station Leader (Australian Antarctic base). Allan has a broad range of experience as a farmer, artist, General Manager, Executive Director, CEO, Project Manager and Company Director across government, private and not-for-profit sectors.

http://www.antarctica.gov.au/news/2012/2012-13-antarctic-station-leaders (viewed March 2018).

3 Indigenous to Indigenous Philosophy

The core-benefits verification is an Aboriginal owned system. This is realised through fellow Aboriginal people respecting Aboriginal knowledge, cultural expression and spirituality rather than using a system that privileges western notions.

To be clear, the *Indigenous-to-Indigenous Standard* is not designed to empower non-Aboriginal third parties to undertake 'audits'.

The beginnings of the standard were born from a conversation relating to a participatory impact assessment of a Tribal rights program in India, in 2016. The philosophy of the standard was developed during 2017 and peer-reviewed towards the end of that year. Concurrently a training course and tools were also developed. The first training course was piloted in a remote Aboriginal community in Cape York, Queensland in May 2018.

As Rowan Foley, Chief Executive Officer of Aboriginal Carbon Foundation says: "If only I had a dollar for every time someone tells me they're working the 'right way' and 'putting Aboriginal people in the driving seat".

AbCF feels that there is still a power differential at play in most Aboriginal Affairs[14] programming. Mark Moran's 2016 book "Serious Whitefella Stuff—when solutions became the problem in Indigenous Affairs" describes the forever shifting political landscape, the well intentioned 'experts' all with their own magic bullet, and the barrage of programs and policies that Aboriginal people have to operate under.

Only recently, after a three-day meeting with 250 key Aboriginal representatives producing a succinct, genuine and poignant 'right political way' called the 'Uluru statement from the Heart'[15] did we see this power differential at play. The statement called for a First Nations' voice in parliament. The statement was widely well received and high spirits were built due to the perceived engagement and support from the Australian federal government. However, this good will was short lived as Prime Minister Turnbull criticised the statement and announced there would not be a place for a constitutional First Nations' voice in parliament after all.

It is within this context, where Aboriginal Australians continue to have programs and policies developed 'for them, but not by them' that we embark on an alternative way for the verification of carbon farming core-benefits.

A key aspect of our core-benefit verification standard is the development of tools and a training package that empower Aboriginal communities and rangers[16] who may have little or no literacy and numeracy skills but are strong in culture, traditional ecological knowledge and networking ability with fellow Aboriginal people.

---

[14] Aboriginal Affairs refers to government policy, legislation and programs relating to Australian Aboriginal people. However the term can be used more broadly and beyond government programming to include other sectors such as NGOs, philanthropists, academia etc.

[15] https://www.referendumcouncil.org.au/sites/default/files/2017-05/Uluru_Statement_From_The_Heart_0.PDF (viewed March 2018).

[16] A person, normally a Traditional Owner, employed to undertake land and sea management programs with a locally based Aboriginal organisation. Historically a ranger position has been employed in national parks and wildlife services with a government agency.

This core-benefits verification standard is founded upon the articles set forth in the United Nations Declaration on the Rights of Indigenous Peoples,[17] which Australia became a signatory to in 2009. The following articles are considered especially relevant:

- Article 4 The right to self-determination and autonomy or self-government in local affairs;
- Article 8 The right not to be subjected to forced assimilation or destruction of their culture;
- Article 8(j)—Traditional Knowledge, Innovations and Practices;
- Article 20 The right to be secure in the enjoyment of their own means of subsistence and development;
- Article 21 The right, without discrimination, to the improvement of their economic and social conditions;
- Article 23 The right to determine and develop priorities and strategies for exercising their right to development. In particular, Indigenous peoples have the right to be actively involved in developing and determining health, housing and other economic and social programmes affecting them and, as far as possible, to administer such programmes through their own institutions;
- Article 26 The right to the lands, territories and resources which they have traditionally owned, occupied or otherwise used or acquired; and
- Article 32 The right to determine and develop priorities and strategies for the development or use of their lands or territories and other resources.

Furthermore, the verification standard embraces a particular concept known within the international development sector as 'south to south'. This process challenges dominant international development paradigms that perpetuate dependency by privileging the knowledge of technical 'experts' from the 'North' in delivering services to the 'South'.

The 'south to south' model embraces the expertise and acknowledges the rich talent of human resources existent within the 'Global South'.[18] As a result, southern 'experts' become the trainers, brokers and capacity builders within the development process. The model acknowledges the power differentials at play between 'North' and 'South' entities. However, with the growing interest of Indigenous carbon farming internationally, our philosophy has been coined 'Indigenous to Indigenous' with the vision that there is great appetite for cultural exchange across countries and has already begun with Canada and New Zealand.[19]

---

[17] http://www.un.org/en/universal-declaration-human-rights/index.html (viewed July 2017).

[18] The Global South is a term used to replace the derogatory terminology of the 'Third World', 'developing countries' and 'less developed countries'. The Global South also assumes many of these countries have had a shared history of colonialism, neo-imperialism, and negative economic and social transformation through which large inequalities in living standards, life expectancy, and access to resources have been maintained. https://en.wikipedia.org/wiki/Global_South (viewed July 2017).

[19] Indigenous-to-Indigenous Climate Cooperation Heads of Agreement with First Nations Energy and Mining Council British Colombia Canada, and Indigenous-to-Indigenous Climate Cooperation

In practice, this 'south to south' philosophy sees all verification of core-benefits conducted by a team of trained Aboriginal experts, including rangers and Traditional Owners from across the projects. This principle prevents the extraction of information by external agencies to be used and interpreted without the understanding of, or any required benefit to, the affected community. It therefore balances the role of third party verification with Aboriginal property rights and a requirement for the people with these responsibilities to have strong cultural and project-based knowledge.

4 AbCF's Role in Influencing Social Responsibility

As an Aboriginal organisation with a charter to build inter-generational wealth for Aboriginal people AbCF sees itself as a catalyst for corporate social responsibility and an influencer in working the 'right way' with Aboriginal people.

Social and economic development and social responsibility intersect through the verification of carbon farming[20] core-benefits. The processes and tools to achieve sustainable development through this opportunity have been developed with much thought and debate to ensure Aboriginal people will be in the leadership position.

For AbCF sustainable development is underpinned by control and self-determination. A unique opportunity is afforded through the carbon farming industry which sells premium priced 'carbon' to the private sector if a project has associated environmental, social and cultural values. As a result, AbCF is in a prime position to educate our partners, for example corporates and Government, about the need for alternative verification processes that genuinely enable Aboriginal ownership and decision making around what and how carbon core-benefits are verified.

Through the investment in carbon farming projects in remote Aboriginal communities it is envisaged that rich and often complex conversations relating to Aboriginal self-determination will influence the private sector's sustainable and socially responsible practices, beyond any initial carbon farming project.

Beyond the philosophical, AbCF has pioneered an investment opportunity called the 'Reducing Carbon Building Communities' Fund that facilitates the inter-generational wealth creation through the trade of carbon credits between corporates and Aboriginal communities. The investment fund also leads crucial industry R&D. AbCF believes that entities investing in Aboriginal carbon and doing business in the 'right way' should be rewarded for their ethical and sustainable practices, such as supporting Native Title claims and Indigenous Land Use Agreements (ILUAs).

A three-tiered financial model, corporates are rewarded most, not only for their good environmental stewardship but for their investment in: local jobs; buying locally

---

Heads of Agreement Hikurangi Enterprises Ltd (HEL) is a charitable company owned by Hikurangi Huataukina Trust New Zealand.

[20]A formal system for either carbon abatement or sequestration of greenhouse gasses through approved methodologies in Australia which is regulated by the government Clean Energy Regulator.

sourced and manufactured products; community engagement; and most importantly support to local Aboriginal people and their land claim aspirations.

In partnering with AbCF, the private and public sectors are committing to an ideology that acknowledges and supports the following:

- an obligation to turn tables on how things have been done in the past, due to the disempowerment and trauma caused by colonialism, at every turn we need to apply an 'empowerment' lens and work in a 'right way'[21];
- limited formal employment prospects exist in remote Aboriginal communities therefore the up-skilling opportunities apparent in any initiative must be realised by people who need them most. For too long, capacity has been developed for Govt workers, NGOs, their staff and consultants at the expense of Aboriginal people who can leverage crucial skills and training for application in other employment settings;
- recognising that by working through a strength-based approach and identifying the 'internal' human assets within a carbon farming project, momentum, ownership and agency is built that isn't dependent on an external entity being the knowledge holder. The fundamental theme here being—people themselves have the answers; and
- building skills of rangers and Traditional Owners to decide what it is that they want to measure without an externally created and driven (M&E) process, prescribed by a corporate or other stakeholder is crucial in building agency around verification.

5 Why Invest in Aboriginal Carbon Farming?

The unpacking of altruism is not an academic exercise for AbCF. It guides us when we ask: "why a corporate, based in the prosperous metropolitan cities of Australia's south would be interested in supporting remote Aboriginal communities?". As former Australian Prime Minister Paul Keating is mythologised for saying "always back a horse called self-interest".

In most cases, decision makers within private corporations have never visited a remote community and the world view of both groups couldn't be further apart. However, AbCF believes that most non-Aboriginal Australians have an awareness and acceptance of the injustice that has befallen Aboriginal people.

It was only ten years ago the Australian Prime Minister Kevin Rudd delivered an apology to Australia's Indigenous peoples in the Parliament of Australia's House of Representatives for the forced removal of Aboriginal children, known as the 'Stolen Generations'. AbCF believes that a growing number of non-Aboriginal people feel a national and personal responsibility to address the atrocities and injustice that have been inflicted on so many Aboriginal people. However, we would be naïve to think this good-will would generate investment alone. To borrow from Bill Clinton's successful 1992 Presidential slogan, "it's the economy, stupid!"

[21] See Indigenous to Indigenous section.

Investing in Aboriginal carbon farming in an Australian context is a 'one-stop' shop on multiple levels. Firstly, as outlined in the above paragraph, many private corporations recognise they have a genuine social responsibility to redressing the wrongs that have befallen Aboriginal Australians. Within the Australian private sector, 'Aboriginal empowerment' is often a priority 'Corporate Social Responsibility' pillar for leading banks and service sector companies. Through investment in Aboriginal carbon a direct connection and rich relationship can be built between executives and Aboriginal people through organised community visits.

There is also a move for all sectors to adopt a 'Reconciliation Action Plan (RAP)'[22] and prioritise procurement from Aboriginal owned businesses. Investing in Aboriginal carbon farming can achieve many of the strategic outcomes typically prioritised in corporate RAPs.

Furthermore, 175 countries signed the 22nd April 2016 Paris Agreement. Offsetting a company's carbon emissions through investment in Aboriginal carbon farming is one concrete way corporates can realise their responsibility in addressing climate change and Australia's Paris Agreement commitments.

Additionally, a corporate can be working towards the Sustainable Development Goals (SDGs) by their investment in Aboriginal carbon farming, which addresses most of the seventeen goals.

Therefore, the opportunity to address both Aboriginal empowerment and redress disadvantage as well as upholding Paris Agreement carbon reduction commitments and advancing the SDGs in one activity makes Aboriginal carbon farming therefore, an attractive package for corporate investment.

6 What Is Carbon Farming?

Carbon farming refers to a carbon accounting methodology that reduces Green House Gas (GHG) emissions, and/or captures and holds carbon in vegetation and soils. Carbon Farming is just like any other agribusiness however the farming commodity produced are 'carbon credits'. In Australia this financial commodity is formally known as Australian Carbon Credit Units (ACCU).

There are approximately 35 methodologies of which have been approved by the Australian Government Clean Energy Regulator (CER)[23] of which savanna burning is one. Savanna burning reduces GHG emissions through cool burns early in the dry season when fires are smaller and patchy (i.e. May, June and July). Without these managed burns, late season wildfires would occur (i.e. August, September and

[22] A RAP is a strategic document that supports an organisation's business plan. It includes practical actions that will drive an organisation's contribution to reconciliation both internally and in the communities in which it operates. To date 800 RAPs have been developed. https://www.reconciliation.org.au/reconciliation-action-plans/ (viewed March 2018).

[23] http://www.cleanenergyregulator.gov.au/ (viewed March 2018).

October), leading to significant amounts of GHG emissions being released into the atmosphere.

The carbon abatement is calculated through SavBAT[24] and sold on either one of two markets (outlined below). Across Northern Australia, there are approximately 78 savanna burning projects of which one third are administered by Aboriginal ranger groups. 4,078,963 ACCU[25] have been produced to date, with 70% of these ACCU being produced by Aboriginal savanna burning projects.

There are two opportunities for the trade of carbon within Australia: (1) through the government auction, known as the Emission Reduction Fund (ERF); or (2) through the private avenue, known as the voluntary market.

Through the government auction the CER buys ACCU (the equivalent of one tonne of carbon dioxide stored or avoided by a project) through the Emissions Reduction Fund (ERF). This is a national scheme introduced by the Australian Government in the *Carbon Credits (Carbon Farming Initiative) Act 2011*.

The ERF places no financial value on the social, cultural or environmental core benefits. The guiding principle is 'lowest cost abatement' to buy the maximum amount of ACCU as cheaply as possible. The CER has administered six auctions commencing in April 2015 on a reverse auction basis with the last auction held in December 2017 buying for an average price $13.08 per tonne of abatement (one ACCU).

The voluntary market provides corporations, institutions and individuals with an opportunity to buy ACCU to offset their carbon footprint and/or meet their carbon neutrally requirements under the National Carbon Offset Scheme.

Many corporations are interested in buying ACCU with environmental, social and cultural core-benefits at approximately $16–$20 per tonne to meet their UN SDGs, Reconciliation Action Plans and/or Corporate Social Responsibility goals.

The values that are contained in the environmental, social and cultural core-benefits attract a premium price in the voluntary market because of their internal and external marketing opportunities, due to the ethical investment angle. Furthermore, ACCU bought by large corporations can act as a catalyst for broader discussions with Aboriginal people around investing in Aboriginal lands for agribusiness, eco-tourism and other development projects.

6.1 Purpose of the Core-Benefit Standard

As mentioned above, if a carbon farming project facilitates environmental, social and cultural values, the carbon being sold can attract a premium price on the vol-

[24]The Savanna Burning Abatement Tool (SavBAT) automates the GIS processes and mathematical equations required to estimate the net abatement for savanna burning projects. https://savbat.net.au/ (viewed October 2017).

[25]West Arnhem Land Fire Abatement project (WALFA) presentation at 2018 Savanna Burning Forum, Darwin.

untary market. The core-benefits standard recognises that Aboriginal people have the knowledge to lead the verification process of carbon farming core-benefits and provides tools and training to enable this to eventuate. The verification of these values, however, gives corporates the confidence that the premium priced carbon being purchased is bona-fide.

This verification process, built on Aboriginal values and informed by 'international development' evaluation best-practice,[26] provides integrity, independence and transparency of Aboriginal carbon farming core-benefits. Although developed originally for savanna burning carbon farming projects, there has been much interest to scale up to other carbon farming methodologies and to other agricultural sectors, such as the beef industry. This verification process is not only relevant to all Indigenous people globally but to any sector that values a peer to peer approach. AbCF is currently exploring what this philosophy would look like when utilised by non-Aboriginal farmers to verify their core benefits of carbon farming.

7 What Are Core-Benefits?

7.1 Core-Benefits Versus Co-benefits

The Kimberly Land Council (2016) first coined the phrase core-benefits. Although some debate still exists as to whether the environmental, social and cultural values of carbon farming should be called co-benefits or core-benefits. Largely the question around the correct terminology is indicative of whether one sees carbon farming as the core purpose or the associated ecological and human values as the most meaningful outcome of any 'project'.

In the context of remote Northern Australia carbon farming is seen as a means to an end rather than an end in itself. The core agenda for Aboriginal peoples across

[26]Core instruments researched in the development of the core-benefits standard were:

(1) OECD Development Assistance Committee (DAC) codes of Efficiency, Effectiveness, Impact, Sustainability and Relevance and 'Principles for Evaluation of Development Assistance' (+Glossary of Terms in Evaluation and Results Based Management) http://www.oecd.org/development/evaluation/2755284.pdf (viewed July 2017)

(2) Australasian Evaluation Society (AES) 'Guidelines for the Ethical Conduct of Evaluations' https://albordedelcaosdotcom.files.wordpress.com/2017/03/aes-guidlines10.pdf (viewed July 2017)

(3) Australian Council for International Development (ACFID) 'Principles and Guidelines for Ethical Research and Evaluation in Development' (April 2016) https://acfid.asn.au/sites/site.acfid/files/resource_document/ethics-guidelines.pdf (viewed July 2017)

(4) BOND UK 'Evidence Principles and checklist' https://www.bond.org.uk/monitoring-and-evaluation/monitoring-and-evaluation-tools (viewed July 2017)

(5) DFAT M&E standards April 2017 https://dfat.gov.au/about-us/publications/Documents/monitoring-evaluation-standards.pdf (viewed July 2017).

the north is getting people back on 'country'.[27] Carbon farming is an industry that can support this aspiration. Savanna burning has created an opportunity for people to return and stay on 'country' due to employment via the ranger program. The income from carbon sales allows communities to expand these programs, keeping culture alive, protecting sacred sites and developing the infrastructure of their 'outstations' or 'homelands'.[28]

There is therefore a significant variation between Aboriginal carbon projects with core-benefits and other carbon projects with co-benefits, where selling carbon for income generation is the primary agenda. In the case of the later all subsequent benefits are welcomed (as the carbon can therefore be traded for a higher price) but they are largely viewed as opportunistic by-products.

For AbCF the key thematic core-benefits of savanna burning are environmental, social and cultural. Examples for each theme are outlined below:

7.2 Environmental Core-Benefits

Decreased carbon emissions and incidence and intensity of wildfires by burning country the 'right way' is one concrete ecological benefit of savanna burning carbon farming. Due to the reduction of wildfires, life (human and animal) and flora are protected. Through the protection of fauna and flora native species, environmental biodiversity is recovered and enhanced where in the past it may have been endangered (Country Needs People 2017). Additionally, due to the intentional management of the land in a 'right way' the numbers of invasives are reducing. Finally, this management of the land by ranger groups increases the control of tourism and negates any negative impact.

[27]The term "country" is often used by people to describe family origins and associations with particular parts of Australia. The relationship with the land, geographically and spiritually, involves an understanding of time, place and cultural connections found within a particular area. https://www.qcaa.qld.edu.au/downloads/approach2/indigenous_res005_0803.pdf (viewed March 2018).

[28]Remote communities were primarily instruments of government organisation. In some cases, different clan groups without any connection, were brought together to live in a central location. Outstations and homelands "provide Aboriginal people with a sense of 'home' and belonging while contributing to their cultural responsibilities of caring for country and managing the natural resources of their land and seas," explains Kim Hill, CEO of the Northern Land Council https://www.nlc.org.au/ (viewed March 2017). Extended families in larger towns frequently visit homelands to spend restorative time there, reconnecting with their culture and traditional country. Source: https://www.creativespirits.info/aboriginalculture/land/aboriginal-homelands-outstations#ixzz59kxsPaJJ (viewed March 2018).

7.3 Social Core-Benefits

The social benefits of savanna burning are plentiful. At the community level, the projects allow for increased social capital as community members come together to work on the project (Ryan et al. 2012). The projects also foster opportunities for women to participate and in some communities, discrete women ranger groups have been established.

The strengthened community-level economy and development of other community projects financed by carbon income has led to an increased sense of control and greater certainty within communities over their finances. Some benefits of this are a reduction in welfare dependence and the development of homelands' infrastructure.

More excitingly however is the confidence-change people feel when involved in a project. Self-worth, sense of purpose as well as an increase in pride in self and others have all been linked to savanna burning projects (Country Needs People 2015, 2017; DPMC; Ryan et al. 2012). For communities that are rebounding from the inter-generational effects of trauma and colonisation, this sense of pride is seen as a crucial catalyst for human empowerment and actualisation. This increase in confidence allows people to put themselves forward for other work with Government, private sector and NGOs. The literature supporting 'social determinants of health' affirms that perception of control over decisions that affect people is one of the most important contributors to emotional wellbeing (Marmot 2011).

As renowned anthropologist Altman (2007) promotes: meaningful work for remotely based Aboriginal Australians must align with traditional interests and values. Savanna burning along with performance and art, hunting, tourism and environmental services provides a secure and meaningful employment opportunity for people living in remote communities.

7.4 Cultural Core-Benefits

The 'holy grail' of savanna burning core-benefits, sits of course, within the cultural dimension. Tradition and culture for Australian Indigenous people are the essence of life itself. The protection of sacred sites through savanna burning and the rediscovery of cultural heritage, hidden in some cases for decades due to European land management practices[29] sees culture being retained and flourishing. The maintenance and 'passing on' of traditional knowledge & language and the education of children by Elders is crucial for individual and cultural identity.

[29]For many years Aboriginal Australian land management practices were forbidden as European 'environmental protection' ideology dominated the thinking around best practice. 30,000 rock art pieces were discovered in 2017, when rangers conducted 'right way burning' in Arnhem Land, the Northern Territory. http://www.abc.net.au/news/2017-07-30/aboriginal-rangers-discover-rock-art-site-in-arnhem-land/8750046 (viewed March 2018).

Improved spiritual wellbeing through the regular completion of cultural obligations to country, increased exercise and physical activity by working on the land, and increased nutrition through access to and availability of traditional foods not only makes culture strong, but also improves health outcomes (Marmot 2011). This is significant for a cultural group whose life expectancy is ten years shorter than non-Indigenous Australians.[30] A report in the Medical Journal of Australia in 2009 entitled 'Healthy country, healthy people: the relationship between Indigenous health status and "caring for country" Burgess et al. (2009) found a "substantial association" between working on country and a number of positive health outcomes such as BMI, blood pressure, type 2 diabetes status, cholesterol, cardiovascular risk and others.

8 Eco-colonialism

Eco-colonialism is a term used largely by First Nations people in Canada and in Africa, to describe the role of conservation organisations and other NGOs who dominate government policy which affects indigenous lands. At the heart of 'eco-colonialism' is a fundamental philosophical clash between Indigenous land owners and the 'green' leadership.

'Achieving economic security for Indigenous peoples within the forest landscapes of northern Canada constitutes a distinct challenge for the whole country. Could this challenge also be an opportunity to develop a new paradigm of "best practices" for sustainable forest resource use? (Chapeskie 1999, p. 39). The debate over how best to use natural resources on Aboriginal lands is on-going and does however, provide opportunities not just challenges.

Many Indigenous people feel that policies and legislation are designed to prioritise conservation, and environmental NGOs and do not place a priority on the sustainable economic development of natural resources by Indigenous people to overcome poverty and lack of local employment opportunities. In effect, this is just another wave of colonial views being imposed on Indigenous people for the electoral or other benefit of non-Indigenous people (Financial Post, 4 January 2018). The argument that conservation NGOs know better how to protect areas and restrict the use of natural resources is often put forward under the banner of 'climate change' or 'wildlife' protection.

Carbon farming in Australia has managed to find common ground and provide tangible benefits to Aboriginal people which is widely supported by conservation and NGO groups. However, the themes around Aboriginal-led verification contained in the Indigenous-to-Indigenous Standard have challenged the euro-centric priority placed on text-heavy planning documents and prescribed indicator banks.

[30]https://www.aihw.gov.au/reports-statistics/health-conditions-disability-deaths/life-expectancy-deaths/overview (viewed March 2018).

9 Empowerment Evaluation Model

There is a growing recognition amongst Australian NGOs working with Aboriginal Australians that evaluation is often yet another form of marginalization and disempowerment. Just as many community development projects have had the reputation of being 'top down' with little or no room for Aboriginal people to develop and implement their own solutions (ACFID ATSI working group 2015) many common evaluation methodologies pose a similar problem of stripping Aboriginal Australians of any investigative agency. First Nations researchers and evaluators have drawn specific attention to the need to decolonise Western research methodologies and evaluation practice by developing an evaluation practice that is "of, for, by and with us" (Kawakami et al. 2007; 222; Tuhiwai Smith 1999).

To quote the Brazilian educational liberationist Paulo Freire "*Any situation in which some men prevent others from engaging in the process of inquiry is one of violence; ... to alienate humans from their own decision making is to change them into objects*"

Similarly, Price et al. (2012) highlight the issue of 'outsiders' going into communities with a predetermined agenda to extract specific 'data' and leaving Aboriginal groups disenfranchised in the process. Tarsilla (2010) discusses that communities need be "active agents in the entire process" and similarly, Cochran et al. (2008) look at the need to move beyond Indigenous communities just being the "objects of study". We need to be most careful in carbon farming so that Aboriginal rangers don't just become the instruments data collection.

'Participatory' evaluation approaches have been accepted best practice in NGO programming for several decades, however these investigative events are still largely reliant on external people to drive. A new conversation is emerging, however. One which talks about an empowerment model whereby Aboriginal Australians are making the decisions about what is measured, what methodology will be developed, and what culturally appropriate tools will be utilized to collect data. Furthermore, Aboriginal Australians make sense of the data collected and share insights and recommendations based on this 'expert' analysis. The move towards the importance of Aboriginal evaluators and researchers taking the lead is slowly increasing.

As passionately expressed by Aboriginal Australian actor Sharena Clanton recently on the Australian Broadcasting Commission's (ABC) prominent current affairs program 'Q&A'[31] non-Aboriginal people have been making decisions on behalf of Aboriginal people for far too long and it was time that Aboriginal people became the architects of their own destiny.

[31] http://www.abc.net.au/tv/qanda/.

10 Why the Need to Create Something Different?

Tried and tested M&E good practice from the international development sector is the foundation of this core-benefits standard. However, we are building rigour and credibility in a way that genuinely sees Aboriginal people driving this work. More important than 'auditing' or 'upward accountability' M&E good practice provides a mechanism for continuous learning for improvement. All M&E good practice therefore has the additional purpose of building skills for people involved in that process.

Through our experiences as community development practitioners with M&E skillsets honed throughout the 'Global South' and remote Australia, we have concluded there is a mismatch of 'good western M&E practice' and genuine ownership and uptake from people who these processes are meant to benefit.

Our challenge is to build good evaluation practice that instils hope and confidence in people's identity and self-worth, something that can be unintentionally diminished when developed and driven externally. As a result, we have chosen to build on what people already do and the successful ingredients around that in an application that is deemed worthwhile, but ultimately driven from within.

Perhaps what is also different here is that we are not auditing a Government grant to track progress or to see the impact of that funding. The core-benefits verification standard is intended to be fully owned and operated by Aboriginal people to enhance their own carbon projects. The work of verification will take its place within a context of community leadership and can be integrated within other cultural activities. For example, verification begins with an invitation and a welcome—a 'smoking ceremony', an ancient Aboriginal tradition to cleanse, ward of bad spirits and celebrate the bringing of people together. The lens applied therefore is one of celebration and good news, not that of an audit.

10.1 Whose Capacity Is Being Built?

At the Australian Council for International Development (ACFID) 2016 conference Fijian woman Emele Duituturaga, Executive Director of the Pacific Island Association of NGOs (PIANGO) pleaded with Australian NGO representatives to stop building the capacity of their own organisations at the expense of local entities. Most ANGOs working in the Pacific context have field offices, staffed primarily with people originating from the INGO headquarters, in key decision-making roles. Similarly, the Australian Peak Organisations Northern Territory (APONT) has also published a guideline titled 'Partnership Principles—NGOs working with Aboriginal organisations and communities'. The principles emphasise the need for non-Aboriginal organisations, supporting them to grow where invited rather than competing with them for funding and service delivery contracts (ACFID ATSI working group 2015).

It is this investment in external staff and infrastructure that smothers genuine local ownership and drive in the 'international development' and remote Aboriginal contexts. Furthermore, this external investment is short lived. There is a common phrase bandied about in remote Aboriginal communities "whitefellas are like Toyotas, you get three years out of them".

10.2 Skills Development

Using a strengths-based approach, project participants and the affected communities will be supported to develop the skills to take on ownership of core-benefits verification. Without ownership, the participants and affected communities will remain dependent on the involvement of external people (Coady International Institute 2012) who are not as well placed to collect, interpret or communicate accurate and meaningful information about the project's core-benefits.

AbCF, Caritas Australia and the Centre for Appropriate Technology[32] (an Aboriginal registered training organisation) has developed a five-day training package on carbon farming, core-benefits identification, monitoring, reporting and verifying. This training sits within an existing vocational 'Conservation and Land Management' training course. The training tools are largely image based and text-light and provide Aboriginal verifiers with the knowledge and instruments for collecting and interpreting qualitative and quantitative data.

Although there are hundreds of rangers involved in savanna burning projects, no 'carbon farming' course has existed until now to expose rangers to the architecture and its workings of their own industry. We must ask ourselves: how do Aboriginal people participate in an industry they know very little about?

10.3 Why not just Use Existing Standards?

During the development of the core-benefits standard, existing international measurement instruments were researched. There are only two or three standards that could be utilised for an Australian carbon farming context: The Gold Standard for Global Goals, The Climate, Community and Biodiversity Standards; and REDD+ Community Monitoring, Reporting and Verification (CMVR). AbCF recognises the enormous amount of thought and work that has gone into the production of these standards. However, they have been developed with a Euro-centric lens.

The existing standards are dozens of pages long, written in Spanish, French, Portuguese, or English, the idiom of the colonisers and therefore require proficiencies in these languages to navigate. Additionally, existing standards all contain a host of

[32]http://cfat.org.au.

prescribed lists of SMART[33] indicators and targets to draw from. AbCF is reluctant to prescribe any 'indicators' however, as we believe this automatically takes agency away from the rangers. The development and identification of specific indicators that the rangers themselves feel most passionate about, is crucial in building the confidence and appetite to drive the verification process.

Critical to the successful verification of core-benefits is measuring the values people feel passionate about yet we propose to do this without indicators. The identification of SMART indicators has become a mainstream approach for the evaluation of program impact in the international development sector and is a key element of the existing verification frameworks. However, SMART indicators are an abstract way of mapping information needs and they require the development of highly precise statements, often in French, Spanish, Portuguese or English. The process of developing or selecting indicators therefore risks alienating remotely located Aboriginal people, who speak English as a third or fourth language.

Furthermore, SMART indicators have become so common that their use is often mistakenly conflated with process rigour. While indicators often serve as a useful road map for identifying the pieces of information that need to be collected the rigour of any research process is primarily built upon the way bias is strategically addressed in the research design. We propose that rigour for our framework will be ensured through triangulation. In the 'right way' core benefits standard we are re-defining triangulation as '*asking the right questions, to the right people, in the right way*'.

AbCF recognises the existing standards are developed to drive good practice. However, the audience or 'end-user' is always an external body. We are trying something new. We believe that enabling Aboriginal people to set the agenda, bring their expertise to bare on the process, and control the information generated will build genuine interest, agency, control and more importantly self-determination.

11 Conclusion

In the development of a core-benefits standard we have made the conscious decision not to 'massage' a western M&E model into an Aboriginal context. Nor are we 'tweaking' existing standards whose souls we consider to be, fundamentally different. For AbCF sustainable development and social responsibility is underpinned by control and self-determination. A unique opportunity is afforded through the carbon farming industry for the private sector and Government to invest in alternative verification processes that genuinely enable Aboriginal ownership and decision making. Therefore, within our scope of influence we call on investors to support initiatives that work in a 'right way' with Aboriginal Australians.

For AbCF developing a core-benefits verification approach invests in the knowledge gain and skills development of Aboriginal people. In the first instance there is a need for rangers and Traditional Owners to understand their savanna burning and

[33] Specific, Measurable, Achievable, Relevant and Time phased.

more broadly the carbon farming industry they operate within. In the development of a core-benefits standard an opportunity has presented itself to build confidence, pride and self-worth in communities where these individual and communal characteristics are in short supply.

There is a small but growing body of evidence examining community development practice in remote Aboriginal community contexts. Hunt (2005) began to look at lessons from international development practice and its potential application in Aboriginal community development. Australian international development NGOs with ‘First Australian’ programs are contributing to this exploration of how to build genuine ownership and control of programming and policy in Aboriginal communities. However, the academic literature contributing to this crucial empowerment theme is in its infancy, largely because the organisations and individuals engaged in this discourse are practitioners and not academics. With this in mind, AbCF has seen an opportunity to test and develop this good practice ideology and plans are afoot to establish a ‘think tank’ to promulgate and make widely accessible examples of ‘right way’ practice.

For AbCF the core-benefits standard draws on a rich political heritage. The pride and confidence gained through leading a core-benefits verification event we envisage and hope, will reinforce someone’s worth and value. The core-benefits work becomes a launching pad for possibility.

In remote Aboriginal Australian communities, where meaningful employment opportunities are scant a ranger with a tool box of traditional knowledge and verification skills becomes an attractive ‘on country’ vocational prospect.

Colonisation of our lands by European governments has actively been resisted for many years. The models, policies and legislation they used were not designed to serve an Indigenous population. Indigenous ways of thinking and working together supports our independence. It is this collective voice that promotes a ubiquitous ‘right way’ in all walks of life, all sectors and all interactions between non-Indigenous and Indigenous people that holds most weight in seeing Indigenous people thrive.

References

ACFID Aboriginal and Torres Strait Islander Working Group (2015) Effective development practice with aboriginal and torres strait islander communities. Practice Note, Australian Council for International Development (ACFID), Canberra

Altman JC (2007) Alleviating poverty in remote indigenous australia: the role of the hybrid economy. Dev Bull 72:47–51. Development Studies Network, Australian National University (ANU), Canberra

Burgess CP, Johnston FH, Berry HL, McDonnell J, Yibarbuk D, Gunabarra C, Mileran A, Bailie RS (2009) Healthy country, healthy people: the relationship between Indigenous health status and “caring for country”. Med J Aust 190:567–572. Australasian Medical Publishing Company, Sydney

Chapeskie A (1999) Northern Homelands. Linking Culture and, Northern Frontier

Coady International Institute (2012) Compendium of tools for assets based community driven development. Unpublished training manual, St Francis Xavier University, Nova Scotia

Cochran et al (2008) Indigenous ways of knowing: implications for participatory research and community. Am J Public Health, 98(1). American Public Health Association (APHA), Washington

Country Needs People (2015) Working for our country, a review of the economic and social benefits of Indigenous land and sea management. Pew Charitable trust and synergies economic consulting. Department for the Prime Minister and Cabinet. Canberra

Country Needs People (2017) Protecting nature for all of us, environmental benefits of Australia's Indigenous Protected Areas and Indigenous rangers. Pew Charitable Trusts. Department for the Prime Minister and Cabinet, Canberra

Hunt J (2005) Capacity development in the international development context: implications for indigenous Australia. Discussion paper no. 278/2005. Centre for aboriginal economic policy research, Australian National University (ANU), Canberra

Kawakami AJ, Aton K, Cram F, Lai MK, Porima L (2007) Improving the practise of evaluation through Indigenous values and methods, decolonizing evaluation practise—returning the gaze from Hawai'i and Aotearoa. In: Brandon P, Smith N (eds) Fundamental issues in evaluation. Guilford Press. New York pp 219–242

Kimberley Land Council (2016) Indigenous savanna carbon industry core-benefits discussion paper. Unpublished. Kimberley Land Council, Broome

Marmot M (2011) Social determinants and the health of Indigenous Australians. Med J Australia. 194(10): 512–513. Australasian Medical Publishing Company, Sydney

Moran M (2016) Serious whitefella stuff—when solutions became the problem n indigenous affairs. Melbourne University Press, Victoria

Price M, McCoy B, Mafi S (2012) Progressing the dialogue about a framework for Aboriginal evaluations: sharing methods and key learnings. Eval J Australas 12(1):32–37. Australasian Evaluation Society, Melbourne

Ryan R, Wilczynski A, Watkins S, Rose J (2012) Assessment of the social outcomes of the working on country program. Final report. Urbis. Department of Sustainability, Environment, Water, Population and Communities, Canberra

Tarsilla M (2010) Inclusiveness and social justice in evaluation: can the transformative agenda really alter the status Quo. J Multi Eval 6(14). Simon Fraser University, Canada

Tuhiwai Smith L (1999) Decolonizing methodologies: research and indigenous peoples. Zed Books, London

Social Responsibility and Sustainability: How Companies and Organizations Understand Their Sustainability Reporting Obligations

L. Folkens and P. Schneider

Abstract Since 2015 exists at EU level the obligation for certain enterprises to prepare a sustainability report. This concerns particularly the DAX 30 organizations, which had to adapt their corporate reporting till 2017, including the adaptation of the existing reporting on social and ecological contents. Siemens, for instance, already presented in 2000 a "Corporate Citizenship Report" (CCR). Beside the CCR there is a wide range of terms for sustainability reporting, such as "CSR-Report", "Sustainable Value Report" or "Sustainability Report". Beside the variation of the report titles, also regarding the focal issues an ambivalent interpretation occurs. A large part of the reporting enterprises orientates itself on the reporting standards of the Global Reporting Initiative (GRI). Also the integration of ecological and social subjects in the annual business report is practiced. The present publication compares the sustainability reports of all DAX 30 enterprises and discusses the content differences. Beside the report titles, it investigates the correlation between titles and content. Scope of the contribution is to find out whether a clear border can be pulled between CSR reports and sustainability report and if so which are the assessment criteria.

Keywords Corporate social responsibility and sustainability
Corporate responsibility report · Sustainability report
Corporate citizenship report · Sustainable value report

1 Introduction

Social responsibility is considered as an ethical framework, in which individuals are accountable for fulfilling their civic duty; the actions of an individual must benefit the whole of society. Social responsibility means sustaining the equilibrium between

L. Folkens · P. Schneider (✉)
University of Applied Sciences Magdeburg-Stendal, Breitscheidstr. 2, 39114 Magdeburg, Germany
e-mail: petra.schneider@hs-magdeburg.de

W. Leal Filho (ed.), *Social Responsibility and Sustainability*, World Sustainability Series, https://doi.org/10.1007/978-3-030-03562-4_9

economic development and the welfare of the society and environment (Palmer 1995; Preston 1997; Emerson 2015).

Social responsibility does not only apply to business organizations, but also to all actors whose actions affect the environment. This responsibility can be passive by avoiding socially dangerous acts or being active by carrying out activities that directly advance social goals (Invernizzi et al. 2017). Social responsibility must be intergenerational, as one generation's actions have consequences for the next. Social responsibility in the form of Corporate Social Responsibility (CSR) is now also playing a key role at company level. Some critics argue that CSR detracts from the fundamental economic role of business; others argue that it is nothing but "green-washing"; others consider it an attempt to force the role of governments over powerful companies (Preston 1997; Emerson 2015). However, social responsibility is much more: the implementation of sustainability at practically all levels. Sustainability is an action principle for the use of resources, in which the preservation of the essential properties, the stability and the natural regeneration capacity of the respective system is in the foreground. The model of sustainable development, visualized in the sustainability triangle, is based on the idea that sustainable development can be achieved through the simultaneous and equal implementation of environmental, economic and social goals that are mutually dependent School/Logica (2011).

Since 2014, large capital companies have been obliged to provide information on the economic, ecological and social effects of their entrepreneurial activities on an annual basis in the form of a sustainability report. Since then, the introduction of this reporting requirement by the European Parliament, which had to be implemented by the relevant companies by 2017 at the latest, has contributed to the fact that more and more companies are dealing with questions of social responsibility and global change. While global players have been producing reports on sustainability for several years already, opportunities, but also risks have recently emerged, especially for small and medium-sized enterprises.

The increasing interdependence of internationally active companies through globalization has led to the fact that today, more than ever before, supply chains are multinational in their orientation and influence each other to a great extent. In 2017, for example, a similar study investigated CSR disclosure in polish-listed companies (Matuszak and Różańska 2017). In a Spanish study on this topic, the different communication quality of CSR reports was examined using four Spanish companies as an example (Baviera-Puig 2015). The interweaving of the terms sustainability and CSR was examined in more detail in another Spanish publication in 2017 (Bosch-Badia et al. 2017). The increasing number of publications on CSR and sustainability reporting illustrates how much the topic has become more relevant.

This work is the first that focuses on the sustainability reporting system of the DAX 30 companies in Germany by comparing them with each other and analyzing their content. The study can thus be seen in the context of various international publications on the topic. Within the past decade, various developments have led to different forms of sustainability reports. One company integrates ecological and social indicators into the annual financial report, while another group of companies prepares a stand-alone sustainability report, which may contain more than 100 pages.

It is interesting to note that the naming of the reports is sometimes handled very differently. In addition to the classic sustainability report, there are many other titles such as corporate responsibility report, corporate citizenship report or sustainable value report. This publication examines whether the title and content of different sustainability reports influence each other or whether they are subject to the design arbitrariness of the respective companies. The aim is to find out whether, for example, a corporate responsibility report takes greater account of social aspects, while a classic sustainability report maintains a balanced focus on the three pillars of economy, ecology and social issues. The procedure is as follows:

Before the theoretical basics are explained, a brief overview of the scientific methods used in this work are given. The theoretical part puts companies in the context of sustainable development and describes the four sustainability challenges they face. Subsequently, approaches to sustainability management and reporting are sketched out before the instrument of sustainability reporting is dealt with in more depth. First of all, a conceptual classification as well as the definition of essential terms takes place. In addition, a historical overview of the development of sustainability reporting since the 1970s is given. Following on from this, the focus is placed on the latest developments in sustainability reporting, with the different dimensions of reporting to be discussed first. Subsequently, the topics of certification, standardization and reporting obligations will be dealt with one after the other. The theoretical part will be rounded off by the presentation of various forms of sustainability reporting. In the practical part, the DAX 30 companies are put at the center of the consideration, as described at the beginning. These are compared with respect to their sustainability reporting systems. Special attention is paid to the different titles of the reports, without neglecting the analysis of differences in content. At the end of this publication is a summary with an implicit outlook.

2 Methodology

The theoretical part of this thesis follows a classical literature search, whereby the attempt was made to present the topic as up-to-date as possible and to include the latest figures.

The methodology used for the practical investigation of the reporting obligations consisted of a background analysis (with a literature review), complemented with a descriptive data analysis. Latter describes a statistical methodology, which sums up a given data or record. In case a total survey or, in general, a data set is available, it is the task of the data analysis to condense the information contained in the individual data and to summarize the essentials. Tables, graphical representations and characteristic values are used for this purpose. The data analysis is of a descriptive nature only (Springer Gabler Verlag 2018).

The survey was conducted in the period from November 2017 to February 2018. It refers to the DAX 30 companies as of December 2017 and is based on data from the GRI database, which was last checked for validity in January 2018.

3 Companies in the Context of Sustainability

One of the most influential institutions in today's society are commercial enterprises. Their actions decisively determine the orientation of an economy (Senge 2011). For this reason, more and more companies are including questions of social responsibility (corporate social responsibility—CSR) and corporate sustainability in their goals and raising the accomplishment of sustainability challenges to the highest management level (Bundesministerium für Umwelt 2007). The goodwill has become a tangible asset and the reputation of a company with regard to its ecological and social orientation has developed into a competitive advantage. "Ethical consumer decisions" are on the increase, as there is a greater awareness of sustainability issues within society than there was a few years ago. Career decisions are also increasingly being made on the basis of companies' sustainability orientation. The best want to enter into an employment relationship with a company that contributes to society and shares their values (Senge 2011).

The goal of corporate sustainability is supported by sustainability management, "*firstly by striving for sustainable organizational development through the integrative consideration of social, ecological and economic goals and secondly by contributing to the sustainable development of the economy and society.*"[1] This definition describes sustainability management as a function.

It can also be considered from an institutional point of view. In this case, it refers to a group of persons or the organizational structure in the company that deals with social and ecological issues and their integration into the course of business (Bundesministerium für Umwelt 2007). The tasks include "*the development and dissemination of qualitative and quantitative as well as temporally defined sustainability goals, the selection of measures and finally the determination of the degree of achievement.*"[2]

In the context of a company's sustainability management, the sustainability strategy is of central importance. It lays down the decision-making premises and creates the basis for the implementation of sustainable organizational development measures (Fifka 2014).

Derived from the three pillars of ecology, social and economic issues and their simultaneous integration into conventional management, four central challenges of sustainability arise. These are shown in the following figure (Fig. 1).

The ecological challenge addresses the question of how a company can reduce the environmental impact caused by its activities (increasing eco-effectiveness). Consequently, the social challenge focuses on the question of how to minimize the socially undesirable effects emanating from a company (increasing socio-effectiveness). The economic challenge for the environmental and social management of enterprises consists of the improvement of eco- and socio-efficiency, which contains the question: "How can environmental protection and social commitment be implemented in a cost-effective, profitable and value-adding manner?" The task of bringing together the three aforementioned challenges and integrating environmental and social man-

[1] cf. Schaltegger et al. (2007), p. 6, translated by the author.

[2] cf. Berthold and Lingenfelder (2014), p. 414, translated by the author.

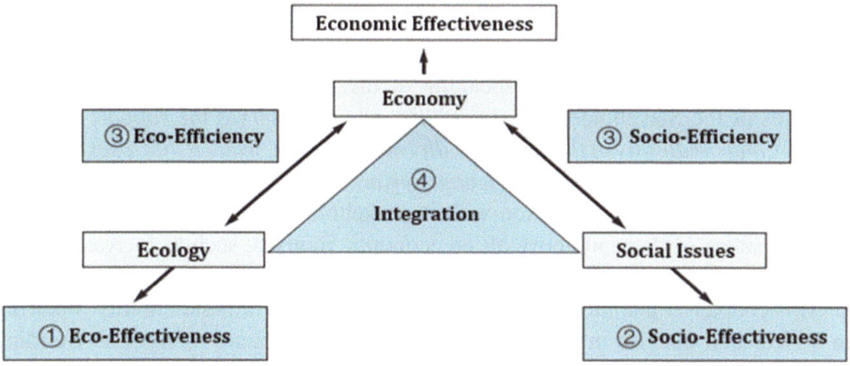

Fig. 1 The four sustainability challenges for companies (cf., Bundesministerium für Umwelt 2007 p. 14, translated by the author)

agement into the conventional economic management, is part of the integration challenge (Bundesministerium für Umwelt 2007).

Sustainability management is based on the entrepreneurial task of meeting these four challenges successfully, using various management approaches. These include concepts, systems and instruments. One of the tools from this compendium is sustainability reporting, which will be explained in more detail below.

4 Sustainability Reporting

Sustainability reporting, as a dialogue instrument moved increasingly in the focus of the public, in the past few years. Beside the annual report, which mostly contain only monetary figures, more and more enterprises considering ecological and social facts in their reporting obligations. In 2014, the European Union made concrete efforts to introduce a reporting requirement that would require large corporations to publish an independent annual sustainability report. While most listed companies already provide extensive annual reports on sustainability issues in a separate report, the discussion on reporting requirements for small and medium-sized enterprises is of the highest relevance (Fifka 2014).

In 2011, 95% of the world's 250 largest companies already issued an independent sustainability report ("stand alone report"). However, sustainability reporting has only become a standard in recent years. In 1999, only 35% of the world's 250 largest companies published such a report. The United Kingdom and Japan lead in international comparison, with 100 and 99% of the 100 largest companies in each country publishing a report, respectively. France and Denmark followed with 94 and 91% respectively. In Germany, 87% of the DAX 30 companies issued a report in 2010 (KPMG 2011; Fifka 2014). Due to the reporting requirement introduced by the

EU Parliament, it can be assumed that the figures within the EU are now also close to 100%

A reason for the increasing meaning of the reporting dues to the increasing demands of the Stakeholder. "*Sustainability reporting serves the support of transparency and credibility of the sustainability management, while it discloses information about measures and results oriented towards the past and the future.*"[3] In addition to improving communication with its stakeholder groups, the focus of efforts for companies is also and above all on economic motives, such as increasing their reputation or brand image and increasing their innovative strength.

This chapter is intended to present the complex field of sustainability reporting in a clear and concise manner using well-founded literature references. Building on the knowledge already described, it is to form the basis of practice-oriented considerations.

4.1 Linking Sustainability Management and Reporting

As has already been shown, instruments for meeting the various sustainability challenges must be integrated into existing corporate management. This is particularly evident in sustainability reporting. It must not be viewed separately from the management system, but as an integrative approach. For example, a sustainability report can be generated from the management system and, conversely, the management system can be optimized through reporting (Fifka 2014).

The main task of management systems is to coordinate and systematize entrepreneurial activities with the help of fixed control and monitoring mechanisms. A management system geared to sustainable development should focus on the implementation and monitoring of corporate social responsibility (Bundesministerium für Umwelt et al. 2014). Since there is currently no uniform management system covering all aspects of CSR, integrated sustainability management incorporates various sustainability-related standards that cover ecological, social and economic aspects in their entirety (Fifka 2014). These include quality and environmental standards (ISO 9001, ISO/TS 16949, VDA, ISO 14001, EMAS regulation) as well as standards of social responsibility (SA 8000, AA 1000), occupational health and safety (OHSAS 18001) (Bundesministerium für Umwelt 2007). One approach that covers a large part of sustainability aspects is ISO 26000, which takes into account important issues of sustainable management such as organizational management, working conditions, the environment, human rights and stakeholder involvement. However, the guideline has so far only served as an orientation guide and does not yet describe a certifiable management system (Fifka 2014).

Audits are carried out in order to check the functionality of the system on a regular basis. This can be certified/validated by an external audit of the management system (Bundesministerium für Umwelt 2007). It should be noted that each of these standards

[3]cf. Berthold and Lingenfelder (2014), p. 415f, translated by the author.

covers implementation, verification and auditing to varying degrees. SA 8000, for example, which aims to comply with minimum social standards, is primarily focused on auditing/certification. On the other hand, ISO 14001 places greater emphasis on environment-oriented management processes and their continuous improvement (Bundesministerium für Umwelt et al. 2014).

Sustainability reporting can contribute to organizational development through the presentation of the company's vision, mission, mission statement and the sustainability strategy. The defined goals, measures and key figures as well as the achieved performance levels and defined improvement potentials can help to optimize the management system. It also serves internal and external stakeholders to communicate sustainability activities. In this way, the coupling of reporting and management systems can contribute to the company's self-promotion, accountability, future orientation and organizational development at the same time. This results in a systematic and strategic development of the company into a sustainable enterprise (Fifka 2014).

4.2 Definition and Conceptual Demarcation

The linking described above is already expressed in the extensive conceptual dissemination of sustainability reporting, according to which "sustainability reporting" in literature often describes the entirety of measuring and disclosing sustainability-relevant information. This definition has been widely accepted, but is only partially correct, as "reporting" actually only describes the disclosure of information. It is therefore part of the higher-level "sustainability accounting". This includes sustainability auditing, sustainability assurance and sustainability reporting (Fifka 2014)". In line with the technical literature, this paper also considers sustainability reporting as a whole of measurement and disclosure.

Furthermore, there are conceptual ambiguities in the designation of sustainability reports. In addition to the sustainability report, there are a number of different terms used in the literature. Widespread are, for example "corporate (social) responsibility report", "corporate citizenship report" or "sustainable value report". On the one hand, this is related to the partly synonymous use of corporate sustainability and CSR in corporate practice, and on the other hand, the different titles can also be attributed to the arbitrary design of companies (Fifka 2014). According to Fifka and Drabble, the content of sustainability and corporate responsibility reports does not vary. "*In short, a sustainability report is nothing more than a corporate responsibility report. So the choice of title is ultimately a "matter of taste"*."[4] Although from a scientific point of view there are differences between these terms, the core function of all of

[4]cf. Fifka (2014), p. 4, translated by the author (Fifka and Drabble 2012).

them is the publication of sustainability relevant indicators and aspects.[5] The term "sustainability report" is mainly used here. This can be defined as follows:

> The sustainability report summarizes the ecological, social and economic dimensions of the company's activities on an annual basis. Both the target groups and the motives may differ: A distinction can be made between market-oriented, management-oriented and public-orientated motives.[6]

Market-oriented motives usually appear in economic sectors in which ecological and social aspects are an important purchasing criterion and in which a company can differentiate itself from its competitors through its sustainability performance. The motive for improving internal processes and structures as well as the motivation and information of employees can be attributed to the management-oriented approach. Companies that are increasingly under public scrutiny due to their industry choose publicity-based motives for reporting. This is the case when the industry has an increased impact on the environment and society (Institut für ökologische Wirtschaftsforschung (IÖW) 2001). This means that the sustainability reporting instrument can be used for both internal and external accounting purposes. The target groups of the report can therefore vary widely and vary from company to company.

4.3 *Origin and Historical Development*

Sustainability reporting dates back to the 1970s, when companies began publishing non-financial information in a structured form. Especially in Western Europe, multinational companies have drawn up so-called social balance sheets for this purpose, which mainly provided information on social issues, such as created jobs or charitable commitments. These companies tried to counteract the increasing criticism of their economic power by voluntarily disclosing information (Abbott and Monsen 1979). This reactive nature of reporting, caused by external pressure, also continued in the 1980s, when ecological concerns were increasingly included in the reporting driven by environmental disasters.[7] In the 1990s, companies saw a potential competitive advantage arising from the company's desire for more social and ecological responsibility and began to proactively approach the subject of reporting (Gray 2001). With the aim of achieving reputational and image benefits (Azzone 1997), environmental aspects were initially placed at the center of attention, and social reports were often replaced by environmental reports (Owen 2008). The uniform integration of social, ecological and economic concerns in a report that went beyond the conventional annual report did not take place until the turn of the millennium. The first

[5]Burckhardt points out that sustainability refers to an entire society and that CSR is about a company's responsibility for sustainable development. Consequently, the organization is the focal point here. (Burckhardt 2013).

[6]cf. Institut für ökologische Wirtschaftsforschung (IÖW) (2001), p. 4ff, translated by the author.

[7]1979: nuclear incident at Harrisburg (USA), 1986: nuclear incident at Chernobyl (Ukraine), 1984: chemical accident at Bophal (India), 1989: oil tanker Exxon Valdez in Alaska (USA).

reports in this form not only represented an extension of the content, but also a conceptual extension to previous types of reports and were, among other things, titled sustainability report, corporate responsibility report or corporate citizenship report (Blankenagel 2007). The conceptual differences have already been addressed above. The first reporting German company was Siemens, which presented a corporate citizenship report in 2000. In the context of the automotive industry, BMW AG is worth mentioning, which published a sustainable value report for the first time in 2008 and continues to adhere to this title today.

4.4 *Developments in Sustainability Reporting*

The still young history of sustainability reporting requires that new developments and trends constantly broaden the context of this topic. The following sections provide an overview of the dimensions of reporting and provide information on how sustainability reporting can be audited and standardized. It also discusses the EU Directive on mandatory reporting. Finally, the most important reporting types are explained.

4.4.1 Dimensions of Sustainability Reporting

Sustainability reporting includes many subject areas that can vary widely from one industry to another. In addition to fields of action that can be directly assigned to one of the three pillars (ecology, economy and social issues), these also include aspects that only partly affect them, but that still affect corporate sustainability in general. In connection with the economic sustainability challenge, the so-called corporate governance, which deals with the management and control structures of a company, can be cited here. In addition, ethics management should also be mentioned here. The implementation of ethical directives includes, for example, the publication of a "code of conduct" or measures to combat corruption (Fifka 2014).

The topics on which companies report are often related to the diverse interests of stakeholders. In order to comply with this, companies could modify or falsify information in their favor in order to give a more positive image. In view of the lack of guidelines on the content, materialities and addressees of sustainability reports, the question arises as to how far truthful information on corporate sustainability information can be guaranteed. The audit/verification of the sustainability report by third parties is an appropriate way of verifying the correctness of the statements made and gaining the trust of stakeholders (Manetti and Becatti 2008).

4.4.2 Certification in Sustainability Reporting

In order to increase the credibility of the information provided, there is, as in the case of classic financial reporting (annual financial statements), the possibility of being audited by an independent auditing firm. Verification is not mandatory in the context of voluntary sustainability reporting. In 2011, however, 46% of the world's 250 largest companies already took advantage of this option and had their reports audited (Gómez et al. 2013).

In the absence of statutory regulations in the area of sustainability reporting, various regulations and auditing standards have evolved over time. Many of the major auditing firms now offer services in the area of sustainability assurance. On the one hand, they can support companies in preparing reports and on the other hand, they can guarantee a review according to firm principles. The International Standard on Assurance Engagements (ISAE) 3000 and the Accountability 1000 Assurance Standard (AA 1000 AS) have established themselves internationally (Gómez et al. 2013).

In addition to this voluntary formalized control by auditing organizations, sustainability reports are also subject to ever-increasing scrutiny by stakeholders such as NGOs and critical media.

Consequently, verification of a report by a third party can underpin its credibility and, at the same time, counteract reputational risk due to spoiled or falsified information. However, the challenge of a company publishing only selected information cannot yet be met (Fifka 2014). "*The certification of information previously selected by the company does not yet permit any significant conclusions to be drawn about a company's sustainability performance. Companies must therefore accept the accusation that they would disclose information which is exclusively beneficial to them, while deterring any negative aspects.*"[8]

This is where the standardization and quantification of sustainability reporting begins.

4.4.3 Standardization of Sustainability Reporting

The selective selection of information is not only a hindrance to correctly assess the sustainability performance of a company, but also makes it difficult to compare sustainability reports. Reporting standards address this problem and provide a framework for the publication of ecological, social and economic performance criteria based on quantitative measurable indicators (Fifka 2014).

This brings significant benefits for both stakeholders and companies. One of the advantages for stakeholders is that standardization increases the credibility of reports. The selection options for companies are limited by a defined reporting framework, which can help to obtain a comprehensive picture of their ecological and social commitment. Furthermore, the comparability of reports from organizations in the same

[8]cf. Fifka (2014), p. 6, translated by the author.

sector is also improved, as standardization generally leads to the publication of the same indicators. As a result, competition from the market economy always tries to meet the sustainability requirements of stakeholders as closely as possible. This in turn is one of the advantages for the companies themselves. They can differentiate themselves from their competitors through good performance and gain market advantages. The high credibility of their reports also improves communication with stakeholders by providing reliable information for collaboration. Here, too, it should be noted that reporting is not only beneficial to the relationship between organizations and their stakeholders, but also, and above all, it improves internal reporting and corporate management. Insights into one's own strengths and potential for improvement can be better identified and documented, which helps the company as a whole to achieve sustainable management (Fifka 2014).

"Despite these advantages, it should not go unmentioned that measuring indicators in particular involves considerable effort and requires technical know-how as well as financial and human resources."[9] Especially in small and medium-sized companies, the resulting additional costs can lead to economic difficulties that should not be disregarded.

4.4.4 Commitment to Sustainability Reporting

Before following on from the comments on standardization in the sustainability report, various forms of presentation are described, the subject of the obligation to report on independent sustainability reporting, which has already been mentioned several times, is briefly discussed. It was adopted by the EU Parliament in 2014 and is intended to oblige companies to produce independent sustainability reports. For this purpose, there are explicit requirements as to which companies are affected by the obligation and which are not. The decision stipulates that three essential criteria must be met. On the one hand, it initially addresses large companies with a balance sheet total of more than EUR 20 million and net sales of more than EUR 40 million per year. Furthermore, only companies with more than 500 employees are considered. Last but not least, the requirement is aimed at organizations of public interest, which means listed groups, credit institutions, insurance companies or other companies declared to be of public interest by the EU Member States. A company is included in the reporting requirement if all the above-mentioned points are fulfilled at the same time. However, there are exceptions to this rule. For example, subsidiaries are not affected by the obligation if the parent company is responsible for reporting. The commitment has been fully effective since 2017. Approximately 6000 organizations across Europe are affected. It should also be noted that although the EU directive does restrict the number of reportable companies, sustainability issues should also be proactively addressed by companies that are not included. The advancing globalization and the ever more complex value chains thus indirectly include other companies as well. In the future, reportable companies will be considering how to integrate sustainability

[9] cf. Fifka (2014), p. 7, translated by the author.

more closely into their supply chains. This applies in particular to the supply industry (Recarbon Deutschland Gmbh 2014).

4.5 Sustainability Reporting Types

The publication of ecological, social and economic aspects can take place in different formats. On the one hand, information can be integrated into traditional financial reporting and published in the form of an annual and management report, and on the other hand it can be based on established reporting standards. Numerous national, international and industry-specific guidelines and standards have been established here, especially in the past 20 years. The reporting guidelines of the Global Reporting Initiative (GRI) have been particularly outstanding. In 2014, 80% of the world's 250 largest companies designed their reports according to the specifications of GRI (KPMG 2011). Other international standards include the UN Global Compact, the EFFAS guidelines and the aforementioned ISO 26000 standard. In Germany, the German Sustainability Code initiated by the German Council for Sustainable Development (RNE) is gaining ground (Berthold and Lingenfelder 2014).

Based on financial reporting, sustainability reports, such as those prepared in accordance with GRI guidelines, are also based on the basic principles of truth, materiality, clarity, consistency, comparability and public access to information. The materiality of selected information is often determined in advance of a sustainability report by means of a materiality analysis and depends, among other things, on the respective business model (Berthold and Lingenfelder 2014).

Various formats can be selected for the publication of a sustainability report. In addition to publication as a print report or PDF document, the medium of the Internet is increasingly becoming the focus of reporting. Web portals allow Stakeholders to select information as required (Fifka 2014).

In terms of content, regardless of the form of reporting followed by a company, essential requirements have emerged that should be taken into account in every sustainability report. These include the presentation of basic strategies and control mechanisms, the identification of key corporate challenges and their materiality, identifying development potential and sustainability goals in the context of sustainable organizational development, and the involvement of stakeholders in the reporting process (Berthold and Lingenfelder 2014).

4.5.1 Integrated Reporting

In Germany, companies have been obliged to integrate non-financial aspects into their annual financial statements and management reports since 2004. With the so-called Balance Sheet Reform Act, the legislature has created a basis for the accountability of large capital companies, according to which these business developments, results of operations and the Group's position must be presented in such a way that a true and fair view is conveyed (Umweltbundesamt 2006). It is stated that non-financial performance indicators, such as environmental and employee information, should be disclosed where they are relevant to understanding the business performance or situation (Braun et al. 2010).

The International Integrated Reporting Council (IIRC) in particular is promoting the integration of sustainability aspects that go beyond statutory obligations. The aim here is to bring together all the information that is essential for the course of business and which includes not only economic, but also social and ecological criteria, in order to obtain a comprehensive company profile (Fifka 2014).

The fundamental integration of sustainability was carried out by 27% of the 250 largest companies in 2011. Two years earlier, only 4% had integrated sustainability information comprehensively into annual and management reports (KPMG 2011).

4.5.2 UN Global Compact

"*The UN Global Compact was presented at the World Economic Forum in Davos in 1999 by then UN Secretary-General Kofi Annan and came into force a year later.*"[10] As a network for organizations of all kinds, it supports companies in the fulfilment of their social responsibility and calls on them to align their activities with a catalogue of values consisting of ten principles. These are based on the four areas of human rights, labour standards, environmental protection and the fight against corruption. As a voluntary platform for stakeholder dialogue, the Global Compact initiates networking and exchange of experience between companies through various local as well as international forums and promotes mutual learning (Bundesministerium für Umwelt et al. 2014).

"*Participants in the UN Global Compact undertake to report annually on their progress in implementing the ten principles. This Communication on Progress (COP) is a notice of a company to its stakeholders. A complete COP describes the actions taken and the results achieved.*" (see footnote 10) It first of all includes a declaration of support from the company. This must be signed by a board member. It is also necessary to specify what measures have been taken to implement the ten principles since the last COP and to what extent the company intends to further advance their future fulfillment. This also includes the definition of target indicators. Finally, the COPs are published on the freely accessible database of the UN Global Compact. There are currently about 12,900 uploaded progress reports (Compact 2014). The UN Global Compact is thus the largest communication platform for socially committed companies (Compact 2015). Around 10,000 COPs were deposited at the beginning of 2014. This reflects the ongoing trend towards reporting. It should be noted that the COPs are a formal requirement of the UN Global Compact and are not checked for quality and accuracy (Bundesministerium für Umwelt et al. 2014).

Organizations publishing a progress report have been classified into two categories since 2011. "GC Active" companies are in constant dialogue with their stakeholders

[10] cf. Berthold and Lingenfelder (2014), p. 418, translated by the author.

and report on all aspects of the UN Global Compact. Companies striving for excellence in terms of sustainability principles will continue to receive the “GC Advanced” rating. To this end, they are guided by various instruments such as the “Blueprint for Corporate Sustainability Leadership” (Compact 2015). This is an action plan for companies, which comprises fifty concrete measures to improve sustainability performance and is divided into three dimensions. The first dimension describes the implementation of the ten principles in the strategy and business activities of companies. Active support in the management of further UN objectives and issues is the second perspective of the plan of action, while the commitment to the Global Compact rounds off the range of dimensions. The blueprint was developed by the “Global Compact Lead Initiative”. Founded in 2011 by Ban Ki-Moon, this is a platform of more than 50 commercial enterprises that aims to contribute to the improvement of overall economic sustainability performance and the further development of corporate sustainability (Berthold and Lingenfelder 2014).

4.5.3 Global Reporting Initiative (GRI)

As a multi-stakeholder network, the Global Reporting Initiative (GRI) has been working closely with the UN Global Compact since its foundation in 2002. The non-profit foundation, based in Amsterdam, has developed a sustainability reporting guideline that is recognized worldwide and is currently used by large corporations to communicate sustainability aspects (Bundesministerium für Umwelt et al. 2014).

The standard is intended to guide organizations of all types and sizes in measuring and reporting environmental, social and economic performance. To this end, various indicators and management approaches in these three areas of action are examined. These are systematically organized, with the aim of achieving above all the goal of standardization to increase transparency, comparability and quality of sustainability reporting. The underlying reporting principles were established in cooperation with numerous stakeholders of the organization and are regularly checked for their topicality (Berthold and Lingenfelder 2014).

Due to a change of application, there are two versions of the Sustainability Reporting Guidelines—the so-called G3.1 and G4 guidelines. The G3.1 guideline introduced in 2011 has been replaced by the G4 standard published in mid-2013, for which a German translation has also been available since November 2013. By 31 December 2015, the G3.1 directive had been completely replaced (Berthold and Lingenfelder 2014).

The new reporting framework promises to be easier to use, in particular through increased compatibility with other standards or the integrated approach to reporting. In addition, greater emphasis is being placed on integrating the supply chain of companies, since ecological and/or social problems also and especially here exist, the holistic approach to which lies within the remit of an advanced sustainability management (Berthold and Lingenfelder 2014).

The G4 guidelines differentiate the quality of a report in two ways, distinguishing between “core” and “comprehensive”. Necessary core indicators must be provided by all organizations reporting in accordance with G4. The general topic areas (General Standard Disclosures) include information on strategy and analysis, organizational profile, identified essential aspects and limits, integration of stakeholders, reporting profile, corporate governance as well as ethics and integrity (Global reporting initiative 2013). A total of 58 indicators are assigned to these seven thematic areas. These indicators must first be examined with regard to their materiality for the reported entity's respective business activities. Organizations are encouraged to involve their stakeholders. It is also expected that a materiality matrix will be drawn up to determine the main content for companies and stakeholders (Global Reporting Initiative 2015). In a second step, the relevant indicators will be reported on. Comprehensive reporting companies must review all 58 indicators for their materiality and, if necessary, report on them. For the simplified report claim (core), however, it is sufficient to look at 34 indicators. The general topics are followed by the Specific Standard Disclosures. These consist of information on the management approach and economic, ecological and social aspects and comprise a total of 92 indicators. (Global reporting initiative 2013) "*These include, for example, economic performance, water consumption, gender equality, child labour, anti-corruption and consumer protection.*"[11] Here, too, it is true that organizations only have to report a reasonable number of key figures if they are relevant to their business activities. The preparation of a report with a simplified claim (core) requires that each subject area considered relevant treats at least one corresponding indicator. If a company wishes to meet the comprehensive requirement, the report must include information on all relevant indicators of an important topic (Berthold and Lingenfelder 2014).

Reports prepared in accordance with GRI guidelines are to be registered on a central online platform of the GRI. There are currently about 29,400 reports published in accordance with GRI guidelines. Furthermore, the database currently lists more than 12,000 reporting organizations (Global Reporting Initiative 2017). In 2015, by comparison, there were 18,300 listed GRI reports and approximately 7300 companies in the GRI database. In 2013, approximately 13,600 reports were registered, confirming the trend of increasing reporting using the GRI guidelines (Bundesministerium für Umwelt et al. 2014).

Verification by an independent audit firm is recommended by the GRI and increases the credibility of the information published.

4.5.4 Reporting According to EFFAS Principles

The European Federation of Financial Analysts Societies (EFFAS) has been developing indicators for environmental, social and corporate governance (KPIs for ESG)[12] since 2007. The current version KPIs for ESG 3.0 has been active since September 2010 (EFFAS 2015). It comprises approximately 130 cross-sector and cross-industry criteria.

[11]cf. Berthold and Lingenfelder (2014), p. 419, translated by the author.

[12]Key Performance Indicators for Environmental Social and Governance Issues.

The complex catalogue of indicators, in contrast to previously considered reporting standards, is aimed exclusively at profit-oriented companies and addresses listed groups in particular. In this way, EFFAS is countering the growing interest in non-financial indicators, which can have an influence on the profitability of a company, also among analysts and investors (Berthold and Lingenfelder 2014).

4.5.5 The German Sustainability Code (DNK)

In recent years, the German Sustainability Code (DNK) has emerged in Germany as a benchmark for the sustainability management of various organizations. It was developed in 2011 by the German Council for Sustainable Development (RNE). In this context, they actively collaborated with various interest groups. The aim is to increase transparency and comparability of sustainability relevant aspects of organizations of different types and sizes.

The DNK is applied on a voluntary basis and comprises 20 reporting criteria. A declaration of conformity provides information on the extent to which an organization complies with these criteria (comply) or, if necessary, also explains why a certain criterion has not been taken into account (explain). Corresponding industry-specific additions can be added by the companies themselves. The declaration of compliance with the DNK can be validated by external audit firms. Furthermore, the Code is deemed to have been complied with if the relevant organization has already prepared a sustainability report based on the model of GRI or EFFAS and if it complies with it to the highest degree. In terms of content, the DNK is therefore also guided by these standards, as well as by the principles of the UN Global Compact, the OECD Guidelines for Multinational Enterprises and the DIN ISO 26000 report guideline (Berthold and Lingenfelder 2014).

Restriction to a compressed number of report criteria increases the user-friendliness of the report. According to RNE, it also favours companies without an already established reporting system and therefore makes it easier for small and medium-sized companies in particular to get started with sustainability management and reporting (Rat für Nachhaltige Entwicklung 2015).

The official database of the DNK currently lists about 277 reporting companies. The list also includes five declarations of conformity issued by companies in the German automotive industry. These include documents from Audi AG, BMW AG, Daimler AG, Porsche AG and Volkswagen AG. All in all, the database primarily lists large corporations and makes explanations of small or medium-sized companies missing. This shows that despite the user friendly and compact reporting standard, comparatively few small and medium-sized companies report according to the model of the DNK (Rat für Nachhaltige Entwicklung 2018).

5 Results and Analysis

The present study compares the DAX 30 companies with respect to their sustainability and corporate responsibility reporting and has the data as of January 2018.

The table below lists all the DAX companies considered. With the help of the GRI database, it was possible to collect essential information for companies' sustainability reporting. Please note that the latest reports registered in the database have been used. This results in deviations in the reference years. Furthermore, this does not mean that a corresponding company has not already published a more recent report. Of relevance here is only whether the report is available in the GRI database.

The table shows that almost all the companies examined have sustainability reporting systems. This is not surprising, as they are all covered by the above-mentioned EU directive. Fresenius is an exception to this rule. In this case, the exception described above occurs after a subsidiary does not have to report if the parent company does so. 25 companies published a stand-alone report in the period under review. Four Companies practiced integrated reporting, although five companies are listed in the diagram below. This is due to the naming of the SAP report, which also suggests integrated reporting.

The GRI database continues to identify two of the 29 companies listed as Non-GRI, which means that the two companies concerned do not report according to GRI standards. These are Adidas and Fresenius Medical Care. All other companies, with the exception of Vonovia, report in accordance with the internationally recognized GRI G4 Guidelines (26 companies). Vonovia already reports according to the latest guidelines, the "GRI Standards", which were published in 2016 and are intended to gradually replace GRI G4. These are only valid until 30 June 2018 (Global Reporting Initiative 2018).

As described above, companies are largely free to decide to what extent (core or comprehensive) they would like to report on their sustainability activities. In the present study it was determined that 8 companies report comprehensively and 19 companies follow the adherence level "core".

In the following, we will go into more detail and compare the contents of different reporting classes with each other. The GRI database was helpful in that it provides a clear overview of the recognized standards to which each reporting company refers in its report. The present study was limited to some important principles in order to increase clarity. These include the UN Sustainable Development Goals (SDGs), the Carbon Disclosure Project (CDP), the UN Global Compact (UNGC), the OECD Guidelines and ISO 26000, and their relevance with regard to the sustainability reporting of the examined companies is now to be defined in accordance with GRI Data Legend (Table 1).

The table above shows that 12 of the 29 listed companies refer to the SDGs. The UN Global Compact is the most frequently referred to as a whole. He is covered by 24 companies, while only three companies include ISO 26000. Measured in terms of frequency, the CDP and the OECD guidelines are in the midfield. The former is represented 18 times, while 15 of the 29 companies refer to the OECD Guidelines.

Table 1 Investigation on sustainability reporting of DAX 30 companies

No.	Name of Company Name of Report (latest on GRI database)	Stand-Alone	Integrated	GRI-Type	Adherence-Level	SDGs	CDP	UNGC	OECD GL	ISO 26000
1	Adidas Adidas Sustainability Progress Report 2015	✓	×	Non	×	×	×	×	×	×
2	Allianz Allianz Group Sustainability Report 2016	✓	×	G4	Core	✓	✓	✓	×	×
3	BASF BASF Report 2016	✓	×	G4	Comprehensive	✓	✓	✓	✓	×
4	Bayer Bayer Annual Report 2016	×	✓	G4	Comprehensive	✓	✓	✓	✓	✓
5	Beiersdorf Sustainability Review 2015	✓	×	G4	Core	×	×	✓	✓	×

(continued)

Table 1 (continued)

No.	Name of Company Name of Report (latest on GRI database)	Stand-Alone	Integrated	GRI-Type	Adherence-Level	SDGs	CDP	UNGC	OECD GL	ISO 26000
6	BMW Sustainable Value Report 2016	✓	×	G4	Comprehensive	✓	✓	✓	✓	×
7	Commerzbank GRI report of Commerzbank	✓	×	G4	Core	✓	×	✓	×	×
8	Continental Sustainability Report 2015	✓	×	G4	Core	×	✓	✓	×	×
9	Daimler Sustainability Report 2016	✓	×	G4	Comprehensive	✓	×	✓	✓	×
10	Deutsche Bank Corporate Responsibility Report 2015	✓	×	G4	Core	×	×	✓	✓	×
11	Deutsche Börse Corporate Report 2015 (annual 2015)	×	✓	G4	Core	×	×	✓	×	×

(continued)

Table 1 (continued)

No.	Name of Company Name of Report (latest on GRI database)	Stand-Alone	Integrated	GRI-Type	Adherence-Level	SDGs	CDP	UNGC	OECD GL	ISO 26000
12	Deutsche Lufthansa Balance 2017	✓	×	G4	Core	×	×	✓	×	×
13	Deutsche Post Corporate Responsibility Report 2015	✓	×	G4	Core	×	✓	✓	✓	×
14	Deutsche Telekom Corporate Responsibility Report 2016	✓	×	G4	Core	✓	✓	✓	✓	×
15	E.ON Sustainability Report 2016	✓	×	G4	Core	✓	✓	✓	×	×
16	Fresenius /	×	×	×	×	×	×	×	×	×
17	Fresenius Medical Care Annual Report 2015	×	✓	Non	×	×	×	×	×	×

(continued)

Table 1 (continued)

No.	Name of Company Name of Report (latest on GRI database)	Stand-Alone	Integrated	GRI-Type	Adherence-Level	SDGs	CDP	UNGC	OECD GL	ISO 26000
18	HeidelbergCement Sustainability Report 2015	✓	×	G4	Core	×	✓	✓	✓	×
19	Henkel Sustainability Report 2016	✓	×	G4	Core	✓	✓	✓	✓	×
20	Infineon Technologies Sustainability at Infineon 2016	✓	×	G4	Core	×	✓	✓	×	×
21	Linde Corporate Responsibility Report 2015	✓	×	G4	Comprehensive	×	✓	✓	✓	×
22	Merck Corporate Responsibility Report 2016	✓	×	G4	Comprehensive	✓	✓	✓	×	×

(continued)

Table 1 (continued)

No.	Name of Company Name of Report (latest on GRI database)	Stand-Alone	Integrated	GRI-Type	Adherence-Level	SDGs	CDP	UNGC	OECD GL	ISO 26000
23	Munich Re Corporate Responsibility Report 2015/2016	✓	×	G4	Core	×	×	✓	✓	×
24	ProSiebenSat. 1 Media Public Value and Sustainability 2015/2016	✓	×	G4	Core	×	✓	×	×	×
25	RWE Our Responsibility 2015	✓	×	G4	Core	×	×	✓	×	×
26	SAP SAP Integrated Report 2016	✓	×	G4	Core	×	✓	✓	✓	×
27	Siemens Sustainability Information 2016	✓	×	G4	Comprehensive	✓	✓	×	✓	✓

(continued)

Table 1 (continued)

No.	Name of Company Name of Report (latest on GRI database)	Stand-Alone	Integrated	GRI-Type	Adherence-Level	SDGs	CDP	UNGC	OECD GL	ISO 26000
28	Thyssenkrupp Integrated Report 2015/2016	×	✓	G4	Core	×	✓	✓	×	×
29	Volkswagen Sustainability Report 2016	✓	×	G4	Comprehensive	✓	✓	✓	✓	✓
30	Vonovia Sustainability Report 2016	✓	×	Standards	Core	×	×	×	×	×
		25	**4**	**G4: 26**	**Core: 19** **Comp.: 8**	**12**	**18**	**24**	**15**	**3**

SDGs "Indicates explicit reference to the UN Sustainable Development Goals (SDGs) in the report. Tracks whether the reporting organization has indicated that the report addresses any of the UN Sustainable Development Goals (SDGs)." (cf. Global Reporting Initiative 2018, p. 13)
CDP "Indicates explicit reference to the organization responding to one of the annual Carbon Disclosure Project (CDP) questionnaires, or participating in an associated CDP project."(ibid)
UNGC "Indicates explicit reference to/use of the United Nations Global Compact and its principles in the report." (ibid)
OECD "Indicates explicit reference to/use of the OECD Guidelines for Multinational Enterprises in the report." (ibid)
ISO 26000 "Indicates explicit reference to/use of the ISO 26000 clauses in the report." (ibid)

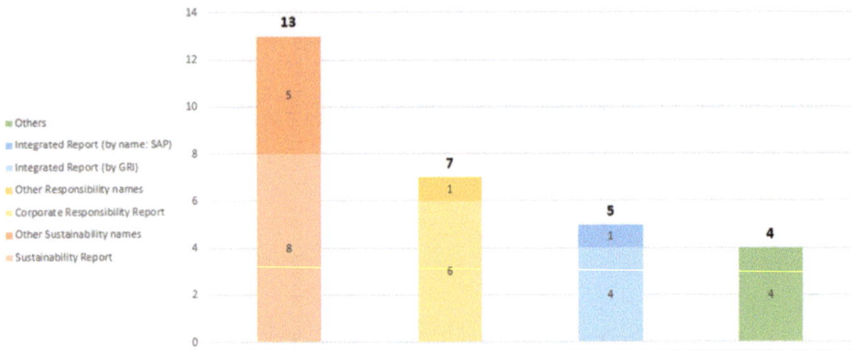

	Sustainability Reports	CR Reports	Integrated Reports	Others
GRI Ad. Level Core	8	5	3	3
GRI Ad. Level Comprehensive	4	2	1	1
SDGs	7	2	1	2
CDP	9	4	3	2
UNGC	10	7	4	3
OECD Guidelines	7	5	2	1
ISO 26000	2	0	1	0
Average of pages	104	147	238	112

Chart 1 Distribution matrix for sustainability and CR reports of the DAX 30 companies

At this point, the diagram below should go into more depth and allow a statement about how the parameters change, the reports are sorted according to their names or references (Chart 1).

Here it can be seen that the classical sustainability reports generally predominate. 13 companies use the term "sustainability" in the title of their report, with eight companies listed as "sustainability reports". The other five companies have names such as "sustainability progress report" or "sustainability review". Of the total of 13 companies, eleven report according to GRI G4, Adidas is listed as Non-GRI and Vonovia reports according to the latest amendment, the "GRI Standards", but in the scope of the report "core". The adherence level "core" is also used by seven other companies, while four companies report extensively ("comprehensive"). These are BMW, Daimler, Siemens and Volkswagen. Here two essences become clear. On the one hand, it can be stated that the major German automobile companies choose to name their sustainability report as such that it is neither an integrated report nor a CR report, but rather a sustainability report, whereby BMW has always used the title "sustainable value report". Secondly, it can be seen that all three companies report comprehensively on sustainability issues. Like Siemens, Allianz, E. ON and Henkel, they also refer to the SDGs. Adidas does not refer to any of the standards included, nor does Vonovia. Of the eleven remaining companies, ten refer in their reports to the UN Global Compact, nine to the CDP, seven to the OECD Guidelines and two to ISO 26000 (Siemens and Volkswagen). Furthermore, it can be said that companies report on an average of about 104 pages. The extremums are Volkswagen (192 pages) and Beiersdorf (39 pages).

Let us now come to those companies which, according to the title, are already trying to fulfil their special social responsibility. All those companies that have the word "responsibility" in the title of their report have been included in this category. There were a total of seven companies, six of which produce a classic corporate responsibility report. RWE chose the title "Our Responsibility" for its report. With Deutsche Bank and Munich Re, two companies from the financial services sector can be mentioned, which in their report refer to the special social responsibility in the title of their report. All companies report according to the GRI G4 standard, with only two of them reporting comprehensively (Linde and Merck). The remaining enterprises have the adherence level "core". Overall, only two companies refer to the SDGs (Deutsche Telekom and Merck). In percentage terms, that's 29 percent. In the case of the sustainability reports (first pillar), 54% (seven out of 13 companies) refer to the SDGs, which could fundamentally lead to the conclusion that companies that refer to sustainability in the title of their report tend to refer to the UN's recognized goals rather than those companies that refer to their reports as corporate responsibility reports. As with sustainability reports, the number of companies that refer to the UN Global Compact is correspondingly high in the category of corporate responsibility reports. While 77% of the former include them in their reporting, all 7 corporate responsibility reports do. However, none of the listed companies follows the guidelines of ISO 26000. Just over half of the seven companies refer to CDP and OECD guidelines. The length of the reports varies from 93 pages (Deutsche Bank) to 211 pages (Deutsche Telekom). The average value is 147 pages, which means that the corporate responsibility reports in this study are 43 pages longer on average than the sustainability reports.

The integrated reports have the largest scope. They are on average 238 pages long, which can be explained by the fact that in addition to the supplementary ecological and social content, the classic contents of a financial report are added. It should be noted that there is a blatant outlier among the five companies in this category. Deutsche Börse operates its integrated reporting on just 56 pages. The remaining reports vary between 224 pages (Fresenius Medical Care) and 344 pages (Bayer). It should also be noted that only four of the five companies in the GRI database are referred to as integrated reports. SAP's report is specified as a stand-alone report, but was included here because of its title (SAP Integrated Report 2016). With the exception of Fresenius Medical Care, all companies report in accordance with the GRI G4 standard, three of which comply with the adherence level "core" and one of which is "comprehensive" (Bayer). Bayer is also the only company that refers to SDGs and ISO 26000. As a Non-GRI company, Fresenius Medical Care does not refer to any of the stated standards. The other four companies are committed to the principles of the UN Global Compact and three of them are also committed to those of the CDP. Two companies refer to the OECD Guidelines. If the respective industrial sectors are included, it becomes clear that two of the five companies are listed as conglomerates (Bayer and Thyssenkrupp), which could be an indication that the application of integrated reporting in this sector is considered practicable.

All companies that cannot be clearly assigned to one of the other three categories have been classified in the "others" category. This applies to BASF, Commerzbank,

Deutsche Lufthansa and ProSiebenSat. 1 Media. According to the GRI database, all four of them have a separate report in accordance with GRI G4. BASF even reports comprehensively, while the remaining three have been classified as "core". None of the companies refers to ISO 26000, while half refers to SDGs and CDP. As with the previous categories, it can be stated that the majority (here 3 companies) include the principles of the UN Global Compact in their reporting. Only Bayer also refers to the OECD guidelines.

6 Conclusion and Outlook

Although there is no scientific clarity as to whether CR and sustainability reports can be fundamentally distinguished, or whether they both ultimately have the same opinions (see Sect. 4.2), the study presented here has shown, using the DAX 30 as an example, that it is quite possible to draw limits in isolated cases. These are partly based on the divergent motives of the companies described in Sect. 4.2, but can also be attributed to sectoral differences. Thus, this publication can contribute to narrowing the conceptual differences between CR and sustainability reporting. It should also be noted that the DAX 30 include various global players who are among the world market leaders in their sectors. As a result, they exert an increased impact and their actions have an influence on several stakeholders. They can be an example of good or bad entrepreneurial practice. The VW exhaust gas scandal has shown which global waves can cause entrepreneurial decisions. The same applies to sustainability reporting. Global players can set standards that small and medium-sized companies can or even have to follow. The study reveals that some of the largest German companies report very differently on social and ecological issues. Thus, this study can also show the divergence between scientific theory and entrepreneurial practice by demonstrating the occasional differences in the sustainability reporting of companies. Above all, this is made clear by the naming of these. As will be shown later in this chapter, the study is one of a number of international papers on the subject, but is the first publication that compares the sustainability reporting of Germany's largest companies using comprehensive data analysis. Various essences could be disclosed.

It has been shown that the distinction between sustainability, corporate responsibility and integrated reports is particularly useful and interesting when sectoral differences are taken into account. This is especially evident in companies from the automotive industry. All four companies produced stand-alone sustainability reports. This is particularly due to the importance of the supply chain in this sector. The sustainable orientation of the supply chain is one of the key challenges facing automotive companies. In addition to increasing the company's reputation, it is above all a matter of making delivery networks as efficient as possible. Suppliers must commit themselves to adhering to sustainability standards vis-à-vis the major automotive companies. For example, Daimler defines the requirements for suppliers with regard to sustainability aspects in "supplier sustainability standards". In this, the company formulates general principles and expectations with regard to labor standards, business ethics and compliance as well as environmental protection and safety along the supply chain. Even at the time of signing the contract, direct suppliers (first-tier suppliers) undertake to adhere to these standards, to communicate them to their employees and ultimately to integrate them into their own supply chain. BMW and Volkswagen are doing the same with their suppliers. Together with other European automotive manufacturers, all three have continued to develop a cooperative called "European Automotive Working Group on Supply Chain Sustainability" to further improve the sustainability of supply chains (Folkens 2015).

The present investigation has also shown that companies in the financial services sector are less likely to use the term "sustainability report", but tend to include social responsibility in the title of the report. This seems to make sense in the respect that this is a more resource-saving sector than the automotive industry. Consideration of ecological aspects is thus being pushed into the background and social concerns are being given much more attention. A third interesting point, within the sectoral view, is the sector of conglomerates. Two out of three companies preferred integrated reporting (Bayer and Thyssenkrupp). No significant conclusions can be drawn for all other sectors, as the remaining DAX 30 companies operate in very different sectors.

Adidas and Fresenius Medical Care are the only companies that do not report according to the GRI standard. As a result, the GRI database does not include any references to established standards such as the UN Global Compact or SDGs in their reports.

Of the other 27 companies, 19 had an adherence level core, while eight reported comprehensive. Three of these eight companies are the three major German automobile groups BMW, Daimler and Volkswagen. The two conglomerates Bayer and Siemens also provide comprehensive reports. This suggests that especially the financially strongest DAX companies include more indicators in their reporting. Furthermore, it can be stated that reports relating to sustainability clearly outweigh those relating to social responsibility.

Although some conclusions have already been drawn, it should be noted once again with reference to Sect. 4.2 that the contents of CR- and sustainability reports are not fundamentally different (Fifka and Drabble 2012) and that the decision in favour of one or another title can also follow the tastes of companies or, as at BMW has historically evolved.

The reference to international standards was also exciting. This has essentially confirmed what was anticipated in the theoretical explanations. It was written there that the UN Global Compact is currently the largest communication platform for socially committed companies. The figures presented in the present study support this view (24 out of 29 companies with reference to UNGC). Furthermore, it was also interesting to note that more than 50% of those companies with "sustainability" in the title of their report referred to the UN's SDGs, whereas those with a corporate responsibility report only did so in 29% of the cases.

Overall, it can be stated that sustainability reporting has increased significantly over the last five years. This can be seen from the figures in the GRI database, according to which 13,600 reports were registered in 2013, 18,300 in 2015 and unbelievable 29,400 GRI reports in 2018. This corresponds to a percentage increase of 116% compared with 2013. As has been described, this development will also affect medium-sized and small companies in the future, as large companies increasingly integrate sustainable development and social responsibility issues into their supply chains. Since the introduction of the reporting obligation by the EU Parliament in 2014, many companies have adapted their reporting accordingly. This study provides an overview of the reporting of the largest German companies and compares them. In doing so, it must be considered in an international context.

In addition to traditional financial reporting, sustainability reporting is increasingly establishing itself in the corporate practice of leading companies. And not only that. As has been pointed out, this trend will continue to develop in such a way that small and medium-sized enterprises will in future also report more frequently on social and ecological aspects of their actions. Politics, science and industry must provide the appropriate solutions and instruments for this purpose. Adapting the company management of small and medium-sized enterprises to the increased requirements can quickly become a real cost factor. If, for example, they want to report according to the GRI reporting criteria, this goes hand in hand with indispensable adjustments to corporate processes in order to ultimately be able to provide the required core indicators. This results in additional costs for the companies, which must be cushioned by increased economic efficiency. This adjustment will vary depending on the industry. Especially in the area of sustainability reporting of small and medium-sized enterprises, innovative concepts are needed in order to positively shape this transformation process for all parties involved.

As such, CR reporting addresses sustainability, particularly the Sustainability Development Goals (SDG) 3 (Good health and well-being), 8 (Decent work and economic growth), 9 (industry, innovation and infrastructure), 11 (Sustainable Cities and Communities), 12 (Responsible consumption and production), and 13 (Climate Action).

References

Abbott WF, Monsen RJ (1979) On the measurement of corporate social responsibility: selfreported disclosures as a measurement of corporate social involvement. S.l.: Acad. Manag J

Azzone G et al (1997) A stakeholder's view of environmental reporting. Long Range Plann

Baviera-Puig A et al (2015) Assessing the communication quality of CSR reports. A case study on four spanish food companies. Sustainability

Berthold, Lingenfelder M (2014) Wirtschaftswissenschaftliches Studium—Zeitschrift für Studium und Forschung, p 414

Blankenagel L (2007) CSR-Berichte als Kommunikationsinstrument der DAX-Unternehmen. VDM Verlag, Saarbrücken

Bosch-Badia M-T, Montllor-Serrats J, Tarrazon-Rodon M-A (2017) Efficiency and sustainability of CSR projects. Sustainability

Braun S, Doerner U, Horst DW, Loew T (2010) Unternehmerische Verantwortung praktisch umsetzen. [Online] [Cited: April 02, 2015.] https://www.pwc.de/de/nachhaltigkeit/assets/Brosch_Nachhaltigkeit_2010.pdf

Bundesministerium für Umwelt (2007) Naturschutz und Reaktorsicherheit; econsense—Forum Nachhaltige Entwicklung der Deutschen Wirtschaft e. V. und Centre for Sustainability Management Leuphana Universität Lüneburg (Hrsg.). Schaltegger S, Herzig C, Kleiber O, Müller J (Authors) (2007) Nachhaltigkeitsmanagement in Unternehmen—Von der Idee zur Praxis: Managementansätze zur Umsetzung von Corporate Social Responsibility und Corporate Sustainability. Berlin, Lüneburg

Bundesministerium für Umwelt, Naturschutz, Bau und Reaktorsicherheit (2014) Gesellschaftliche Verantwortung von Unternehmen—Eine Orientierungshilfe für Kernthemen und Handlungsfelder des Leitfadens DIN ISO 26000. Berlin: s.n.

Burckhardt G (2013) Corporate social responsibility—Mythen und Maßnahmen. Springer Gabler, Bonn

EFFAS (2015) KPIs for ESG 3.0. [Online] 2010. [Cited: April 01, 2015.] http://www.effas-esg.com/wpcontent/uploads/2011/07/KPIs_for_ESG_3_0_Final.pdf

Emerson J (2015) The blended value proposition (PDF). California management review. Retrieved 20 July 2015

Fifka MS (2014) CSR und Reporting: Nachhaltigkeits- und CSR-Berichterstattung verstehen und erfolgreich umsetzen. Springer Gabler Verlag, Heidelberg

Fifka MS, Drabble M (2012) Focus and standardization of sustainability reporting—a comparative study of the United Kingdom and Finland. S.l.: Bus Strategy Env

Folkens L (2015) Nachhaltigkeitsberichterstattung im Kontext des Nachhaltigkeitsmanagements von Unternehmen der Automobilzulieferindustrie. Hochschule Magdeburg-Stendal, Magdeburg

Global reporting initiative (2013) G4 Leitlinien zur Nachhaltigkeitsberichterstattung

Global Reporting Initiative (2015) Materiality check. [Online] [Cited: March 26, 2015.] https://www.globalreporting.org/reporting/report-services/Pages/materialitydisclosures-service.aspx

Global Reporting Initiative (2017) Database. [Online] [Cited: December 12, 2017.] https://database.globalreporting.org/

Global Reporting Initiative (2018) Sustainability disclosure database—data legend. [Online] [Cited: February 17, 2018.] https://www.globalreporting.org/SiteCollectionDocuments/GRI-Data-Legend-Sustainability-Disclosure-Database-Profiling.pdf

Gómez JM, Lang C, Wohlgemuth V (2013) IT-gestütztes Ressourcen- und Energiemanagement. Springer Vieweg Verlag, Heidelberg

Gray R (2001) Thirty years of social accounting, reporting and auditing: What (if anything) have we learnt?. A European Review, Business Ethics

Institut für ökologische Wirtschaftsforschung (IÖW) (2001) Der Nachhaltigkeitsbericht. [Online] [Cited: Februar 14, 2018.] https://www.ioew.de/fileadmin/_migrated/tx_ukioewdb/Der_Nachhaltigkeisbericht.pdf

Invernizzi DC, Locatelli G, Brookes NJ (2017) Managing social challenges in the nuclear decommissioning industry: a responsible approach towards better performance. Int J Proj Manag Soc Responsib. Management of Megaprojects. 2017-10-01, 35(7):1350–1364. https://doi.org/10.1016/j.ijproman.2016.12.002

KPMG (2011) International survey of corporate responsibility reporting. Amsterdam

Manetti G, Becatti L (2008) Assurance services for sustainability reports: standards and empirical evidence

Matuszak Ł, Różańska E (2017) CSR Disclosure in polish-listed companies in the light of directive 2014/95/EU requirements: empirical evidence. Sustainability

Owen D (2008) Chronicles of wasted time? A personal reflection on the current state of, and future prospects for, social and environmental accounting research. A.l.: Acc Auditing and Accountability J

Palmer K (1995) Tightening environmental standards: the benefit-cost or the no-cost paradigm? J Econ Perspect 9(4):119–132. https://doi.org/10.1257/jep.9.4.119.JSTOR2138393

Preston LE (1997) The social-financial performance relationship (PDF) Business and society. 36(4):419. https://doi.org/10.1177/000765039703600406. Retrieved 20 July 2015

Rat für Nachhaltige Entwicklung (2015) Der Deutsche Nachhaltigkeitskodex—Maßstab für nachhaltiges Wirtschaften

Rat für Nachhaltige Entwicklung (2018) Der Deutsche Nachhaltigkeitskodex—Datenbank. [Online] 2018. [Cited: Februar 15, 2018.] http://www.deutscher-nachhaltigkeitskodex.de/de/datenbank/dnk-datenbank.html

Recarbon Deutschland Gmbh (2014) Webinar—Verpflichtende Nachhaltigkeitsberichterstattung. [Online] [Cited: März 12, 2015.] https://www.youtube.com/watch?v=6ZGAVOFASz8

Schaltegger S, Herzig C, Klinke T (2007) Umweltschutz der Wirtschaft. [Online] [Cited: April 17, 2015.] http://www2.leuphana.de/umanagement/csm/content/nama/downloads/download_publikationen/SchalteggerHerzigKlinke.pdf

Senge PM (2011) Die notwendige Revolution. Carl-Auer, New York

SMI EBS Business School/Logica (2011) Excellence in supply chain sustainability, Studie der EBS business school wiesbaden und logica deutschland GmbH & Co KG

Springer Gabler Verlag (2018) Gabler Wirtschaftslexikon. Datenanalyse. [Online] [Cited: Januar 08, 2018.] http://wirtschaftslexikon.gabler.de/Archiv/1823/datenanalyse-v10.html

Umweltbundesamt (2006) Lagebericht zur Lageberichterstattung. [Online] [Cited: April 02, 2015.] http://www.umweltbundesamt.de/sites/default/files/medien/publikation/long/3236.pdf

UN Global Compact (2014) Database. [Online] [Cited: Januar 15, 2014.] www.unglobalcompact.org/ParticipantsAndStakeholders/index.html

UN Global Compact (2015) About us. [Online] [Cited: March 30, 2015.] http://www.unglobalcompact.org/AboutTheGC/index.html

UN Global Compact (2015) Fortschrittsmitteilungen. [Online] [Cited: March 30, 2015.] https://www.unglobalcompact.org/Languages/german/fortschrittsmitteilungen.html

Continuous Application of Preventive Environmental Strategies as a Way to Introduce Social Responsibility in Companies

Jurgis Kazimieras Staniškis and Eglė Staniškienė

Abstract The social sustainability challenge is still receiving relatively little attention but in future the social aspects will become a more important as a result of the growing expectations and demands addressed to business enterprises by employees and external stakeholders. Profit- oriented business companies run primarily for economic purposes, therefore environmental and social issues in most cases are confronted with the goals of making profit or at least minimising operating costs. As usual environmental and social aspects from management and organizational point of view are dealt separately. The aim of the research to investigate the integration challenge of ecological and social dimensions in the context of economic activity by implementation of different preventive environmental strategies and programs in the companies. The research method is based on development and implementation of special training and preventive innovations generation methodology for company's management and employees. There are many preventive environmental strategies, for instance, pollution prevention, waste minimization, cleaner production, etc. For a particular investigation, the resource efficient and cleaner production (RECP) strategy which applies to industrial processes, products and services was mainly introduced. This is because RECP has a potential to achieve production efficiency through optimization of productive use of natural resources to minimize the adverse impacts on environment and to reduce risks to employees and communities. Sustainable development strategy requires behavioural change by individuals as well. Therefore, the employees engagement is crucial. At the same time, resource efficient and cleaner production concept besides the potential mentioned above provides the possibilities to support their personal development. The methodology for preventive environmental innovations generation and implementation was introduced in more than 150 industrial companies and economic, environmental and social benefits were identified and assessed. Companies, which implemented a successful preventive environ-

J. K. Staniškis (✉) · E. Staniškienė
Kaunas University of Technology, 44239 Kaunas, Gedimino 50, Lithuania
e-mail: jurgis.staniskis@ktu.lt

E. Staniškienė
e-mail: egle.staniskiene@ktu.lt

W. Leal Filho (ed.), *Social Responsibility and Sustainability*, World Sustainability Series, https://doi.org/10.1007/978-3-030-03562-4_10

mental programs benefitted by significantly lowering operating costs, enhancing the safety of their employees, improving product quality and productivity, reducing environmental liability, improving public image, and chances for broader cooperation. Besides that, companies managed to implement employees'engagement strategy by (1) communicating the professional impact that employees contribution had on the company, (2) enabling cross-functional communication and idea exchange. From another point of view, these programs helped financiers secure better clients and make more profitable investments and assists government agencies in their pursuit of a safer and healthier environment.

Keywords Social responsibility · Business company
Preventive environmental strategies · Resource efficient and cleaner production
Sustainable development

1 Introduction

Society is facing many environmental challenges, including the depletion of natural resources, pollution, climate change, the collapse of whole ecosystem. These changes dramatically increase threats to human security, the health and well-being of society. Environmental and social responsibility is a precondition to the prosperity of human beings and important aspect of social responsibility. Social responsibility according ISO 26000:2010 is the responsibility of an organization for the impacts, of its decisions and activities on society and environment, through transparent and ethical behaviour that (International organization for standardization 2010):

- Contributes to sustainable development, including quality of life;
- Incorporate social and environmental considerations in its decision making process;
- Accountable for the impacts on society and environment;
- Is in compliance with applicable law and is consistent with international norms of behaviour;
- Social responsibility is practised in its relationships with stakeholders and takes into account their interests (ISO 26000 2010).

Social aspects require that companies act responsibly toward consumers, investors and the government by motivating employees, environmental perspective is at the heart of sustainability issues, therefore integration of social and environmental dimensions is critical for companies to be more innovative, efficient and effective (Asif et al. 2013; Martinez-Conesa et al. 2017).

During the last decade more and more industrial companies respond to ecological issues. Investigations show that the motives for corporate greening are: competitive advantage, regulatory compliance, ethical concerns, stakeholder pressures, and top management initiatives (Bansal and Roth 2000; Christensen et al. 2014). There is no clear evidence that smaller companies are not necessarily less socially and environmentally responsible than large companies. Smaller firms, on the other one hand, are challenged by shortage of human and financial resources, but in reality, they “silently” engaged by implementing a wide range of tools and practices to uphold social and environmental responsibility in their business operations (Baumann-Pauly et al. 2013; Wickert et al. 2016).

In the past 20–25 years’ new ideas have emerged to reduce emissions to the environment at the source and at the same time to solve some social issues. These waste minimization, pollution prevention and resource efficient and cleaner production strategies appeared to be necessary to reduce the enormous costs of clean-up actions, certainly from the moment that the polluter pays principle was brought into legislation. A preventive approach means that environmental problems are addressed before they arise when choices are made concerning processes, raw materials, design, transportation, services, and more. Such an approach effectively addresses the wasting of natural resources since pollution not only leads to environmental degradation but is also a sign of inefficient production processes or management.

As usual, there is a need of extra investments to replace waste generating industries with new clean technology facilities. However, large gains can often be achieved by simple “good housekeeping” measures, improved inventory control, better maintenance, repairing leaks, simple equipment or process modification, installing measurement and control systems, and improving enterprise and its process control.

UNEP and UNIDO define resource efficient and cleaner production (RECP) as a continuous application of preventive environmental strategies to processes, products, and services to increase efficiency and reduce risks to humans and the environment. RECP works specifically:

- Production Efficiency—through optimization of productive use of natural resources (materials, energy, water) at all stages of the production cycle;
- Environmental Management—through minimization of the adverse impacts of industrial production systems on nature and the environment;
- Human development—through minimization of risks to people the communities, and support to their development (UNEP DTI 2012).

The key importance for resource efficient and cleaner production promotion and implementation is capacity building which should enable:

- Further human development by using workplace policy and addressing important social concerns;
- A common understanding of the RECP concept;
- Incorporation of RECP concepts in policy frameworks;
- Integration into enterprise policies and operations;
- Indicators to measure progress in RECP implementation;
- Providing information about both the technology involved and the environmental management and social tools needed.

The second most important condition for sustained RECP is when it becomes part of the management through a formal company environmental, quality social responsibility management approach. An environmental management system provides a decision-making structure and action plan to support continuous environmental improvements, such as the implementation of RECP. If a company has already established an environmental management system, the RECP can be an effective tool for focusing attention on specific environmental and social problems. If, on the other hand, the company establishes RECP procedures first, this can provide the foundations of an environmental management system (Kliopova and Staniškis 2006; Staniskis and Katiliute 2017).

Experience of RECP programs in Lithuania demonstrated that a long-term training programs emphasising on-the-job training are the most effective way to create domestic professional capacities. Besides that, the only cost and time efficient way of fulfilling this need is to adopt a train-the-trainer approach. The paper presents the methodology of the human development and training in a workplace, including social responsibility and based on the detailed structure of RECP assessment system.

2 Raising Awareness and Building Competency for Social Responsibility by RECP Assessment System

Despite the potential benefits of RECP programs the dissemination process is quite slow, which could be caused by many difficulties and problems. Such barriers could be divided into four groups: (I) policy and market, (II) financial and economic, (III) technical and informational, (IV) managerial and organizational. Besides barriers mentioned above, such programs are seldom implemented systematically with a lack of RECP continuity, i.e. options are implemented but not monitored and extended. Moreover, the analysis of existing RECP assessment systems revealed the lack of a more comprehensive RECP method, as well as a lack of information and clear description of the activities that should take place in each step of RECP program (Silva et al. 2013).

The comparative analysis of the selected nine methods (Silva et al. 2013) has showed that all of them have the same basic steps: development of environmental and social policy, organization and planning, assessment/audit, identification of options for improvement, evaluation of options, implementation and monitoring and review, which also resemble the main phases of Deming cycle or so-called PDCA (Plan-Do-Correct-Act). Based on studied methods authors developed standard method where most part of improvements suggested and explained are related on the Plan phase of the PDCA cycle. The explanation is that "all major decisions taken in this stage have a direct influence on the continuation of the remaining cycle phases, and when a good Plan is established it is more likely that the other phases will also have good implementation success. Therefore, it can be expected that some benefits of a RECP program will be maximized and at the same time consumed by each activity will be reduced by applying this proposed method". This is true, but not always, in other words – there are more conditions for a success.

To make sound business plan for RECP innovation, the information on the environmental and social aspects of the relevant product, raw material or constituent part of the process should be gathered. This information should consider the environmental effects not only of the production phase and product life cycle but also of extracting and transporting the alternative raw materials and of treating any avoidable waste. Innovations with significant capital costs will require more detailed analysis. There are a number of factors that make RECP costs and benefits difficult to calculate for many business plans. Many of those costs are hidden or probabilistic, although the risks are real and it is difficult to predict the event occurrence date from past experience. Total Cost Analysis suggests some approaches to calculating indirect and probabilistic costs so that their full impact can be included in economic feasibility assessments. Innovation business plan should provide the basis for obtaining funding for RECP project. Such type projects should not be sold on their technical merits alone; a clear description of both tangible and intangible benefits can help a proposed project obtain funding. There will probably be other projects, such as expanding production capacity or moving into new product lines, that will compete with RECP projects for funding. Some companies may have difficulty raising funds internally for capital investment. The company has to look for outside financing, for instance, private sector financing includes bank loans and other conventional sources of financing. Revolving financial funds that provides soft loans for short pay-back projects are extremely attractive for RECP innovations financing.

The RECP projects does not end with implementation. Track of its effectiveness versus the claims made—technical, economic, environmental, social should follow. Short description of the RECP assessment, including building capacity for social responsibility:

The recognised need to RECP: Top management decision, policy statement, consensus building. Understanding the benefits of social and environmental responsibility should start at the top of the organization;

First step—planning and organisation: Name the assessment leader and group, define goals. It would be useful to focus initial efforts on some parts of an organisation which are more interested and receptive in taking actions on social responsibility;

Second step—pre–assessment: Collect data on all air, water, and solid waste emissions and releases, identifying their sources and quantifying the true costs of pollution control, treatment and waste disposal; site visits, signing priorities to process, operations, and materials; Identify a need for changes in decision making process and governance that would promote more freedom and motivation to suggest new approaches and ideas;

Third step—write Program plan: contact external groups from outside the company including surrounding community; define the objectives in quantitative terms and target dates; identify potential obstacles into four broad categories: economic, technical, social, regulatory, and institutional/human; define cost-benefit analysis procedures, develop schedule addressing some social responsibility issues in the short term and some over a longer period of time;

Fourth step—assessment: define additional staff to comprise detailed assessment team(s) with relatively specific focus; review pre–assessment data, organise a thorough site review and interview workers; prepare material and energy balances as a means of analysing pollution sources, their environmental and social impact and opportunities for eliminating them;

Fifth step—define RECP innovation options: generate options encouraging creativity and independent thinking of each assessment team member, use brainstorming sessions; screen options by procedure (for example, weighted sum method) when the assessment team selects the options that appear the best after examining each of them, taking account of social responsibility when conducting operations for the organisation;

Sixth step—feasibility analysis: prioritized list of RECP options/innovations should be examined to determine which are technically, environmentally, economically and socially feasible and selected for implementation;

Seventh step—assessment report: write and review the report, which should be the basis for innovation(s) business plan preparation; report has to be reviewed by people affected by the proposed project(s) in order to eliminate inaccuracies or misunderstandings;

Eight step—business plan of RECP innovation: content: company description, feasibility data, procurement of the equipment, investment analysis (payback period, net present value, Internal rate of return, risk analysis, sensitivity analysis), company's economic „health" analysis (profit-loss statement, balance sheet, economic ratio analysis), project implementation schedule, supervision and progress monitoring. Plan should confirm that the principles of social responsibility are adequately applied in its governance and reflected in plan structure;

Ninth step—financing: select the source of external capital: commercial bank, shareholders, leasing companies, revolving facility, environmental fund, etc., adapt the RECP project business plan to banking institution requirements;

Tenth step—implementation: procure and install all the equipment listed in RECP project business plan; at the same time install extra measuring devices for materials, energy savings and social responsibility documentation. Indicators should be quantitative or qualitative, and able to compare and demonstrate changes over time. Our experience has confirmed that if any financing of RECP is to be provided, a revolving facility is a more appropriate mechanism. Such facility may offer low interest rates, grace periods or other incentives to borrowers. But it is not necessarily easier to get financing from a revolving facility than from a commercial bank. A revolving facility might even ask for additional information or participation of borrower in specific, for instance RECP programs or other support systems. They do not solve the problem of unprepared applicants who are not yet able to formulate a good proposal (Staniškis et al. 2010).

Eleventh step—monitoring: measure progress against goals, assess the degree to which goals are being met; use normalized factors, watch for shifts of wastes to other media, carefully assess toxicity, not just quantity produced, focus on social responsibility performance, including reviews at appropriate intervals, benchmarking and obtaining feedback from stakeholders;

Successfully implemented innovation project: resource efficient and cleaner production is an ongoing effort that will be best maintained with the establishment of awareness program and performed by personnel in the production area. Top management have to demonstrate continuing commitment to RECP program by conducting annual reviews, which should be communicated to all employees and social partners through written announcements and meetings.

The most important element of the resource efficient and cleaner production program is training. The training program should include all levels of personnel within the company. The main goal is to make each employee aware of waste generation, its impact on the site, regional and global environment, social responsibility and ways how negative environmental and social impact of processes, products and wastes can be reduced or prevented. More detailed training should be provided to new employees after they have been on the job for a few weeks. Specialized training sessions on environmental and social policy, procedures, and techniques should be provided to staff when their job scope is in the areas where assessment is under planning.

Effective communication between managers and employees is a critical requirement for maintaining a successful program. Employees ideas should be always welcome and good suggestions should be put into practice and recognized.

3 Methodology for Human Development and Workplace Learning on RECP

Rapid change in the world of work and in society, fast development of information and communication technology, and changes in the organization of work have aroused the value of learning throughout the working career. There is a broad extent and diversity of workplace learning methods. One of the few holistic models is presented by Biggs in his 3-P model of learning (Biggs 1999). This model refers to the three basic components of the learning phenomenon: presage, process and product. There are two presage factors, such as prior knowledge and motivation, and as teaching methods and assessment. The central part in learning is process component which describes how students approach learning, i.e. either deep or surface level. The product part refers to learning outcomes, for instance, knowledge, skills and attitudes. Biggs' model is developed in the context of school learning and cannot directly be applied to workplace learning as such, but the basic model structure (presage, process and product) fits well with the concept of workplace learning. A very similar model for workplace learning was developed by Marsick et al. (2011), where the same three main components are presented, although the "presage" named after "inputs".

Tynjala (2010) have adopted 3-P model for workplace learning. First, the socio-cultural environment in a wide sense, as a defining the possibilities and constraints of workplace learning, was introduced. Another modification is introduction of an additional learner's interpretation of presage factor between the presage and process. And finally, the label "student factors" was replaced by "learner factors" because learners at the workplace are seldom in a student position. Besides that, the factors of learning context, the process and product components in the model of workplace learning are also very different from school learning processes (Tynjala 2013; Tynjala 2010).

The RECP assessment system (Fig. 1) and workplace learning methodology structure (Fig. 2) clearly shows that the process part is the most important and at the same time—most complicated. RECP assessment system is complicated, requiring technical experience, environmental and social responsibility knowledge, a basic understanding of economics and financial engineering. The learning process is inter-active and combines classroom studies, group work, in-company project work, in-company advice and data search in international data bases. The learning is up with interaction between "out—of-company teaching" and group work and separate "in-company project". Participants from universities, consulting companies and governmental agencies take part not only in classroom group work and plenary sessions but in the "in-company project" work as well. The objective of the learning program is to train a group of approximately 30–40 experts in RECP assessment and to create successful examples of the application of RECP on company level. Alter the learning program the trainees should be able to execute similar RECP projects in industry on their own. Thus, the trainees later on will become the key players in disseminating and implementing RECP innovations wider in industry. The group of trainees comprise: 2–3 specialists from approximately ten companies, representatives from universities, industry associations, consultancy, local government, environmental protection departments, NGOs. They have to be selected on personal commitment for RECP as well as educational background to compose an interdisciplinary group of trainees.

4 Conclusions and Considerations

The first version of cleaner production assessment system has been developed by the Norwegian Society of Chartered Engineers (NIF) in co-operation with the Society of Engineers of Poland (Nedenes 1994). During the next fifteen years this methodology was improved and elaborated as a RECP assessment system by many authors (Khan 2008; Environment Canada 2009; UNEP DTIE 2012). The version presented above (Fig. 1) was developed and implemented by experts from the Institute of Environmental Engineering at Kaunas University of Technology (Staniškis et al. 2010). In the period (1993–2016) only in Lithuania more than 150 companies have participated in CP/RECP program and have generated more than 250 preventive innovations with average payback period less than two years. Within increased environmental awareness, strong competition and the need to improve production efficiency, especially with regard to material efficiency, the cost of tracking and tracing material flows throughout the company are by far outweighed by the improvement potentials identified and realised.

In such cases, non-financial support services are expected to be offered by other business service providers (BDS). BDS can facilitate possible partnership with investors but only when there is sufficient data, records and information on the business and the investment proposal. For each RECP investment project, a supervision

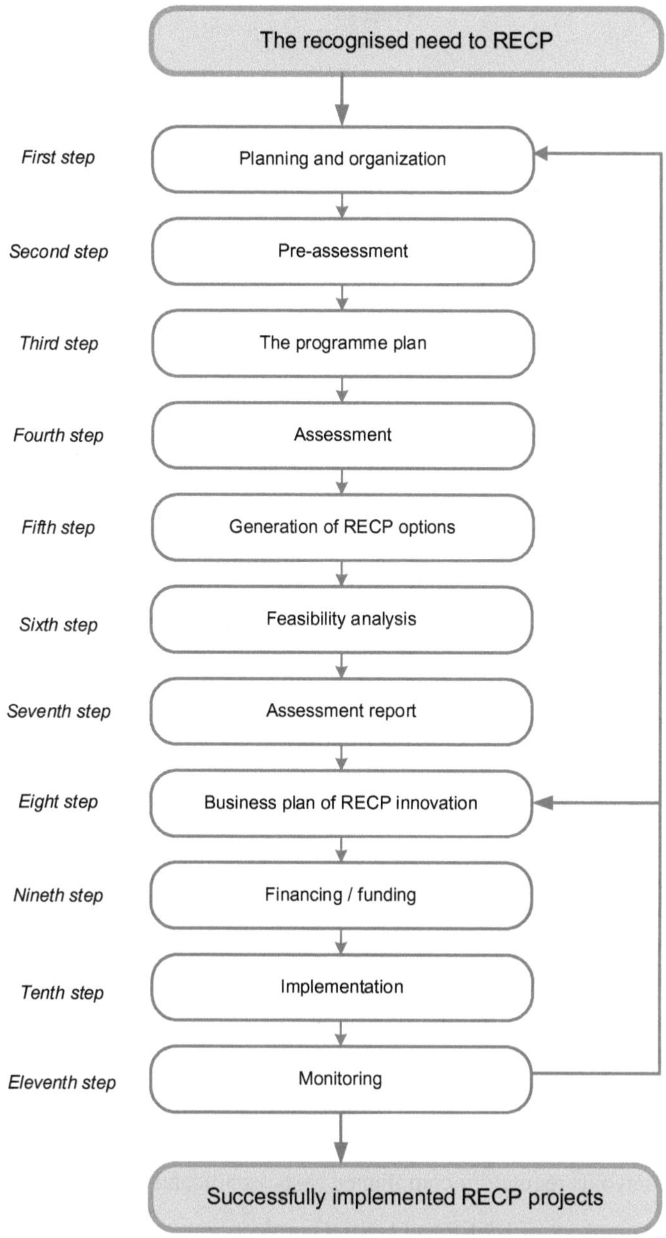

Fig. 1 The system for resource efficient and cleaner production assessment

and monitoring plan is prepared as part of the loan application The company has to provide BDS with a possibility to inspect goods, sites, factories, installations and

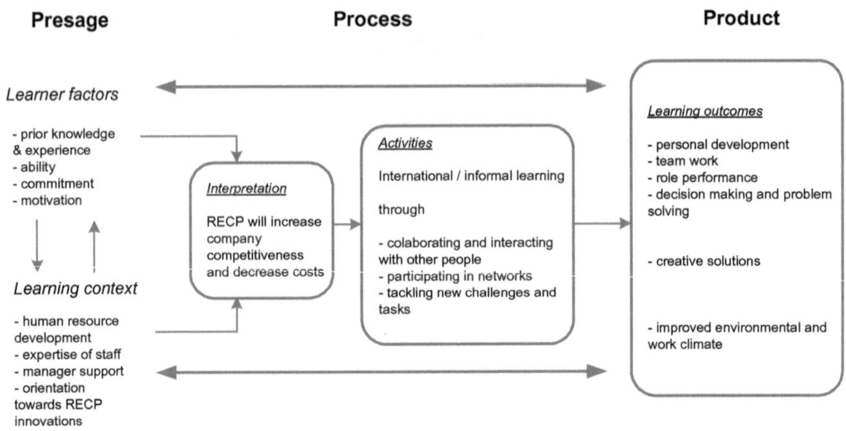

Fig. 2 The structure of human development and workplace learning methodology for RECP

construction sites included in the investment project, any related documentation, and to supervise the implementation of the project. Financial and physical performance indicators are specified in the loan agreements. Environmental performance indicators (EPIs) monitor the company's effectiveness and efficiency of resource management. This applies mainly to physical resources (materials, etc.).

For each project, the realised savings are verified and compared with the expected savings. A standardised reporting format is provided, with a focus on savings in energy use, water use, use of chemicals, etc. The environmental effects of each project are verified. This verification should document the reduction in emissions and wastes and the reduction in inputs (water, energy, chemicals, etc.). The mentioned standardised format ensures the management of all RECP innovation data in a special database which is being continuously update.

More than five hundred RECP experts have been trained by the methodology, presented above (Figs. 2, 3) It could be concluded that implemented innovations:

- Increases profitability, lowers production costs;
- Provides a rapid return on any capital or operating investments required;
- Leads to a more efficient use of energy, natural resources and raw materials;
- Increases staff motivation through reduced worker risks;
- Reduces the risk of environmental accidents;
- Reduces/avoids regulatory compliance costs significantly (Staniškis 2010).

In the framework of UNEP and UNIDO projects, our experience was transferred to several countries in Africa, Central America, South East Asia, China, Central and Eastern Europe, Russia. Some key **conclusions** concerning RECP workplace learning and implementation:

- RECP is not one-time event, it is continuous process, but to ensure the follow-up of such activities at company level after the project ends it is difficult;

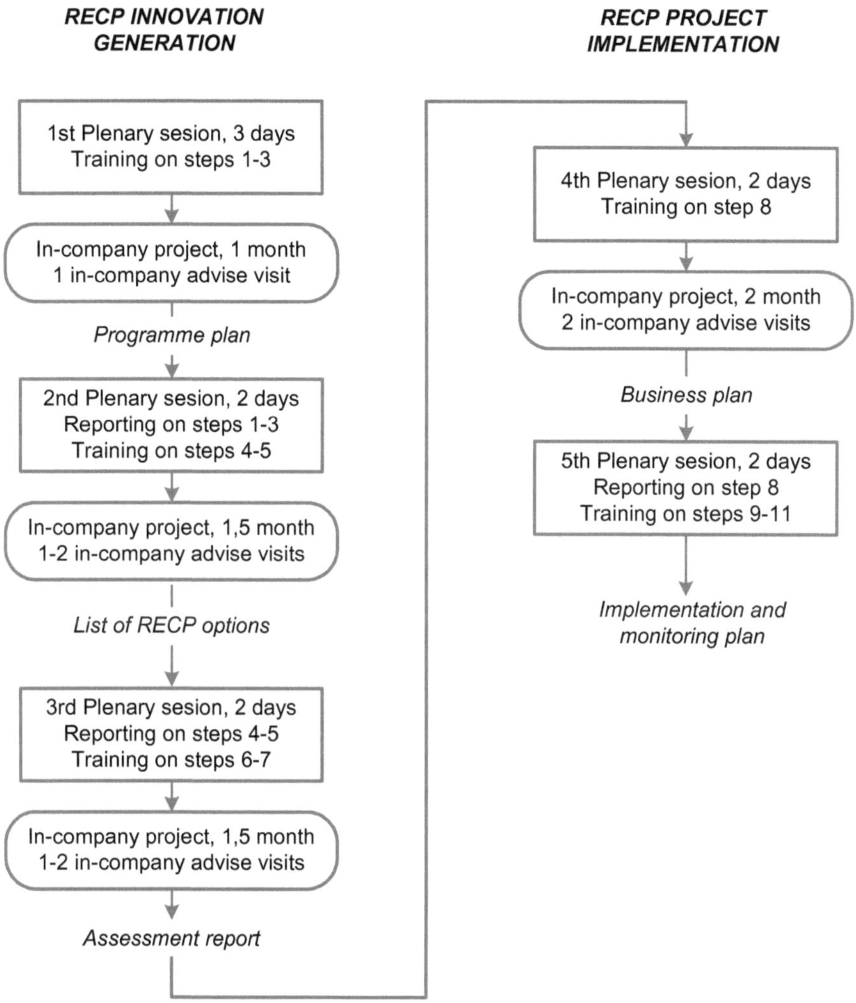

Fig. 3 Process algorithm for human development and workplace training

- Barriers to RECP mostly related to human rather than technical factors. The end-of-pipe approach is well known and accepted by industry and engineers. The existing government policies and regulations often favor end-of-pipe solutions. There is a lack of communication between those in charge of production processes and those manage the wastes that are generated. Managers and workers, who know that the factory is inefficient and wasteful, are not rewarded for suggesting improvements;
- Learning/training program comprising social responsibility aspects are playing a key role in the introduction of RECP strategy in industry. Stakeholder's involve-

ment in RECP learning program is the prerequisite for local adaptation and for the continued success of the program.

- The best number of companies in the program group should be not less than ten companies from different branches. When companies of the same branch—confidentiality issues arise; when companies from different branches—possibilities for experience and ideas transfer to different branch occur;
- Selecting suitable people for a learning program has proved to be an important issue. Environmental specialists are usually very interested in the issues but often possess a peripheral position in the companies. They are often not in the best position to carry out the changes in management and production which are looked for. Broad experience has showed that RECP project leaders in learning program should preferably be recruited from production leaders, not environmental staff members;
- Governments should stimulate relevant non-governmental organizations (industrial and engineering associations, universities) to run modified workplace learning program for production managers, operators, industrial advisors, university teachers, governmental officers.

An overall experience would be that it is often not fruitful to generalize among workplace learning programs or among countries. The situation in the companies/countries is rapidly changes and differs markedly, i.e. local circumstances and considerations and, most significantly, the persons involved will define the outcome. RECP a workplace learning/training program is more likely to be successful when carried in companies closely linked to real issues within and outside the organization. Therefore, further research might explore the impact of standards on social responsibility management and the differences in the influence of standards and management structures at global and local level.

Acknowledgement This research was funded by a grant (No. S-MIP-17-122) from the Research Council of Lithuania.

References

Asif M, Searcy C, Zutsi A, Fisscher OAM (2013) An integrated management systems approach to corporate social responsibility. J Clean Prod 56:7–17

Bansal P, Roth K (2000) Why companies go green: a model of ecological responsiveness. Acad Manag J 43(4):717–736

Baunann-Pauly D, Wickert C, Spence LJ, Schrer (2013) Organizing corporate social responsibility in small and large firms: size matters. J Bus Ethics 115:693–703

Biggs J (1999) Teaching for quality learning at university. What the students does. Society for Research into Higher Education and Open University Press, Buckhingham

Christensen JL, Mackey A, Whetten D (2014) Taking responsibility for corporate social responsibility: the role of leaders in creating, implementing, sustaining, or avoiding socially responsible firm behaviors. Acad Manag Perspect 28(2):164–176

Environment Canada (2009) Pollution prevention handbook. http:/www.ec.gc.ca/planp2-p2plan/default.asp?lang=En8n=56875F44-1. Last Accessed 27 Mar 2018

International organization for standardization (2010) ISO 26000: 2010 (en) Guidance on social responsibility

Khan Z (2008) Cleaner production: an economical option for ISO certification in developing countries. J Clean Prod 16(1):22–27

Kliopova I, Staniškis JK (2006) The evaluation of cleaner production performance in Lithuanian industries. J Clean Prod 14(18):1561–1575

Marsick V, Watkins K, O'Connor BN (2011) Researching workplace learning in the United States. The SAGE handbook of workplace learning, 198–209

Martinez-Conesa I, Soto-Acosta P, Palacios-Manzano M (2017) Corporate social responsibility and its effect on innovation and firm performance: an empirical research in SMEs. J Clean Prod 142:2374–2383

Nedenes OS (1994) The Norwegian industrial transfer of know-how programs on waste minimization/cleaner production to Central and Eastern European countries. In: Material of invitational expert seminar. Introducing cleaner production in Eastern Europe. Lund University, Sweden, pp 13–19

Silva DAL, Deai I, de Castro MAS, Ometto AR (2013) Quality tools applied to cleaner production programs: a first approach toward a new methodology. J Clean Prod 47:174–187

Staniskis JK, Katiliute E (2017) Systems approach to resource efficient and cleaner production solutions: method and implementation. In: sustainability through innovation in product life cycle design. Springer, Singapore, pp 385–398

Staniškis JK, Stasiškienė Ž, Kliopova I, Varžinskas V (2010) Sustainable innovations in Lithuanian industry: development and implementation. Kaunas, Technologija, p 460

Tynjala P (2010) Workplace learning in transformation. Keynote presentation at the EARLI Learning and professional development SIG conference. Germany, Munich, pp 25–27

Tynjala P (2013) Toward a 3-P model of workplace learning: a literature review. Vocations and Learning. Springer. pp 11–36

UNEP DTIE (United Nations Environmental Program, Division of Technology, Industry, and Economics (2012) Understanding Resource Efficient and Cleaner Production. http://www.uneptie.org/PC/cp. Last Accessed 15 Mar 2018

Wickert C, Scherer AG, Spence LJ (2016) Walking and talking corporate social responsibility: implications of firm size and organizational cost. J Manage Stud 53(7):1169–1196

Jurgis Kazimieras Staniškis is research manager of the Institute of Environmental Engineering, Kaunas University of Technology, full member of Lithuanian Academy of Sciences, member of the United Nations Independent Group of Scientists for Global Sustainable Development Reporting. Active for over two decades on sustainable development strategies and their application in industry, universities and governmental institutions in Lithuania, Central and Eastern Europe, Africa, Latin America and Asia. Author of 12 books, more than 250 publications and 140 inventions.

Eglė Staniškienė is an investigator in Sustainable Management research group, School of Economics and Business, Kaunas University of Technology. Active on sustainable university, sustainable development, quality management systems, higher education quality research and interdisciplinary research. Author of more than 40 publications.

International organization for standardization (2010) ISO 26000: 2010 (en) Guidance on social responsibility
[illegible] (2008) Cleaner production as a [illegible] option for ISO certification in developing countries. J Clean Prod [illegible]
Kliopova I, Staniškis JK (2006) The evaluation of cleaner production performance in Lithuanian industries. J Clean Prod 14(18):[illegible]

Managing Incomplementarity: Implementing Social Responsibility in Companies

Simon Dabrowski and Florian Lottermoser

Abstract Publicly listed companies nowadays face contradictory expectations from their institutionalised environments concerning social responsibility and sustainability (SRS) on the one hand, and financial performance on the other. From the perspective of the financial logic, the economic performance of social responsibility is uncertain, whereas sustainability views the financial logic as obstructive to achieving long-term regenerative capacity of economic and non-economic resources. Thus, enacting both institutional logics in decision-making and strategy can create goal-conflicts or incomplementarity within organisations. Incomplementarity is an obstacle to the implementation of SRS. Therefore, incomplementarity needs to be resolved and managed in order to ensure organisational reproduction. This paper presents an empirical analysis of the organisational social responsibility strategies dealing with the incomplementary institutional logics in German publicly listed companies. The results are based on 14 qualitative management interviews in three German publicly listed companies (DAX30) and emerge from a current research project granted by the German Research Foundation (DFG). Organisational strategies as patterns of managing incomplementarity internally are sketched which can be shown to be interwoven with institutional intentions as part of institutional work. Finally, an integrated framework is proposed for analysing the management of incomplementarity as an interplay between social responsibility strategies and institutional intentions that can serve to assess the different outcomes of social responsibility and sustainability.

Keywords Institutional incomplementarity · Social responsibility
Financial logic · Sociology

S. Dabrowski
Institute of Sociology, Martin-Luther-University Halle-Wittenberg, Emil-Abderhalden-Straße 26-27, 06108 Halle (Saale), Germany
e-mail: simon.dabrowski@soziologie.uni-halle.de

F. Lottermoser (✉)
Faculty of Business, Economics and Social Sciences, University of Hamburg, Sedanstraße 19, 20146 Hamburg, Germany
e-mail: florian.lottermoser@wiso.uni-hamburg.de

W. Leal Filho (ed.), *Social Responsibility and Sustainability*, World Sustainability Series, https://doi.org/10.1007/978-3-030-03562-4_11

1 Introduction: Institutional Incomplementarity of Social Responsibility

In their decision-making and associated organisational structures, how do listed companies deal with two seemingly incomplementary institutional logics: the principle of social responsibility and sustainability (or SRS for short) on the one hand and the financial logic of the shareholder value principle on the other hand?

The literature discusses the possibility of institutional incomplementarity when companies face the decision whether and how to implement SRS. There is a potential clash between the logic of SRS and the logic of shareholder value of the financial markets, i.e. these two logics exhibit institutional incomplementarity (Beckert 2010; Beyer et al. 2007; Hiß 2014; Senge 2018). On the one hand, SRS require the company to adopt social, ecological and financial institutions, i.e. the expectations of its stakeholders in society. On the other hand, social and ecological commitment requires the company to make investments whose financial contribution is often difficult to measure so that the economic rationality of the commitment remains unclear (Curbach 2009, p. 160f; Perrini et al. 2010, p. 191; Senge 2007, p. 54). It is particularly against the background of increasing management orientation towards the financial markets (Beyer 2006; Münch 2008, p. 163) that a company's SRS activities prove to be a phenomenon that appears to stand in contradiction to the traditional business objectives, which gives rise to institutional incomplementarity: Generally speaking, organisations face a multitude of institutional expectations, which may be aligned or run counter to each other (Friedland and Alford 1991; Kraatz and Block 2008; Selznick 1949). Institutional diversity is associated with conflicting goals, fragmentation, incoherence and organisational instability (Kraatz and Block 2008, p. 243). SRS activities can create a clash between ecological, social and financial institutions that follow diverging logics and confront the company with opposing expectations which nevertheless need to be satisfied in order to ensure legitimacy and other resources required for the company's reproduction in their social environments (Meyer and Rowan 1977; Kraatz and Block 2008, p. 247ff). At first glance, the SRS logic and the financial logic do not appear to complement each other. In this context, 'institutional complementarity' refers to a situation where at least two institutions mutually enhance each other. Such a situation may arise either if the institutions are similar and thus reinforce each other or if they are different in such a way that they can compensate for each other's deficiencies (Crouch 2005, p. 359f; Thornton and Ocasio 1999, 2008). Conversely, institutional incomplementary exists if at least two institutions reduce the effectiveness of each other.

In order to implement SRS, companies need to resolve and manage incomplementarity. It is assumed that companies resolve any institutional incomplementarity by means of the strategy of financialising SRS. Financialisation here refers to "the increasing role of financial motives, financial markets, financial actors and financial institutions in the operation of the domestic and international economies" (Epstein 2005, p. 3). Research has shown how in in the Anglo-American context SRS often becomes co-opted with the predominant economic and financial logics (Banerjee

2008; Fleming and Jones 2013). Other work on corporate disclosure (Milne and Gray 2013) and environmental or sustainability accounting (Hiß 2013; Lohmann 2009; Malsch 2013) has emphasized how SRS is aligned with corporate goals of profitability (and thus with the financial logic), producing selectivity concerning the SRS issues and problems tackled and attended to (Senge 2015). Both in concrete organizational decision processes and in relation to the environment there is a tendency towards greater opacity and/or simplified information, rather than the proclaimed organizational accountability, transparency and insight (Christensen and Cornelissen 2015; Garsten and de Montoya 2008). The argument has been made that the practices of loose coupling or decoupling (Meyer and Rowan 1977; Orton and Weick 1990) constitute typical organisational responses to multiple and potentially conflicting institutional logics: In the practical implementation of SRS, a decoupling between the activity structure and the formal structure is conceivable to the effect that social and ecological activities may be publicly communicated in order to conform with the institutional expectations, yet in actual fact, for the sake of efficiency, these activities are performed only to a limited if any extent (Hasse 2006; p. 152f; Hiß 2006, p. 165ff). As a consequence, numerous companies' SRS efforts have mainly been perceived as superficial and symbolic formal structures or façades while core activities continue in a "business as usual" mode (Boiral 2007; Banerjee 2008; Fleming and Jones 2013; Cho et al. 2015). Loose coupling or decoupling may thus reinforce the impression of SRS activities as "fake" measures with potentially detrimental effects on the legitimacy and trust ascribed by relevant stakeholders.

This paper clarifies that it is not the strategy of financialisation alone which contributes to the resolution of institutional incomplementarity and thus the implementation of SRS activities. Instead, incomplementarity is overcome only by a combination of financialisation with a strategy of cooperative allocation and institutionally integrative intentions among the company's management. From the institutional perspective of institutional logics and institutional work, this argument is supported by an empirical comparison of three of the largest listed companies in Germany. In 14 management interviews covering SRS projects, we identify the strategies that these companies have found to resolve institutional incomplementarity: on the one hand, the organisational strategies, and on the other hand, the institutional intentions of the managers involved. The patterns that emerge from the organisational strategies, the institutional management intentions and the standing of the three companies in sustainability rankings suggest that cooperative allocation strategies and institutionally integrative management intentions can foster the resolution of institutional incomplementarity in the implementation of SRS.

In the following, we will initially provide an overview of the methods of data collection and analysis. We then present the identified organisational strategies and institutional management intentions. Based on that, the strategical patterns we found in the three companies are described, from which we can ultimately derive conclusions regarding the relevance of strategies and intentions for the resolution of institutional incomplementarity.

Adidas	Deutsche Börse	Lufthansa
Allianz	Deutsche Post	Merck
BASF	Deutsche Telekom	Munich RE
BMW	E.ON	ProSiebenSat.1
Bayer	Fresenius Medical Care	RWE
Beiersdorf	Fresenius	SAP
Commerzbank	HeidelbergCement	Siemens
Continental	Henkel	Thyssenkrupp
Daimler	Infineon	Volkswagen
Deutsche Bank	Linde	Vonovia

Fig. 1 DAX 30 index composition (as of April 15, 2016). *Source* www.dax-indices.com

2 Data and Methodology

The empirical analysis is based on a qualitative comparison of the strategies employed by three of the largest German listed companies (DAX30) to resolve institutional incomplementarity. Figure 1 shows an overview of the DAX30 companies (as of April 15, 2016). The data basis comprises 14 management interviews covering corporate projects on SRS. The interviews were conducted between July 2015 and August 2016. The companies and the participants were assured anonymity.

The analysis serves to identify different strategies used to resolve incomplementarity in the companies. Therefore, the focus of the sample selection was on the distinctiveness of the companies with respect to the solution of incomplementarity. All companies in the DAX 30 state that they are committed to SRS topics. Since the analysis does not claim quantitative representativeness and research access to the DAX30 companies is limited, the analysis focuses on three companies that perform differently in sustainability performance rankings. Based on the assumption that strategic solutions of institutional incomplementarity favor SRS implementation, a better ranking should at least partially speak for the overcoming of institutional incomplementarity in the respective company. The three selected companies clearly differ in their position in sustainability performance rankings. Company 1 is ranked in the middle. Company 2 is highly ranked. Company 3 could not qualify for rankings. The ranking positions refer to two sustainability performance rankings in 2015 and 2016. Due to the anonymity assurance to the companies, the rankings at this point cannot be named.

Figure 2 provides an overview of the projects mentioned and discussed in the interviews that took place in the three companies. All managers interviewed were involved in company decisions concerning SRS, specifically the implementation or further development of such projects. The managers were affiliated with the following company divisions: social responsibility and sustainability, investor relations, (management) accounting, human resources, business strategy, the environmental office, compliance, and public relations/communications.

The transcribed interviews regarding the corporate projects on SRS were analysed by a mix of methods. Organisational strategies were identified in a two-fold

Company 1	Company 2	Company 3
development of a social indicator	electromobility in the corporate fleet	cooperation project with environmental organisations
implementation of the human rights index	internal offer of ecological mobility	materiality analysis
ecological product range	monetary consolidation of social, ecological and financial indicators	further development of the investor relations office for socially responsible investors
resource efficiency, reduction of resource use	further development of an integrated sustainability report	further development of integrated and separate sustainability reports
internal waste reduction	further development of non-financial sustainability indicators	negotiation of the budget for social responsibility, establishment of a budget for sustainability
sustainability analysis of the monetary influence of social / ecological activities	further development of the sustainability department	
establishment of a sustainability report and reporting structures		
establishment of organisational sustainability structures, sustainability committee, sustainability department		

Fig. 2 Projects in the three companies that were analysed

process. First, all data were analyzed by coding text passages topically. Then they were interpreted according to the principles of hermeneutical qualitative interpretation (Reichertz 2016) in order to reconstruct the orientations and meanings elaborated by participants in the process of talking about their practices. The interpretation process was conducted both inductively as well deductively by referring to the analytical framework sketched below. Interviewees' statements were assumed to represent instantiations of the logics of SRS and/or the financial market which were addressed and related to each other by the participants in terms of concrete projects, positions, communication channels, goals, reporting or decision processes; participants were thus expected to handle both logics, the logic of SRS and the financial market logic, by referring to, and actualizing, such organisational structures in the communicative process of the interview. Through case comparison, typical patterns of dealing with these logics were identified and condensed.

To assess the management intentions, the interviews were condensed and categorised in a three-stage structured analysis of content (Mayring 2014). The material was first coded and given a preliminary structure using the MAXQDA software, then it was structured more precisely in a separate analysis for each project and transferred to the final categories. The categories were defined by a combination of deductive and inductive means. The intentions were worked out deductively and were then, using the constellations of them that underlay the individual projects, inductively consolidated into three categories, which we refer to as institutionally integrative intentions, institutionally selective intentions, and institutionally indifferent intentions, respectively.

3 Organisational Strategies

The organisational strategies refer to the 'how', i.e. the different ways of managing institutional incomplementarity within organisations. In order to make this visible, we need to posit some basic assumptions concerning the internal operations of organisations. In line with traditional organisational theory, sociological systems theory (Luhmann 2011; Seidl and Becker 2006) views organisations as networks of interlinked decisions. Decisions intendedly or non-intendedly condensate particular organisational structures such as vertical and horizontal communication channels, positions, procedures or formal or informal rules in general, all of which can serve as premises informing further decisions that again impinge upon structures etc. Decisions again selectively draw upon heterogeneous societal higher-order principles—institutional logics (Friedland and Alford 1991; Thornton and Ocasio 2008)—that shape organizational decisions and actions cognitively, evaluatively and normatively; they also serve as (legitimatory) accounts towards internal and external observers. Corporations usually draw upon diverse institutional, especially economic logics (Kraatz and Block 2008) such as the financial one but also more ambiguous principles such as the SRS logic. It can be assumed that the necessity to implement both logics creates decision problems as both represent incomplementary expectations that organisations are expected to satisfy in one way or another; yet decisions must be made to ensure organisational continuity—and this requires viable solutions by means of decisions. We term these solutions strategies.

A *decision problem* that could be reconstructed in most of our interviews was the very basic problem of creating connectivity or even attention for SRS-related decisions and topics in other relevant departments and levels such as the board. Most of the time, this problem implies a legitimacy deficit of SRS primarily inside but also, regarding the board's task to present and to 'sell' strategies, outside the firm. This problem could be interpreted as an empirical manifestation of the theoretical problem of incomplementarity: with its introduction or rising importance, the SRS logic challenges the worldviews and routines shaped by established economic, especially financial logics; one immediate reaction to this may be organizational members' indifference.

The most prominent solution to this problem was to frame SRS topics such as sustainable products or also employee satisfaction in a terminology, concepts and devices referring to *financial markets* and their discourses; this strategy was actively pursued especially by SRS managers. SRS issues were translated (Sahlin and Wedlin 2008) into particular, yet standardized, publicly justifiable formats (Thévenot 2007) in order to increase the chance for further decisions: Interviewees described how they created comparability and commensurability (Espeland and Stevens 1998) between

SRS-related and economic, especially financial values by relating them to financial metrics and key indicators such as gross margin or EBIT. Similarly, decision pressure for SRS-related activities at board level could be built up by provoking the 'risks' of urgent and inevitable disclosure requirements of financial rankings or indices deemed relevant such as the Dow Jones Sustainability Index, indirectly reflecting investors' expectations. Although not always directly referring to financial markets, most of the participants routinely used their language, reflecting hegemonic concepts of corporate governance: shareholders' scrutiny did not act by its actuality but by its potentiality. By linking SRS to financial concepts and ideas, decision situations were framed by financial valuation devices such as performance indicators and accounting techniques or external rankings. Therefore, one could speak of a *financialisation* of SRS (Chiapello 2015; Hiß 2013). SRS activities were more likely to be selected and pursued when they are financialized. Through this strategy, attention and decidability could be created for SRS issues, which, howeverbecame redefined and reduced in scope and complexity at the same time.

Another strategy was to *allocate* CR-related decisions inside the organisation to other departments or positions. Here we can distinguish two sub-strategies: Decision problems were either delegated or other organisational actors were involved cooperatively. *Delegation* means that competences for decision were shifted horizontally or vertically. Some managers for example ascribed the relevant authority for the solution to other departments, thus delimiting their own domain and responsibilities towards others. In another company, environmental initiatives taken by local organisational members at the shop-floor were retrospectively entangled with the companies' SRS strategy while keeping this project at the local level. Generally, in a talk mode, organisational members at all levels were, addressed as socially and environmentally responsible subjects, contributing to the company's economic and SRS-related development, and being able to balance such expectations. *Cooperation* refers to the practice of involving other departments or actors into developing a solution to a decision problem, which was often done in SRS councils or committees. Here, different expectations and rationalities represented by selected departments and boundary-spanners were drawn upon and arranged in a heterarchical way (e.g., Stark 2011) which often facilitated compromises. Cooperation and delegation as instances of the allocation strategy have in common that by distributing competences for decisions heterogeneous and incomplementary expectations on different levels could be monitored and accounted for; consequently, solutions became more complex, ambiguous and time-consuming.

Some organisations also sought to *influence and manipulate their respective environments* in order to achieve more favourable conditions for their operations. One company for example attempted to create reporting standards and indicators that could fit its own requirements. Another company, through its cooperation with stakeholders such as NGOs, tried to align their interests with those of the company. Furthermore, not only this, but also other companies selected more or less actively which stakeholders would be included into cooperation and thus be considered relevant, determining which external actors would be classified as 'stakeholders'.

The *loose coupling* between certain organisational elements, such as departments or activities/decisions (Orton and Weick 1990) is the fourth strategy we could identify. Different departments and SRS-related positions in particular were kept separate and only indirectly or occasionally linked—for example, in one company, in 'obvious cases' being considered SRS-related while the SRS management merely played on informal relations. In some companies, the links of SRS-related decisions and programs at corporate level and subsidiary or local level were weak or missing, making translations between these activities difficult. Moreover, SRS projects that had been initiated, often did not relate to the company's core activity at all, yet were formally signified as part of the SRS strategy, sometimes addressed in the interviews as being part of 'charity'. Loose coupling in many companies allowed strategic decisions in a 'business as usual'-way while marginalizing SRS or assigning it a largely ceremonial role (Meyer and Rowan 1977).

The organisational strategies presented here enable organisations to reproduce themselves in a pluralistic environment also shaped by the financial and the SRS logic. The solutions to the decision problems arising through incomplementarity often lead to selectivity and short-term reactivity as to which SRS issues were attended to, prioritizing a slightly modified 'business as usual'—yet at the same time opening corporate decision-making to SRS issues.

4 Management Intentions

We analyse the actor level of the institutional management intentions from the perspective of the sociological concept of institutional work (Lawrence and Suddaby 2006), which emerged as an actor-centred framework within sociological neo-institutionalism, building upon earlier neo-institutionalist work (DiMaggio 1988; Oliver 1991, 1992). Institutional work provides a suitable theoretical basis for our analysis because it focuses on the institutional intentions of actors within organisational processes (Lawrence et al. 2009, p. 7ff): intentions of organisational actors that are geared towards "creating, maintaining and disrupting institutions" (Lawrence and Suddaby 2006, p. 215). Institutional work refers to such purposeful, institution-oriented actions (ibid.). The premise that is central to the analysis of institutions holds that institutional effects in organisational processes are the consequence of institutional work on the part of individual and collective actors (Lawrence and Suddaby 2006, p. 219; Senge and Dombrowski 2015, p. 400). As an institutional concept, institutional work is in accordance with the core assumption of embedded agency (Holm 1995; Seo and Creed 2002; Greenwood and Suddaby 2006) that actors who influence institutions in turn use institutions for orientation and are limited by them (Lawrence and Suddaby 2006, p. 220; Lawrence et al. 2011, p. 52; Zietsma and Lawrence 2010, p. 189).

For our analysis of SRS, we are specifically interested in the institutional intentions of the management. In contrast to institutionalised organisational routines that actors perceive and follow according to objective facts (Zucker 1977), management decisions on SRS in particular in the case of institutional incomplementarity require dedicated solutions that result from the actors' reflective processes (Greenwood et al. 2015, p. 328; Jennings and Zandbergen 1995; Seo and Creed 2002), which means that the decisions can be attributed to the institutional intentions of the managers. The intentions correspond to the 'creating' and 'maintaining' of social, ecological and financial institutions that are implemented in corporate projects, strategies and goals. We may conceive of various constellations of intentions: The intentions behind a sustainability decision may be directed either at a single institution or at several institutions simultaneously. Furthermore, the intention need not primarily aim at the institution that is directly affected by the decision. This variant may occur if institutional intentions have unintended consequences, for example in that the decision has an impact on institutions other than the one it is addressed to (Lawrence et al. 2009, p. 11). Moreover, besides institutional intentions, institutional decisions may also be driven by individual intentions that follow an actor's personal interest (Senge 2011, p. 160), so that institutional effects may result from individual interests—in which case one would not speak of institutional work (Lawrence et al. 2009, p. 13).

The contentual analysis of the interview transcripts allowed us to identify three intentional constellations among the management: institutionally integrative intentions, institutionally selective intentions, and institutionally indifferent intentions. If a SRS decision is driven by an *institutionally integrative intention*, the decision aims for the integration of social or ecological institutions with financial institutions. Such integrative approaches may either seek to combine the institutions in a common goal or project, or they may promote the corporate institutions independently of each other. An *institutionally selective intention* behind a SRS decision in turn strives for one-sided support for the social, ecological or financial institution in question. This is the case for example if a social sustainability decision is based on either a social or a financial intention, whose integration is however not sought. Finally, an *institutionally indifferent intention* is directed towards organisational or solitary managerial objectives that do not concern social, ecological and financial institutions. Evidently, the institutional intentions behind social and ecological management decisions may, but need not, coincide with the potential SRS effects of the decisions, seeing that the intentions may deviate from the institutional effects.

Institutionally integrative intentions and institutionally selective intentions thus comprise social, ecological and financial management aims, which may be pursued either jointly ('integratively') or individually ('selectively'). At a more detailed level, the three types of intentions concern the following sets of issues:

Social intentions refer to the improvement of working conditions, social benefits and incentives for employees, health and safety conditions, employee satisfaction and involvement, as well as the opportunity to support charities (e.g. in the form of in-kind donations) and more generally to make a contribution to society. The category furthermore comprises the social translation also of ecological strategies and measures, as well as the general impetus that such strategies and measures provide for future social decisions and activities to be undertaken by the company.

Ecological intentions aim for the improvement and development of ecological product features, the reduction of the company's use of water, energy, and other resources, compliance with environmental standards, carbon emission reduction and compensation, other contributions to climate protection, waste reduction. Additional measures include the offer for employees to enjoy ecological services such as public transport subsidies, and the gathering and exchange of ecological know-how with external stakeholders such as environmental experts. Another ecological intention consists in the general incentive function of ecological strategies and measures for future ecological decisions and activities by the company, which includes the possibility to develop and establish new instruments to gauge the company's ecological performance. As with social intentions, this also includes the monetarisation of ecological activities in order to generate management interest in ecological issues so that they can achieve more prominence within the company.

Financial intentions finally refer to the monetary influence of social and ecological strategies and measures generally on the improvement of indicators such as profitability, productivity, efficiency, sales, product quality, customer loyalty and employer branding to the extent that it entails benefits on the labour market. Financial intentions also concern the fundamental ability to recognise and calculate any such monetary influence because that will help to legitimise future measures in terms of a financial return. Further financial intentions include innovative impulses for the core business, inhouse R&D, in particular the development of new products and services, the acquisition of additional funding on the financial markets and thus a reduction in the cost of capital, other cost savings, increased sales through greater differentiation from the competition, improved reputation, the incorporation of employee loyalty in the calculation of a business case and more generally the use of strategies and measures for SRS as a basis for business decisions.

Institutionally indifferent intentions are geared towards either organisational or solitary (individual) interests, and they are independent of social, ecological or financial institutions. From a more detailed perspective, organisationally oriented intentions aim to initiate and to improve organisational processes and structures, by which we mean the fundamental procedural and structural conditions, regardless of whether they support sustainability-oriented activities. Thus, the focus is on the organisational goal, which could for example be to improve the organisational processes and structures, the mechanisms of control, coordination and monitoring, the internal and external instruments of communication, and thereby also to improve the organisational options for decision, strategy and action, as well as to reduce business risk. This means that to establish a SRS department may also be an organisationally oriented intention if it aims at the company structure, e.g. to improve the coordination of all SRS-related issues within the company. Likewise, this category may also comprise SRS indicators if they are primarily intended as a tool of external communication or to better monitor the company's supply chain. We may also speak of an organisationally oriented intention if manager want to externally position their company as a role model among the competition or if a team chooses a SRS-related project for the primary reason that such a project is most likely to achieve success within the organisation. In both cases, i.e. the external role model and the internal success of a project, a SRS project is contentually realised, yet the intention focuses not on the social or ecological activity but rather on the external or internal success.

A further organisational intention consists in the formal fulfilment of externally set requirements, such as ensuring the company's compliance with the applicable laws, specific quotas in areas related to SRS, and generally meeting the expectations of external stakeholders. The integration of stimuli from external stakeholders into the organisational strategies, processes and structures in turn serves to improve these very same organisational strategies, processes and structures.

Solitarily oriented intentions, by contrast, mostly aim at the characteristics of a person's own function or position within the company. Among them we may count the meaningfulness and effectiveness of someone's job, knowing that what they do carries a positive image, the achievement of personal freedom of decision, receiving recognition, being able to apply one's skills, reducing one's risk of errors on the job. These intentions also include the fundamental individual performance of one's duties and the opportunity to make a valuable contribution to society in the course of one's job, to do good, but also to meet external expectations that the person's private environment may have regarding the job, for example the expectation to be doing sustainability-oriented work, which an ecologically minded spouse may express at least implicitly.

5 Strategical Patterns

Comparing the three organisations according to institutional management intentions on the one hand and organisational strategies on the other in organisational decisions made it possible to identify certain patterns, i.e. distinct constellations of institutional intentions and organisational strategies across organisational decisions. Furthermore, as an additional criterion of comparison, the positions of these companies in sustainability performance rankings were used as a reference. We regard rankings as instantiations of external legitimacy criteria within an organisational field, i.e. as devices that through their employment of specific categories and evaluative and ordering mechanisms reflect institutionally anchored ideas and criteria of appropriateness which in turn serve as cognitive and normative orientations for organisations. Thereby, rankings potentially contribute to the diffusion and institutionalisation of ideas and models within a field (Espeland and Sauder 2007; Heintz 2010). It can further be assumed that these rankings also guide the decisions of organisationally relevant external stakeholders such as financial investors or also NGOs.

According to their position in external sustainability performance rankings, the three companies have been assigned very different performance categories. While company 1 ranked in the middle and in one ranking didn't qualify, company 2 was one of the leading companies in the rankings included; company 3 however didn't qualify in any of the rankings and had even been explicitly excluded from the ranking.

Across all three companies and all decisions, the financialisation strategy occurred most often as a solution to a decision problem caused by incomplementarity. Among the strategies, it was the allocation strategies and, by tendency, the loose coupling strategy which were the most discriminatory. In company 1 and—less dominantly—

company 3 decisions were often delegated (as part of the allocation strategy) according to (mostly formally defined) vertical (hierarchical) or horizontal communication channels. Hence, interviewees reported that in decisions relating to SRS, operational competencies and responsibilities—and thus agency—tended to be shifted to other positions once formally defined competences and 'zones of indifference' (Barnard 1974) of the positions concerned had been crossed by the respective issue. In company 2 however, allocation was practiced more through involving different expectations and rationalities of positions, hence in a cooperative and heterarchical mode. Here, finding solutions to decision problems was described by the managers as drawing upon the differences of orientations and expectations within the organisations mostly in formally established intermediary councils (such as a sustainability council), but also in less formal organisational everyday interactions. The intermediary councils consisted of selected representatives of departments deemed relevant—such as controlling, sustainability or investor relations—activating and arranging the respective departmental and organisational priorities and goals that the aspired solution had to account for.

The differences in allocation strategies could further be supported by looking at loose coupling which mostly occurred in company 1, but also in company 3 where activities and departments often were not relating to each other or were marginalized from organisational core activities and decisions—whereas in company 2 technical-economic core activities and SRS activities were intrinsically linked.

These patterns in strategies coincided with certain constellations of institutional management intentions pursued in decision situations. In company 2, most decisions characterized by financialization and cooperative strategies were predominantly grounded in an institutionally integrative intention, aiming to merge the SRS and financial logics. It was attempted to integrate SRS topics into the operative core activities and the strategy as a whole mostly referring to projected cost reductions, product innovations or win-win-constellations that could be generated by integrating both logics. In comparison with the other two companies, company 2 performed the best in influential sustainability rankings. In company 1, a more mixed constellation of institutionally integrative and also selective intentions was predominant—combined with delegative and financializing strategies. As already mentioned, corporate performance rankings assigned the company a middle position. In company 3, the financializing and delegative strategies in decisions could be attributed mostly to an institutionally indifferent intention referring to intentions related to other institutions than the financial or SRS institutions—here especially to other organisationally relevant objectives.

To conclude, the comparison of three organisations could reveal that different patterns or constellations of organisational strategies and management intentions make a difference for organisational solutions of institutional incomplementarity. The latter in turn also seem to be related to different positions in SRS rankings. A combination of financializing and cooperative strategies on the basis of institutionally integrative management intentions goes along with a higher position in corporate rankings. The constellation of financializing and delegative strategies with selective or even indifferent intentions however was related to less favourable ranking positions.

6 Conclusion and Integrated Framework

For the implementation of SRS, institutional incomplementarity is an obstacle that needs to be resolved and managed. The empirical analysis of the organisational strategies and the institutional management intentions in three of the largest listed companies in Germany has shown how these companies deal with institutional incomplementarity in cases where there is a clash between, on the one hand, the principle of SRS and, on the other hand, the principle of shareholder value. It is typically assumed that incomplementary decision situations at the intersection of SRS and shareholder value are resolved through the financialisation of the former area, where financialisation means "the increasing role of financial motives, financial markets, financial actors and financial institutions in the operation of the domestic and international economies" (Epstein 2005, p. 3). By contrast, our analysis has shown that the resolution of institutional incomplementarity and thus the implementation of SRS activities

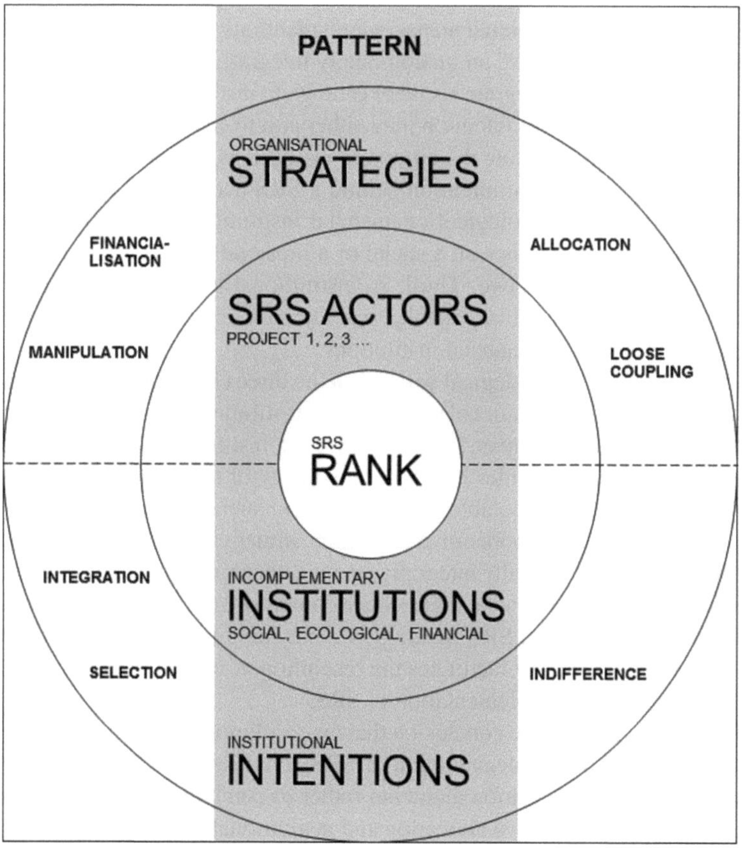

Fig. 3 An integrated analytical framework for institutional incomplementarity

is not achieved by financialisation alone. Instead, only a combination of financialisation with a strategy of cooperative allocation and institutionally integrative intentions by the company's management can serve to overcome incomplementarity. Our findings can be summarised as follows:

(1) On the organisational level, we could identify four strategies. As expected, financialisation plays a central role among them: Companies translate SRS into financial concepts and thus create decision situations that are framed in terms of financial indicators—the financialisation of SRS (Chiapello 2015; Hiß 2013). A second strategy was to allocate decisions related to SRS to other departments or positions within the organisation. Two sub-strategies can be distinguished: Decisions were either delegated (the decision competence was shifted horizontally or vertically), or cooperation with other organisational actors was sought to find a joint solution to the decision problem. Additional strategies were the deliberate manipulation of the organisation's environment and loose coupling between certain organisational elements such as departments or activities (Orton and Weick 1990).
(2) Our analysis of institutional management intentions yielded three constellations of such intentions: First, an institutionally integrative intention refers to a SRS decision aiming to integrate social or ecological institutions with financial institutions. An integrative intention may either aim to combine the institutions in a common goal or promote the corporate institutions independently. Second, an institutionally selective intention behind a SRS decision seeks unilateral support for the social, ecological or financial institution in question, e.g. if such a decision is based either on a social or a financial intention but no attempt is made to integrate the two. Third, an institutionally indifferent intention aims at organisational or solitary managerial objectives that have nothing to do with social, ecological or financial institutions.
(3) The comparison of strategical patterns in the three companies suggests that certain patterns of organisational strategies and institutional management intentions may facilitate the company's overcoming the institutional incomplementarity. While all three companies featured the strategy of financialisation, company 2 is distinguished by a notable pattern: Here, financialisation tended to occur in combination with a cooperative allocation strategy, which was predominantly based on an institutionally integrative management intention that sought to combine SRS with the financial logic. Since company 2 outperformed the other two companies in external SRS rankings, we may assume that the strategic pattern found in that company facilitates the resolution of incomplementarity and thus the organisational implementation of SRS.
(4) This result suggests the conclusion that the resolution of institutional incomplementarity and the implementation of SRS activities are not attributable to the strategy of financialisation alone but rather to combining financialisation with a strategy of cooperative allocation and institutionally integrative management intentions.

Based on these findings, we are now able to propose an integrated framework for analysing the management of incomplementarity as an interplay between organisational strategies and institutional intentions. The actor-based framework can help to assess the different outcomes of SRS. To see how a company deals with institutional incomplementarity, consider the three levels of analysis depicted in Fig. 3: On the first level, the company's performance in terms of SRS is assessed, for example by comparing its position in published rankings to the competition (inner ring). On the second level of analysis, the institutional incomplementarity to be analysed can be defined on the basis of the actors involved in actual SRS projects, for each of which the institutionally incomplementary aims concerned are stated (second ring). Finally, on the third analytical level, for each individual actor the organisational strategies and institutional intentions that are being applied to the SRS projects are worked out (outer ring). All three levels of analysis are considered jointly in order to identify a potential pattern. Any such pattern can be used to substantiate whether and to what extent the way in which the company internally deals with institutional incomplementarity can enhance its SRS performance.

References

Banerjee SB (2008) Corporate social responsibility: the good, the bad and the ugly. Crit Sociol 34:51–79

Barnard C (1974) The functions of the executive. Harvard University Press, Harvard

Beckert J (2010) Sind Unternehmen sozial verantwortlich? In: Schumann O, Brink A, Beschorner T (eds) Unternehmensethik. Forschungsperspektiven zur Verhältnisbestimmung von Unternehmen und Gesellschaft. Metropolis, Marburg, pp 107–122

Beyer J (2006) Vom "kooperativen Kapitalismus" zum Finanzmarktkapitalismus—eine Ursachenanalyse. In: Brinkmann U, Krenn K, Schief S (eds) Endspiel des kooperativen Kapitalismus? VS, Opladen, pp 35–57

Beyer J, Hiß S, Wielgohs J (2007) Unternehmen zwischen Aktionärsinteresse und sozialer Verantwortung. Berliner Debatte Initial 18(4/5):3–5

Boiral O (2007) Corporate greening through ISO 14001: a rational myth? Organ Sci 18:127–146

Chiapello E (2015) Financialisation of valuation. Human Stud 38:13–35

Cho CH, Laine M, Roberts RW, Rodrigue M (2015) Organized hypocrisy, organizational façades, and sustainability reporting. Account Organ Soc 40:78–94

Christensen LT, Cornelissen J (2015) Organizational Transparency as myth and metaphor. Eur J Soc Theory 18:132–149

Crouch C (2005) Three meanings of complementarity. Socio-Econ Rev 3:359–363

Curbach J (2009) Die Corporate-Social-Responsibility-Bewegung. VS, Wiesbaden

DiMaggio PJ (1988) Interest and agency in institutional theory. In: Zucker LG (ed) Institutional patterns and organizations. Ballinger, Cambridge, pp 3–21

Epstein GA (2005) Financialization and the world economy. Edward Elgar, Cheltenham

Espeland WN, Sauder M (2007) Rankings and reactivity: How public measures recreate social worlds. Am J Soc 113:1–40

Espeland WN, Stevens ML (1998) Commensuration as a social process. Annu Rev Soc 24:313–342

Fleming P, Jones MT (2013) The end of corporate social responsibility: crisis & critique. Sage, London

Friedland R, Alford RR (1991) Bringing Society Back In: Symbols, Practices, and Institutional Contradictions. In: Powell WW, DiMaggio PJ (eds) Symbols, practices, and institutional contradictions. The new institutionalism in organizational analysis. University of Chicago Press, Chicago, pp 232–266

Garsten C, de Montoya ML (eds) (2008) Transparency in a new global order. Edward Elgar, Cheltenham

Greenwood R, Suddaby R (2006) Institutional entrepreneurship in mature fields: the big five accounting firms. Acad Manag J 49:27–48

Greenwood R, Jennings PD, Hinings B (2015) Sustainability and organizational change. An institutional perspective. In: Henderson R, Gulati R, Tushman M (eds) Leading sustainable change. An organizational perspective. University Press, Oxford, pp 323–355

Hasse R (2006) Der Neo-Institutionalismus als makrosoziologische Kulturtheorie. In: Senge K, Hellmann K-U (eds) Einführung in den Neo-Institutionalismus. VS, Wiesbaden, pp 150–159

Heintz B (2010) Numerische Differenz. Überlegungen zu einer Soziologie des (quantitativen) Vergleichs. Zeitschrift für Soziologie 39:162–181

Hiß S (2006) Warum übernehmen Unternehmen gesellschaftliche Verantwortung? Ein soziologischer Erklärungsversuch. Campus, Frankfurt/New York

Hiß S (2013) The politics of financialization of sustainability. Competition and Change 17:234–247

Hiß S (2014) Was bleibt von der Nachhaltigkeit nach ihrer Finanzialisierung? In: Heires M, Nölke A (eds) Politische Ökonomie der Finanzialisierung. SpringerVS, Wiesbaden, pp 211–224

Holm P (1995) The dynamics of institutionalization: transformation processes in norwegian fisheries. Adm Sci Q 40:398–422

Jennings PD, Zandbergen PA (1995) Ecologically sustainable organizations: an institutional approach. Acad Manage Rev 20(4):1015–1052

Kraatz MS, Block ES (2008) Organizational implications of institutional pluralism. In: Greenwood R, Olivier C, Suddaby R, Sahlin K (eds) The SAGE handbook of organizational institutionalism. Sage, London, pp 243–275

Lawrence TB, Suddaby R (2006) Institutions and institutional work. In: Clegg St. R, Hardy C, Lawrence TB, Nord WR (eds) The Sage handbook of organization studies. Sage, London, pp 215–254

Lawrence TB, Suddaby R, Leca B (2009) (Eds) Institutional work. Cambridge University Press, Cambridge

Lawrence TB, Suddaby R, Leca B (2011) Institutional work: refocusing institutional studies of organization. J Manage Inq 20:52–58

Lohmann L (2009) Toward a different debate in environmental accounting: the cases of carbon and cost–benefit. Account Organ Soc 34:499–534

Luhmann N (2011) Organisation und Entscheidung. Springer VS, Wiesbaden

Malsch B (2013) Politicizing the expertise of the accounting industry in the realm of corporate social responsibility. Account Organ Soc 38:149–168

Mayring P (2014). Qualitative content analysis. theoretical foundation, basic procedures and software solution. Klagenfurt

Meyer JW, Rowan B (1977) Institutionalized organizations. Formal structure as myth and ceremony. Am J Soc 83:340–363

Milne MJ, Gray R (2013) W(h)ither ecology? the triple bottom line, the global reporting initiative, and corporate sustainability reporting. J Bus Ethics 118(1):13–29

Münch R (2008) Jenseits der Sozialpartnerschaft. Die Konstruktion der sozialen Verantwortung von Unternehmen in der Weltgesellschaft“. In Maurer A, Schimank U (eds) Die Gesellschaft der Unternehmen—die Unternehmen der Gesellschaft. VS, Wiesbaden, pp 163–187

Oliver C (1991) Strategic responses to institutional processes. Acad Manage Rev 16(1):145–179

Oliver C (1992) The antecedents of deinstitutionalization. Organ Stud 13:563–588

Orton JD, Weick KE (1990) Loosely coupled systems: a reconceptualization. Acad Manag Rev 15:203–223

Perrini F, Castaldo S, Misani N, Tencati A (2010) The relationship between corporate responsibility and brand loyalty in retailing: the mediating role of trust. In: Smith NC, Bhattacharya CB, Vogel D, Levine DI (eds) Global challenges in responsible business. Cambridge University Press, Cambridge, pp 191–214

Reichertz J (2016) Qualitative und interpretative Sozialforschung. Eine Einladung. VS Springer, Wiesbaden

Sahlin K, Wedlin L (2008) Circulating ideas: imitation, translation and editing. In: Greenwood R, Olivier C, Suddaby R, Sahlin K (eds) The SAGE handbook of organizational institutionalism. SAGE, London, pp 218–242

Seidl D, Becker KH (2006) Organizations as distinction generating and processing systems: Niklas Luhmann's contribution to organization studies. Organization 13:9–35

Selznick P (1949) TVA and the grass roots: a study in the sociology of formal organization. University of California Press, Berkeley

Senge K (2007) Ist corporate social responsibility ein ökonomischer Wert?. Berl Debatte Init (18)4/5:47–55

Senge K (2011) Das neue am neo-institutionalismus. Der Neo-Institutionalismus im Kontext der Organisationswissenschaft, VS, Wiesbaden

Senge K (2015) Zur Translation und Finanzialisierung von Corporate Social Responsibility. In: Graf A, Möller Ch (eds) Bildung, Macht, Eliten. Campus, Frankfurt/M, pp 266–286

Senge K (2018 forthcoming) Corporate Social Responsibility als Legitimationsprinzip von Unternehmen in der World Polity. Springer VS, Wiesbaden

Senge K, Dombrowski S, Lawrence TB (2015) Suddaby, Roy: Institutions and Institutional Work. In: Kühl S (eds). Schlüsselwerke der Organisationssoziologie. VS, Wiesbaden, pp 400–402

Seo M-G, Creed WED (2002) Institutional contradictions, praxis, and institutional change: a dialectical perspective. Acad Manage Rev 27(2):222–247

Stark D (2011) The sense of dissonance. Princeton University Press, Princeton, NJ, Accounts of Worth in Economic Life

Thévenot L (2007) The plurality of cognitive formats and engagements. moving between the familiar and the public. Eur J Soc Theory 10:409–423

Thornton P, Ocasio W (1999) Institutional logics and the historical contingency of power in the higher education publishing industry, 1958–1990. Am J Soc 105(3):801–843

Thornton P, Ocasio W (2008) Institutional logics. In: Greenwood R, Olivier C, Suddaby R, Sahlin K (eds) The SAGE Handbook of Organizational Institutionalism. SAGE, London, pp 76–98

Zietsma C, Lawrence TB (2010) Institutional work in the transformation of an organizational field: the interplay of boundary work and practice work. Adm Sci Q 55:189–221

Zucker LG (1977) The role of institutionalization in cultural persistence. Am Sociol Rev 42(5):726–743

Part II
Case Studies

Sustainability and Green Project Management Skills: An Exploratory Study in the Construction Industry in Dubai

Hala Al-Qassab, Alberto Paucar-Caceres, Gillian Wright and Rosane Pagano

Abstract Organisations appreciate green project management leads to competitive advantage and expect project managers to be effective in this respect. This research emphasizes the identification and evaluation of skills that project managers require to make the activity of project management safe for the environment while also ensuring that it does not contribute negatively to natural surroundings. This should be of interest to executives and project managers working in the construction industry. This paper provides understanding of how project managers react to this requirement and evaluates their skills. The literature indicates that effective green project managers have a strong profile of schedule management skills, stakeholder communication and human resources management. The research approach is based on a survey of 33 project managers in the United Arab Emirates country of Dubai, followed by semi-structured interviews to delve further into the survey responses. Our findings indicate awareness of green project management is in its infancy, the focus is budget compliance. Stakeholder management and communications management skills are not considered as important as problem solving or schedule management. Fluctuating prices of materials and difficulty in sourcing sub-contractors hinder sustainable project management. While some organisations train and develop project managers, helping them acquire green project management skills, many do not and a dearth of industry standards further hampers green project management. Since this exploratory study dealt with a small sample (30+) the biggest limitation of the study was that the sample population was not reflective of the wider population. Though the construction industry in the United Arab Emirates country of Dubai, the loca-

H. Al-Qassab · A. Paucar-Caceres (✉) · G. Wright · R. Pagano
Manchester Metropolitan University Business School, All Saints Campus, Oxford Road, Manchester M15 6BH, UK
e-mail: a.paucar@mmu.ac.u

H. Al-Qassab
e-mail: hala_alqassab@hotmail.co.uk

G. Wright
e-mail: g.wright@mmu.ac.uk

R. Pagano
e-mail: r.pagano@mmu.ac.uk

W. Leal Filho (ed.), *Social Responsibility and Sustainability*, World Sustainability Series, https://doi.org/10.1007/978-3-030-03562-4_12

tion for this fieldwork is a thriving multi-billion dollar industry, the research could benefit from verification across a wider geographical area. A further area for future research is to investigate the differences between project managers in their attitudes to green project management. The initial findings should be of interest to executives and green project managers working in the construction industry wishing to become more involved in green practices and be more environmentally aware. This paper provides some empirical evidence as to how project managers react to this requirement and evaluates their skills.

Keywords Project management · Green management · Construction industry Dubai · Developing countries

1 Introduction

Increased concern for the environment, climate change, and the need to minimise the negative impact on the stakeholders and the communities in which organisations operate drive the increasing emphasis on green project management (Yang et al. 2016). Most organisations have also come to view green project management as a means of gaining a competitive advantage in the market place. It seems increasingly evident that to manage the complexity of projects whilst addressing environmental demands, project managers are expected to possess the necessary skills to be considered as effective project manager's that have credible green credentials.

Sustainability is a notion which appreciates that development should be an activity through which needs of the present generation can be fulfilled, but does not jeopardise the ability of future generations to meet their desires (Hwang and Ng 2013; Silvius et al. 2012). Maltzman and Shirley (2016) argued that successful project management has traditionally meant delivering a product or service on time and within budget constraints, using resources (people and materials) in an optimal way, satisfying the needs of the customer and the supplying organization.

With the increasing attention being given to sustainable development and growing awareness of environmental concerns, especially global climate change, the allure of becoming green is inspiring project managers to include environmental goals in their activities, and to seek guidance on how to assess the sustainability of construction projects (Zhang et al. 2014). Suggestions for how new companies and individuals can 'go green' can be found almost everywhere, therefore, it is essential for project managers to possess the essential, relevant, skills and knowledge to make the task of project management sustainable and less harmful to the environment.

Sustainable project management has emerged as an important topic for companies because of loss in biodiversity, rising pollution levels, changing climates, etc. Moreover, the concept of sustainability is also related to a great extent to the image and reputation of companies in their markets. Today, for project managers, it has become essential that they have the skills, talent, knowledge and abilities through which they can make this task as safe as possible for the environment (Banihashemi et al. 2017;

Robichaud and Anantatmula 2011). We aim to gain an understanding of the views of project managers concerning sustainability and the nature of the essential skills and knowledge they should possess.

This issue is especially important in developing countries where the construction industry is becoming the most important sector in the economy, this research is based in Dubai which is one such region. Furthermore, the construction sector has a major impact on the environment as commercial and residential buildings combined are responsible for between 20 and 40% of the world's energy consumption and these values are rising steadily every year (Hwang and Tan 2012). Therefore, with rising awareness of sustainable development it is crucial for project managers to ensure that they possess essential skills and talents through which they can manage resources efficiently and effectively, while ensuring that the natural environment is safe.

This research emphasises the identification and evaluation of skills that project managers require to make the activity of project management safe for the environment while also ensuring that it does not contribute negatively to natural surroundings. This should be of interest to executives and project managers working in the construction industry. For the purpose of determining which skills are required by sustainable project managers and also to evaluate their perspectives these skills in their live projects, the following research questions were formulated:

- What are the various components that support the concept of sustainability in project management?
- What are the sustainability risks related to time and quality management?
- What are the skills and abilities that project managers require to accommodate sustainability?
- In what ways can tasks of project management be made more sustainable?

This article is organised as follows: after this introduction, in Sect. 2 we outlined the key issues of green building practice and green management found in recent literature emphasizing the importance of these practices in project management. In Sect. 3 discuss the different elements in Green project management and sustainability. The research methods are briefly describe in Sect. 4 and the results presented in Sect. 5. We discuss the results in Sect. 6. Finally, some conclusions and agenda for further research are suggested in Sect. 7.

2 Green Management: Green Buildings Practice and Sustainability in Project Management

Frank (2002, cited in Hwang and Ng 2013: 272) determined that the project manager has direct influence over 34–47% of project success, establishing the importance of their role with respect to competitiveness. Furthermore, there is a crucial relationship between project success and construction project management competencies Hwang and Ng (2013). Therefore, if a business organisation has not determined criteria for a competent project manager, then it would not be able to survive (Eskerod and Huemann 2013; Hwang and Ng 2013).

Conversely these authors suggest that negative impacts of project management activities have increased considerably and therefore it has become imperative that project managers have skills and talents in green management. Evaluation of green buildings has been undertaken with an expectation of their improved performance (Ding 2008).

The importance of project management to competitiveness has been established, Silvius et al. (2012), with project management a critical competence identified as central to the management of change crucial in an organisation Silvius et al. (2012). Early research suggested that sustainability should be an integral part of this project management contribution and that utmost attention must be paid to this by the project manager and other authorities Newman (1999). Shen et al. emphasized that incorporating sustainability into project management processes can contribute to competitive advantages and economic benefits while reducing harm to the environment and society. Emphasis on the importance of sustainability has increased and pressure on organisations to model their operations in a manner through which their negative impact on the environment can be reduced has grown.

There are many ways through which the activities of project management could be made sustainable, such as making optimal use of the local resources that are available and reducing waste (Sears et al. 2015). The first step towards this is to establish a clear vision and framework for sustainability that robustly defines the business case and the second step should be communicate the importance of the vision and educate management team. There are clear calls in previous research for project managers to possess the essential skills and knowledge which enable them to act as a green project manager (Silvius et al. 2012; Sears et al. 2015).

3 Elements to be Considered in Green Project Management and Sustainability

Although there has been previous research examining the essential skills and knowledge of project managers in a general context, little has been undertaken to focus on specific context of green construction and to address the necessary skills required to implement green projects. From the literature, we built a base for our research exploring: (a) the financial elements that both hinder and support green project management; (b) the skills that are required of a green project manager; and (c) the different risk and quality related issues that are faced by the construction industry. The following section discusses these elements.

3.1 *Financial Elements in Green Project Management and Sustainability*

Sustainable buildings can now be constructed as about the same cost as traditionally constructed buildings but can save over 30 per cent in operating costs. From the fiscal perspective, such buildings are highly cost-effective and this understanding has led to the creation of different sustainable buildings. This has helped in saving substantial corporate funds through the reduction in overall operational and maintenance costs, thus creating financial stability whilst often reducing taxpayer burden, also saving money for taxpayers (Glavinich 2008).

Whilst good project management and cost management software could have the major impact on demonstrating the benefits of various green building measurements, having good information about costs, benefits, and return on investment can be critical to keeping good green measures under consideration, instead of losing them to strictly financial considerations.

3.2 *Skills Required by Sustainable Project Managers*

Project managers should possess knowledge of the technical aspects of the industry which enable them to understand the project being delivered (Hwang and Ng 2013) and construction management needs a unique skills and experience beside the managerial skills and knowledge to improve feasibility and allow construction managers to complete their jobs within the allocated budget and requirements (cited in Hwang and Ng 2013).

Specifically, construction industry project managers need to be able to: apply green building strategies to support reduction in energy costs whilst minimizing carbon output; coordinate and manage the large logistic sites (Abudi 2010); develop and implement programs such as safety, environment protection, and quality control. Manage construction wastes in compliance with LEED's goals. Furthermore, they also need to be able to work with supervisory personnel, contractors, designers and owners to undertake a constructive discussion on issues such as complaints, procedures and construction problems (Shen et al. 2010). This is supported by Azim and Gale (2010) who identified a similar profile: human resources, scheduling; resource allocation; risk management; budgeting; time and team management; change management; conflict management; leadership skills; communication abilities; foresight; and delegating tasks. This confirms the findings of Campbell (2009), that good communication management and stakeholder management enhance manager's performance and enables them to cooperate better with employees and other stakeholders. However, other research has identified that in practice, other issues may take priority, that cost, communication and schedule management are considered more critical than other management knowledge areas (Hwang and Ng 2013).

A further benefit of good communication and with advanced human resources knowledge can decrease the level of conflict, reduce errors and misunderstanding between team members, and ensure the high levels of commitment, productivity (Daft 2017; Brewster et al. 2012; Dainty and Loosemore 2012). Appropriately skilled project managers will be much in demand; a core reason to require highly skilled and knowledgeable project managers is the competitive marketplace wherein both companies and individuals are competing for a unique position and chance to grow (Westland 2013).

3.3 Risks Involved in Sustainable Construction Projects

Sustainable construction techniques provide an ethical and practical response to issues of environmental impact and resources consumption. Kibert articulated seven principle of sustainable construction, which would ideally inform project managers and decision- making during each phase of the design and construction process, continuing throughout the building's entire life cycle. These principles are: reduce resources, reuse resources, use recyclable resources, protect natural, eliminate toxics, apply life-cycle costing, and focus on quality. However, there are different risks in the construction sector that are in direct a relationship with the quality and time management problems that project managers could face and these need to be taken into account when understanding project managers, their drivers and skills (Kubba 2010). These risks are:

- *Certification risk*: the simplest risk element in green construction because most the construction companies look for validation of their business projects so as to seek permission from officials to continue with their task (Kosaroglu and Hunt 2009).
- *Performance risk*: related directly to the skills and knowledge that project managers possess therefore it is crucial to be familiar with details in the project to avoid such risks (Hicks 1996). According to Hwang and Ng (2013) risks due to insufficient knowledge or unfamiliarity with green techniques, materials and process could affect performance, outcome and project delivery.
- *Quality and Safety*: related to construction projects Bing et al. (2005) noted that for project managers, it is imperative that quality is maintained, ensuring the safety of employees.
- *People management*: project managers need to properly supervise and motivate people in order that they do not feel isolated or cheated in any way, as this could jeopardize the project (Banaitiene and Banaitis 2010).

4 Research Method

This study was conducted in 2016 over a period of six months. The research approach is based on a survey of 33 project managers in the United Arab Emirates country of Dubai, followed by semi-structured interviews to delve further into the survey responses. The initial survey comprised three elements: the financial elements that hinder or support green project management; the quality and other risks encountered by green project managers and data on the skills required by green project managers.

The data collection instrument included 5-point Likert scales consisting 1 = strongly agree to 5 = strong disagree. From the responses, nine open-ended questions were developed for the interview schedule and participants were encouraged to discuss elements of their careers as green project managers.

The interviews were subject to a thematic analysis that was informed by the initial survey and the conceptual base developed from the literature. The initial sampling frame comprised 60 project managers in the construction industry in the United Arab Emirates, 33 completed the scoping survey and eight agreed to participate in an interview.

5 Results

The interviews were based on a combination of the literature review and a scoping survey of a small sample of project managers. This section presents the outcome of this exploratory survey (Table 1), followed by the insights into financial, time and quality risks and green project management skills. In our report of the interviews, comments made by interviewees are presented in italics.

Scoping of the study
Financial issues

The initial survey identified attitudes to the importance of financial issues, risk and green project management.

Regarding financial elements the majority of project managers (84%) agreed that keeping within budget is important to ensure the sustainability of the project. However, there was also considerable agreement that keepings costs low (50%), coming in under budget (70%) and making cost savings are insufficient to ensure sustainability. Most agreed that waste reduction is important, though this was not as high as those agreeing that keeping within budget is important.

Time and Quality Management

The summary results of the scoping study are presented in Table 1. Responses indicate that sourcing low cost high quality products is not the biggest problem faced with regard to ensuring the high quality of project deliverables, but rather that organisations face many challenges in sourcing qualified green contractors and sub-contractors.

Table 1 Attitudes to green project management

	Agree/strongly agree (%)	Uncertain	Disagree/strongly disagree (%)
Financial elements of sustainability			
Keeping to budget most important	85	–	15
Smaller budget with robust pre-planning more important than keeping within budget	40	10%	50
Within budget alone insufficient for a sustainable project	50	–	50
Being under budget and making savings ensures long-term sustainability	70	10%	20
Waste reduction and efficient resource use for sustainability more important than being within-budget	70	10%	20
Quality risks			
Difficulty in sourcing and selecting sub-contractors	60	10%	30
Difficulty in obtaining reasonably priced green material and equipment	24	6%	70
Using a variety of planning techniques	60	10%	30
Unforeseen quality issues and challenges	65	5%	30
Difficulty in comprehending specifications in a green project contract	70	9%	21
Green project management skills			
Schedule Management	60	–	40

(continued)

Table 1 (continued)

	Agree/strongly agree (%)	Uncertain	Disagree/strongly disagree (%)
Stakeholder Management	20	5%	65
Effective Human Resources Management	42	3%	55
Efficient Human Resources Management	70	–	30
Communication Management	30	–	70
Problem Solving	60	–	40
Strong analytical skills	60	–	40
Effective decision making less important than other skills	60	–	40
Team working skills are more important than other skills	67	–	33
Problem solving not as important other skills	35	–	65

Furthermore, it is also evident that other challenges faced by project managers including the complexity of contract, which have a direct impact on the quality of the deliverables, as well as the different techniques and standards that are used are important. Other quality risks that were cited in the survey include unforeseen and unplanned risks that can pose an immense threat to the long-term sustainability of the project.

Essential Skills for Green Project Managers

In terms of the skills that a green project manager should possess, respondents believed that schedule management and the ability to work within a budget is important, while stakeholder management and communications management was the least important. It is also clear that the majority considered that problem solving and analytical skills are important while less valued decision-making skills compared to team-working, schedule management, human resources management, stakeholder management and communications management and problem-solving.

Attitudes to Green Project Management

Participants were familiar with the term 'green project management', and though there was a general feeling that it is something of a 'buzz word' and said that it has various meanings to different people. However, overall, they interpreted it as "*a type of project management philosophy that strives to minimize the negative impact on the environment and stakeholders, while increasing the efficiencies, reducing costs.*" The reasons for the importance of green project management varied. Some referred to the competitive advantage it can bring by demonstrating to clients what outcomes they are able to provide compared to the competition. There was acknowledgement that this is a good thing in its own right, it means that impact on the environment is kept to a minimum. Others commented on the mutual benefits to the organisation and their customers.

Implementing green project management

There were various approaches to implementing green project management. Some referred to structured learning from previous projects, seeking areas that can be improved *we continuously review our policies and procedures with the aim of increasing sustainability*. Others favour hiring employees with skills were aligned to the green project management philosophy whilst providing on-going training for managers within the company to acquire these skills. There was also, however, acknowledgement that some organisations are not addressing the issue at all.

Financial influences

The inability to source low cost high quality materials was considered to hinder green project management, while *pre-construction project planning often helps to reduce much wastage and costs.* Fluctuating raw material prices was a concern *is the biggest problem we face* along with difficulty to source skilled labour *at reasonable prices,* though this was mitigated through pre-construction project planning. Another consideration was the changing scope of a project *without adequate budgets* which was noted as a significant hindrance.

Risks

Price fluctuations and changes to the scope of the project were considered major risks though *we try our best to protect again raw material fluctuations and schedule changes.* There was acknowledgement of the importance of stakeholder communication with respect to risk, *we state to the client that the environment is given priority and that we need flexibility, however, it was also acknowledged that this flexibility is not always forthcoming.* Risk mitigation was an important issue, participants mitigate against many risks and *these change according to the different projects and the project scope.*

Green Project Management Skills

Whilst some respondents believed that it is very important that they had green project management skills in order for them to progress through their organisation, it was also indicated that there was a feeling that these skills had to be acquired *in order to succeed in a different organisation if they wanted to move to a new job*. Others suggested that it would be useful only if they wanted to move to a different organisation where this is valued, but they only needed to *acquire whatever skills their organisation believed was important to become a more 'green project management' oriented company.*

In discussion of training, there was a core emphasis on the traditional project management skills, *we are continuously provided with feedback on our problem solving and critical thinking skills and monitored and measured on our ability to stay on schedule and within budget.* In another organisation, it was reported that the respondent was provided with much support and training on *how to effectively manage my team, keep conflicts to a minimum, problem solve and stay within budget.* For another, the organisations focus was on *finding creative ways to stay within budget.* A more positive report was of training that *provided me with training on how to keep human resources related and other costs low, while also reducing wastage and staying on schedule.*

Almost all participants were negative about the possibility of undertaking green project management training in their own time or funding it themselves, though one was contemplating this because *from a career perspective both within the company and externally* it would be very beneficial and would add value to his resume. In organisations that are not engaged with green project management, it was seen as important to introduce the idea and change the culture within the organisation in order to prepare it for sustainability. There was support for the provision of more focused training and development and the introduction of performance management, rewards and recognition programmes *around the acquisition and application of green project management skills.*

6 Discussion

There is considerable alignment with the move towards the awareness and understanding of green project management in previous research of Hwang and Ng (2013). At a time when most organisations are now designing their operations in a manner that would be least harmful to the environment and help with preservation and conservation, instead of destruction (Othman and Sirbadhoo 2009), it is evident that the respondents have a fairly accurate understanding of the concept. However, it was also evident that respondents were divided in their opinions about the role played by budgetary management and quality in the overall green project management process.

6.1 Financial Aspects of Green Project Management

Whilst the majority of the respondents agreed that staying within the allocated budget was of high importance in their respective organisations, there were differences in opinion on how such an objective should be achieved. While it was the view of Glavinich (2008) that keeping costs low through green buildings, saving corporate and tax payer funds, was important, respondents were of a different opinion, feeling that whilst staying on budget was important, using different techniques such as pre-construction project planning, and keeping costs low were less so. Thus we postulate that this implies that while it is important to stay within the budget, costs should not be cut to the extent that there would be a surplus. Furthermore, it was also interesting to note that most respondents believed that staying within budget alone was sufficient to ensure the sustainability of the project.

Thus, leaving no room for any assumptions, which high levels of cost cutting could result in low quality end products could lead to a lack of sustainability. However, again, respondents were more concerned about the staying within budget and considered it to be more important that the efficient use of resources. Thus, the results suggest that staying within budget is given a higher priority than the actual impact that such activities could have on the environment. There is a strong element of learning from all aspects of a project when it is completed, even though there remains a strong emphasis on budget control. The learning is seen as a way to improve practices and processes.

This indicates that organisations are focused on optimising their operations and processes in order to conform with a green project management conceptual framework. Furthermore, it is also important to note that when asked what financial elements or constraints supported or hindered green project management, one of the interviewees stated that the biggest challenge was sourcing high quality raw materials at a low cost and that pre-construction project planning helped to eliminate this problem

6.2 *Time and Quality Risks*

This research is congruent with some previous findings, though it also diverges in other respects. One of the biggest challenges faced by green project managers is the inability to source and select sub-contractors who provide green services (Hwang and Ng 2013; Hwang and Tan 2012), and this is confirmed in this research indicating a significant challenge in this developing economy. Moreover, the different planning techniques used throughout the project management process is a significant risk to quality as is the likelihood that unforeseen issues, challenges and obstacles arise (Hwang and Ng 2013). Another further challenge and risk to quality is the inability to easily comprehend the many specifications in the contracts (Bing et al. 2005).

Our study confirms these risks faced in green project management, indicating that the situation in other countries is echoed here. However, this research differs from the results of other studies in that unlike elsewhere, this study found that there is not a difficulty in sourcing unlike in the case of many other countries and the observations made by, reasonably priced, high quality green materials and equipment (Hwang and Ng 2013; Hwang and Tan 2012). A further difference noted in our results is that price fluctuations and the changing scope of the projects are threats to quality faced by green project managers.

6.3 *Essential Skills for Green Project Managers*

Previous studies highlighted several different skills essential for green project managers, including the ability to schedule, manage and coordinate the project; to effectively and efficiently manage human resources, to effective manage stakeholders and communicate with them in a manner that ensures the smooth flow of the project (Hwang and Ng 2013). Other competencies have been identified as essential for green project managers to successfully overcome many of the challenges faced by them during the project management process. These are skills in problem solving, decision making, analytical thinking and team working were in order. This research emphasised especially, the need for skills in schedule management skills and efficient human resources management. Interestingly in this research, the ability to manage stakeholder's communication was highlighted as more important than decision-making and team-working skills. However, it was also clear that respondents believed that problem-solving skills were much more important than stakeholder management, communications management and schedule management skills.

Our findings agree that effective human resources management skills are essential; agreeing with (2010) believed that it is more important to be able to effectively manage human resources than to effectively manage them. While this may seem counter-intuitive, we postulated that in the construction industry (as in many other industries), labour costs make up a significant portion of the overall cost for the organisation and therefore managing the labour resources efficiently is a top priority compared to effectively managing them. On the other hand, it can also be argued, that if project managers are able to manage and motivate employees effectively in a manner that elicits high levels of commitment, engagement and performance, error rates will drop, and productivity increase, thus indicating that effective human resources management rather than efficient human resources management is more beneficial (Daft 2017; Brewster et al. 2012; Dainty and Loosemore 2012).

While most interviewees had received some form of training and development in green project management, the type of training differed between companies. This suggests that there are no clear overall and companies establish their own ad hoc development programmes based on their views of essential skills. The general view was that it would be beneficial for them to acquire more skills to establish them as green project managers and if provided with the opportunity to make improvements within their organisations, they would re-evaluate the current training initiatives and focus more on specific areas that need further enhancing.

Though we found considerable consistency in our findings with the extant literature, there were some subtle differences suggesting that there may be contextual inferences for green project management. So while the industry may have evolved similarly to developed countries in some aspects, it is not as advanced in others and is continuing to change and develop.

7 Conclusion

Green Project Management continues to gain prominence across the world as concerns for the environment and the need to reduce the impact on the environment increases. At the same time, the need to maximise benefits for stakeholders is an increasing priority. It is evident from the literature that many organisations have adopted a green project management philosophy, as they believe that it can provide them with a competitive edge in the market place. Furthermore, it is also evident that in the context of this research in the developing area of the United Arab Emirates, Green Project Management is embraced, but unlike in many developed economies, in the Dubai, the concept of Green Project Management is still evolving and undergoing many changes.

7.1 Implications of the Study

Our research demonstrates that staying within the budget is considered to be a top priority in most organisations and that the inability to source sufficient numbers of contractors and sub-contractors who provide green services, the use of multiple techniques and the complexities of green contracts often leads to quality-related risks and challenges in projects. Previous research found that some of the most important skills that a green project manager should possess included the effective schedule management, stakeholder management, communications management and human resources management Hwang and Ng (2013). However, in this context, it is evident that schedule management and the efficient management of human resources are considered to be more important than stakeholder and communications management.

It was also clear that effective human resources management skills were not considered to be as important as efficient human resources management skills. We postulate that the scarcity and high cost of skilled labour could by the driving force behind this finding. Problem solving, analytical, decision-making, and effective team-working skills are essential competencies for a green project manager (Hwang and Ng 2013), because these skills help the project manager to overcome their challenges and manage risks effectively. However, in this context, we found that team-working and decision-making skills were not considered to be as valuable as problem-solving skills. Furthermore, problem-solving skills were considered second only to schedule management for green project managers.

Our research has indicated that while there were no standards for the training and development of green project managers, most organisations have a process to look back at past projects and learn from them. Furthermore, in some organisations project managers were also provided with ongoing feedback on their critical thinking and problem solving skills as well as their ability to manage schedules and stay within budget. Team, conflict and human resources management related training programmes are provided by many organisations. However, it is also important to note that there are organisations that do not provide any such training for project managers and have not adopted a green project management philosophy. Yet, even companies that do not have a green project management philosophy expect their managers to stay within budget.

It is also apparent that mitigating against fluctuations in raw materials and labour costs, through pre-construction project planning helps to increase the sustainability of projects. Furthermore, unlike in many other economic and industrial contexts, it was evident that sourcing high quality raw materials at a reasonable cost was not difficult, even though there are frequent fluctuations in their costs. Therefore, while green project management exists in this area, the conceptual notion and implementation of green project management is still evolving and therefore the focus of organisations is to stay within budget. While there are several different quality-related risks that are experienced by organisations, the inability to source sub-contractors that provide green services is considered to be one of the most significant challenges faced by the industry.

7.2 Main Lessons from the Study

It is evident that while most organisations strive to train and develop their project managers and help them to acquire green project management skills, there are no set standards in the industry and there are also companies that do not provide any training even though their managers are expected to stringently adhere to budgetary constraints. Based on the outcomes of this study, and looking at the extant literature, we summarise key points in our findings, and provide directional recommendations to address these challenges.

Pre-Construction Project Planning Pre-construction project planning allows for contingency planning and the more efficient allocation of resources, which in turn will help the organisation to have better control of its raw material, labour costs, and mitigate against price fluctuations (Dainty and Loosemore 2012). It also enables project managers to look ahead and identify potential risks before contracts and agreements are signed, thus ensuring that identifiable financial, quality and other risks are taken into consideration when estimating budgets, scheduling, carrying out scope management and other related activities (Dainty and Loosemore 2012).

Green Project Management Culture Cultivating the environment for a green project management culture within the organisation would also help to synergise the processes and procedures within the company and enable the organisation to focus on different areas such as the quality of deliverables as well as the budgetary constraints on the project. Crucially, such a culture would help to increase the emphasis on the acquisition of skills and abilities for employees which supports green project management (Maylor 2017).

Training and Development Project managers should be provided with ongoing training and skills development in schedule, stakeholder, communications, human resources, problem solving, critical thinking, decision making and team working, and these should be the consideration of their performance (Chapman and Ward 2011; Hwang and Ng 2013; Maylor 2017). They should be provided with ongoing feedback on their progress and a knowledge base should be created with the aim of learning from past projects to continuously improve organisational processes and procedures in green project management (Chapman and Ward 2011; Maylor 2017).

7.3 Limitations of the Study

There was inability to survey and interview a wider sample of participants. Since only 33 of respondents actually participated in this research study, we felt that the biggest limitation of the study was that the sample did not reflect the wider population as extensively as we would have hoped.

Although the location for this fieldwork is a thriving multi-billion dollar industry—the construction industry in the United Arab Emirates country of Dubai—the research could benefit from verification across a wider geographical area.

There are some interesting contradictions in relation to the literature, and although these are explicable in the findings, and indeed, we sought to identify uniqueness and difference in the area, this makes repetition and verification in similar locations an interesting basis of future research.

A further area for future research is to investigate the differences between project managers in their attitudes to green project management. We suggest that length of experience, and experience of different locations or industries could have had an impact on the manner in which they responded to the issues addressed in this research.

The aim of the research was to gain an understanding of the different skills that were essential for green project managers. Therefore, to gain multiple perspectives from recruitment agencies and industry experts would also be potentially beneficial.

References

Abudi G (2010) Project managers need leadership skills. Retrieved from http://www.projectsmart.co.uk/project-managers-need-leadership-skills.php. Accessed on 10 May 2014

Azim S, Gale A (2010) The importance of soft skills in complex projects. Int J Manag Proj Bus 3(3):387–401

Banaitiene N, Banaitis A (2010) Risk management in construction projects. Retrieved from http://cdn.intechopen.com/pdfs-wm/38973.pdf. Accessed on 13 May 2014

Banihashemi S, Hosseini M, Golizadeh H, Sankaran S (2017) Critical success factors (CSFs) for integration of sustainability into construction project management practices in developing countries. Int J Proj Manage 35(6):1103–1119

Bing L, Akintoye A, Edwards J, HardCastle C (2005) The allocation of risk in PPP/PFI construction projects in the UK. Int J Proj Manage 23(1):25–35
Brewster C, Sparrow P, Vernon G, Houldsworth E (2012) International human resource management, 3rd edn. Chartered Institute of Personnel and Development (CIPD), London
Campbell M (2009) Communication skills for project managers. AMACOM Division of American Management Association, New York
Chapman C, Ward S (2011) How to manage project opportunity and risk: why uncertainty management can be a much better approach than risk management, 3rd edn. Wiley, Hoboken
Daft R (2017) Management, 12th edn. Cengage Learning, Boston
Dainty A, Loosemore M (2012) Human resource management in construction, 2nd edn. Routledge, New York
Ding CKD (2008) Sustainable construction—The role of environmental assessment tools. J Environ Manag 86(3):451–464
Eskerod P, Huemann M (2013) Product development tensions: exploring contrasting styles of project management. Int J Manag Proj Bus 6(1):36–50
Glavinich ET (2008) Contractors guide to green building construction: management, project delivery, documentation, and risk reduction. Wiley, New York
Hicks ER (1996) Experiential learning in a postgraduate project management programme. Education + Training 38(3):28–38
Hwang B, Ng W (2013) Project management knowledge and skills for green construction: overcoming challenges. Int J Proj Manag 31(8):272–284
Hwang B, Tan J (2012) Green building project management: obstacles and solutions for sustainable development. Sustain Dev 20(5):335–349
Kosaroglu M, Hunt AR (2009) New product development projects and project manager skill sets in the telecommunications industry. Int J Manag Proj Bus 2(2):308–317
Kubba S (2010) Green construction project management and cost oversight. Butterworth-Heinemann, London
Maltzman R, Shirley D (2016) Driving project, program, and portfolio success: the sustainability wheel. CRC Press, New York
Maylor H (2017) Project management, 4th edn. Pearson Education
Newman WGP (1999) Sustainability and cities: extending the metabolism model. Landscape Urban Plan 44(4):219–226
Othman EAA, Sirbadhoo N (2009) An innovative partnership framework for sustainable development of rural areas in South Africa: The role of project management firms. J Eng Des Technol 7(3):243–263
Robichaud L, Anantatmula V (2011) Greening project management practices for sustainable construction. J Manage Eng 27(1):48–57
Sears K, Sears A, Clough H, Rounds J, Segner R (2015) Construction project management: a practical guide to field construction management, 6th edn. Wiley
Silvius G, Schipper R, Planko J, Brink J, Kohler A (2012) Sustainability and project management. Routledge
Westland J (2013) 7 Critical skills for project managers. Retrieved from http://projectmanager.com.au/7-critical-skills-for-project-managers. Accessed on 10 May 2014
Yang R, Zou P, Wang J (2016) Modelling stakeholder-associated risk networks in green building projects. Int J Proj Manage 34(1):66–81
Zhang X, Wu Y, Shen L, Skitmore M (2014) A prototype system dynamic model for assessing the sustainability of construction projects. Int J Proj Manage 32(1):66–76

Engaging Employees in Corporate Social Responsibility Projects—A Case Study from the Lufthansa Group Showcasing Experiences and Lessons Gathered in Kenya, Rwanda, Nigeria and Columbia

Jens M. Unger and Johannes M. Luetz

Abstract Social responsibility and sustainable development are concepts of growing interest to the corporate sector, including in the so-called developing world. Although the benefits are manifold and typically include mutual learning, innovation, humanitarian engagement, intercultural exchange, new markets, and effective theory-praxis integration that is aligned with the idea of giving back to the community there is a dearth of research studies featuring the practical outworking of actual cases in a comprehensive, analytical and critical manner. To address this gap in the literature the study employs an exploratory methodological approach examining the Impact Week, a Lufthansa Group CSR program aimed at progressing entrepreneurship, innovation, and intercultural exchange that systematically engages employees as part of an internal qualification program. Data from projects in Kenya, Rwanda, Nigeria and Columbia showed long-term local effects both on a program as well as an individual level of analysis. Findings further indicate that engaging employees in CSR practice has the potential of making organizations more meaningful and thus positively impacting organizational culture and attracting new talents. A human centric as opposed to a utilitarian orientation among stakeholders emerged as an important factor for the longevity and sustainability of CSR practice. Results are relevant to stakeholders interested in learning how social responsibility and sustainability may be enhanced in organizational practice.

Keywords Sustainability · Social responsibility · Lufthansa Group
Entrepreneurship · Innovation · Social capital · Impact week

J. M. Unger (✉)
Lufthansa Group, Lufthansa Aviation Center, 60546 Frankfurt, Germany
e-mail: jens.unger@dlh.de

J. M. Luetz
CHC Higher Education, Brisbane, Carindale, QLD 4152, Australia
e-mail: jluetz@chc.edu.au; j.luetz@unsw.edu.au

J. M. Luetz
University of New South Wales (UNSW), Sydney, Australia

W. Leal Filho (ed.), *Social Responsibility and Sustainability*, World Sustainability Series, https://doi.org/10.1007/978-3-030-03562-4_13

1 Introduction

Taking the example of the Lufthansa Group sponsored Impact Week (IW), this study seeks to better understand what makes 'Corporate Social Responsibility' (CSR) projects successful and sustainable. The IW is a qualification program now in its third year since inception that aims at fostering entrepreneurship, innovation, and intercultural exchange in countries and communities in which the Lufthansa Group operates. The company systematically engages employees in the IW as part of an internal qualification program. The study examines the sustainable impact of the program, in particular long-term and self-sustainable local effects as well as effects on employees and potential effects on a company level.

Section 1.1 offers a literary overview of social responsibility in contemporary business practice. Section 1.2 details the paper's specific intended contribution to scholarship.

1.1 Corporate Social Responsibility: A Short Review of the Literature

Over recent years it has become increasingly accepted and even expected that a company's obligations reach well beyond its immediate profiteers or shareholders (Boatright 2012; Cole 2005; Higgins 2009), Hence in discussing CSR, it seems more expedient to emphasise a company's 'stakeholders' rather than 'shareholders' (Dunfee 2009), namely all those persons or legal, paralegal or non-legal[1] entities that have a legitimate 'stake' or interest in the sustained wellbeing both of the company itself, and the 'externalised' wellbeing (or greater good) that flows from the company to society and the world at large (Cole 2005, pp. 26–29). Consequently, CSR opens up more holistic perspectives (Boatright 2012; Crane et al. 2009). Moreover, given that the list of stakeholders with a vested interest in matters pertaining to a company can be quite extensive, discussing 'stakeholders' (rather than merely 'shareholders') typically also introduces an analytical level of heightened diversity and complexity (Aras and Crowther 2010). Evidently, the list of 'stakeholders' who are in some way influenced by (or can influence) a company's actions and decisions may be quite extensive and may include actors and non-actors such as employees, customers, investors, suppliers, consultants, community organizations, government, neighbourhoods, and more broadly, other species and the environment at large, among others (Cole 2005, pp. 26–29; Crane et al. 2009; Moscardo et al. 2013).

According to Dahlsrud (2006), there is both "uncertainty as to how CSR should be defined" (p. 1), and "an abundance of definitions, which are, according to Van Marrewijk (2003), often biased toward specific interests" (Dahlsrud 2006, p. 1). Hence Dahlsrud posits that "it is not possible to develop an unbiased definition" (p. 2,

[1] https://en.oxforddictionaries.com/definition/us/non-legal.

attributed to Berger and Luckmann 1966). Even so, his analysis of 37 definitions of CSR identifies five dimensions that can be discovered (in whole or in part) in all definitional approaches: The stakeholder, social, economic, voluntariness, and environmental dimension (Dahlsrud 2006, p. 5). His analysis concludes that "the challenge for business is not so much to define CSR, as it is to understand how CSR is socially constructed in a specific context and how to take this into account when business strategies are developed." (p. 6).

Given that CSR largely functions within a self-regulatory operational framework[2] whereby businesses monitor and ensure their own compliance with the law, ethical and environmental standards and national and international norms (White 2008; Rasche et al. 2017), critics have questioned both the sometimes lofty societal expectations of CSR as "unrealistic or ill founded" (Henderson 2001, p. 155), and/or have even challenged the "underlying conceptual need for corporate social responsibility (CSR) in the first place" (van Oosterhout and Heugens 2009, p. 1). Further, CSR has been famously dismissed by the neoclassical economist Friedman (1970) on the grounds that "[t]he social responsibility of business is to increase its profits" (p. 17; 2007). Other criticisms include the charges that CSR in practice promotes the so-called 'greenwashing'[3] or 'window dressing' of unsustainable business practices (Fallon 2013, p. 48; White 2008), or may even be employed as a means to pre-empt the watchdog function of national governments seeking to regulate the actions of powerful multinational corporations (Shamir 2011; White 2008).[4]

The mining industry lends itself as an example. While mining is widely conceived as 'unsustainable' in the sense that non-renewable mineral resources can only be extracted once, thus making the action of their extraction literally 'unsustainable' (in addition to charges relating to environmental degradation, human displacement processes and dispossession of indigenous land etc.),[5] recent years have seen the mining industry actively engage in public relations campaigns to reposition its operations in the public consciousness as 'sustainable' (Gilberthorpe et al. 2016; Padel and Das 2010; Whitmore 2006). According to Whitmore (2006), by adopting and mainstreaming the neologist oxymoron[6] "sustainable mining" (p. 309), the mining industry has created the impression "that there is a 'new, sustainable mining' which is different from the old, bad practices of the past." (p. 309). However, Whitmore's (2006) analysis leads him to conclude that "[a]ttempts by the mining industry to greenwash itself as a new, improved, sustainable industry simply will not wash […] Those on the other side of the bulldozer can easily see that the emperor is naked, and the more he insists it is not so, the further away an honest dialogue is." (p. 313).

---

[2]See Higgins (2009) for a discussion on "The rise of soft regulation" (pp. 204–207).

[3]According to White (2008), "[t]he phenomenon of 'greenwashing' has been incorporated into most companies' operational practices in one way or another. Greenwashing refers to putting a particular corporate 'spin' on environmental issues and problems. Much of it has to do with image-making, and hence it is heavily tied up with public relations" (pp. 262–263).

[4]See Crane et al. (2009) (Part 3) for an overview of critical discourses relating to CSR.

[5]See Frynas (2005), Gilberthorpe and Banks (2012), Trebeck (2008), and O'Faircheallaigh and Ali (2009) for examples and discourses.

[6]See Benson and Kirsch (2010) for other examples of corporate oxymorons.

In short, CSR can be open to the criticism that powerful corporate actors with strong business self-interests may be using 'Corporate Social Responsibility' as a front to put their own social and/or sustainability 'spin' on their principal raison d'être: Profit maximisation (Friedman 1970, 2007; Lenz et al. 2017). Nevertheless, despite this criticism, overall Boatright (2012) suggests that in today's corporate environment, "CSR has become a virtual industry, with most large corporations proclaiming long lists of activities." (p. 292). This is also reflected in the attempted standardisation of CSR by the International Organization for Standardization (ISO), which has developed ISO 26000[7] from the collective input of 500 experts, albeit conceding that its implementation is de facto limited to self-regulation, given that this standard "provides guidance rather than requirements" (ISO n.d., para 3): "ISO 26000 is not a management system standard. It does not contain requirements and, as such, cannot be used for certification." (ISO 2010, p. 8).

1.2 Intended Paper Contribution to Research and Scholarship

In synthesis, despite the significance of CSR in contemporary business practice (Crane et al. 2009), there is little comprehensive, analytic study of how companies actually engage and invest in social projects, which remains "highly ambiguous" (van Oosterhout and Heugens 2009, p. 6). Hence increasing our understanding about the effects and success factors of such projects in relation to stakeholders, including the company itself, is therefore relevant for scholars and practitioners. This offers the promise of more equitable CSR implementation practices, in addition to fostering future research on the topic.

This paper discusses CSR through the lens of a Lufthansa Group (LHG) case study that showcases experiences and lessons gathered in Kenya, Rwanda, Nigeria and Columbia. In doing so, this paper extends previous research in the following ways: First, based on the inherent CSR conflicts discussed in Sect. 1.1, we present and investigate a special case of CSR activity, namely the combination of CSR with a qualification program for employees. Second, we contribute to the stakeholder view of CSR by analysing multiple perspectives and outcomes with special focus on the sustainability of such programs. Third, applying a case study approach, we aim at generating insights from in-depth analysis to inform future research and practise.

The paper is organized as follows: Sect. 2 introduces the historical evolution and intent of the LHG Impact Week (IW) initiative. Section 3 elaborates the paper's case study methodological approach and presents data analytical approaches employed. Section 4 presents the data: It first provides an overview of all IWs conducted to date. Starting with the inception of the IW using a pilot in Kenya, Nairobi, the section then provides contextual descriptions, qualitative and quantitative data where available for each IW location in chronological order. Thereafter, the section synthesises sustain-

[7]https://www.iso.org/iso-26000-social-responsibility.html.

able outcomes. The discussion in Sect. 5 analyses main findings and looks at selected key stakeholders that were affected in some way by this CSR initiative. Section 6 recapitulates the main findings of this paper with a concise concluding synthesis that distils both the main lessons learned and offers hypotheses and a shortlist of recommendations for future similar projects.

2 Historical Evolution and Intent of the Lufthansa Group Impact Weeks

In this section, we describe the concept and intent of the IW. We intentionally start with the macro perspective—the Lufthansa Group CSR strategy and the overall context of the IW (Sect. 2.1). We then describe the micro level—the evolution, concept, procedure and intentions of the IW (Sect. 2.2).

2.1 Macro Perspectives: Lufthansa Group CSR Strategy and Overall IW Contextual Fit

As part of its corporate strategy, the LHG commits itself to creating added value for customers, employees and investors and to take on responsibility for environmental and societal issues:

> Corporate responsibility, that is to say sustainable and responsible entrepreneurial practice, is an integral part of our corporate strategy. It means that we are committed to creating added value for our customers, employees and investors and to meeting our responsibilities toward the environment and society. (LHG n.d., para. 1)

This ambition also means including social responsibility and corporate citizenship within the company's sustainability agenda and "as a company [to] actively engage in numerous social concerns" (LHG n.d., para. 2). In 2015, when the United Nations' member countries adopted the 2030 Agenda for Sustainable Development and its 17 sustainable development goals,[8] the LHG became "the first signatory of the UN Global Compact from the ranks of the aviation industry" (LHG n.d., para. 7). Apart from the company's concern with climate and environmental responsibility "[…] the Lufthansa Group is particularly committed to the areas of education and enabling through the help alliance, its charitable organization." (LHG n.d., para. 8). With its strategic focus on "Education & Enabling" the *help alliance*[9] supports projects around the world and thereby collaborates closely with competent local partners around development cooperation. Originally founded in 1999 by a group of 13 Lufthansa employees, the *help alliance* today marks the central pillar of the

[8] https://www.un.org/sustainabledevelopment/.

[9] https://www.helpalliance.org/?lang=en.

Lufthansa Group's corporate citizenship activities. The engagement and involvement of many different employees of the LHG continues to be the core element of the organization, e.g. every project is coordinated by a LHG employee on a voluntary basis. Based on this strong employee-driven background, the *help alliance* is constantly developing into a more professional organization. Hence 2016 marked the transformation from a registered association to a non-profit making private limited company and the integration into the LHG.

2.2 *Micro Perspectives: Evolution, Concept, Procedure and Intentions of the LHG IW*

The LHG engagement in the IW,[10] a combination of CSR engagement and personnel development, started in 2016 as an initiative of the lead author of this paper, who spearheaded the program in his role as one of the leading change managers responsible for the cultural transformation of the company. The IW is the first program within the LHG to systematically combine strategic CSR with an internal training of key business skills for employees, including senior managers, project leaders, and team leaders. The initiative was enabled by three developments: (1) the professionalisation of the *help alliance* with its strategic focus on education and enablement; (2) a growing demand on the part of employees to engage themselves in meaningful, social projects; and (3) an increased market requirement reported by the business units for more customer centricity, innovation, and speed to market. The IW had the potential to offer answers in respect of all three demands. First, the actual implementation of the IW included the integration in the Lufthansa Group CSR strategy, specifically into the portfolio of the *help alliance*. Second, it required the integration of the IW into the training portfolio of the LHG, qualifying employees and leaders in relevant business skills. Third, the IW was designed as the combination of a CSR project fostering entrepreneurship and innovation in underprivileged regions of the world and as an internal personnel development measure to qualify employees and leaders for the digital and cultural transformation of the company.

Impact Week (IW) Idea and Inception: The IW was first launched in 2015 as an initiative of Kirche in Aktion (Church in Action),[11] a Frankfurt-based organization, which runs projects all over the world with the clearly expressed aim of improving living conditions all around the globe. Its motto is "to create a piece of heaven on earth".[12] The IW teaches people how to create business models out of existing problems and offers them tools and knowledge to found new businesses and make a living. Figure 1 displays the four phases of the IW model since the Lufthansa engagement commenced in 2016, including (1) set-up, (2) implementation, (3) follow-up support after the IW, and (4) long-term sustainability.

[10]http://impactweek.net/.

[11]www.kircheinaktion.de.

[12]http://www.kircheinaktion.de/donate/.

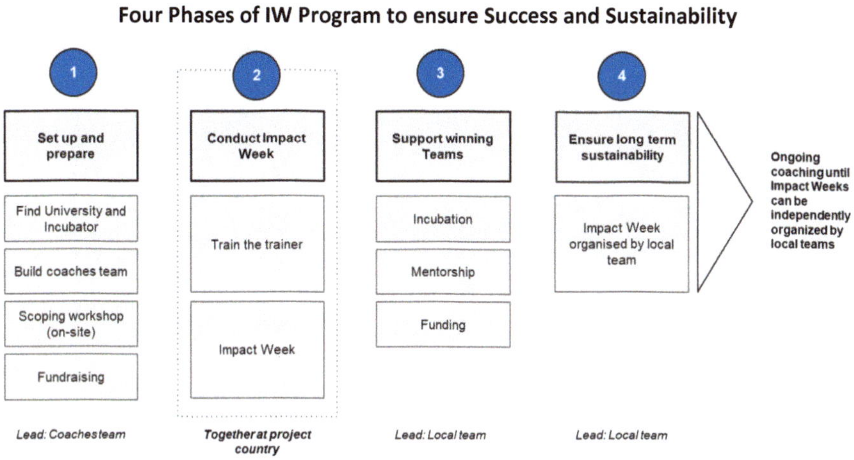

Fig. 1 The four phases of the IW model: (1) set-up; (2) implementation; (3) follow-up; (4) long-term sustainability

Each IW includes four phases. During the preparation stage (Phase 1) the IW organizers identify a local partner institution and incubator, find sponsors and recruit a team of professional and corporate volunteers. Professional volunteers are recruited via social media and networks. Thereafter (Phase 2), the IW is implemented. Since 2016, the IW itself has consisted of two parts. The first part is a two-day training or so-called train the trainer (TTT), which qualifies faculty members of local higher education institutes as well as employees from the business world (corporate volunteers) to coach students to develop new business models. The TTT component is a key sustainability element of the program: Through the qualification of professors and lecturers, university institutions are being enabled to run their own programs without external input (see Gielnik et al. 2015). In the TTT, faculty members and corporate volunteers learn how to apply the Design Thinking approach, an innovation process (Grots and Creuznacher 2016) that "is in large part a response to the increasing complexity of modern technology and modern business" (Kolko 2015, para. 1) and that many companies like IBM and GE have adopted in order to simplify and humanise their product development processes.[13] In the educational sector Design Thinking is often linked to business and innovation studies and has become popular at Stanford University. In the second part of the IW, immediately following the TTT, faculty members and corporate employees apply their new skills in a four-day workshop format with students from the respective university (Fig. 2).

More specifically, they coach students to identify business opportunities, develop solutions, build prototypes, test their prototypes with potential users, and pitch their business models and products in front of a jury of local and international experts (see Figs. 3, 4, 5 and 6 depicting exemplary pictures of the Design Thinking phases

[13]https://hbr.org/2015/09/design-thinking-comes-of-age.

Fig. 2 Photo exemplar showing a group of students, faculty and international volunteers who participated in a four-day workshop (Part 2 of the IW) at Kigali, Rwanda, in 2017

taken during the IWs in Kigali, Rwanda). Five winning teams are selected who get access to an incubator and receive prize money to fund and further develop their initial business ideas. After the IW, professors, lecturers, corporate and professional volunteers are available to support the teams by offering mentoring.

Besides the Lufthansa Group with the *help alliance*, Lufthansa German Airlines, and Brussels Airlines, a number of other sponsoring organizations also contributed to the IWs over the years, including SAP, Accenture, Price Waterhouse Cooper, design@business, and German Society for International Cooperation (GIZ), among others.

In line with the general strategic targets for LHG social responsibility and corporate citizenship activities, the specific goals for the IW were defined along three dimensions: Social impact, employee impact, and business impact (Fig. 7).

The aspired social impact of the program is to foster entrepreneurship, innovation and intercultural exchange. On the individual level, it affects the local stakeholders at the IW locations. Faculty as participants of the participants of the TTT component are expected to be able to coach and facilitate innovation processes. Students as participants in the workshop are expected to improve their entrepreneurial mindset and innovation skills. On the institutional level, the IW is expected to enable local universities to carry out independent IWs and to increase student start-up rates. Finally, due to the multinational nature of the IWs, the initiative is expected to increase participants' intercultural competence and thereby to contribute to a better international understanding and integration between participating nations.

Fig. 3 IW Kigali, Rwanda (2017): Interviewing potential customers to identify specific needs and business opportunities

Fig. 4 IW Kigali, Rwanda (2017): Mastering complexity and developing solutions through visualisation and clustering

Fig. 5 IW Kigali, Rwanda (2017): Building prototypes to collect customer feedback

Fig. 6 Award ceremony IW Kigali, Rwanda (2017): Student participants with Design Thinking certificates

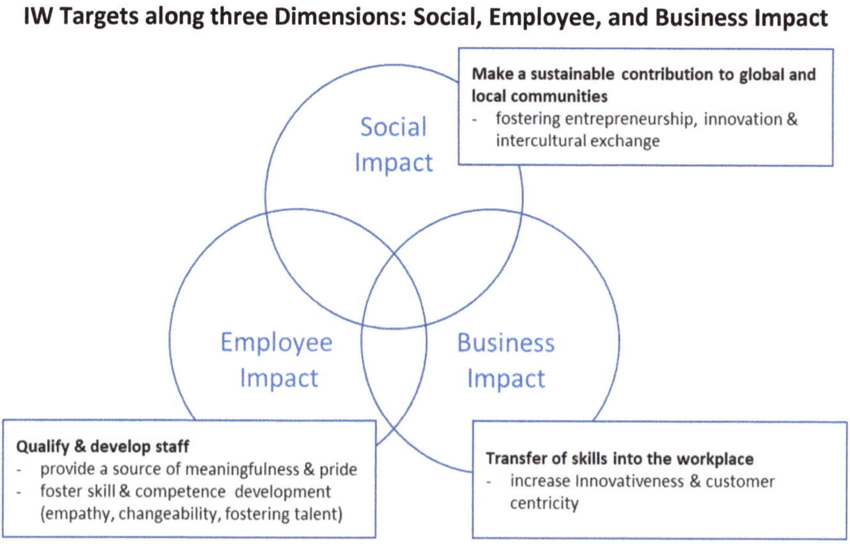

Fig. 7 In line with the Lufthansa Group CSR framework, IW targets include specific targets contributing to social, employee, and business impact

The immediate anticipated impact on employees can be characterised as the acquisition of new skills and an entrepreneurial mind-set, as well as intercultural competence. In the long run, the program is also expected to have positive effects on employee engagement and on employer branding.

Finally, the main desired business-related outcome of the IWs is for program participants to apply and further develop in daily business operations those new skills and techniques that were acquired during the program, as well as to nurture new networks with innovation experts from other companies. Figure 8 summarises the main anticipated long-term effects of the IW.

3 Case Study Methodological Design Features and Approaches

In our study we report data collected between July 2016 and April 2018. We chose a case study methodological design in order to portray a holistic view of the IW program. Case study approaches offer important benefits for research into fields of investigation characterised by complex multidisciplinary and multicausal interrelationships (Bryman 2016; Johnson and Christensen 2017; Punch 2014). Benefits of the case study methodology are elaborated in Sect. 3.1 followed by a short synthesis of data analytical approaches (Sect. 3.2).

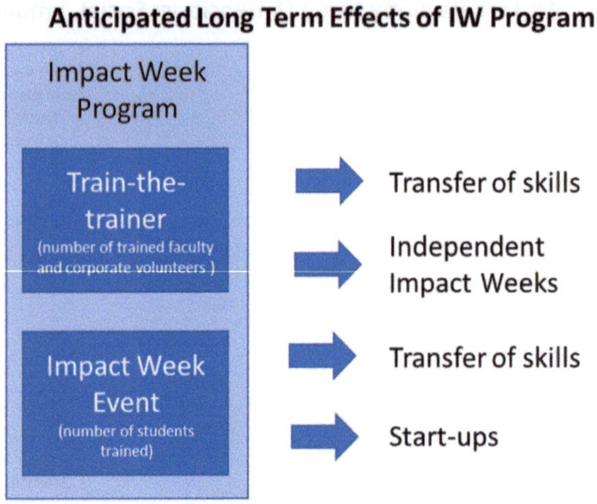

Fig. 8 IW Program elements and desired long-term outcomes

3.1 Case Study Methodological Appropriateness: A Short Overview Discussion

Expressed in simple language, "[t]he basic case study entails the detailed and intensive analysis of a single case." (Bryman 2016, p. 60). Relatedly and importantly, a case study "has a holistic focus, aiming to preserve and understand the wholeness and unity of the case." (Punch 2014, p. 120). Moreover, a case study can offer particular promise where there is a need for in-depth evaluation, assessment or analysis (Stake 1995). According to Creswell (2014, p. 14),

> Case studies are a design of inquiry found in many fields, especially evaluation, in which the researcher develops an in-depth analysis of a case, often a program, event, activity, process, or one or more individuals. Cases are bounded by time and activity, and researchers collect detailed information using a variety of data collection procedures over a sustained period of time. (attributed to Stake 1995 and Yin 2009, 2012)

Further, cases may be comprised by a multiplicity of defining features that are conceived and studied within a bounded context and that may not be similarly captured, comprehended and described by alternative research methodological approaches. They are particularly suited in situations where knowledge is shallow, fragmented, incomplete, non-existent.

> All data relevant to the case are gathered, and all available data are organized in terms of the case. The case study method gives a unitary character to the data being studied by interrelating a variety of facts to a single case. It also provides an opportunity for the intensive analysis of many specific details that are often overlooked with other methods. (Punch 2014, p. 121)

Even though case studies are open to the criticism that they are "only one case, so how can we generalise?" (Punch 2014, p. 122), their perceived benefits outweigh this potential weakness for at least five important reasons: (1) There are situations where generalisability may not constitute the primary study intention but "the case may be so important, interesting or misunderstood that it deserves study in its own right" (Punch 2014, p. 122); (2) Nevertheless, dissimilarities should also not be overstated: "Clearly, every case that can be studied is in some respects unique. But every case is also, in some respects, similar to other cases." (Punch 2014, p. 123); (3) Case studies are consistently employed as effective teaching tools (Bryman 2016; Johnson and Christensen 2017) in situations where "historical cases are studied in great detail and are used to train managers, doctors, lawyers, and so on" (Punch 2014, p. 124) to prepare them to deal with future scenarios; (4) Case studies enable both in-depth analyses of complex circumstances and contexts and are therefore well-suited to facilitate comparative analyses across "two or more cases" (Bryman 2016, p. 67) through "cross-case analysis" (Johnson and Christensen 2017, p. 437); (5) Finally, the capacity of the case study design to engage and manage complex research contexts surpasses the capacity of other research designs (Punch 2014), wherefore case studies are an effective means for uncovering and conceptually describing opportunities for future research:

> [...] only the in-depth case study can provide understanding of the important aspects of a new and persistently problematic research area. This is particularly true when complex social behaviour is involved, as is the case in much social science research. Discovering the important features, developing an understanding of them and conceptualising them for further study, is often best achieved through the case study strategy. Following this line of argument, it may be that too much research has tried to go straight to measurement and quantitative mapping without a fuller understanding of the phenomena and processes involved that are best achieved by case studies. (Punch 2014, p. 124)

In short, "[s]ome of the best-known studies in sociology are based on this kind of design" (Bryman 2016, p. 60).

3.2 Synthesis and Data Analytical Approaches

For the reasons mentioned, the case study design seemed to be well-suited to engage and describe the complex sociocultural and ethnoeconomic realities encountered in the course of the IW project planning and implementation. An overview of data analytical approaches is provided next.

In our case study approach, we move from description of implementation, providing facts and figures, to analysis and interpretation. Specifically, the case study approach includes the following steps: On a descriptive level providing qualitative and quantitative data we illustrate how the IW concept was put into practice and which and how many stakeholders were included. As already mentioned (Sect. 3.1), this case study comprises predominantly qualitative data analytical approaches. Notwithstanding, although the IW was not designed as a quantitative training evaluation, we add some supplementary numerical data for the IW locations Bogota, Kigali, and Lagos, which were collected from short online questionnaires. Approximately one week after the IWs, all roles involved in the IW Bogota and Kigali (professional volunteers, participants of the TTT, as well as students) received a link. In Lagos only professional volunteers and coaches from the LHG received the link. The short questionnaire included open questions like improvements that participants thought should be made in future IWs, a comment field to briefly describe how they personally experienced the IW, and a scaled question from 1 to 10 to indicate the likelihood with which they would recommend the IW to other people (these data were subsequently converted into the Net Promoter Score (NPS), with possible scores from −100 to 100 [see Grisaffe 2007]). In total 73 participants filled out the questionnaire. In Kigali participants additionally filled out a paper-pencil questionnaire after the IW event asking for their main employment status (74 respondents). Approximately five months later participants were contacted again in a follow up telephone interview done by the GIZ (53 respondents).

Further insights are gained by means of disparate sources, including participant observations throughout the programs, and verbatim personal statements of different stakeholders from sources such as social media blogs, emails, or personal conversations, etc. In a next step of our approach we report specific outcomes of each IW and describe to what extent the initial goals of the IW were met. A special focus is placed on sustainable effects of the IW program. In doing so, we also describe outcomes that were somewhat unexpected and exceeded the expectation of the organizers. In our analysis, we report learnings and analyse success factors of the IW on a macro as well as on a micro level. In a final step, we conduct comparative analyses across the different IWs and inductively offer interpretations and hypotheses (theory generation). The last two steps will be undertaken in the discussion (Sect. 5), which analyses the descriptive results and exposition of outcomes presented next.

4 In-Country Implementation and Outcomes

Section 4.1 provides an overview of data collated from all IWs conducted to date. Starting with the inception of the IW using a pilot in Kenya, Nairobi, we provide contextual descriptions, and qualitative and quantitative data where available for each IW location in chronological order (Sect. 4.2). Thereafter, we present course of events and outcomes following early IW programs, including sustainable outcomes (Sect. 4.3).

4.1 Descriptive Data About the Impact Weeks

The LHG engagement in the IW started in 2016 in Nairobi after the concept had been extended to include a train the trainer (TTT). Table 1 displays all IWs conducted since 2015, including overall number of participants, organizers, professional volunteers involved, coaches trained, and number of students participating at each location. Since the first IW 619 students or young people have participated in the IWs, ranging

Table 1 Statistics of the Impact Weeks (IW) from 2015–2018

	First IW	Lufthansa sponsored IWs				Independent IWs		Σ
	Nairobi 2015	*Nairobi 2016*[a]	Bogota 2017	Kigali 2017	Lagos 2017	Nairobi 2017	Kibera 2018	
Organizers	4	2	3	3	2	3	2	19
Lead trainer(s)	n.a.	1	1	2	1	1	1	7
Professional volunteers or multipliers[b]	20	7	5	6	10	11[b]	2[b]	61
Coaches trained								
Local faculty/multipliers		20	24	22	26		15	107
Corporate volunteers								
(incl. LHG)	n.a.	11 (5)	11 (6)	12 (9)	16	n.a.	1	51
Students/youth trained	100	122	75	100	144	56	22	619

[a]beginning of Lufthansa Group (LHG) sponsoring and engagement in the IW
[b]from previous IWs

from 22 in Kibera to 144 in Lagos. In total, 107 faculty members or other multipliers have been trained as Design Thinking coaches (20 in Nairobi, 24 in Colombia, 22 in Kigali, 26 in Lagos), together with 51 corporate volunteers, including 36 employees of the Lufthansa Group (LHG). Nineteen organizers helped set up the IWs, and seven lead trainers carried out the TTT and facilitated the whole IW event.

4.2 Implementation and Outcomes of the IWs in Kenya, Colombia, Rwanda, Nigeria

Each Impact Week (IW) was initiated by a former IW participant and had a local partner in the hosting country. The concept and schedule was the same for every location and was implemented as planned with only minor variations across countries.

IW Nairobi, Kenya (1–7 July 2016): The first year of the IW at LHG began in 2016 in Kenya, Nairobi. Partner university was the Africa Nazarene University (ANU),[14] a private university on the outskirts of Nairobi that is based on Christian values, promotes social change and entrepreneurial spirit (Vision: "what starts here transforms the World"). Departments involved were the School of Business and the School of Science and Technology. Nairobi provides good opportunities for start-up activities. There is a vibrant start-up scene as well as several incubators available for

[14]http://www.anu.ac.ke/.

entrepreneurs. Even so, about 42% of the Kenyan population of approximately 44 million people still live below the poverty line.[15] Together, the strong need for job creation in the country, a favourable infrastructure in Nairobi, and the local Lufthansa business engagement were motivating factors for the IW in Nairobi. The IW Kenya served as the lighthouse and pilot project at the LHG for the combination of CSR with a qualification for business skill. An internal announcement of the IW Kenya 2016 with five places available for LHG employees resulted in 31 colleagues applying for participation. A total of 29 managers explicitly supported the participation of their employees. Based on several diversity criteria five employees from different business units were ultimately selected for participation. At the beginning of the IW program, 20 local faculty members, 6 participants from other companies, and the 5 LHG employees were trained as Design Thinking coaches by 7 professional volunteers during a two-day TTT that preceded the IW. Following this TTT schedule, trainers coached teams of students as they facilitated the process from identifying needs, developing solutions, testing prototypes and developing business models. To offer inspirational input, local speakers, opinion leaders and role models were invited who talked about their experiences as entrepreneurs.

A qualitative evaluation of the IW indicated positive results on behalf of the partners in Kenya as well as the LHG participants concerning learning outcomes and application of skills in business (e.g. all participants reported to have applied their new skills in business, e.g. by facilitating Design Thinking workshops). The positive outcomes prepared the ground for the decision to make the IW an integral part of both the Lufthansa Group (LHG) CSR strategy as well as the internal training portfolio. Supported by a diverse range of sponsors, including the *help alliance*, LHG, GIZ (German Society for International Cooperation), and SAP, among others, participants of the IW 2016 began to scale the IW program and to look for new partners, which resulted in three new IW locations being identified in 2017: Bogota, Kigali, and Lagos.

IW Bogota, Colombia (19–29 September 2017): Co-operation partner in Bogota was La Universidad de Bogota Jorge Tadeo Lozano (UTADEO).[16] Following the vision “Innovamos: nuestra esencia” (“we innovate: it is our essence”), the university actively promotes research, innovation and creativity in all academic areas. The IW was hosted by the Faculty of Arts and Design together with the Faculty of Social Sciences, and the School of Engineering. The IW came to Colombia in the year after the peace accord ended the longest running armed conflict in the country and launched the peace process to reintegrate former fighters and foster reconciliation.[17] The IW Colombia was thus initiated to bring well-known tools to a country that persists in a process of significant transformation, to support the peace process, and to allow people to develop sustainable solutions that fit their local culture. Content areas for the development of business models during the IW were chosen by UTADEO prior to the IW. Students from various age groups were preselected by UTADEO

[15]https://www.unicef.org/kenya/overview_4616.html.

[16]https://www.utadeo.edu.co/es.

[17]http://www.bbc.com/news/world-latin-america-38096179.

Fig. 9 Winning team "Justo Gusto", IW Bogota, Colombia (19–29 September 2017), with a business idea to give farmers direct access to the market without intermediaries and to provide city people with organic products

for participation in the IW and split into groups to solve problems pertaining to the following topics: Education, fair trade, tourism, regional development, public policy. In total 35 participants (24 locals, 6 LHG employees, 5 from other companies) were trained as Design Thinking coaches who during the IW event built tandems to coach 75 students (most of them in their final year before graduation) to develop business ideas. The student teams developed 14 business models. Among the teams that were selected by the jury as winners during the IW award ceremony was "Team Parceria", which created a board game to bring politics back to the people by making it fun, non-discriminative, and accessible. Figure 9 shows the first winner, team "Justo Gusto" with a fair-trade business model who formally registered a business after the IW.

Altogether 35 IW participants took part in the online survey (17 were students, 14 coaches, and 4 professional volunteers). The description of the IW experience commonly included statements such as "life changing" (corporate volunteer); "Impact week was a totally changing life experience, a new way to work in every little project" (student); "Impact week was a door for different cultures to get to know themselves, learn from each other, and understand how innovation is being tackled in different countries" (professional volunteer); "Impact Week was an intercultural, interdisciplinary and personal exchange" (lecturer, local coach). Over 94% indicated that they would highly recommend the IW to others (NPS = 91).

IW Kigali, Rwanda (3–13 October 2017). The IW Kigali was hosted by the Technical Highschool of the SOS Children's Villages in Kigali[18] and was locally supported by the GIZ. The partnership between SOS, GIZ and the IW was based on a shared vision of increasing youth employment and a strong belief in the important role of education and qualification for the future of the country. Main local organizer of the IW was the president of the alumni network of the Technical High School, together with the high school director. Among the topics that the Technical High School chose for the IW to tackle were e-commerce, agriculture and health care. The 34 participants of the TTT included teachers of the Technical High School, volunteers from NGOs, a team from the Impact Hub, an incubator in Kigali, and the corporate volunteers, including 9 employees from the LHG. Three former participants of the IW Nairobi, Kenya (2016), were also trained by six professional volunteers as participants during the TTT process. In the actual IW event 100 students and alumni from the Technical High School participated (see Figs. 2, 3, 4, 5 and 6). Among the winning teams was "Team FLASH" that is still working on their idea to solve the problem of electrical supply through on call or chat support.

In Kigali, 25 participants took part in the online questionnaire (15 coaches, 7 students, 3 professional volunteers). Most common suggestions for improvement were the wish for more information and better communication prior to the IW, more role clarity, and time for individual preparation. Respondents described their IW experience as follows: "Impact week was truly an amazing learning experience that taught me that there is no limitations to my ideas and how to develop them, there are only plateaus and you must not stay there, you must go beyond them" (local coach); "putting ideas together and then build them and sharing ideas with others" (student); "Impact Week was an amazing experience I wouldn't want to miss. Major flood of insights from all sides with the best team ever! Motivation generator!", "Impact Week was a real life changing experience!" (both corporate volunteers); "Impact Week was a unique experience that will have a lasting effect on my further life" (professional volunteer). Of the participants in Kigali 83, 61% indicated that they would highly recommend the IW (NPS = 83).

Furthermore, surveys conducted by GIZ (Table 2) showed that the majority of participants did not change their student status. However, in respect of employment status there was a positive increase: The percentage of those "employed" rose from 3 to 9%, and the percentage of those identifying as "own account worker" rose from 3 to 13%.

IW Lagos, Nigeria (20–26 October 2017): The IW Lagos was hosted by the University of Lagos (UNILAG)[19] and the School of Entrepreneurship at which an incubator was launched just in time for the IW. Having the biggest national economy on the African continent, Nigeria harbours many start-ups and entrepreneurs. To date the economy strongly depends on its oil reservoirs. The country is therefore exploring new sources of income offering potential opportunities for entrepreneurial activities. Because of a perception of some negative press about Nigeria in Western media,

[18]http://www.soshgths.org.

[19]https://unilag.edu.ng.

Table 2 Employment status of participants contacted during (Baseline) and five months after (Tracer) the IW Kigali, Rwanda (5–12 October 2017)

Main Employment Status				
	Baseline respondents	% (rounded)	Tracer respondents	% (rounded)
Employed	2	3	5	9
Own-account worker	2	3	7	13
Running a business	5	7	2	4
Contributing to a family business	2	3	2	4
Member of a producer's cooperative	0	0	0	0
Internship	0	0	1	2
Unemployed	18	24	8	15
Studying (I'm a student)	43	58	28	53
No answer	2	3	0	0
Total	74	100	53	100

(*Data* GIZ)

the IW was also an opportunity to experience and get to know the country and its people from a different angle. The topics UNILAG chose for the IW were: Tourism, finance, technology, education, commerce, and health. Ten professional volunteers from 5 different nations trained 26 faculty members and 16 corporate volunteers from the LHG to be qualified Design Thinking coaches. The 16 corporate volunteers also included 6 local LHG employees and partners. 144 students took part in the IW event (Fig. 10). Among the winning teams was "Easyfare" who tackled the problem of slow and insecure payments on public transport by introducing a new cashless service.

In Lagos 13 participants (4 professional volunteers, 9 coaches) took part in the online survey. Most common suggestions for improvement were to receive more time or an extra day for the TTT process, more time to reflect, and more rest. Respondents described their experience in similar ways: "Impact Week was an unforgettable experience, packed with Design Thinking learnings as well as learnings about the Nigerian people and culture. I left with a full heart, good memories, and new friends.", "It [the experience] was encouraging at first and exciting in many ways, and now looking back I'm sure there isn't going to be something alike. I'm sure this is something I'm going to remember lifelong, not only because of the people, the coaches, the country, the climate, the conditions, the food […] but of what has made me grow as a person and as a professional." (both professional volunteers); "Impact Week was

Fig. 10 Impact Week Lagos, Nigeria (20–26 October 2017): Biggest impact week with 144 students and 42 faculty members

one of the most intense and rewarding events I have ever experienced and broadened my horizon in so many ways I could have never imagined", "A once in a lifetime experience, absolutely amazing and mind-changing." (both corporate volunteers). Of all participants in Lagos, 100% indicated that they would recommend the IW (NPS = 100).

4.3 Sustainable Outcomes and Unexpected Results of IW Programs

Figure 11 depicts all IWs held since 2015 and illustrates sustainable effects of investments made in early IWs, in particular in Kenya. Comparing Fig. 11 and Table 1 reveals that several independent IW initiatives have organically emerged from inputs made during the LHG sponsored IWs. These spin-off IW activities were (and continue to be) independent of Lufthansa sponsoring and are thus classified as 'sustainability effects'. For the purposes of this paper, these 'sustainability effects' are an important effect of Lufthansa's CSR engagement. Relevant outcomes are discussed next.

After IWs 2015 and 2016 in Nairobi, Kenya, faculty of the ANU conducted a prototype IW with 40 students from a local high school without support from any external sources. Thereafter, ANU conducted their first independent IW (1–3

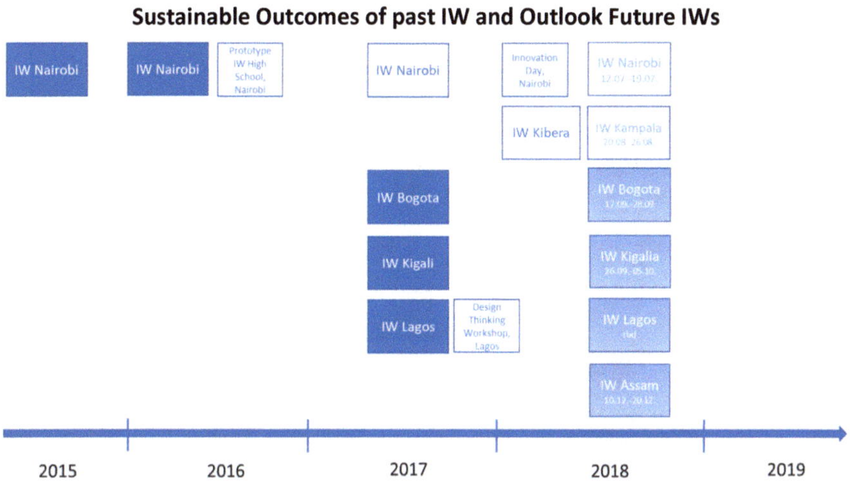

Fig. 11 Past Lufthansa supported and sponsored IWs (blue rectangles), independent IWs or similar events (white rectangles), and future IWs planned for later in 2018 (light blue rectangles)

November 2017) on their premises following a one-day refresher for the faculty members. In the three-day IW event 8 faculty members and 3 trained coaches from IW Rwanda supported 56 students to develop business models (Figs. 12, 13 and 14). ANU also engaged in 7 similar workshop events applying the Design Thinking approach (e.g., during the Innovation Day, the biggest entrepreneurship forum in Kenya where faculty members facilitated a Design Thinking workshop with 20 participants). Altogether 10 faculty members took the lead in organizing the different events and training others. The next independent IW in Nairobi is planned to take place in July 2018 (Figs. 15 and 16).

Most recently, another independently organized IW became the first-ever IW to be offered in a slum area, namely in Kibera, Kenya (16–20 April 2018), considered to be "the biggest slum in Africa and one of the biggest in the world." (Kibera UK, para. 1) (Figs. 17, 18, 19 and 20). IW Kibera was organized by 29-year-old Caleb Aringa. Caleb, born and raised in Kibera, had participated in the IW Nairobi 2016 as a non-student. He also took part in the IW Kigali where he was trained as a coach. Two faculty members of Africa Nazarene University, Nairobi, supported Caleb and conducted the TTT and the IW moderation. The two faculty members had also taken part in the IW Kigali where they had intensified their coaching skills.

IW Kibera started with a 2-day TTT of 15 Kibera residents known as community leaders who were between 25 and 35 years old. The majority were small and micro business owners or so called "husslers", people who do different small jobs as they come along. The IW event took place in the Town Centre Kibera and had 22 participants aged between 17 and 25 years, all residents of Kibera, with the majority still attending school or being apprentices. Lufthansa representation was limited to a single former participant at IW Nairobi who joined the IW Kibera.

Fig. 12 Independent ANU organized Impact Week in Nairobi, Kenya (1–3 November 2017): Interviewing potential customers to identify specific needs and business opportunities

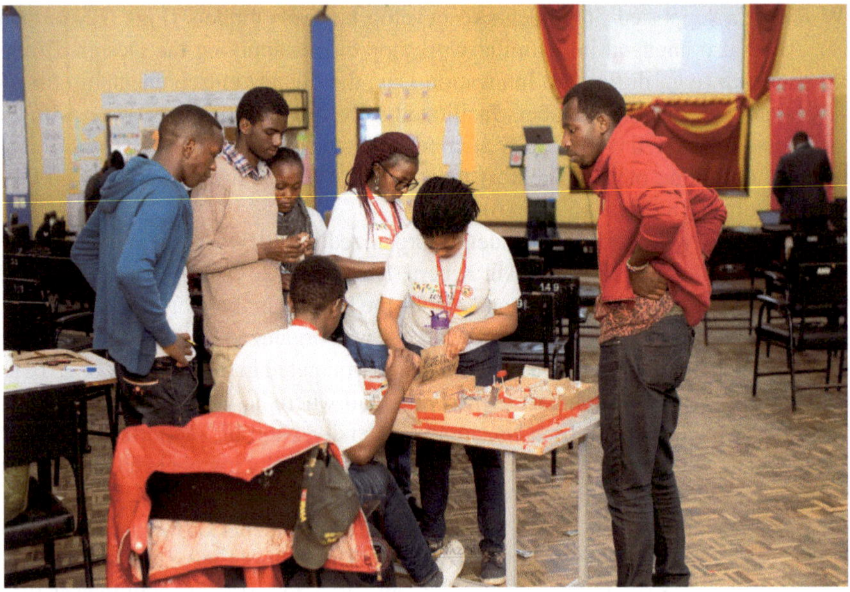

Fig. 13 Building prototypes in order to collect customer feedback at independent IW Nairobi, Kenya (1–3 November 2017)

Fig. 14 Presenting business model and prototype at independent IW Nairobi, Kenya (1–3 November 2017)

Fig. 15 Team coaches including participants from IW Kigali receiving coach certificates at award ceremony during independent IW Nairobi, Kenya (1–3 November 2017)

Fig. 16 Celebrating the success of the first independent IW in Nairobi, Kenya (1–3 November 2017)

Fig. 17 Participants of IW Kibera in front of the Community Centre in downtown Kibera, Kenya's biggest informal settlement (16–20 April 2018)

Fig. 18 Manager of Kibera Community Centre at IW Kibera, Kenya (16–20 April 2018)

Fig. 19 Participants inside Kibera Town Centre at IW Kibera, Kenya (16–20 April 2018)

Over the course of the different IWs, both expected and unexpected multiplier and snowball effects could be observed. Professional and corporate volunteers were

Fig. 20 Team Impact Week Lagos from 8 nationalities and 18 companies: "One team, one goal—that's all that matters" (quote professional volunteer)

somewhat expected to step up into leadership roles after their first participation in an IW. This was indeed the case. All lead organizers of the IW Bogota, Kigali, and Lagos had been engaged in the IW Nairobi 2016 either as corporate or professional volunteers. Of the five Lufthansa volunteers in IW Nairobi 2016, four co-organized an IW event in 2017. What came as an unexpected result was the unplanned exchange between African IW locations and that a similar snowball effect could be observed as with the professional and corporate volunteers. Two participants of IW Nairobi 2016 applied for participation in the TTT in Kigali via the official online registration form. In the end three places were sponsored for Kenyan participants who further enhanced their skills in Kigali. One month later, IW Kigali sent three participants to Kenya to support IW Nairobi 2017. IW Kigali 2018 on the other hand will once again be supported by two Kenyans who—after five IWs and eight related workshops—have reached professional skill levels.

Moreover, the spin-off to other Design Thinking formats and events exceeded most expectations and—with one event in Lagos—could not only be observed in Kenya. Finally, IW Kibera in and of itself, initiated, organized and conducted locally, was an unexpected result that could be traced back to IW Nairobi 2016. The LHG volunteer who joined IW Kibera described the IW Kibera as the most effective and intense she has joined. A number of circumstances are likely to have played a role in this. First, the IW was small in size, allowing more personal interaction between group members. Second, all participants had the same background of being born and raised in the slums, thus creating a sense of togetherness and intimacy right from the start.

Third, local IW leaders with their clear-cut messages may have contributed to an even deeper understanding on behalf of the participants. Finally, and perhaps most importantly, there was an urgent felt need on the part of participants to learn and innovate in order to improve the living conditions of their communities. Records of daily participant feedback further support these observations and indicate what impact the program has made. Several sample verbatim responses are reproduced next:

> "Everyone showed who they really are and everyone was open and free and we really collaborated. I wished we could have more time. I plan to change the community with that."
>
> "I was very skeptical in the beginning but this really came out to be a new level of fulfillment concerning the learning experience and team work."
>
> "We really created a family. Now we are together."
>
> "We are encouraged to collaborate but in daily life we work alone most of the time. So, finding time to actually collaborate is really great and I loved how everyone encouraged me and helped me going."
>
> "I have an own business and there I can use it immediately. I will use it for our community center. This training has really taken me to another level."

The sustainability of the IW concept on a program level of analysis is also underpinned by accounts of local stakeholders, thus indicating long-term effects on an individual level of analysis:

> [I] attended the Impact Week in Nairobi which lead to a life changing experience: Design Thinking helped to change my life and enabled me to improve the circumstances in my community. Design Thinking allows even the simplest of ideas to develop into a business. Thanks to the Impact Week in Kenya I am now running a community project, and are constantly exchanging my learnings with other former participants who managed to also start their own business. (participant IW Nairobi 2016)
>
> As a result of the uptake of the Design Thinking approach, my students and I have been able to come up with an innovative modus operandi for facilitating the learning of programming fundamentals. We instituted tutorials that have shown a quicker uptake for programming and a heightened interest in learning. I have also re-addressed how I require class presentations to be conducted, and I now encourage presentations beyond the use of power-point so as to nurture innovation and critical thinking. (faculty member, IW Nairobi 2016)

5 Discussion and Interpretation of Results

Since the engagement of the Lufthansa Group (LHG) in the Impact Week (IW) project commenced in 2016, a total of 28 professional volunteers, 51 corporate volunteers, 92 faculty members and 519 students participated in the IWs. The first long-term goal of the IW was to generate independent IWs carried out by local universities. The second long-term goal was to increase the student start-up rate at the respective universities. Both goals represent the desired social impact of the IW. From an internal corporate angle, employees and the business were supposed to benefit from the IW. The desired effects on corporate volunteers were engagement and acquisition of new skills (employee impact). As another long-term effect, employees were expected to apply their new skills in business at their individual workplace (business impact).

In this section, we address long-term effects of the IW and explore factors that enabled or prevented desired outcomes. Further, we analyse the internal corporate perspective of the IW and identify general conditions that are favourable for CSR (Sect. 5.1). In a final step, we address limitations and avenues for future research (Sect. 5.2).

5.1 Analysis of IW Impact: Social, Employee, and Business Impact

The CSR engagement in the IW Nairobi 2016 showed a long-term effect, namely it generated the first independent IW in Kenya in 2017 carried out by local faculty. Additionally, the second independent IW is now scheduled to take place in July 2018. There were also various spin-off activities by faculty members that included elements of the IW, thereby further demonstrating that the skills acquired during the train the trainer (TTT) are being continuingly applied. An unexpected long-term effect was the IW Kibera, which was the first IW that took place in a slum area outside the formal educational sector and with different target groups. A second series of IWs is scheduled or planned in Bogota, Kigali, and Lagos. Assam in India will be added as a new IW location. Kampala is planned as a new locally organized IW and spin off of IW Kibera. All things considered, the IW concept appears to be a successful concept that is being locally adopted (and adapted) across different cultures. The study demonstrates lasting effects on both the individual as well as the institutional level, indicating that the IW concept can generate sustainable results.

A comparative analysis across all locations where IWs were organized reveals important success factors for the effects described above. First, the IW setting fostered a working climate of shared ownership and mutual learning between volunteers and locals. Corporate volunteers and faculty members came to the TTT as participants and learners. They both faced the same challenges and later assumed the same role as coaches of student teams. In the majority of cases, student teams were coached by a tandem of local faculty members and corporate volunteers.

Second, the intercultural exchange was an explicit component of the IW mission, thus putting a strong emphasis on intercultural learning and relationship building. Again, the key was to have a common goal and to learn and work together.

Third, the Design Thinking approach appeared to be particularly suitable for safeguarding against "western knowledge" being imposed on other cultures. The Design Thinking approach—a human centric approach—merely offered a framework and a process. The volunteers facilitated this process, created the setting, asked questions, and supported their teams. Contents, ideas, and solutions were developed by the students themselves and thereby gave maximum room for the local and cultural perspective.

Fourth, the involvement of multiple stakeholders ensured that the IW mission was broadly supported by numerous partners with dissimilar yet complementary strengths. In short, all volunteers were committed to the goal to foster entrepreneurship, innovation, and intercultural exchange through Design Thinking (Fig. 20).

Fifth, former participants stepping into leadership positions ensured the sustained emergence of local multipliers, thus securing and perpetuating the longevity and sustainability of IW benefits. The exchange of coaches between Kenya and Rwanda was particularly critical in giving individuals the opportunity to further perfect their skills and emerge as local leaders.

Finally, lasting relationships and a cross-company community have emerged from the individual IWs, facilitating both ongoing local support and the organization of future IWs. In summary, the IW concept has generated long-term social impact, is accepted by local stakeholders, and appears to be a self-sustainable concept. We now turn to the internal corporate perspective of CSR.

The IW is now in its third year and continues to be a part of the Lufthansa Group (LHG) CSR activities. Given conflicts of interest and ongoing cost cutting and efficiency programs that are integral parts of big corporations, the longevity of such CSR initiatives cannot be taken for granted (Hubbard et al. 2017). The combination of external social program and internal qualification for the business as applied in the IW offers a way out of this "CSR dilemma". In the case of the LHG, the IW trained employees in critical business skills that are relevant for the implementation of the company's innovation and digitalization strategy. Participants later applied their skills in business, e.g. by organizing or facilitating innovation and product development processes or by training others and passing on their new skills. Moreover, IW participants gained new perspectives and mastered intercultural challenges, thereby increasing their adaptive capacities, a competence that is believed to be a critical success factor in today's business world marked by complexity and uncertainty (Keith et al. 2016; Unger et al. 2011). In summary, from a business perspective learning was the main selling point for management and the most notable return on investment for the company.[20] We therefore suggest that CSR activities include employee engagement and offer learnings that are closely aligned with overarching strategic business needs.

As discussed in Sect. 1, CSR has been open to the criticisms of 'greenwashing' and its self-regulatory 'check-box' approach. The setting and the development of the IW appears to have largely prevented such practices to occur and may thus offer some general principles and lessons for good CSR practice. First, the IW started as an initiative of a single employee followed and supported by a small group of dedicated internal promoters who helped further implement the concept. This organic employee driven approach of social engagement, e.g. via the *help alliance* (which again was also founded by Lufthansa employees), has a long tradition in the Lufthansa Group

[20]Former IW participants have acted as multipliers and first points of contact for the Design Thinking approach in their respective business units supporting a range of innovation and digitalization programs. The IW is now an established training program that is being announced annually in the company's management news.

and reflects a desire from within to take on responsibility, particularly at those locations where the company operates. Second, all internal stakeholders participated on a voluntary basis. Step by step a variety of stakeholders, including from the ranks of senior management, began to support the program either by engaging their employees, sponsoring flight tickets, prize money or other material goods. None of these stakeholders were instructed or ordered to do so but acted out of their own conviction and volition. Third, IWs were set up in a way that empowered organizers and participants to plan and implement the IW program as independent projects without interference from other stakeholders. Finally and importantly, our study suggests that corporate organizations can most successfully counteract 'greenwashing' effects and enable the organic evolution of CSR activities and employee engagement if ensuing activities can occur without any entanglement and interference of corporate politics.

5.2 Limitations and Future Research

As with all research, the current study has limitations and shortcomings. First, the study was not designed as a training evaluation and therefore does not allow causal inferences between measures and subsequent effects. For properly investigating effects and causalities future research should evaluate the IW program following a randomised control group design (see Gielnik et al. 2015). However, a number of indicators including time-lagged effects reported in the study suggest long-term effects of the IW measures. Second, not all IWs included the same set of data for the analysis. Moreover, some data were incomplete. This prevented a more thorough comparative analysis across different IW locations.

A third limitation pertains to the concept and implementation of the IW itself. Although there was some proof of entrepreneurial activity following the IW program, there was limited evidence of actual start-ups following the IWs (Table 2). Given the complex processes involved in founding businesses, the IWs are more likely to raise students' general interest in entrepreneurship and provide them with necessary tools and skills. One way to increase the likelihood of actual business creation is by means of a better and more targeted selection of participants (e.g. a stronger focus on talent or on individuals who are already intent on starting a business).

On a broader level, the IW program could be extended to also include a start-up training, guiding individuals with a concrete business idea through the process of registration and founding a business (Fig. 21). Such start-up trainings have previously been shown to be effective in increasing local start-up rates (Gielnik et al. 2015). Figure 21 also proposes a basic theoretical model from awareness and increased self-efficacy and skills to the intention to start a business to actual implementation and entrepreneurial activity. Future studies could combine IW and start-up training and test the assumed effects included in the model.

From the perspective of corporate stakeholders, the following questions should be addressed: Can employee engagement in CSR projects increase engagement at work? Do participants of such projects experience higher levels of meaningfulness

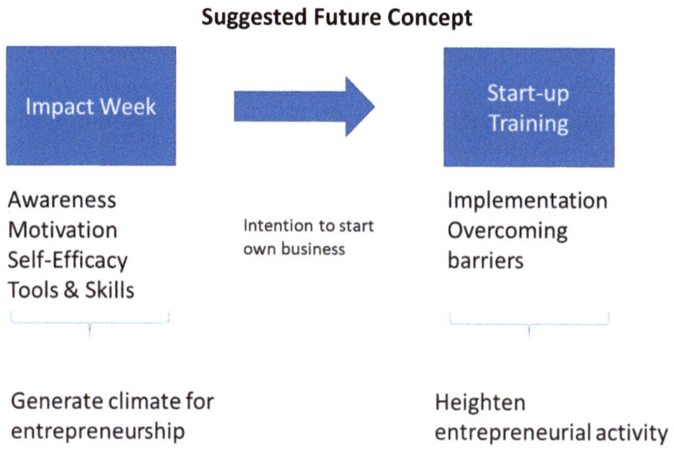

Fig. 21 Theoretical model from generating a climate for entrepreneurship to increasing entrepreneurial activity suggesting an extension of the IW to include start-up trainings

and purpose in life? (cf. von Devivere 2018) Does this affect motivation at work, and if so, in what ways? Other business-related impacts that were not addressed in this study include effects on employer branding. Does CSR engagement increase the attractiveness as employer and for whom (e.g. for the millennial generation with an assumed stronger desire for meaningfulness at work), and/or does it represent a competitive advantage in a company's 'war for talent'? (Bhattacharya et al. 2008).

Furthermore, emerging themes from our study suggest fertile opportunities for future research that explores the role of 'deep personal impact', 'longing for meaningfulness', and 'sense of togetherness' across countries, cultures and/or corporations. For example, there was strong agreement from among all stakeholder groups that the IWs were "life changing" events or even a "once in a lifetime experience". Taken together, data presented in our study appear to culminate in one simple finding, namely that human centricity is a core success factor for initiatives that help mobilize diverse stakeholders, maintain commitment over time and thus seem to increase the potential for sustainable effects.

Human centricity in the corporate world may be described as putting at the center the human component as opposed to the conventional utilitarian view in organizations (see Janssen 2016). Researchers for future trends are already observing a gradual shift of organizations from a narrow profit to a broader human centric orientation, which places at the center the needs of customers, employees and society at large.[21] In the literature, however, there is insufficient theorizing and conceptualizing about human centricity. Future research should address this gap by answering the following questions: Is sustainable CSR practice one component of human centric organizations? What else constitutes human centricity in organizational practice? How does human centricity evolve? Can it be promoted and with what effects?

[21] https://goodplace.org/human-centricity-der-neue-trend-fuer-unternehmen/.

In sum and based on the findings of our case study, we argue that human centricity and CSR practice (especially when actively engaging multiple stakeholders) play a critical role in making organizations meaningful—a topic that has recently become the subject of exploration amongst organizational research and that is likely to become even more relevant in the future:

> We often take as given that an organization's purpose is to produce economic value; and, although economic value can often add to social value, sometimes it does not. This disjuncture raises the question of meaningfulness. The meaningfulness of an organization is its expression of purpose, values or worth. It involves a sense of significance that goes beyond material success or profitability; rather, it highlights how organizations can play a larger and more positive role in the world. It is an approach embraced by the next generation of workers, the millennials (in their mid-to-late 30 s), who often focus on making a positive difference in the world and a contribution to society – with organizations, not in spite of them.[22](cf. von Devivere 2018, p. 181)

6 Summary and Conclusions

The study was motivated to better understand success factors and sustainable outcomes of CSR activities and to contribute to the stakeholder view of CSR in the literature. Seven projects in four countries with over 800 people involved were analyzed following a multiple case study approach. Findings showed long-term social impact in the local communities on an individual as well as an institutional level, thereby demonstrating the sustainability of the CSR program from the perspective of the local beneficiaries. The study further indicates long-term impact on professional and corporate volunteers (employee impact). A unique feature of the analyzed program was the systematic combination of CSR with the qualification of employees in relevant business skills. Such combinations offer distinct advantages for CSR by creating almost immediate returns on investments for the business (business impact), minimizing potential conflicts of interest and thus maximizing the longevity of CSR initiatives. The voluntary nature of stakeholder engagement as well as the alignment of multiple stakeholders towards one shared, desirable goal emerged as critical elements for CSR to be successful. In principle, we propose that human centricity (as opposed to a narrow customer centricity) will increasingly become a discriminating factor in a company creating deep impact and fostering a sense of meaningfulness, both inside and outside the company.

Acknowledgements Our special thanks go to all professional and corporate volunteers as well as all local hosts for dedicating their time and effort to the IW programs. Grateful acknowledgement is made to Michael Huebl, co-founder of the IW, to Church in Action for all logistic support and to all sponsors and partners who made this project and research possible by providing funds and expertise. Our special thanks go to Julia Wenzel for providing important qualitative content for the study and for her feedback on earlier versions of the manuscript as well as to Jan Wessel for the evaluation of

---

[22]Mary Ann Glynn, 2016 Program Chair, Academy of Management; http://aom.org/Meetings/annualmeeting/2016/AOM-2016-Theme--Making-Organizations-Meaningful.aspx/.

IW Kigali. We also thank all photographers including Thomas Ade, Jason Terschueren, and Jens Stapelfeldt for their photographic material. Finally, we wish to thank Kirsty Andersen for her copy-editorial support, and Susanne Kotysch, Monika Ruehl, Gregor Preis for providing information on Lufthansa Group CSR and the help alliance.

References

Aras G, Crowther D (eds) (2010) A handbook of corporate governance and social responsibility. Gower, Surrey, UK

Benson P, Kirsch S (2010) Corporate oxymorons. Dialect Anthropol 34:45–48. https://doi.org/10.1007/s10624-009-9112-y

Berger P, Luckmann T (1966) The social construction of reality: a treatise in the sociology of knowledge. Penguin, London

Bhattacharya CB, Sen S, Korschun D (2008) Using corporate social responsibility to win the war for talent. MIT Sloan Manag Rev 49(2). https://ssrn.com/abstract=2333549. Accessed 31 May 2018

Boatright JR (2012) Ethics and the conduct of business, 7th edn. Pearson, Upper Saddle River, NJ

Bryman A (2016) Social research methods, 5th edn. Oxford University Press, Oxford, UK

Cole K (2005) Management: theory and practice. Pearson Education Australia, Frenchs Forest, NSW

Crane A, Matten D, McWilliams A, Moon J, Siegel DS (2009) The Oxford handbook of corporate social responsibility. Oxford University Press, Oxford, UK

Creswell JW (2014) Research design: qualitative, quantitative and mixed methods approaches. Sage, Thousand Oaks, CA

Dahlsrud A (2006) How corporate social responsibility is defined: An analysis of 37 definitions. Corp Soc Responsib Environ Manag 15:1–13. https://doi.org/10.1002/csr.132

Dunfee TW (2009) Stakeholder theory: managing corporate social responsibility in a multiple actor context. In: Crane A, Matten D, McWilliams A, Moon J, Siegel DS (eds) The Oxford handbook of corporate social responsibility. Oxford University Press, Oxford, UK, pp 346–362

Fallon W (2013) Sustainability, stakeholders and the nature of the firm. In: Wells M et al (eds) Sustainability in Australian business: principles and practice. Wiley, Milton, QLD, pp 35–66

Friedman M (1970) The social responsibility of business is to increase its profits. New York Times Magazine, p 17

Friedman M (2007) The social responsibility of business is to increase its profits. In: Zimmerli WC, Holzinger M, Richter K (eds) Corporate ethics and corporate governance. Springer, Berlin, pp 173–178

Frynas JG (2005) The false developmental promise of corporate social responsibility: evidence from multinational oil companies. Int Aff 81(3):581–598. https://doi.org/10.1111/inta.2005.81.issue-3

Gielnik MM, Frese M, Kahara-Kawuki A, Katono IW, Kyejjusa S, Munene J et al (2015) Action and action-regulation in entrepreneurship: evaluating a student training for promoting entrepreneurship. Acad Manag Learn Educ 14(1):69–94. https://doi.org/10.5465/amle.2012.0107

Gilberthorpe E, Banks G (2012) Development on whose terms?: CSR discourse and social realities in Papua New Guinea's extractive industries sector. Resour Policy 37(2):185–193. https://doi.org/10.1016/j.resourpol.2011.09.005

Gilberthorpe E, Agol D, Gegg T (2016) 'Sustainable mining'? Corporate social responsibility, migration and livelihood choices in Zambia. J Dev Stud 52(11):1517–1532. https://doi.org/10.1080/00220388.2016.1189534

Grisaffe DB (2007) Questions about the ultimate question: conceptual considerations in evaluating Reichheld's net promoter score (NPS). J Consum Satisfaction Dissatisfaction Complaining Behav 20:36–53

Grots A, Creuznacher I (2016) Design thinking: process or culture? In: Brenner W, Uebernickel F (eds) Design thinking for innovation. Springer, Cham

Henderson D (2001) Misguided virtue: false notions of corporate social responsibility. Institute of Economic Affairs, London

Higgins W (2009) Standardizing corporate social responsibility. In: Bubna-Litic D (ed) Spirituality and corporate social responsibility: interpenetrating worlds. Gower, Surrey, UK, pp 199–212

Hubbard TD, Christensen DM, Graffin SD (2017) Higher highs and lower lows: The role of corporate social responsibility in CEO dismissal. Strateg Manag J 38(11):2255–2265. https://doi.org/10.1002/smj.2646

ISO—International Organization for Standardization (n.d.) ISO 26000—Social responsibility. Resource document. Author. https://www.iso.org/iso-26000-social-responsibility.html. Accessed 15 April 2018

ISO—International Organization for Standardization (2010) Discovering ISO 26000: guidance on social responsibility. Genève, Switzerland: Author. https://www.iso.org/files/live/sites/isoorg/files/archive/pdf/en/discovering_iso_26000.pdf. Accessed 15 April 2018

Janssen B (2016) Die stille revolution: Führen mit Sinn und Menschlichkeit. Ariston, Germany

Johnson RB, Christensen L (2017) Educational research: quantitative, qualitative and mixed methods approaches, 6th edn. Sage, Thousand Oaks, CA

Keith N, Unger JM, Rauch A, Frese M (2016) Informal learning and entrepreneurial success: a longitudinal study of deliberate practice among small business owners. Appl Psychol 65(3):515–540

Kolko J (2015) Design thinking comes of age: organizational culture. Harvard Business Review (HBR). Online news article. https://hbr.org/2015/09/design-thinking-comes-of-age. Accessed 28 Apr 2018

Lenz I, Wetzel HA, Hammerschmidt M (2017) Can doing good lead to doing poorly? Firm value implications of CSR in the face of CSI. J Acad Mark Sci 45(5):677–697. https://doi.org/10.1007/s11747-016-0510-9

LHG—Lufthansa Group (n.d.) Corporate responsibility within the Lufthansa group. Resource document. Author. https://www.lufthansagroup.com/en/responsibility/corporate-responsibility.html. Accessed 23 Apr 2018

Moscardo G, Lamberton G, Wells G (2013) Sustainability in Australian business: principles and practice. Wiley, Milton, QLD

O'Faircheallaigh C, Ali S (eds) (2009) Earth matters: indigenous peoples, the extractive industry and corporate social responsibility. Greenleaf, London

Padel F, Das S (2010) Cultural genocide and the rhetoric of sustainable mining in East India. Contemp South Asia 18(3):333–341. https://doi.org/10.1080/09584935.2010.503871

Punch KF (2014) Introduction to social research: quantitative and qualitative approaches, 3rd edn. Sage, Thousand Oaks, CA

Rasche A, Morsing M, Moon J (eds) (2017) Corporate social responsibility: strategy, communication, governance. Cambridge University Press, UK

Shamir R (2011) Socially responsible private regulation: world-culture or world-capitalism? Law Soc Rev 45(2):313–336. https://doi.org/10.1111/j.1540-5893.2011.00439.x

Stake RE (1995) The art of case study research. Sage, Thousand Oaks, CA

Trebeck K (2008) Corporate social responsibility and democratisation: opportunities and obstacles. In: O'Faircheallaigh C, Ali S (eds) Earth matters: indigenous peoples, the extractive industry and corporate social responsibility. Greenleaf, London, pp 8–23

Unger JM, Rauch A, Frese M, Rosenbusch N (2011) Human capital and entrepreneurial success: a meta-analytical review. J Bus Ventur 26(3):341–358

Van Marrewijk M (2003) Concepts and definitions of CSR and corporate sustainability: between agency and communion. J Bus Ethics 44:95–105

van Oosterhout J, Heugens P (2009) Much ado about nothing: a conceptual critique of corporate social responsibility. In: Crane A, Matten D, McWilliams A, Moon J, Siegel DS (eds) The Oxford handbook of corporate social responsibility. Oxford University Press, Oxford, UK, pp 197–226

von Devivere B (2018) Meaningful Work: Viktor Frankl's Legacy for the 21st Century. Springer, Cham, Switzerland. https://doi.org/10.1007/978-3-319-89791-2
White R (2008) Crimes against nature: environmental criminology and ecological justice. Willan, Portland, OR
Whitmore A (2006) The emperor's new clothes: sustainable mining? J Cleaner Prod 14(3–4):309–314. https://doi.org/10.1016/j.jclepro.2004.10.005
Yin RK (2009) Case study research: design and methods, 4th edn. Sage, Thousand Oaks, CA
Yin RK (2012) Applications of case study research, 3rd edn. Sage, Thousand Oaks, CA

Dr. Jens M. Unger is Senior Manager for cultural transformation at the Lufthansa Group and independent organizational consultant and researcher. Dr. Unger has assumed various leadership and management positions at the Lufthansa Group in organizational development and learning. In his research he has specialized on the psychology of entrepreneurship and success factors of small and micro business Africa. Past research includes projects on learning processes in organizations, cross-cultural comparisons in entrepreneurship, meta-analysis on the effects of human and social capital, and studies on psychological success factors of small and micro business owners in Germany, Zimbabwe, South-Africa, Zambia, Peru, Russia, and China.

Dr. Johannes M. Luetz is Senior Lecturer, Postgraduate Coordinator and Research Chair at CHC Higher Education in Brisbane, Australia, and Adjunct Lecturer at the University of New South Wales (UNSW) in Sydney in the School of Social Sciences, where he also earned his Ph.D. in Environmental Policy and Management with a thesis on forced human migration. Dr. Luetz has worked extensively with World Vision International on research projects raising awareness of the growing effects of climate change on poor and vulnerable communities in Asia, Africa and Latin America, and the need to meaningfully address vulnerabilities through praxis-informed education approaches that work in the real world. His previous international development work experience includes the role of Lecturer in Development Studies at UNSW, Sydney.

Sustainability Governance in Traditional Crafts Communities: A Project Proposition

Walter Leal Filho and Lez Rayman-Bacchus

Abstract This overview paper describes the relevance of sustainability governance to traditional crafts communities in developing economies. It highlights the importance of this work, both as a livelihood and as way of life, for millions of people. Some theoretical lenses for thinking about traditional craft communities are presented, as well as the practical challenges such communities face, especially in terms of potentially meeting sustainable development ambitions. The paper also describes a research design, comprising research aims, and the use of case studies, outlining how to address the research needed in this central field.

Keywords Sustainability · Crafts · Skills · Stateholders · Developing countries

1 Introduction

This paper describes a research proposal aimed at promoting sustainability governance thinking and practices among traditional crafts communities in developing economies. It comprises three parts: Background, Methodology, and Conclusions. Background outlines the importance of traditional crafts, as a manifestation of living cultural heritage, and recognised by UNESCO. It further shows how important this activity is culturally and economically to most developing economies. Set in the context of sustainable development, the need for sustainable governance in living traditional crafts is highlighted. Methodology describes the design to address the research needed in this central field. Specific research aims are presented, followed

W. Leal Filho (✉)
Faculty of Life Sciences, Research and Transfer Centre "Sustainable Development and Climate Change Management", Hamburg University of Applied Sciences, Ulmenliet 20, 21033 Hamburg, Germany
e-mail: walter.leal2@haw-hamburg.de

L. Rayman-Bacchus
University of Winchester Business School, Winchester SO22 5HT, UK
e-mail: lezmichael@gmail.com

W. Leal Filho (ed.), *Social Responsibility and Sustainability*, World Sustainability Series, https://doi.org/10.1007/978-3-030-03562-4_14

by a description of what data needs to be collected, by which means, and how such a research project should be governed. Conclusions follow.

2 Background

In recent decades, the meaning of 'heritage' has evolved considerably. As expressed in several UNESCO conventions, cultural heritage no longer ends at tangible artifacts, such as monuments and objects, but also includes (intangible) knowledge, skills, practices, traditions and other living expressions.[1] Such heritage provides individuals and communities a sense of place, belonging, and continuity, cherished as an inheritance from ancestors, and as a cultural compass to future generations. The production and consumption of traditional crafts,[2] is one manifestation of cultural heritage, and valued the world over, not only culturally but also economically. Indeed the production and sale of traditional crafts provide the only livelihood for millions of people throughout the developing world. Still, there is a lack of data on the nature and extent of traditional craft, by country. The traditional crafts sector is not clearly defined. While some craft production involves the full-time use of specialist manual skills, such as hand weaving of carpets, some may use machinery for some parts of the production process. Further, in many economies traditional craft work is informal, often carried out in rural areas by workers otherwise occupied on agricultural production. As agricultural work is seasonal, these workers may also engage in artisanal work.[3]

Considered as 'living heritage', traditional craft work offers an opportunity to challenge the tendency to reify all cultural heritage, including the intangible.[4] Such thinking overlooks that culture is alive and evolves as society develops; practices change or disappear as the material and social needs of communities change. Set within the context of sustainable development, the notion of 'living' further invokes a need to broaden the understanding of 'livelihood'; from a narrow focus on communities overcoming poverty to one of pursuing a *sustainable* livelihood. This means communities addressing factors that help or hinder the capacity of craft communities in making a living that is not only economically viable, but at the same time socially

[1]Convention Concerning the Protection of the World Cultural and Natural Heritage (1972); Universal Declaration on Cultural Diversity (2001); Convention Safeguarding Intangible Cultural Heritage (2003); Convention on the Protection and Promotion of the Diversity of Cultural Expressions (2005).

[2]Traditional craft are artifacts used in everyday life, made of natural materials, typically made by hand using traditional skills and techniques, often by artisans, sometimes involving small scale production, and carried out in certain areas. See for example http://www.aiweb.or.jp/english/dento/html/howto1.html.

[3]Yang et al. (2018), report on a 'Lack of availability of sufficient data', in their study 'Preservation of Cultural Heritage Embodied in Traditional Crafts in the Developing Countries: A Case Study of Pakistani Handicraft Industry'.

[4]Kurin (2004, 74), notes that 'the UNESCO 2003 convention tends to reduce intangible cultural heritage to a list … atómistically …conceived'.

inclusive and ecologically responsible.[5] Factors likely to encourage communities to move toward a sustainable livelihood approach include their experiencing the economic consequences of causing environmental harm (over-consumption of natural resources, pollution and waste, cost, regulation) and demand for social change, both from within communities (against the exploitation of labour, for better working conditions, income) and beyond (respect for human rights, inclusion of the vulnerable, investigative journalism, regulation). In this move, the design, function, production, and how to eventually dispose of traditional craft products, demands innovation in both materials used and in working practices, if these communities and their traditions are to survive and realise the potential of emerging and new market opportunities. Ignoring changing market demands (for ethical or sustainable products) and social norms (including heightened awareness of the fragility of the environment) risks both social costs and economic failure for traditional craft communities.[6]

As a primary driver for the emergence, conservation and development of viable communities, a 'living cultural heritage' view emphasises not only acknowledgement of the spatial interdependence between communities and their environment, but also the importance of securing continuity across future generations, both cultural and economic. The cultural and economic viability of traditional craft communities, both in present and future terms, must be assessed in the context of the pressures of financial survival, weak local environmental and social governance, weak regulatory enforcement, and the influence of international supply chains.

Fostering a sustainable livelihood approach to the production and consumption of traditional crafts unavoidably raises sustainable governance challenges, in particular around: mitigating harmful impacts on natural resources (e.g., use of raw materials, pollution, waste); creating quality of life pathways (e.g., access to welfare and education, scope for inclusion and integration); and harnessing economic value for the common good (training of labour supply, fostering innovation capability, access to finance, and tax policies). Yet mainstream policy prescriptions and the global hegemony of commercial priorities continue to encourage a notion of sustainable development underpinned by economic rationality.

This rationality drives tensions between the pursuit of sustainable (in practice commonly interpreted as economic) development, and the mobilisation of traditional crafts (both as a livelihood and a way of life), with the result that social and environmental values may be compromised, including rights to decent working conditions and human dignity, and environmental degradation. These tensions are strong, especially in developing countries where traditional craft work is critical, providing *the* livelihood of communities, and nationally recognised by many developing economies for its significant aggregate economic contribution, including providing employment, and cultural value.

---

[5]Chambers and Conway (1991) defines a sustainable livelihood as one able to 'cope with and recover from stress and shocks, maintain or enhance its capabilities and assets,...provide ...opportunities for the next generation, and ...contribute ...benefits to other livelihoods at the local and global levels...'.

[6]Liebl and Roy (2004).

The importance of sustainability governance[7] for cultural heritage is reflected in the aims of various multilateral bodies, some concerned with the preservation or conservation of heritage for all humanity (e.g., UNESCO), others concerned with protecting ownership (intellectual property) rights (e.g. TRIPS/GI/AO).[8] As noted, expressions of cultural heritage, both tangible and intangible, provide individuals and communities not only a sense of identity, but also a means of development. Regulatory protection of such cultural and intellectual resources is essential for securing not only economic freedom, but also the advancement of social justice and an ecological ethic.[9] Indeed, observers highlight that (sustainable) development should enable the fulfilment of the citizen, wherein human dignity is central to economic development, and warn that social justice failures cripple substantive freedoms (capabilities) and potential socio-economic contribution.[10]

Given the rural, localised, and informal nature of traditional craft communities in developing economies, there may or may not be appropriate formal institutions to support local development strategies. Such context suggests any analysis should include understanding the drivers of *local* development. This is likely to involve understanding the roles of: [1] existing traditions of local governance (processes, structures, rules); [2] social capital (social relations and networks); and [3] wider social and political institutions (territorial, national, international, multilateral); and [4] the character of the wider national policy landscape (e.g. political culture and degree of autonomy). Policy studies suggest promoting local development policies involve utilising local particularities (incl. resources and capabilities, cooperation), supporting entrepreneurial activity, and facilitating market access. Such policies should also embrace sustainable development goals (help reduce poverty, be socially and economically inclusive, environmentally responsible, and economic viability).[11] While a focus on the local is critical, there is an attendant danger of sliding into regressive localism, facilitating intolerance and regional inequalities.[12]

The production and consumption of traditional crafts provides multiscalar challenges for sustainable governance, between national and state policy makers, regional and local government, and traditional craft communities. Helpfully, the UN Sustainable Development Goals (SDGs) provides nations with an action framework, for

---

[7]Sustainability governance is the performance of governments towards achieving sustainability, democracy, the rule of law, social integration and welfare.

[8]WTO Agreement on Trade-Related Aspects of Intellectual Property Rights (TRIPS), negotiated during the 1986–94 Uruguay Round, introduced intellectual property rules into the multilateral trading system for the first time; GI (Geographical Indications); AO (Appellations of Origin. [*World Intellectual Property Organization*].

[9]Finger and Schuler (eds.) (2004).

[10]Sen (1999), Simons (2013), Robeyns (2005), ASEAN (2002), http://asean.org/?static_post=development-for-human-dignity [accessed May 8th, 2018].

[11]Begin-Gillis et al. (2014).

[12]Trigilia (2001). Grabher (1993), Financial Times (2017), https://www.ft.com/content/ce35e7f6-6d71-11e7-bfeb-33fe0c5b7eaa.

example as has India.[13] Moreover, the Sustainable Governance Index (SGI) provides a useful measure of the relative performance of 149 nations, in terms of their performance in delivering the SDGs,[14] highlighting the performance of individual nations.[15] The SDGs and the SGI provide a useful background for this proposal, helping to locate the sustainable governance achievements and potential of any given nation state.

There is a paucity of research exploring the nature of sustainability governance practice in traditional craft communities in developing economies, and which go beyond the above noted economic rationality. Further, there is a need to understand the tensions between the need for economic development and mobilising living cultural heritage in the context of sustainable development. Such understanding would help in developing more nuanced sustainability governance principles/frameworks and practices (i.e. a framework that does not privilege economic viability), but which is also environmentally ethical, socially relevant and culturally informed. In particular, this knowledge may help in examining the sustainability governance practices which are suitable to the crafts sector, and in guiding initiatives for influencing practices, and for encouraging and promoting local development strategies that influence national policy making. The findings and policy proposals emerging from this study should inform similar studies and interventions in other regions where traditional crafts form an important part of developing economies, economically, socially, environmentally.

3 Sustainability in the Crafts Sector: Some Questions to Be Addressed and a Research Proposition

One key question which is often posed is: how can the handicraft sector in developing countries be made more sustainable? In order to address this question we need a better understand of the challenges to sustainable governance faced by traditional craft communities in the context of developing economies. In particular, we need to examine sustainability governance principles and practices through a close examination of the particularities of traditional craft areas, including the materials and technologies employed (i.e., pottery, woodworking, metal, textile, and involving manual production), and the treatment of living cultural heritage, as an everyday livelihood. Drawing on such understanding, one can characterize sustainability governance, propose strategies, and undertake capacity-building activities that effectively contribute to strengthening sustainability governance in the traditional craft communities.

[13]India Sustainable Development Goals (SDGs), Targets, CSS, Interventions, Nodal and other Ministries (08.06.2016), http://in.one.un.org/page/sustainable-development-goals/, available at http://niti.gov.in/writereaddata/files/SDGsV20-Mapping080616-DG_0.pdf [accessed May 7th, 2018].

[14]Sustainable Governance Index provides evaluations of national policy (economic, environment, social), progress on democracy, and governance, http://www.sgi-network.org/2017/About.

[15]Sustainable Development Index and Dashboard: a global report (2016). The SGI score gives a ranking out of 149 nations, and the individual score: India—110/149 [48.4]; Pakistan—115/149 [45.7]; Bangladesh—118/149 [44.4].

This paper proposes the need to research the sustainability of traditional crafts sectors, comprising the following aims:

(a) to identify and understand the main sustainability governance arrangements (policies, institutions, strategies) and practices, in place for the protection and development of living heritage as a livelihood and way of life within the specified artisanal communities, assessing the extent to which environmental, social, and economic concerns are embedded within said policies, institutions, strategies and practices.
(b) to better understand the tensions inherent in the mobilisation of living heritage in the context of sustainable development, including: tensions between looking back and looking ahead (e.g. between histories and traditional ways of working against aspirations, attitudes to preservation and innovation, resilience and adaptation, conflicts between institutions and actors).
(c) to gain insight to the preservation of human dignity in the context of development: examining how individuals and their communities experience development (e.g., hopes, fears, opportunities, threats), their assessment of their quality of life, set in the context of their social structures and institutions, and to understand local conceptions and likely tensions between the rights of individuals and of communities.[16]

Drawing on the understanding resulting from (a), (b), (c), a research project may seek to effect the beginnings of a sustained positive change in the life of traditional craft communities.

3.1 Methodology

The research methodology used needs to be exploratory in nature, combining an extensive review of the available literature, with a set of 3 concrete thematic case studies of traditional craft communities, in differing developing countries, focused on three of the four traditional craft technological categories: pottery, woodworking, metal, textile, and involving manual production. The research team should include project partners, each being a local subject expert familiar with the local context of each case study. The research work may be divided into work packages, reflecting a progression from data collection, through analysis, policy implications, and practical implementation (e.g. training workshops).

[16]The authors are minded of the fate of the Kerpalti slum in Delhi, involving tensions between the government and the Kerpalti community. The government determined that Kerpalti is a slum, occupying valuable developable land in the city, and they should rehouse this community in modern housing within a new development in the same location. The Kerpalti community regard themselves as a community of artisans and not as (individual) slum dwellers. This conflict has forced many of Kerpalti artisans to question their identity, with some rejecting the government assessment outright, while others acquiesce and accept their fate.

A critical review of studies will identify and reflect on the key parameters considered important to communities successfully pursuing living heritage work as a livelihood. These include: level of participation in decision making, provision of educational opportunities, income generation strategies (community consumption vs. tourism); the role of handicraft as decent work in local communities, in raising living standards, improving security and securing identity, and promoting wellbeing, equity and empowerment; tensions between development and mobilisation of heritage, and preserving human dignity.

Thus, duly informed and fully taking into account the objectives around sustainable governance, the research proposition may deploy a combination of data collection approaches in each location, i.e.:

(a) stakeholder engagement,
(b) case studies,
(c) participatory capacity enhancement, and
(d) knowledge sharing and co-creation.

This combination of approaches will ensure the project meets the best interests of the final beneficiaries, traditional craft communities. Case studies enable understanding of complex social phenomena in context; a participatory strategy opens the way to methods of changing mindsets, and combined with capacity enhancement and knowledge sharing, will provide for long term positive impact on economic development and social welfare. The approach will also consider ways of strengthening those broader institutional responses (policy dialogue and policy making) that shape local community attitudes toward traditional craft production, and also look at ways of strengthening craft workshop practices, as they affect sustainability governance elements.

The project partners should collect field data through interviews with traditional craftspeople, community leaders, and craft associations, selective use of focus groups, and by reviewing relevant reports, policy documents and local records and audits. Data may also be collected on sustainability impacts: economic (incl., employment numbers, tourism income), social (incl., equity, opportunity), and environmental (incl., waste, recycling, energy employed), and the management of these impacts, and any existing organising governance frameworks, both formal and informal. Mindful that such frameworks tend toward economic rationality, ignoring aesthetic, pragmatic and reflective practices, further data will be collected that inform understanding of the intertwining between cultural practices and traditions with sustainable development theory and policy schemas. In order to identify appropriate data, there will be a systematic determination of indicators of competing valuations and interpretations of cultural heritage and of human rights, and the particularities of local histories and institutions, faith and moralities. Said determination will arise from interviews, focus group discussions, and close examination of existing frameworks and indicators, including gaps and overlaps, (e.g., indicators of development and of conservation; assessing dignity through a rights based approach to development.

Ultimately, the research proposition will contribute to the state of knowledge on the means by which sustainable governance in traditional craft communities can be strengthened, and how a sustainable development oriented approach may promote a continuous transmission through subsequent generations, of craft knowledge and skills informed by environmental ethics, sense of social justice, and strengthened dignity and sense of identity.

All the objectives and the associated activities within the project are strongly interrelated and employ a wide range of methods and approaches in order to achieve the overall aim of the project.

3.2 Project Approach and Governance

The proposition uses approaches sensitive to the cultural proclivities of the participant countries, to be chosen based on (1) their presence on the ODA list of approved developing economies and legitimate beneficiaries of development assistance; (2) that they show weak performances on the Sustainable Governance Index, falling below 100 out of 149 nations and therefore providing scope for improvement; and (3) have in common similar histories and democratic institutions, but may also have important cultural differences (developed differing histories, have differing languages, and religions). The particular traditional craft communities should be selected based on their scale, the range of handicrafts produced, and the economic challenges accompanying these localities.

What is learnt from these countries should provide insights to sustainable governance in other nations, in both rural and urban contexts, as well as being relevant to other concerns other than tangible and intangible heritage (replicability of the research).

The stakeholders relevant to the project, and who will also act as beneficiaries as part of the research proposition, are:

- local actors represented by craftsmen and craftswomen working with pottery, woodworking, metal, textile, and involving manual production
- local NGOs working with traditional craft regions and firms
- national actors: managers of tourism sites
- national heritage associations
- local cities and municipalities which manage the areas where handicrafts are sold
- International organisations with expertise in heritage (e.g. UNESCO, World Heritage Forum).

The management and governance of the project should be ensured by means of a project management board (PMB) comprising the work package leaders and target country partners. The technical excellence will be ensured by means of a project advisory board (PAB) composed of representatives of relevant heritage-related organisations/NGOs and relevant traditional craft associations, and institutions working with sustainable development, who will provide ad hoc advice to the project.

4 Importance of the Research Proposition

The importance of the study is based on the following contributions:

(a) Better understanding of how traditional craft communities in a developing economy operate within a sustainable development framework, i.e., the extent to which there is a sustainability governance framework in use, and if so its effectiveness.
(b) Appreciation of how cultural context influences sustainability governance attitudes and practices.
(c) Stronger insights to the complex relationships between living cultural heritage and human dignity with sustainable development.
(d) Improve practices and attitudes to sustainable development in the economic mobilization of living cultural heritage, including the protection of human dignity, in particular traditional crafts communities in developing economies.

Furthermore, the project should advance our understanding of the relevance and importance of traditional craft communities and their work as conduits of history, culture and heritage, and the potential role played by sustainable development thinking in shaping and passing these on to future generations.

The results deriving from the research proposition will be useful in three main ways:

i. wider use in other contexts of the principles and practices of sustainable governance in support of traditional crafts
ii. in support of policies seeking to reduce the vulnerabilities of traditional crafts sectors
iii. in meeting the demand for more robust information on the complex relationship between sustainable development and the exercise of living heritage work, here exemplified by traditional craft production communities, immersed within particular cultural contexts and set in developing economies.

Data and experiences from the project, combined with conceptual reflection, may be used to characterise the challenges faced by the communities investigated. These insights may be used to suggest relevant measures and practices to overcoming the identified challenges, and to promote, develop, and preserve the role of traditional crafts as living cultural heritage, valued for not only for their economic impacts but equally regarded for their social value. Further, these insights should help develop and strengthen the practical wisdom of all beneficiaries on the interdependence of humanity and the natural environment. The methods proposed in this project may be replicated in other countries and regions, throughout the developing world.

5 Conclusions

There is a need to identify and explore the key issues that pose substantial challenges to the preservation and development of traditional crafts as living heritages. This is especially so in developing countries. As this research proposition shows, insights gained may be replicable, and hence may assist traditional craft communities in developing economies to take advantage of the many opportunities that sustainable governance offers them.

As to opportunities for further work, we need novel research approaches that address the need for deeper understanding of how sustainability thinking may help traditional crafts work, for example in terms of models of sustainable livelihoods. At the same time, there is also need to better understand the operational everyday (sustainability) challenges craftsmen and craftswomen face, seen from their perspective. Here we also need to identify the key challenges to the preservation of traditional crafts as living heritages. The integration of sustainable development theory and everyday practices in tradition crafts should inform both, contributing to not only more relevant and useful theory but also to more sustainable practices in the production and consumption of traditional crafts. Studies such as that proposed here, replicated in other geographical regions, should go a long way toward improving not only the economic performance of the traditional crafts, but also their social and environmental performance, globally.

References

ASEAN Welcome address by Rodolfo C. Severino, Secretary-General of the Association of Southeast Asian Nations, at the High-Level Conference on Social Development in the National Development Agenda organized by the ASEAN Secretariat and the World Bank, Jakarta, 16 Jan 2002

Begin-Gillis M, Sam KL, Yule A (2014) Creating a strategic plan for local economic development: a guide prepared by the Federation of Canadian Municipalities—Caribbean Local Economic Development Project

Chambers R, Conway GR (1991) Sustainable rural livelihoods: practical concepts for the 21st century. Institute of Development Studies DP 296, 1991. University of Sussex, Brighton

Financial Times, 24 July 2017 'Danger of new localism driving urban inequality'

Finger JM, Schuler P (eds) (2004) Poor people's knowledge: promoting intellectual property in developing countries. The World Bank and Oxford University Press, Washington DC, USA

Grabher G (1993) The embedded firm. Routledge, London

Kurin R (2004) Safeguarding intangible cultural heritage in the 2003 UNESCO convention: a critical appraisal. Mus Int no. 221-222, 5(1–2):66–77. Blackwell, Oxford

Liebl M, Roy T (2004) Handmade in India: traditional craft skills in a changing world. In: Finger JM, Schuler, P (eds) Poor people's knowledge: promoting intellectual property in developing countries. The World Bank and Oxford University Press, Washington DC, USA, pp 53–73

Robeyns I (2005) The capability approach: a theoretical survey, J Human Dev 6(1):93–114. Development for Human Dignity

Sen A (1999) Development as freedom, Random House, New York. Nussbaum N.C, 2011, Creating capabilities; the human development approach, Harvard University Press, London

Simons P (2013) Human dignity central to economic development. Koers—Bull Christ Sch 78(1), Art #424
Sustainable Development Index and Dashboard: a global report (2016)
Trigilia C (2001) Social capital and local development. Eur J Soc Theor 4(4):427–442
Yang Y, Shafi M, Song X, Yang R (2018) Preservation of cultural heritage embodied in traditional crafts in the developing countries: a case study of Pakistani handicraft industry. Sustainability 10

Trade Tradition and Sustainable Development: A Health Promotion Experience

Glória Lúcia Alves Figueiredo, Carlos Henrique Gomes Martins, Jaqueline Lopes Damasceno and Márcio Ronan Policarpo

Abstract Rapid urbanization is exerting pressure on various areas, including the industrial sector, to adopt sustainable practices. To be sustainable, a company must be concerned not only with economic growth, but also with environmentally and socially ethical attitudes. This paper describes a local initiative that allies commercial tradition and sustainable development. The economy of Franca, a city in Brazil, is based on the footwear industry, which has gained international recognition. The companies established in Franca have to meet a series of predefined requirements regarding small- and large-scale production and waste management. Compliance allows a company to use a Provenance Indicator (IP) logo together with a code and a QR Code in the shoes it produces. The QR Code directs the consumer to information about the origin of the product and facilities, employees, and social responsibility and environmental care actions of the company. All the data are available as numbers and images that can be consulted before, during, and after the purchase. In addition to promoting environmental preservation, this initiative adds value to the product (shoes) because it can positively change the image of the company. A non-governmental operational entity registers the companies that are interested in the initiative. In an increasingly competitive market, a company having the IP logo on its products attests to its commitment to quality and to social and environmental responsibility, but it is the consumer that controls this responsibility.

Keywords Health promotion · Industry · Sustainability

G. L. A. Figueiredo (✉) · C. H. G. Martins · J. L. Damasceno · M. R. Policarpo
University of Franca, Franca, SP 144040-600, Brazil
e-mail: gloria.figueiredo@unifran.edu.br

C. H. G. Martins
e-mail: carlos.martins@unifran.edu.br

J. L. Damasceno
e-mail: jaquelinelopes.bio@hotmail.com

M. R. Policarpo
e-mail: marcioa4arquitetura@gmail.com

W. Leal Filho (ed.), *Social Responsibility and Sustainability*, World Sustainability Series, https://doi.org/10.1007/978-3-030-03562-4_15

1 Introduction

Footwear production is part of our history, our culture, and we are recognized by it.

Brigagão do Couto

Given the ancient and current global sustainability needs, involvement and participation of different society sectors are crucial for the population to adopt preventive and health-promoting practices that will not only deal with environmental issues, but also promote quality of life in more general and extended ways. According to Buss (2000) and Czeresnia (1999), these are the premises of Health Promotion that have been discussed since the 1980s.

On the basis of International Conference Charters, Health Promotion is a set of strategies that promote both individual and collective health. Intra- and inter-sectorial articulation and cooperation characterize these strategies, which involve broad participation and social control (Brasil 2015). Health Promotion acknowledges existing policies and technologies and aims to ensure everyone's right to equity and quality of life by reducing the vulnerabilities and health risks resulting from social, economic, political, cultural, and environmental determiners.

According to Figueiredo et al. (2017), right to health and right to the city are interconnected: by ensuring that the population has access to healthy urban spaces, it is possible to reduce iniquities and to guarantee that the positive effects of urbanization will also be enjoyed by disadvantaged and non-priority groups. From 1986 to 2016, it became evident that these rights were present in the formulation of policies and in the agenda of social movements at different levels. Collective construction of local identity agendas effectively revealed more assertive decisions about the city in which the population wishes to live.

In Brazil, cities where the economy increasingly depends on industries not only contribute to economic indicators, but also influence environmental sustainability and health indicators, the so-called economic health indicators. Unemployment in the industrial sector is a complex issue to deal with more thoroughly: it modifies not only the family dynamics and childcare, but also the epidemiological profile of diseases. Nevertheless, locally speaking, successful production and improvements in the quality of life must walk side by side. For example, the population of Franca, a city in Brazil, relies on the success of its footwear industry to improve its inhabitants' quality of life.

High production costs and constant oscillations in the footwear sector impair the competitiveness of a company in the market. This has led the footwear industry to seek alternatives that add value to their products. In Franca, the footwear industry has bet on more assertive strategies and actions concerned with sustainability. Industries in Brazil generally depend on national public policies and incentives to grow. Therefore, in this article we will focus on the sustainability actions taken by the footwear industry union and directed to the population and to the urban spaces of Franca. This is our first attempt to bring industrial and academic activities together, especially with respect to sustainability issues involved in footwear production. With a view to improving our knowledge in the field of Health Promotion, we aim to discuss a strategy that combines the commercial tradition of footwear production with sustainable economic development.

2 Industry and Sustainability

Factors such as drastic climate changes, imminent decline of nonrenewable energy resources, and loss of biological diversity (Weiss and Bradley 2001), among others, can reduce the resilience of our ecosystem and impair its capacity to meet human needs (Beddoe et al. 2009). Driven by easy and fast access to information, the society has become increasingly concerned about the environment, which has boosted the demand for sustainable products. The environment is now a more frequent issue in the relation between consumers and companies, and environmental issues are taken into consideration from the moment a product is designed until the moment it is disposed of after its useful life.

Global awareness of the ecological crisis has raised questions about the way the society is organized, especially in terms of economic development. Indeed, economic development can no longer be understood only as a synonym of growth to justify the unlimited exploitation of environmental resources. It is now essential to worry about how economic development affects the society (Scotto et al. 2011) and whether production processes are sustainable.

Coined in forestry, the concept of sustainability originally meant never reaping more than the forest could produce in a new growth cycle (Wiersum 1995). For Kuhlman and Farrington (2010), sustainability can be defined as maintaining well-being over a long, perhaps even indefinite period. This largely covers the environmental dimension of our survival. Ideally, the potential impact of a sustainable development policy, program, or project should lead to greater well-being and to a positive or at least neutral effect on the overall state of natural resources, to preserve the future of environment (Pearce et al. 1989; Kuhlman and Farrington 2010; Torresi et al. 2010).

When it comes to sustainable solutions, some societal responses stand out: maintaining sustainable food supply and a steady and non-rising population and implementing forest policies, agricultural technologies, and carbon neutral communities, for example (Weiss and Bradley 2001; Diamond 2005; Beddoe et al. 2009; Van der Schoor and Scholtens 2015).

The potential global collapse requires a new ecological adaptation, which causes the society to change how it thinks and acts and demands a general realignment of the way people view and interact with the environment (Gunderson and Holling 2002). Thus, what may have begun with specific actions around the world has resulted in a new perspective on how to live without harming the environment, one of the basic principles of sustainable development.

Sustainable economic development should guarantee the permanence of the productive capacity of the natural environment for human needs. Worldwide, sustainable development is widely accepted as being underpinned by environmental, social, and economic factors (Haines et al. 2012). In this context, the concept of sustainable development defined in the Brundtland Report of the World Commission on Environment and Development (WCED) of the United Nations in 1987 stands out: sustainable development meets the needs of the present without compromising the possibility that future generations will meet their own needs. This definition emphasizes intra- and inter-generational equity and denotes that every adult and child has the right to the opportunity of developing freely in a stabilized society where their basic needs are met while the environment is protected (Chang et al. 2018).

The concept of sustainability is broad and comprehensive (Merico 2009). However, at the heart of social and historical needs, the concepts of development and sustainability will guide actions more or less effectively depending on how well these concepts can respond to the population's needs.

3 Methodology

This is an exploratory, cross-sectional, qualitative study that describes a strategy developed from the perspective of economic and sustainable development. It involves footwear companies established in the city of Franca, Brazil. A documentary analysis was conducted, and the study was complemented by an interview with the president of the Footwear Industry Union of Franca (SindiFranca). Data were collected from February to July 2018.

The history of Franca began in the eighteenth century. People on their way to find gold in the interior of Brazil stopped there to stock up before continuing the long journey. The first experiences with footwear were shoe and harness repairs and the manufacture of other leather artifacts. These experiences were later boosted when immigrants arrived and the first footwear industry was founded.

In Franca, footwear production came along with industrialization in the 1920s. This tradition generated development and wealth and promoted regional integration. In 2017, Franca produced 27 million pairs of shoes, which represented 21.23% of the national footwear production. Franca is currently one of the main shoe manufacturing hubs in Brazil, especially with respect to men and leather footwear. There are 1383 small, medium-sized, and large industries in the city. The footwear industry effectively contributes to generating jobs both in the city and the region, increasing the citizens' income, promoting foreign exchange through exports, and supplying the national market with products of good added value. In 2016, people employed by the footwear industry corresponded to 29.7% of the total population of Franca (SindiFranca 2018).

Franca is located in the interior of the state of São Paulo, in Brazil. In 2017, the population was estimated at 347,237 inhabitants; the territorial area was 605,679 km^2; the population density was 526.09 inhabitants/km^2; the infant mortality rate was 9.73 per thousand live births; life expectancy was 75.9 years; the Gross Domestic Product per capita was R$ 24,679.09; and the Human Development Index was 0.780 (IBGE 2017).

In Franca, the schooling rate is 98.2% among inhabitants aged from 6 to 14 years (Atlas Brasil 2013). The city has five higher education institutions, three and two of which are public and private, respectively. In the health sector, the population relies on comprehensive care made up of basic health units, Family Health Strategy units, and emergency units. All these units belong to the Brazilian Unified Health System (SUS), which guarantees the constitutional right of free access to consultations, examinations, medication, and hospital or outpatient surgical interventions to the entire population.

4 Results and Analysis

4.1 Footwear Manufacture

Footwear production determined by raw material (leather, synthetic material, or fabric), segment (athletic footwear, safety footwear, men's and women's casual or social footwear, etc.), and fierce competition is complex. Manufacturing footwear made of synthetic materials leads to higher productivity and requires lower degree of production complexity as compared to leather footwear production, which in turn demands many semi-handcraft procedures and meets with serious automation difficulties (Espíndula 2009). Table 1 describes the six main steps in leather footwear production.

In Franca, there are 1383 footwear industries. To keep this number of factories operating and producing almost 30 million pairs of shoes (SindiFranca 2018), a high amount of raw material is necessary, and the volume of waste produced at the end of the manufacturing process is large. Each production stage generates effluents, emissions, and waste that are mostly recyclable.

To reduce the amount of industrial waste, specialists advocate reusing said waste. When skin is transformed into leather and then into footwear and artifacts, over 40% of the raw material employed in the operation is discarded as waste. Chromium is present in garbage and footwear industry waste and is classified as Class I hazardous waste due to its toxicity (Alves and Barbosa 2013).

Eco-efficiency is a Sustainable Development strategy that refers to the production of more goods while employing fewer resources and bearing social equity in mind. The productive sector has gradually been incorporating eco-efficiency into its activities. Briefly, Lemos and Nascimento (1999) describe eco-efficiency as practices that reduce the consumption of materials and energy; add value to goods and services; decrease the dispersion of toxic substances; intensify the recycling of materials; promote the sustainable use of renewable resources, and prolong the durability of products (Vieira and Barbosa 2011).

Table 1 Steps and activities in leather footwear manufacture

Step	Activity description
Designing/molding	Identifying sources, analyzing and interpreting fashion trends, and behaviors, and tailoring production
Cutting	Factories can outsource this step because many professionals have the necessary equipment (hydraulic cutting rocker). Laser cutting, water jetting, and air compressing are not usually outsourced
Sewing/backstitching	Joining the pieces that were cut and marked in the previous step by sewing or stitching. Factories can also outsource this step provided the professional has the stitching machine
Assembling	Collating the upper in the mold for attachment to the mounting insole. This stage also comprises placement of toes or cuirass, buttresses, cambers (when leather is shaped into the desired way, usually in boots), interlace, and others
Outsole	Mounting insoles, which can be nailed, glued, vulcanized, or sewn. In this stage, the soles and upper leather are also sanded
Finishing	Taking the footwear out of the mold and going through the final touches including placement of linings and heels, painting, waxing, boxing, and others

Source The authors

Eco-design and cleaner production has been a trend in footwear production in Franca. This type of production focuses on sustainability and is under accelerated expansion. Eco-design and cleaner production are the result of increasing consumer environmental awareness: today, consumers want to know what the product is made of and how it is manufactured. In this scenario, ecological footwear is modifying opinions, market trends, concepts, and manufacturing and distribution processes.

Large footwear industries in Franca already rely on ecological outsole and linings and have been seeking solutions that minimize their environmental impact. This constitutes a vision of planetary insertion in which the success of a business can no longer be measured by financial profit or shared appreciation only, but also by environmental responsibility and future commitment. Some industries produce a line of footwear that reuses everything that has been created in previous editions (modeling, products, and materials) and transforms them into new projects through updates and changes, following current trends. These industries focus on 100% sustainable production (Freeway 2017; Abicalçados 2018).

On a national basis and also in Franca, the use of local materials reflects "Brazilianness" and is linked to the idea of environmental sustainability. Examples of local materials are vegetable fibers and wood, which can be a design advantage and increase the competitiveness of the national product. This advantage has been termed "Cara Brasil" and has been the focus of strategic actions that will be adopted by the footwear industry in the coming years. Footwear design relies on eco-design, incorporates Brazilian cultural materials and characteristics, and focuses on environmental sustainability by addressing issues such as the recycling and reuse of components (Guidolin et al. 2010).

4.2 Economic and Sustainable Development in the Footwear Industry

Sustainability has a globalized character, which evokes profound transformations in the ways the population thinks, produces, and consumes. Incorporation of the environmental variable has been crucial for the survival of organizations of the leather footwear sector: for these organizations to become competitive in the foreign trade scenario, they will have to meet the increasingly stricter international norms regarding the environmental impact of their products. On the basis of tools that help to monitor and to eliminate environmental problems, companies will be able to establish themselves in the market. Thus, the Brazilian footwear industry opted to strengthen its position though environmental qualification (Gatelli et al. 2010).

According to Almeida (1998), for a company to be sustainable, it must seek eco-efficiency in all its actions and decisions, so that it can increase its production both quantitatively and qualitatively while polluting less and using fewer natural resources. The company that is in favor of the sustainability principles must still be socially responsible because it is immersed in a social environment where it exerts influence at the same time that it is influenced by said environment.

Production in the shoe sector has undergone internationalization since the late 1960s. De Oliveira et al. (2013) pointed out that investment in product development and innovation, sustainable production processes, waste reuse within companies, and reduction in energy consumption is increasingly present in the industries of Franca.

Carloni and Pires (2017) analyzed the sustainable development practices in small, medium-sized, and large industries in the city of Franca. In the case of the emission of pollutants, 74% of the companies reported taking actions to control and/or reduce residues. For gaseous residues, the industries have used filters and filtration technologies and appropriate residue-based machines. In the case of solid residues, the companies have increased the use of raw materials, recycled residues, and discharged residues in legalized landfills. Concerning the issues related to social progress, 100% of the industries stated that they have respected and adopted local labor laws and agreements when they hire or outsource human resources for the company, but only 60% of the industries reported actions for socioeconomic progress (training, positions, and transparent actions). The majority of the companies (80%) reported addressing issues related to staff health and safety. In the interest of sustainability, 53% of the industries claimed that they have managed the product life cycle along the entire production chain and have worried about the raw material and its disposal.

Companies have sought sustainable solutions to differentiate their products. In this sense, today some companies already develop projects or some activity related

Fig. 1 IP Franca Seal *Source* Couromoda web page (Available at: https://www.couromoda.com/noticias/ler/indicacao-de-procedencia-de-franca-conta-com-novo-logotipo/. Accessed in August, 2018)

to sustainability and environmental concern. In a survey of these initiatives, Silva et al. (2015) described the project "Sustainable Production—Happier Tomorrow" ("Produção Sustentável— Amanhã mais feliz), which shows to consumers how the companies deal with environmental responsibility by separating and discarding the solid waste generated during their industrial activity. In addition, there are companies that have been awarded the label "Sustainable origin", a certification given by the Brazilian Association of Footwear Industries (Abicalçados) and the Brazilian Association of Leather, Footwear, and Artifact Components Companies (Assintecal) in partnership with the Sustainability Laboratory (LASSU) of the University of São Paulo (USP) and the Massachusetts Institute of Technology (MIT). This certification follows a scale that attests how much the company has already incorporated sustainability actions aligned with four established pillars—environmental, economic, social, and cultural pillars—in its processes. Recyclable packaging, sustainability report, solid waste management program, environmental management system (EMS), and sustainable footwear lines are other practices that have been described by the companies.

In this article, we highlight SindiFranca's initiative, which combines the historical context of the city, marked by its tradition in footwear manufacture and adoption of sustainable measures during this activity. Besides adding more value to the product, this initiative constitutes an additional commercial tool in promoting the products of well-known quality that have been manufactured in the city of Franca for almost two centuries. The Franca footwear industry registered this initiative with the National Institute of Industrial Property (INPI) under the name Provenance Indication of Franca (IP Franca), which was officially launched in 2017. This is the first Provenance Indication that has been used in the economic sector in the State of São Paulo and the first Provenance Indication that has been used in the footwear manufacture segment in the world (Fig. 1).

All the footwear that is produced in the industries located in the city of Franca may use the IP Franca seal as long as they accept assuming responsibilities for competitive differentials that rest on the five sustainability pillars and comply with the Use Regulations and Control Standards, to be administered by SindiFranca and by two bodies, the Regulatory Council and the Administrative Council.

Table 2 Sustainability pillars and industry classification

Pillars	Diamond	Gold
Manufacture origin	Franca	Franca
Management/production	Institute and own program	Institute or own program
Raw materials	Leather	Leather and/or synthetic
Social action	Pro-Child[a] and own action	Pro-Child[a]
Environmental management	Recycling and legal disposal via SindiFranca	Legal disposal via SindiFranca

Source The authors
[a]Education program for children and teenagers

IP Franca is based on the premise of product traceability; that is, the product can be monitored and controlled through participation of the society, via digital access. Each shoe can have its origin visited by means of a code and a QR Code that will direct the consumer to information about the manufacturing company: data on facilities, employees, social work, environmental care, and e-commerce will be available through numbers and images that can be consulted before, during, and after purchase. The consumer can follow the footwear production process and see the entire history, dedication, and compliance with environmental responsibility assumed by the industry. IP Franca aims not only to value the footwear industry, but also to add value to the product for the final consumer.

The industries that are interested in using the IP Franca seal must apply for accreditation with SindiFranca. The accreditation is valid for three years. After this period expires, the company must request renewal and go through the requirements of the Regulation again. The footwear product could be a target for male, female, or infant consumers and should contain a label or tag with information about its characteristics (material, product composition). For certification purposes, the SindiFranca Board will classify the industry into two categories, diamond or gold (Table 2), depending on their compliance with the five pillars.

The Manufacture Origin contemplated by IP Franca prizes a product of quality that is manufactured by the hands of shoemakers in the city of Franca and is partly handcrafted.

According to the requirements of Use Regulations and Control Standards, Management/Production recognizes and values the best level of industry management and production, including a company having its own improvement programs.

As for Raw Materials, leather is more valued than synthetic materials.

In terms of Social Action, industries are expected to contribute to social projects that improve the quality of life of the people living in Franca. In terms of partnerships between the city of Franca and the footwear industry entrepreneurs, the Footwear City Institutes and the Pro-Child Institute are worth mentioning. The latter institute was created in 1995 with the purpose of eradicating child labor. This institute aims to support actions that promote the education of children and adolescents by preparing them to exercise citizenship and qualifying them for work. The institute offers scholarships for computer courses, foreign language courses, sports practice (soccer, ballet), and training in mechanics, among others (Barbosa et al. 2012).

Environmental Management is concerned with the disposal of industrial waste in the footwear manufacture sector. In this context, the Eco-Efficient Collection Project was created, which aims to manage and to dispose of all the industrial waste generated by the sector in accordance with the legislation.

The industrial waste in Franca has a peculiar nature and requires specific monitoring during its collection and disposal. SindiFranca, which is one of the institutions that are responsible for managing the waste generated by the footwear industry in Franca, was one of the supporters of the Municipal Plan for Integrated Solid Waste Management with a view to the National Waste Policy. As a result of this Plan, the São Paulo State Environmental Company (CETESB) licensed the city's landfill and allowed it to receive the waste produced by the footwear industry in the municipality. This landfill also serves as a parameter for other regions.

5 Conclusion

Environmental perspectives are troubling. On both local and global scales, solving or mitigating environmental issues implies implementing deep social and economic changes that constitute a challenge for current technologies, practices, policies, and ways of thinking. To remain in the market and to participate actively and sustainably in economic development, companies will have to consider global, integrated, and long-term perspectives that will require new knowledge and achievements.

Regarding the footwear production, sustainability issues constitute a challenge in terms of competitiveness, management, and waste disposal. The shoe industry in the city of Franca became aware of the fact that the commercial impact of their sustainability actions could determine their maintenance in the market. Therefore, the companies started to search for a competitive advantage and to promote viable actions. The IP Franca seal resulted from these actions and represents the combination of quality and tradition with social and environmental responsibility through sustainable choices.

Human prosperity cannot be detached from environmental protection. This novel attempt to bring the footwear industry and sustainability together may be important to ally economic development with academia to produce knowledge.

With a view to advancing knowledge, sustainable economic development must be considered under the premises of Health Promotion. It is necessary to seek intersectorial solutions and the participation of society, which moves and controls the cycle of product consumption. Access to information provides knowledge about the product to be acquired, but it is the population that will decide whether to purchase it. Entrepreneurs need to take care of their product, whereas consumers need to worry about their health for this partnership to work.

Acknowledgements We thank the Coordination for the Improvement of Higher Education Personnel (CAPES) for granting the doctorate scholarship to Jaqueline L. Damasceno. We also thank the president of SindiFranca José Carlos Brigagão do Couto for the interview.

References

Abicalçados (2018) Associação Brasileira das Indústrias de Calçados. http://www.abicalcados.com.br/noticia/em-busca-da-sustentabilidade-industria-da-moda-brasileira-ja-se-prepara-para-2018-com-as-inspiracoes-do-salao-inspiramais (Last Accessed 28 July 2018)

Almeida RO (1998) Ecoeficiência e as empresas do terceiro milênio. Tendência do Trabalho www.perspectiva.com.br/leitura/ (Last Accessed 26 July 2018)

Alves VC, Barbosa AS (2013) Management practices of leather footwear industry in Franca-SP. Gest Prod 20(4):883–898

Atlas Brasil (2013) http://www.atlasbrasil.org.br/2013/pt/perfil_m/franca_sp (Last Accessed 26 July 2018)

Barbosa AS, Alves EA, Brago Filho H (2012) Corporate social responsibility in the footwear industry in São Paulo (Brazil): an analysis of the industrial clusters of Franca and Birigui. G&DR 8(3):183–197

Beddoe R, Costanza R, Farley J, Garza E, Kent J, Kubiszewski I, Martinez L, McCowen T, Murphy K, Myers N, Ogden Z, Stapleton K, Woodward J (2009) Overcoming systemic roadblocks to sustainability: The evolutionary redesign of worldviews, institutions, and technologies. PNAS 106(8):2483–2489

Brasil (2015) Ministério da Saúde. Política Nacional de Promoção da Saúde: revisão da Portaria MS/GM nº 687, de 30 de março de 2006. Brasília: MS, p 36

Buss PM (2000) Health promotion and quality of life. Ciênc Saúde Colet 5(1):163–177

Carloni AR, Pires DF (2017) Actions and perceptions of sustainability in large footwear industries of Franca/SP. Rev Eletrôn Adm 16(2):296–321

Chang YJ, Lehmann A, Winter L, Finkbeiner M (2018) Application options of the sustainable child development index (SCDI)—assessing the status of sustainable development and establishing social impact pathways. Int J Environ Res Public Health 15(7):E1391

Czeresnia D (1999) The concept of health and the difference between prevention and promotion. Cad Saúde Pública 15(4):701–709

De Oliveira AFL, Bianchini VK, Queiroz GA, Do Amaral Razzino C, Da Silva MC (2013) Sustentabilidade: uma análise estratégica para o setor calçadista. XXXIII Encontro nacional de engenharia de produção. Salvador, Brasil, pp 1–14

Diamond J (2005) Collapse: how societies choose to fail or succeed, 1st edn. Viking Press, New York, pp 293–304

Espíndula FM (2009) Processo Produtivo dos Calçados. Available in http://www.artigos.com/usuario/fmespindula (Last Accessed 24 July 2018)

Figueiredo GLA, Martins CHG, Damasceno JL, Castro GG, Mainegra AB, Akerman M (2017) Right to the city, right to health: what are the interconnections? Cien Saude Colet 22(12):3821–3830

Freeway (2017) http://www.blogfreeway.com.br (Last Accessed 26 July 2018)

Gatelli E, Dal Col Zeve CM, Sikilero CB (2010) Impacto ambiental da cadeia produtiva do setor calçadista do Vale do Rio dos Sinos. In: XXX Encontro nacional de engenharia de produção, São Carlos, Brasil

Guidolin SM, Da Costa ACR, Da Rocha ERP (2010) Indústria calçadista e estratégias de fortalecimento da competitividade. Calçados (BNDES Setorial) 31:147–184

Gunderson LH, Holling CS (2002) Gunderson LH, Holling CS (eds) Panarchy: understanding transformation in human and natural system, 1st edn. Island Press, Washington, DC, p 308

Haines A, Alleyne G, Kickbusch I, Dora C (2012) From the earth summit to Rio+20: integration of health and sustainable development. Lancet 379:2189–2197

IBGE (2017) Instituto Brasileiro de Geografia e Estatística. https://cidades.ibge.gov.br (Last Accessed 25 July 2018)
Kuhlman T, Farrington J (2010) What is sustainability? Sustainability 2(11):3436–3448
Lemos AD, Nascimento LFA (1999) produção mais limpa como geradora de inovação e competitividade. Rev Adm Contemp 3(1):23–46
Merico LFK (2009) Economia e sustentabilidade: o que é, como se faz. Loyola, São Paulo
Pearce DW, Markandya A, Barbier EP (1989) Blueprint for a green economy, 1st edn. Earthscan Publications, London, UK, p 108
Scotto G, Carvalho IC de M, Guimarães LB (2011) Desenvolvimento Sustentável, 6th edn. Petrópolis, RJ: Vozes, Coleção Conceitos Fundamentais
Silva AH, Moraes CAM, Modolo RCE (2015) Avaliação ambiental do setor calçadista e a aplicação da análise de ciclo de vida: Uma abordagem geral. VI Fórum Internacional de Resíduos Sólidos, São José dos Campos, Brasil
SindiFranca (2018) Sindicato da Indústria de Calçados de Franca http://www.sindifranca.org.br/estatisticas.html (Last Accessed 26 July 2018)
Torresi SIC, Pardini VL, Ferreira VF (2010) O que é sustentabilidade? (Editorial). Quim Nova 33(1):5
Van der Schoor T, Scholtens B (2015) Power to the people: Local community initiatives and the transition to sustainable energy. Renew Sust Energ Rev 43:666–675
Vieira EL, Barbosa AS (2011). Práticas tradicionais e de ecoeficiência na indústria de calçados no Brasil. Revista Eletrônica: Tempo - Técnica – Território 2(1):30–50
Weiss H, Bradley RS (2001) What drives societal collapse? Science 291:609–610
Wiersum KF (1995) 200 Years of sustainability in forestry: lessons from history. Environ Manag 19:321–329

Glória Lúcia Alves Figueiredo obtained her Ph.D. and M.Sc. degrees in Public Health Nursing at the Ribeirão Preto College of Nursing of the University of São Paulo (EERP/USP) and graduated in Nursing from the same school. She is currently a researcher and the vice-coordinator of the Postgraduate Program in Health Promotion of the University of Franca (UNIFRAN) and is a Professor in the Undergraduate Nursing Course at UNIFRAN. She is also the Leader of the Nucleus of Studies and Research into Health Promotion / NEP-Health Promotion / UNIFRAN / CNPq.

Carlos Henrique Gomes Martins obtained his Ph.D. degree in Applied Microbiology and his M.Sc. degree in Clinical Analysis at the São Paulo State University "Júlio de Mesquita Filho"—UNESP. He graduated in Biomedicine from the Faculty of Health Sciences "Barão de Mauá". He is currently a researcher and the coordinator of the Postgraduate Program in Health Promotion and a researcher of the Graduate Program in Sciences of the University of Franca (UNIFRAN). He is also a Professor of the Undergraduate Course in Biomedicine, Dentistry, and Nursing at UNIFRAN. CNPq Productivity Research Grant—level 2.

Jaqueline Lopes Damasceno obtained her B.Sc. degree in Biomedicine and her M.Sc. degree (funded by the São Paulo Research Foundation—FAPESP) with emphasis on mutagenesis at the University of Franca (UNIFRAN). She is currently a Ph.D. student in the Postgraduate Program in Health Promotion of the same institution (funded by Coordination for the Improvement of Higher Level Personnel [CAPES]).

Márcio Ronan Policarpo obtained his degree in Architecture and Urbanism at the University of Franca (UNIFRAN). He is an active Architect and Urbanist in the city of Franca and suburbs. He is also a Professor of the Undergraduate Course in Architecture and Urbanism at UNIFRAN. He is currently working to obtain his M.Sc. degree in Health Promotion at UNIFRAN.

Social Responsibility Versus Sustainable Development in United Nations Policy Documents: A Meta-analytical Review of Key Terms in Human Development Reports

Johannes M. Luetz and Mohamed Walid

Abstract Social responsibility (SR) and sustainable development (SD) are dissimilar yet complementary concepts. Over recent decades their increase in popularity has seen the two terms become firmly integrated within international development policy discourse. Nevertheless, even though both terms are intertwined and cannot be meaningfully discussed in isolation, there is a paucity of research that addresses the interrelationships of the two terms in human development and policy discourse. To address this gap in the literature, this research employs an inductive and exploratory methodological approach. Conducting a systematic keyword search and expert literature meta-analysis in all 25 United Nations (UN) Human Development Reports (HDRs) published to date from 1990 to 2016, the study investigates what prioritisation the UN ascribes to 'economic', 'sustainable' and 'social' development. Keyword analysis reveals that 'economic' perspectives dominate 'sustainability' and 'social' perspectives by a factor of 2 and 4.67 respectively. In synthesis, the UN remains espoused to 'economic' development as its primordial panacea for poverty reduction, which it increasingly advocates under the guise of 'sustainable' development. Relatedly and importantly, UN HDRs continuingly advocate 'economic growth' as a solution instead of identifying it as a problem. This study extends previous research by focusing expressly on the intersection of economic development, environmental sustainability, and social responsibility as a fertile space for inquiry. The research proposes a Triple Bottom Line (TBL) approach as a bridging notion for a more holistic human development agenda.

J. M. Luetz (✉)
CHC Higher Education, Brisbane, Carindale, QLD 4152, Australia
e-mail: jluetz@chc.edu.au; j.luetz@unsw.edu.au

J. M. Luetz
University of New South Wales (UNSW), Sydney, Australia

M. Walid
London School of Economics, London, UK
e-mail: M.Walid-Lotfy@lse.ac.uk

W. Leal Filho (ed.), *Social Responsibility and Sustainability*, World Sustainability Series, https://doi.org/10.1007/978-3-030-03562-4_16

Keywords Social development · Sustainable development
Economic development · Social policy · Social responsibility
United nations (UN) · Human development reports (HDRs)
Development discourse · Policy analysis
United nations development programme (UNDP)

> When the last tree is cut down, the last fish eaten, and the last stream poisoned, you will realize that you cannot eat money —Native American saying. (In: *Oxford Dictionary of Proverbs*, Simpson and Speake 2009)

> A transition to sustainability cannot be achieved if our economic system is not radically changed, simply because limitless economic growth is impossible within a limited planet. (Pacheco et al. 2018, p. 238; cf. Ripple et al. 2017)

> Sustainability requires putting the environment and society above the market. To make progress towards sustainable societies requires policies based on the inclusion of all races and cultures, equity and solidarity among societies, and cooperation among governments. The first step in meeting this challenge is repairing the environment and society. (Fisher and Ponniah 2003, pp. 127, 128, cited in Ledwith 2005, p. 175)

1 Defining Social Responsibility and Sustainable Development

This introduction comprises four parts. Section 1.1 introduces the study motivation and sketches its broader literary context. The next two sections explore the theoretical and conceptual evolution of 'Sustainable Development' (SD) (Sect. 1.2) and 'Social Responsibility' (SR) (Sect. 1.3). Section 1.4 offers a short synthesis of how the two concepts may be merged or theorised concurrently.

1.1 Study Motivation

This study was motivated by a desire to better understand United Nations (UN) policy discourse at a meta-analytical level. More specifically, the study sought to investigate what prioritisation the UN ascribes to so-called 'economic', 'sustainable' and 'social' development in its global flagship Human Development Reports (HDRs), published (almost) annually since 1990. In short, this meta-analytical study of all 25 UN HDRs published to date promised to offer new policy perspectives on the role and place of different development foci in "the most influential [literature] in the field of development" (Telleria 2017, p. 2143, attributed to McNeill 2007). As such, this study offers auspicious benefits of macro-analysis that may promote a more holistic agenda for so-called 'human development', including in areas of social responsibility and sustainable development. The following literary overview will set the stage contextually for this study's interdisciplinary investigation and discussion.

Since the inception of the first HDR in 1990 (UNDP 1990), all the way to the most recently published HDR (UNDP 2016), 'social development' did not enjoy a prime seat at the table of global development discourse—unlike economic-related concepts such as 'economic growth' or 'economic development'. For instance, the 1996 HDR clearly identifies 'human development' as the end, with 'economic growth' constituting the corresponding means (UNDP 1996). However, scrutinising this situation more carefully highlights a key concern, namely that the consequence of focusing primarily on 'economic growth' as the all-powerful engine to drive 'human development' has spawned rising inequality and social injustice (Hoekstra and Wiedmann 2014; Hickel 2015; Oxfam 2015; Simms 2008; WWF 2016; Salleh 2016). Further, obsession with 'economic growth' has also been identified "as a major cause of environmental crisis" (Pacheco et al. 2018, p. 238; cf. EOD 2017; Ripple et al. 2017). This disparate development emphasis is rather clearly reflected in the data of this meta-analytical inquiry.

Economic growth has been classically about boosting the gross national/domestic product (GNP/GDP) (Dang and Pheng 2015). Nonetheless, this seems to be a rather narrow goal of development following a reductionist approach. A reductionist approach in that sense is one that reduces 'human' development with its multi-dimensions (social, environmental, economic—among others), to predominantly 'economic' concerns, thereby creating unbalanced and unsound development that fails to inspire and promote equality, justice and sustainability (Todaro and Smith 2003).

> To maximize income growth, environmental considerations were left to languish on the sidelines; the standard of living was often allowed to slide; large inequalities between classes, regions, and genders were ignored; and poverty was tolerated more than it should have been in the rush to generate maximum growth. (Basu 2000, p. 64)

Looking retrospectively, it seems that world politics has been geared towards finding quick fix solutions—through economic stimulants—thus creating inequitable development outcomes. By consequence, the social aspect of development in the equation of sustainable development has been missing or at the very least undermined due to excessive attention being directed to economic aspects of development. Theoretically, social responsibility should serve as a good reminder to policy-makers to ensure that balance is maintained among the pillars of sustainability (social, environmental, economic) (Fig. 4). Social responsibility is about integrating the different pillars for the sake of creating a more balanced welfare society (Carroll 1979). From that angle, social responsibility can be viewed not only as a concept *of* development but also as a criterion *for* development. Hence, although social development has remained side-lined and stymied by economic growth, it has the potential to claim a more significant role in global human development politics.

To begin this discussion, it is crucial to establish a clear understanding of the meaning and parameters of both Social Responsibility (SR) and Sustainable Development (SD), each in their own right, before attempting to bring them together. It should be clarified, however, that in our attempt to find clear-cut definitions for both concepts, it seems that neither of them have a commonly accepted definition so far

(Ebner and Baumgartner 2008). Nonetheless, there is no shortage of definitions for either of them independently. In one study, Dahlsrud (2005, 2008) identified 37 different definitions of SR/CSR, and admittedly, his study did not capture all available SR/CSR conceptualisations. Similarly, Ebner and Baumgartner (2008) attempted to gather data on both concepts from the fields of economics and business management. They ended up with a non-exhaustive compilation of 43 articles, 13 of which were on SD, and 30 on SR/CSR. To that extent, it seems there is an ever-growing interest in researching both concepts, albeit without reaching a point of satiation in terms of theorising them together.

1.2 Sustainable Development (SD)

We start with SD as it was conceptualised in official text. The *Brundtland Report*, published in 1987, famously advanced the following conceptual definition of SD:

1. Sustainable development is development that meets the needs of the present without compromising the ability of future generations to meet their own needs. It contains within it two key concepts:
   - the concept of 'needs', in particular the essential needs of the world's poor, to which overriding priority should be given; and
   - the idea of limitations imposed by the state of technology and social organization on the environment's ability to meet present and future needs.
2. Thus the goals of economic and social development must be defined in terms of sustainability in all countries - developed or developing, market-oriented or centrally planned (WCED 1987, p. 41).

SD has different connotations (Fig. 1). Even so, while the term is often enlisted to refer to environmental contexts, its most frequently used definition, that of the *Brundtland Report* (WCED 1987), did not overly accentuate the environment, although it implied its vitality in sustaining life by alluding to the future. One definition that buttresses the importance of the environment within the sustainability paradigm of the UN was published in a World Bank paper. It states that SD will improve quality of life while lowering the intensity of natural resource use for future generations (Munasinghe and Lutz 1991).

Aside from mentioning the environment in this definition, of further interest is the use of the term 'quality of life', which was still relatively new in the realm of policy-making and SD. The Australian Government (1992) followed the same path taken by the World Bank after making a direct link between ecological SD and the increase in the 'total quality of life'. The introduction of 'quality of life' into the dictionary of SD signified the importance of the 'social pillar', hence opening a window of opportunity for SR to bond with SD.

Author	Definition	Dimensions & Orientation
Brundtland Report (World Commission on Environment and Development [WCED] 1987, pp. 16, 41)	"1. Sustainable development is development that meets the needs of the present without compromising the ability of future generations to meet their own needs."	Generic
MacMillan (1988)	It is the kind of economically-driven development that continues based on its exploitation of environmental resources jeopardizing the environment	Economic > Environmental
Pearce (1988)	SD translates to per capita utility that increases over time	Economic > Economic
Braat (1991)	SD combines two notions: economic development & ecological sustainability, both leading to maximising welfare, which can be ecologically sustained	Economic-Environmental > Social
Holmberg (Ed., 1992)	SD is a result of interaction between trade-offs between biological, economic and social systems	Complex (involving the three pillars interactively)
O'Riordan and Yaeger (1994)	Managing economic growth within the bounds of natural replenishable systems	Economic > Environmental
Winograd (1995)	SD is about satisfying human needs without compromising the environment	Socio-Economic > Environmental
Choucri (1997)	Managing social demands without eroding life properties or social cohesion	Social > Environmental-Social

Fig. 1 Brief overview of selected SD models. The arrow sign > indicates the intentional direction of one of the three pillars (economic, social, environmental) to the other

1.3 Social Responsibility (SR)

According to Carroll (1979), social responsibility (SR) is about the interaction of different responsibilities for the sake of realising the expectations of a welfare society. Prior to Carroll's definition, SR was identified as a set of economic and legal duties companies have towards society (McGuire 1963). Our research finds that SR is the primitive concept from which CSR, Corporate Social Responsibility, emerged. For that reason, both terms are used interchangeably.

With more than 40 different labels for CSR programmes and numerous CSR synonyms available (Visser 2008), a brief overview of selected CSR models is presented below (Fig. 2). Carroll (1979) specifies four components for CSR: Economic, legal, ethical, and discretionary/philanthropic expectations that the society has of an enterprise. Carroll's model was arguably the first comprehensive and most popular in the domain of SR/CSR. Empirical work using factor analysis on a survey to 241

Author	Definition	Dimensions
M. Friedman (1962, 1970, 2007)	The sole social responsibility of businesses is the use of resources to increase profit as long as they engage with honesty and fairness	Organisational, Societal
Carroll (1979)	Interaction between different categories of social responsibilities	Economic, Legal, Ethical, Discretionary
Wartick and Cochran (1985)	"The underlying interaction among the principles of social responsibility, the process of social responsiveness and the policies developed to address social issues" (p. 758)	Economic, Legal, Ethical, Discretionary
Wood (1991)	"A Business organization's configuration of principles of social responsibility, processes of social responsiveness, and policies, programs, and observable outcomes as they relate to the firm's societal relationship" (p. 693)	Institutional, Organisational, Individual
Dahlsrud (2008)	"CSR is viewed as a social construction and [...] it is not possible to develop an unbiased definition" (p. 2). However, existing definitions of CSR can be categorised into five primary dimensions (p. 4)	Environmental, Social, Economic, Stakeholders, Voluntariness
International Organization for Standardization (ISO 2010)	Social responsibility comprises seven core subjects (p. 9)	Organisational governance, human rights, labour practices, the environment, fair operating practices, consumer issues, community involvement and development

Fig. 2 Non-exhaustive overview of selected SR/CSR models and conceptual approaches

executives found that the relative weight of each component is: Economic = 3.5; legal = 2.5; ethical = 2.2; discretionary/philanthropic = 1.3 (Carroll 2016). His survey reflects the perception of executives in terms of their prioritisation, indicating which of these components are most important to them. Perhaps unsurprisingly, economic motivation is accorded the greatest prioritisation.

The findings of the survey resonate with Friedman (1962) as the defender of neoclassical economics:

> There is one and only one social responsibility of business - to use its resources and engage in activities to increase its profits so long as it stays within the rules of the game, which is to say, engages in open and free competition, without deception or fraud. (p. 112)

Others were even more dramatic in their disfavour of SR being conceived as socially-driven or environmentally-inclined. In his 1958 *Harvard Business Review* article, Theodore Levitt spoke of "The Dangers of Social Responsibility." In brief, his position was that the primary concern of businesses is to maximise profitability while

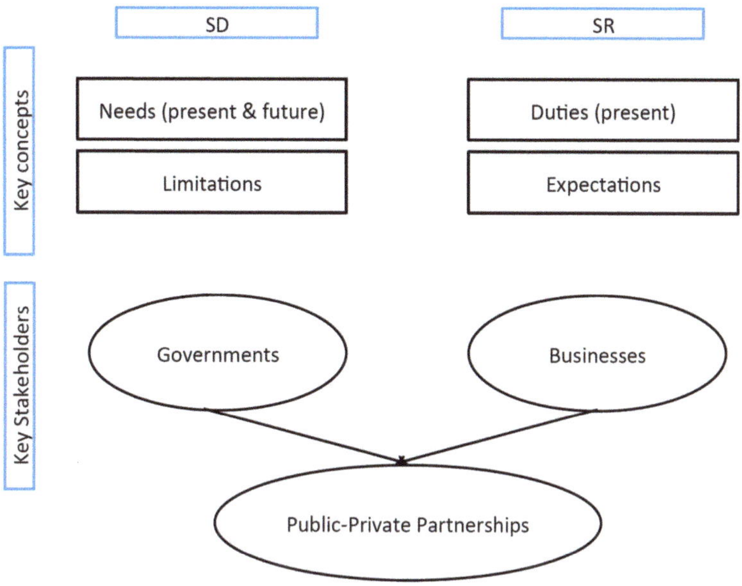

Fig. 3 Illustration by authors. Key concepts behind SD and SR and principal macro stakeholder of each model

engaging in civility (e.g., honesty and good faith). One can easily see the backlash between neoclassical economic and SR theorists. Some key concepts behind SD and SR, as well as the primary macro stakeholders of each model, are illustrated by Fig. 3.

1.4 Toward Synthesis

The definitions of SD and SR show that they had different departure points. SD was initiated as an environmental concept. The modern foundational structure of SD evolved between the years 1972 (United Nations Conference on the Human Environment in Stockholm) and 1992 (United Nations Conference on Environment and Development in Rio de Janeiro). The origin of the term SD itself can, however be traced back to the eighteenth century. It was then used exclusively in forestry-related issues. At the time, in order to guarantee the long-lasting growth of tree populations, a certain number of trees were allowed to be cut down. Thanks to the Club of Rome[1] following its report "Limits of Growth" (Meadows et al. 1972), the heretofore unchallenged normative 'economic growth' supremacy was progressively questioned (Hoekstra and Wiedmann 2014; Oxfam 2015; Pacheco et al. 2018; Simms 2008; WWF 2016). In contrast to SD, SR had a socio-economic starting point. It was

[1]http://www.donellameadows.org/wp-content/userfiles/Limits-to-Growth-digital-scan-version.pdf.

geared towards prioritising and rationalising the interests of businesses (starting with the economic motivation) in a way that can cater for communities' development. The framework of SR was formed in the 1950s. Its focus is supposed to ensure that actions align with societal values (Douglas et al. 2004). Even so, whether those societal values are set by a dominant particular corporate interest or another influential stakeholder remains a critical question to ask while talking about societal values.

2 Bringing Social Responsibility and Sustainable Development Together

Having discussed some popular definitions and conceptualisations of SD and SR, we now explore ways of bringing them together in one comprehensive model. Through thought leaders such as Howard R. Bowen the two concepts SD and SR started finding common ground. In his landmark book *Social Responsibilities of the Businessman*, Bowen (1953) asked the question: "what responsibilities to society may businessmen reasonably be expected to assume?" (p. xi). His thinking stemmed from his keen realisation that the largest businesses in the United States at his time held immense power and influence over people's lives. It sparked ways of thinking about how SD and SR/CSR may be related.

One definition by *SustainAbility* (2004) answers Bowen's question and explores the implied SR/CSR and SD relationship: CSR is "an approach to business that embodies transparency and ethical behaviour, respect for stakeholder groups and a commitment to add economic, social and environmental value" (p. 4). Unlike several other definitions that focused majorly, if not exclusively, on economic or social values, this one placed environmental values on an equal footing with the other factors. Indeed, SD has evolved to embrace more and similar pillars to those of SR/CSR but the aforementioned definition shows how the environmental dimension is gaining traction within the CSR paradigm, at least at the conceptual level. Cranfield School of Management (2017) identified the motif behind CSR and SD in saying that they are about,

> enabling companies to incorporate creation of social and environmental, as well as economic, value into core strategy and operations. This improves management of business risks and opportunities whilst enhancing long-term social and environmental sustainability. (para 1.)

After all, if businesses are expected to exert positive energy in the direction of sustaining environmental resources, they ought to be motivated to do so. Therefore, the definition explicitly mentions the positive outcome business management might get in return.

Leaving SR and the efforts of different stakeholders (businesses and not-for-profit organisations) to include the environment within SR-related definitions, we now look at the situation at the SD front. As mentioned erstwhile, SD began as an environmentally-driven concept. It later went on to incorporate economic and social pillars (Mubarak 2016), both of which are integral to SR, particularly the economic pillar.

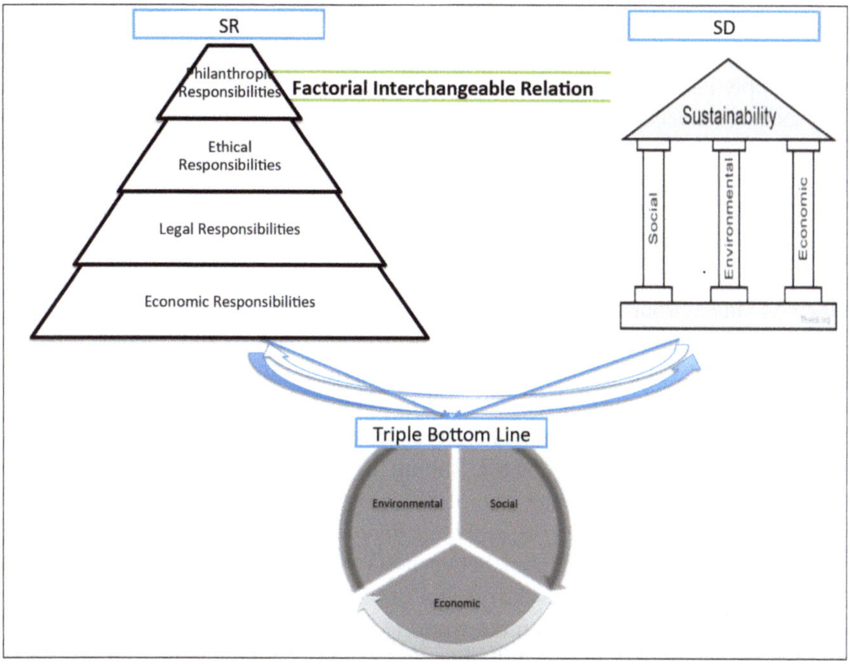

Fig. 4 Illustration by authors. SR and SD entertain a 'factorial interchangeable relation'

Our theoretical literary research leads to two major findings. First, SR and SD entertain a 'factorial interchangeable relation' (Fig. 4). While they both had a different starting point with a distinct goal in mind, they ended up with the three same factors (economic, social, environmental), even if to varying degrees. This is the result of the interactive nature of challenges the world increasingly faces (Elkington 1997), narrowing the gap among the pillars/steps of both concepts. Second, the unique relationship SR and SD enjoy together at the conceptual level may have been accelerated by the Triple Bottom Line (TBL) framework. The aim of the TBL was to redirect business interests from the fully-fledged focus on classical bottom line (represented in revenue) to a broader and more inclusive bottom line (that recognises and incorporates the environment and society). Incidentally, the term TBL was first introduced by John Elkington in his (1997) book, *Cannibals with Forks: The Triple Bottom Line of 21st Century Business*.

Expressed in simple language, the TBL concept comprises three so-called 'bottom lines'. The first, financial bottom-line is considered the traditional bottom line. It is about expanding market share and maximising revenue. Businesses that focus on the second bottom line are often measured by how low their employee turnover is. Treating employees in an ethical and fair manner with equitable compensation is the major success indicator in this second segment of TBL. The third strand in TBL is the environmental bottom-line. This entails ensuring that companies and working people are aware of their environmental footprint. Businesses, in this case, take all necessary measures to reduce their negative environmental impact throughout the entire supply chain. A business or a working person who abides by the TBL is one who seeks a balance between profitability, social responsibility and environmental awareness (Atu 2013). It is of importance to state that TBL is a concept that can apply to any kind of business, governmental institution, not-for-profit organisation, household, or individual. This is what SR or 'social footprint' entails from a holistic perspective (Henriques 2010).

Retrospectively, we can see that all three TBL foci originated from different disciplines; SD in environmental contexts, SR in management, and TBL in business. The multiplicity of sources reflects the broad spectrum of interests and possible relations these concepts have interchangeably. Furthermore, from a chronological perspective, as TBL was the last of these concepts to emerge, we can safely assume that it has been influenced *by* them. Therefore, we propose TBL as a bridging notion that conjoins elements of SD and SR, linking the environment, social and economic components. In fact, in chapter one of his book, Elkington (1997) makes a clear reference to the flagship 1987 *Brundtland Report* (WCED 1987), thereby inviting the possibility for querying the link between SD and TBL. Further, Savitz and Weber (2006) state that TBL "captures the essence of sustainability by measuring the impact of an organization's activities on the world [...] including both its profitability and shareholder values and its social, human and environmental capital." (p. 2). Relatedly, some authors and scholars even argue that the Sustainable Development Goals (SDGs) may be a case of TBL in policy (Cheeseman 2015). Official statements of CEOs on UN platforms also emphasise this point. For instance, Sijbesma (2015), CEO of Royal DSM, recently stated that his company had adopted the TBL approach, seeking to achieve the SDGs by targeting the three dimensions of people, planet and profit.

3 Research Contribution, Design and Methodological Considerations

This paper builds on previous research by means of an inductive research approach that combines a systematic keyword search carried out within key UN policy documents with exploratory analysis of selected key passages. In this way, this research combines quantitative and qualitative components that are distinct and yet complementary. This mixed methods approach (Punch 2014, pp. 301–326) uses an 'exploratory design' paradigm that seems appropriately suited to simultaneously encapsulate exploratory breadth and analytical depth (Creswell 2014; Creswell and Plano Clark 2011).

3.1 Quantitative Systematic UN HDR Keyword Research

Quantitative data used in this paper are derived from systematic keyword searches conducted within key United Nations (UN) policy documents. Methodological approaches were based on past similar systematic keyword research within UN literature, namely: (1) comparative analyses of the concepts 'sustainable development' and 'environmental sustainability' (Walid and Luetz 2018); and (2) critical analyses of the conceptual presentation of 'the poor' as either 'active' OR 'passive' stakeholders in development (Luetz et al. 2019). Following the antecedent of these studies, this research was similarly limited in scope to all 25 English language UN Human Development Reports (HDRs) published to date (from 1990 to 2016), which were chosen "because of their global influence and appeal. Such focus could also be viewed as a limitation"[2] (Walid and Luetz 2018, p. 805).

HDRs are published annually[3] by the United Nations Development Programme (UNDP) and are considered to be important international milestone reports for global development. According to Telleria (2017), HDRs are "the most influential [literature] in the field of development in recent decades" (p. 2143, attributed to McNeill 2007). Reports are quite comprehensive and range in length from 130 pages (UNDP 1991) to 440 pages (UNDP 2006), with 236 pages as the average report size. With a total of 25 HDRs published to date, the total sample size of all reports combined comprises a cumulative 5896 pages overall. For the purposes of this research study, this body of work represented a suitable data set.

In keeping with the TBL focus of this paper, keyword searches concentrated on the three domains (1) economic, (2) environmental, and (3) social. To gauge the overall level of significance that the UN ascribes to these three domains, this research queried all 25 reports for nine search phrases of interest according to TBL emphasis: (1) "economic development", "economic growth", "economic justice"; (2) "sustainable development", "environmental sustainability", "environmental justice"; (3) "social development", "social responsibility", "social justice". These phrases were intentionally chosen to direct the focus of this research on nouns (substantives) that are narrowly aligned with the TBL emphasis of this paper.

Given their currency in development literature, these nine phrases seemed to be well suited to encapsulate the TBL emphasis of this paper, as exemplified by the following HDR inclusions or references: "economic development" (Griffin 1989, cited in UNDP 1992, p. 109); "economic growth" (Friedman 2005, cited in UNDP 2009, p. 125); "economic justice" (Phelps 1973, cited in UNDP 1993, p. 113); "sustainable development" (UNCTAD 2014, cited in UNDP 2015, p. 197); "environmental sustainability" (UNDP 2015, pp. 250–253); "environmental justice" (Sze and London 2008, cited in UNDP 2011, p. 115); "social development" (Sen 2007, cited in UNDP 2013, p. 136); "social responsibility" (Hopkins 1999, cited in UNDP 1999, p. 115); "social justice" (Nussbaum 2003, cited in UNDP 2016, p. 181).

---

[2]This limitation in scope is elaborated in Sect. 6.

[3]There are two exceptions: Over the two-year period 2007–2008 there was only one report published (UNDP 2007), and no HDR was published in 2012.

To obtain the data, Adobe Acrobat PRO DC (Creative Cloud 2018) was used to carry out an 'Advanced Search', querying each phrase as 'whole words only' within all 25 UN HDRs, which were accessed from the UN.[4] All occurrences were subsequently collated and analysed through multiple review cycles. The guiding motif of this research was to gain a better understanding about the prevalence and evolutionary usage of key development terminology in influential United Nations policy documents (McNeill 2007). In point of fact, the UN seems to be self-assured of the influence of its HDRs as evidenced by its UNDP publication entitled *Ideas, Innovation, Impact: How Human Development Reports Influence Change* (Sidhu et al. n.d.).

The research design was prepared during the months November 2017 to January 2018. This was followed by data collection and analysis during the months February and March 2018. Multiple rounds of data collection and analysis ensured "test-retest reliability" (Punch 2014), with "construct validity" (pp. 238, 240) being ensured by means of including three distinct keyword search strings for each TBL emphasis under investigation. Finally, comprehensive sampling ensured that the findings of this meta-analytical study were not obscured by 'publication bias' (Rothstein et al. 2006). Stegenga (2011) posits that "numerous decisions must be made when performing a meta-analysis which allow wide latitude for subjective idiosyncrasies to influence its outcome" (p. 497), including through "publication bias: papers which show statistically significant positive findings are more likely to be published than papers that have null or negative findings" (p. 502). Hence, the possibility of 'publication bias' in meta-analytical research implies that "[t]hose carrying out systematic reviews need to ensure they conduct as comprehensive a search as possible to help avoid introducing bias into their review." (Rothstein et al. 2006, p. 51). Therefore, selecting all 25 HDRs published to date as the 'all-inclusive' and "unbiased sampling frame for meta-analyses" (Rothstein et al. 2006, p. 40) safeguarded the synthesised findings of this meta-analytical research design from being affected and skewed by 'publication bias' (Rothstein et al. 2006, pp. 49–72).

In synthesis, this meta-analytical study of all 25 UN HDRs published to date promised to offer new policy perspectives on the role and place of different development foci in "the most influential [literature] in the field of development" (Telleria 2017, p. 2143, attributed to McNeill 2007). As such, this study offers auspicious benefits of macro-analysis that may promote a more holistic agenda for so-called 'human development', including in areas of social responsibility and sustainable development.

3.2 *Research Contribution*

As previously alluded to in the introductory part, which discussed the study motivation (Sect. 1.1), this paper extends previous research by focusing expressly on the

[4]UN HDRs were accessed online at: http://hdr.undp.org/en/global-reports.

intersection of economic growth, environmental sustainability and social responsibility as a fertile space for inquiry. The research contributes to a more holistic and equitable human development discourse in key UN policy documents. It opens fresh meta-analytical perspectives (Bryman 2016, p. 692) that shape TBL-related priorities in seminal UN development discourse. By offering inclusive and integrated perspectives, the study makes an important contribution to the literature, given that "conventional wisdom and academic disciplines falsely treat the economy as separate from society, or falsely treat the economy and society as separate from ecology. This silo thinking is unhelpful to social change." (Salleh 2016, p. 1).

4 Results and Key Findings

Several key findings emerged from the research, which have been synthesised and consolidated below.

4.1 Quantitative Results Derived from Systematic UN HDR Keyword Review

Quantitative data obtained from a systematic keyword search carried out in all 25 UN Human Development Reports (HDRs) published to date (Sect. 3.1) raised several interesting perspectives.

Figure 5 contains the raw data: The first column reflects the HDR year of publication and its corresponding report title; the subsequent colour-coded columns reflect the number of times the following search strings appeared in each report: 'economic development' (second column); 'economic growth' (third column); 'economic justice' (fourth column); 'sustainable development' (fifth column); 'environmental sustainability' (sixth column); 'environmental justice' (seventh column); 'social development' (eighth column); 'social responsibility' (ninth column), 'social justice' (tenth column). Columns are colour-coded according to their foci: Economic (blue), environmental (green), and social (orange).

The stacked frequency graphs (Figs. 6, 7 and 8) reflect alternative views of this data by combining the values of the colour-coded columns together according to their focus area: The economic emphasis (blue) emerges as the most dominant perspective as reflected by Fig. 6; the environmental emphasis (green) emerges as the second-most dominant perspective as reflected by Fig. 7; the social perspective (orange) emerges as the weakest emphasis as reflected by Fig. 8.

Figure 9 reflects a perspective where the sums of the three colour-coded columns have been added together as absolute values, namely: 'economic growth' + 'economic development' + 'economic justice' = Economic Combined (blue); 'sustainable development' + 'environmental sustainability' + 'environmental justice' = Envi-

Human Development Reports (HDRs) with year of publication	Economic Develop-ment	Economic Growth	Economic Justice	Sustainable Develop-ment	Environ-mental Sustaina-bility	Environ-mental Justice	Social Develop-ment	Social Respon-sibility	Social Justice
1990 - Concept and Measurement of Human Development	7	55	1	3	0	0	5	0	0
1991 - Financing Human Development	1	48	0	3	0	0	4	1	1
1992 - Global Dimensions of Human Development	22	44	0	26	0	0	5	0	0
1993 - People's Participation	8	24	1	9	0	0	8	0	2
1994 - New Dimensions of Human Security	10	27	0	15	1	0	13	0	0
1995 - Gender and Human Development	9	46	1	9	2	0	18	2	3
1996 - Economic Growth and Human Development	12	243	0	5	1	0	10	0	1
1997 - Human Development to Eradicate Poverty	4	55	1	7	0	0	14	0	1
1998 - Consumption for Human Development	5	23	0	25	9	0	8	1	1
1999 - Globalization with a Human Face	2	30	0	4	4	0	14	3	0
2000 - Human Rights and Human Development	4	29	0	3	0	0	8	1	7
2001 - Making New Technologies Work for Human Development	9	21	0	7	1	0	0	3	1
2002 - Deepening Democracy in a Fragmented World	15	32	1	7	4	0	7	10	8
2003 - Millennium Development Goals: A Compact Among Nations to End Human Poverty	16	154	0	45	46	0	12	1	6
2004 - Cultural Liberty in Today's Diverse World	12	25	0	8	2	0	13	0	5
2005 - International Cooperation at a Crossroads: Aid, Trade and Security in an Unequal World	17	63	0	6	2	0	3	0	21
2006 - Beyond Scarcity: Power, Poverty and the Global Water Crisis	10	39	0	26	15	2	12	0	13
2007/2008 - Fighting Climate Change: Human Solidarity in a Divided World	5	60	0	29	10	0	2	0	20
2009 - Overcoming Barriers: Human Mobility and Development	10	12	0	6	0	0	1	1	1
2010 - The Real Wealth of Nations: Pathways to Human Development	10	96	0	5	11	0	8	0	1
2011 - Sustainability and Equity: A Better Future for All	9	21	1	32	40	12	0	0	4
2013 - The Rise of the South: Human Progress in a Diverse World	9	56	0	8	6	0	4	0	4
2014 - Sustaining Human Progress: Reducing Vulnerabilities and Building Resilience	9	25	0	31	3	0	2	1	3
2015 - Work for Human Development	4	22	0	80	24	0	3	1	4
2016 - Human Development for Everyone	9	21	0	141	16	0	6	3	7
Total number of occurrences (all reports combined)	**228**	**1271**	**6**	**540**	**197**	**14**	**180**	**28**	**114**

Fig. 5 Frequency table of 25 UN HDRs published from 1990 to 2016. In terms of the number of occurrences in all reports, the concept 'economic growth' clearly emerges as the most dominant emphasis. Further, the concept 'sustainable development' outnumbers the concept 'social development' by a factor of

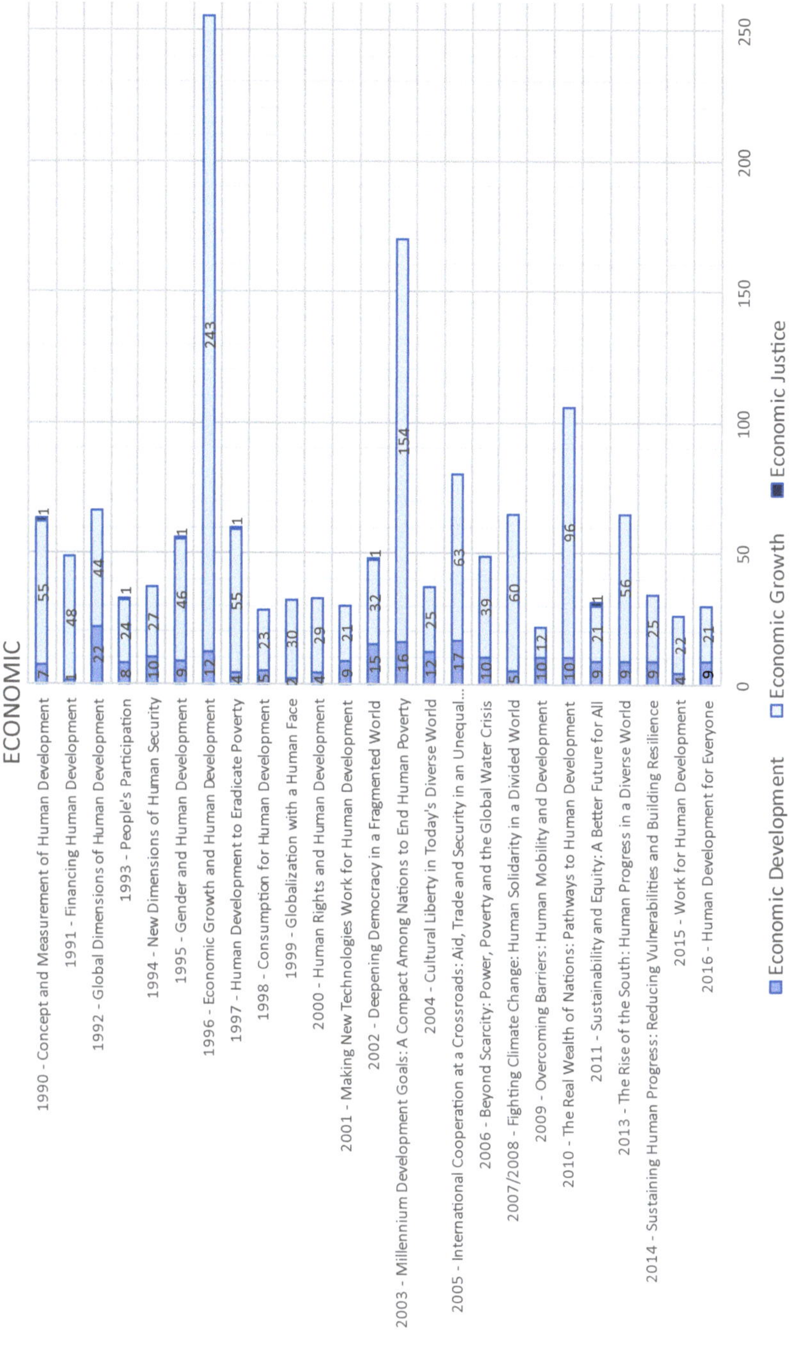

Fig. 6 Economic perspective: Frequency graph of 25 UN HDRs published from 1990 to 2016

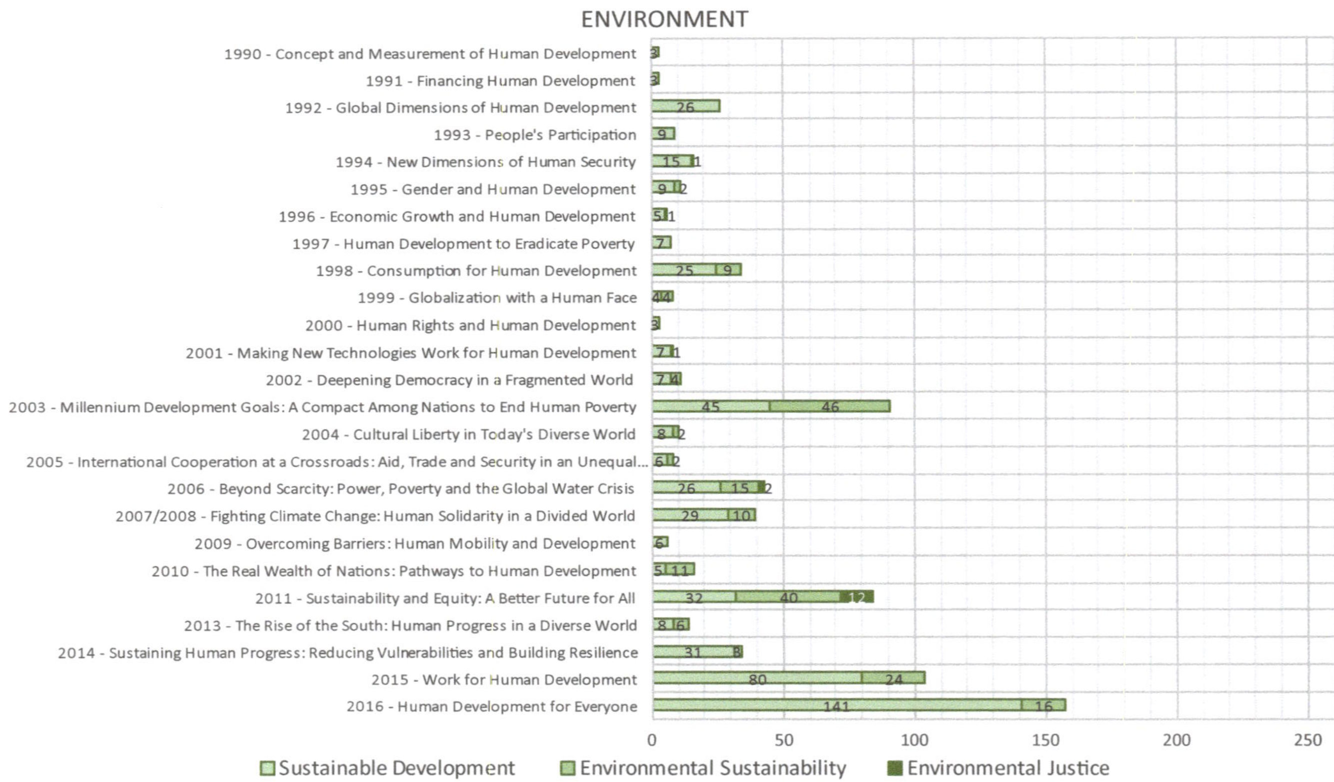

Fig. 7 Environmental perspective: Frequency graph of 25 UN HDRs published from 1990 to 2016

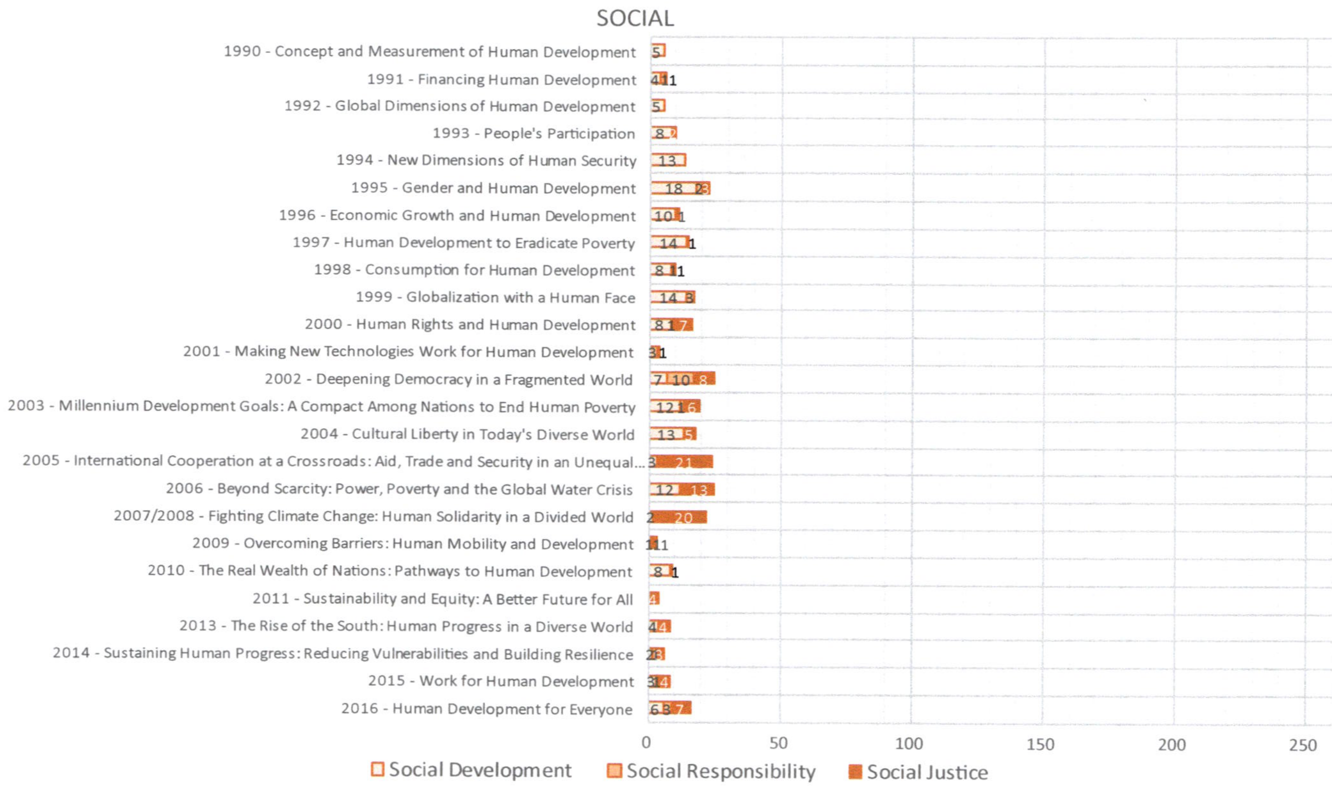

Fig. 8 Social perspective: Frequency graph of 25 UN HDRs published from 1990 to 2016

ronment Combined (green); 'social development' + 'social responsibility' + 'social justice' = Social Combined (orange).

Figure 10 reflects the same values expressed as percentage figures.

Figure 11 reflects a time series according to HDR report number (1–25) and year of publication (1990–2016).

4.2 Synthesis of Quantitative Results Derived from Systematic UN HDR Keyword Reviews

Quantitative data obtained from systematic keyword searches carried out in all 25 UN Human Development Reports (HDRs) published to date (Sect. 3.1) clearly reflect a predominance of 'economic' perspectives in UN HDR development discourse. Further, there is a clear sense that over recent years 'environmental' or 'sustainability' considerations have gained currency. Finally, emphasis on 'social' development is overall and continuingly weak.

5 Discussion, Analysis, Synthesis: A Discourse on Social Developmental Perspectives Is Urgently Needed

Quantitative analysis shows that 'economic growth' (EG) surpasses all other concepts and not only its counterpart, 'economic development' (ED). This finding reflects the number one mantra and priority-of-priorities for the world: Economic growth! It is followed by 'sustainable development' (SD). Considering all 25 HDRs together, 'economic' terms occurred 1505 times, 'sustainability'-linked language occurred 751 times, and 'social' terms occurred 322 times (Fig. 5). Expressed differently, 'economic' perspectives dominate 'sustainability' and 'social' perspectives by a factor of 2 and 4.67 respectively.

Noting the recent rise in currency linked to SD and a gradual weakening over time of ED (Figs. 6 and 7) suggests the possibility that recent years have simply shifted the semantic terminological emphasis. While an outright 'economic' emphasis was still politically *en vogue* in the 1990s (Figs. 5, 6 and 9), as exemplified by the 1996 UN HDR entitled *Economic Growth and Human Development* (UNDP 1996), more recently so-called 'sustainable' (economic) development may have simply taken over as the more 'politically correct' phraseology (Figs. 5, 7 and 9), as exemplified by the 2011 UN HDR entitled *Sustainability and Equity: A Better Future for All* (UNDP 2011). This is especially accentuated in the UN's so-called 'Sustainable'[5] Development Goals (SDGs) (UN-DESA 2015). Hence, the question arises whether the UN's previous emphasis on 'economic' growth/development is still the same neoliberal

[5]cf. https://sustainabledevelopment.un.org/sdgs; https://www.un.org/sustainabledevelopment/.

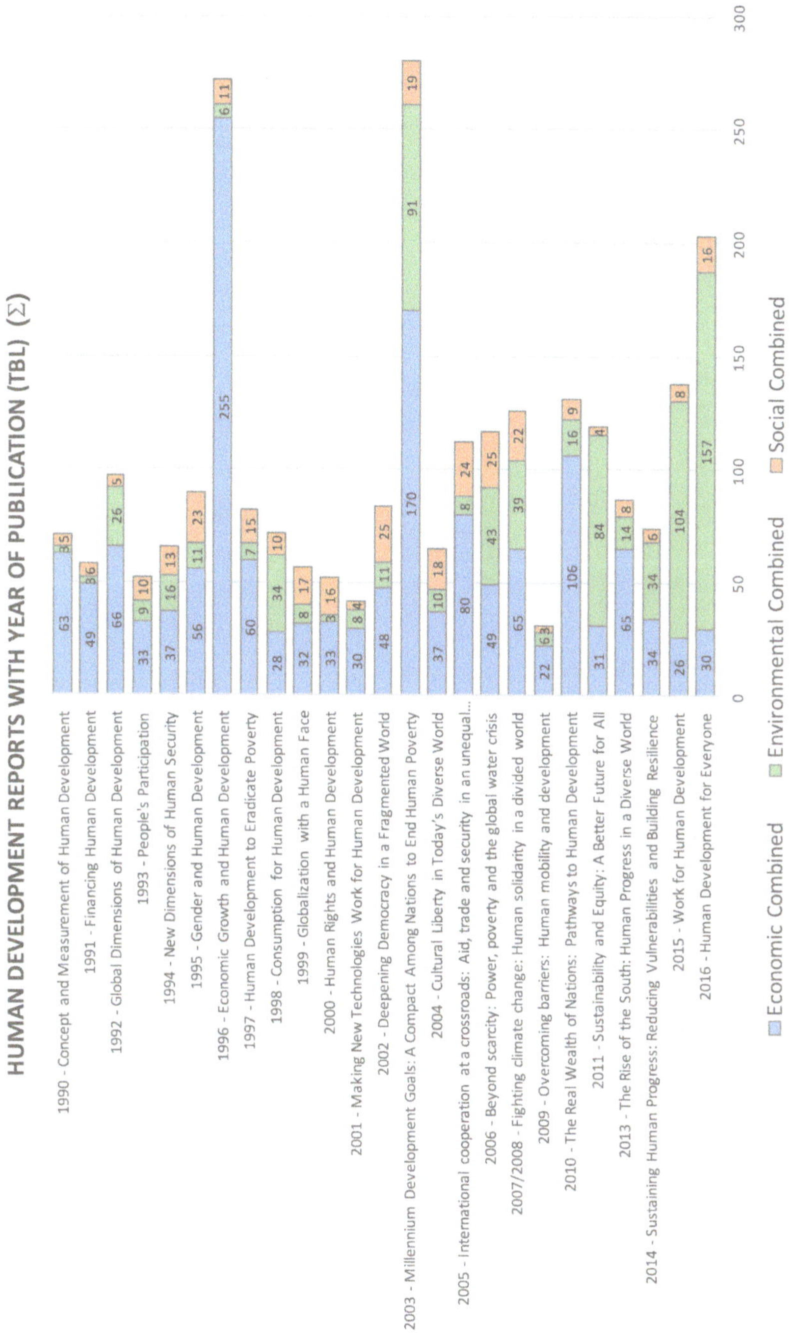

Fig. 9 Combined (Absolute values): Frequency graph of 25 UN HDRs published from 1990 to 2016

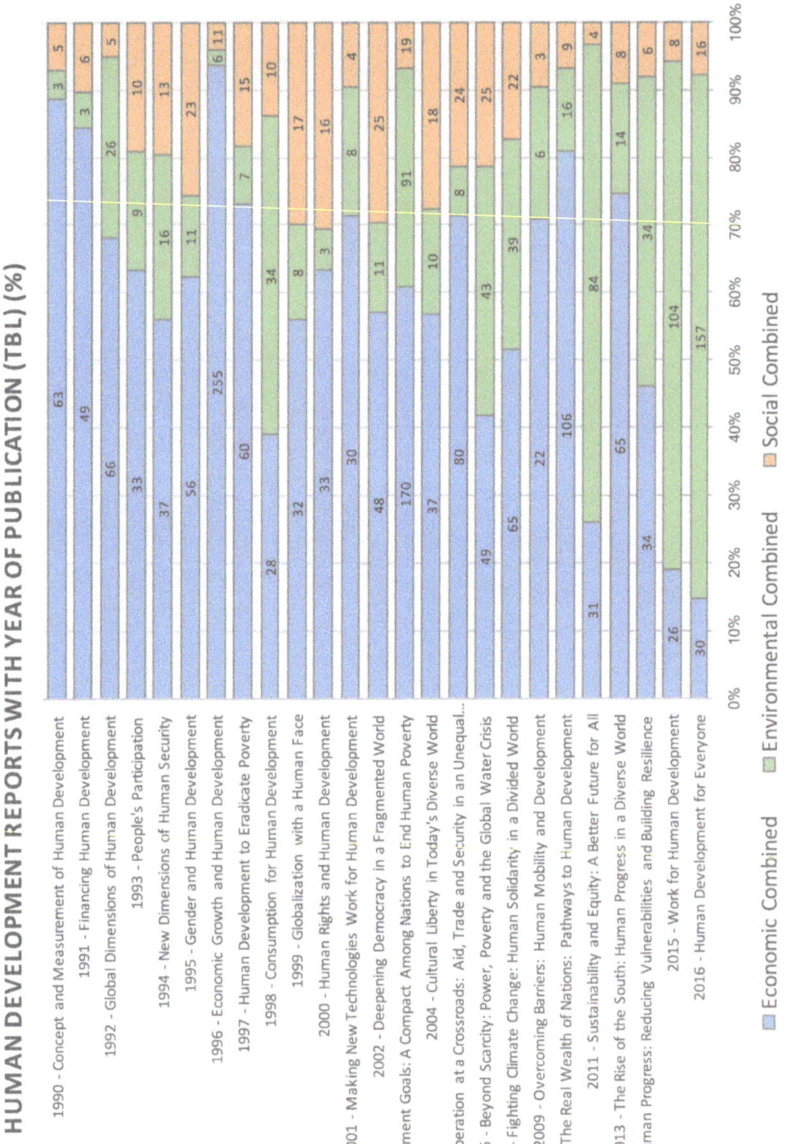

Fig. 10 Combined (Percentage figures): Frequency graph of 25 UN HDRs published from 1990 to 2016

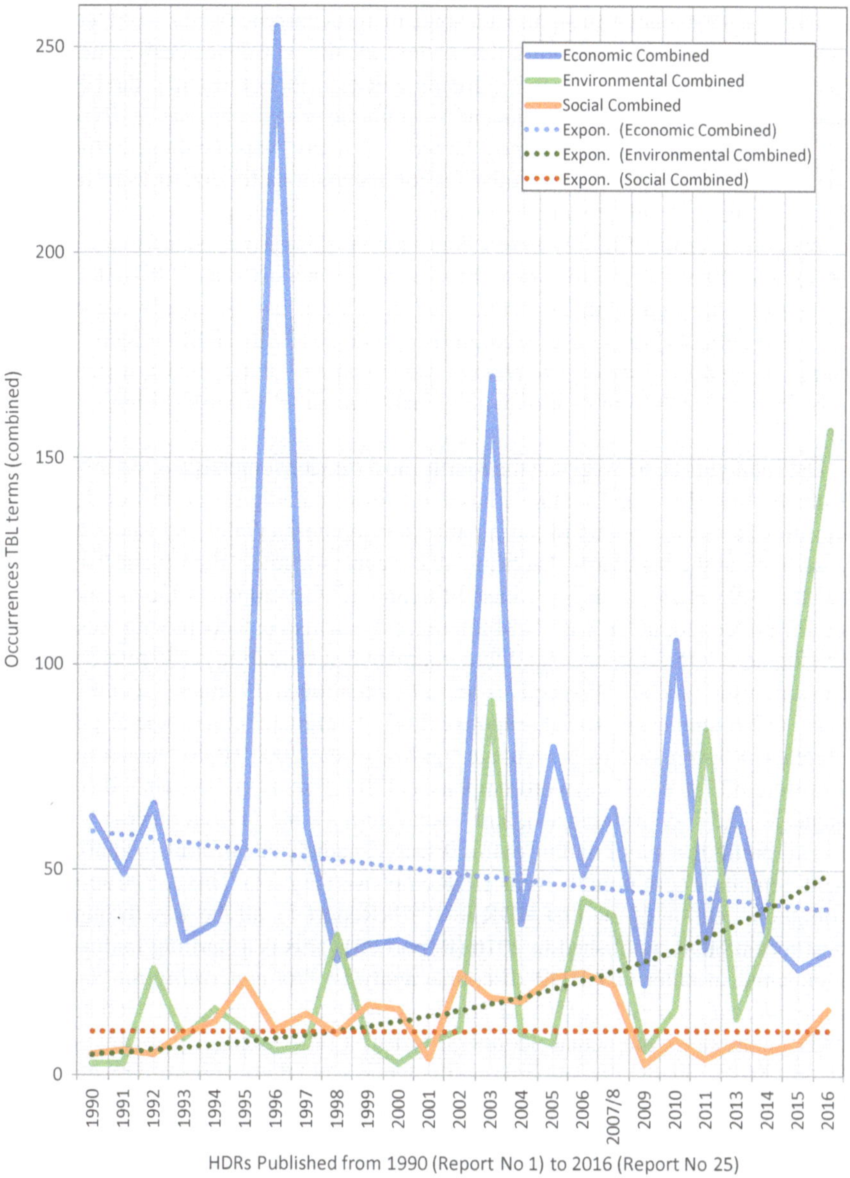

Fig. 11 Time series for 25 HDRs published from 1990 (Report No 1) to 2016 (Report No 25): The time series shows a weakening of the currency of economic terms over time, and an increase in currency for environmental/sustainability perspectives. Social perspectives are overall and continuingly weak

development agenda, which is now merely dressed up in HDRs as a *nouveau* neologistic emphasis on so-called 'sustainable' development/growth? (Telleria 2017).

The year 1996 marked a significant spike in the occurrence of the word 'economic growth' (Fig. 6). It is noteworthy that at the time the world was busily pursuing an agenda on eradicating poverty through economic growth. For instance, the 1996 HDR stated, "Human development is the end—economic growth a means." (UNDP 1996, p. 1). In the same report, 190 times the word 'poverty' was mentioned alongside a narrative on economically-rationalised recommendations for countries to follow *en route* to 'human development'. Incidentally, two years with particularly low counts of 'economic' terms (Fig. 11) were 1998 (which followed the Asian Financial Crisis 1997), and 2009 (which followed the Global Financial Crisis 2007–2008). Even so, any corollary enlightenment that may have been gained clearly did not have any lasting effect (Fig. 11), as exemplified by the 2010 UN HDR entitled *The Real Wealth of Nations: Pathways to Human Development* (UNDP 2010), which swiftly reconfirmed the UN's infatuation with global 'economic' priorities (Figs. 5, 6, 9 and 10).

Data analyses also reveal that the fourth-most frequent emphasis is 'environmental sustainability' (Fig. 5). That likely has to do with a paradigm shift in world political priorities across the pillars of sustainable development (Walid and Luetz 2018). A closer look at the bar charts that bring all concepts together (Figs. 9 and 10) reveals that the 2003 HDR on the so-called Millennium[6] Development Goals once again prioritised 'economic growth' (154 references), but also in tandem with 'sustainable development' (45 references) and 'environmental sustainability' (46 references). At the same time 'social' perspectives are only mentioned 19 times: 'social development' (12 references), 'social responsibility' (1 reference), and 'social justice' (6 references). Expressed as percentage figures, in the 2003 HDR 'economic' terms constitute 60.7% of all keyword emphasis (170 references), 'sustainability'-linked language 32.5% (91 references), and 'social' terms 6.8% (19 references).

This points to a major finding of this research, namely that overall, 'social development' remains a continuingly weak HDR emphasis (Fig. 11). This can be appreciated from the inception of the first HDR in 1990 (Report 1), all the way through to the most recent HDR, published in 2016 (Report 25). This is a startling finding, which is mirrored in other UN HDR discourse analysis: "*By concealing the antagonistic/conflictual dimension of social issues*—poverty, inequality, and exclusion—the UNDP naturalises the actual neoliberal order." (Telleria 2017, p. 2143, emphasis added). More specifically and poignantly, while UN HDRs appear to recognise the rise in social inequality in the world, they clearly fail to juxtapose this awareness with any corresponding discourse on 'social development', 'social responsibility', or 'social justice'. Tellingly, the HDR published in 2011 and entitled *Sustainability and Equity: A Better Future for All*, does not mention the concepts 'social development' and/or 'social responsibility' even once (Fig. 5), and its only four references to 'social justice' amount to little more than platitudinal verbiage.[7]

---

[6]http://www.un.org/millenniumgoals.

[7]All four inclusions of 'social justice' in UNDP (2011) are tokenistic references: (1) "The 2011 Human Development Report offers important new contributions to the global dialogue on this challenge, showing how sustainability is inextricably linked to basic questions of equity that is,

In international development discourse, "social justice aims to create equal worth, equal rights, opportunities for all and the elimination of inequalities reinforced by poverty." (Commission for Social Justice 1994, cited in Ledwith 2005, p. xv). Hence the conspicuous absence of the 'social' dimension in HDR development discourse hints at deeper normative issues surrounding inequality. Given that UN HDRs have recurrently addressed the increase in inequality in the world, it has to be asked why the UN is unable or unwilling to offer any compelling corresponding discourse on 'social justice'?

For instance, the 1992 HDR acknowledged in blunt language a definitive UN awareness of:

> disturbing new analysis of the global distribution of income and opportunities—demonstrating that income disparities have in recent years widened dramatically. In 1960, the richest 20% of the world's population had incomes 30 times greater than the poorest 20%. By 1990, the richest 20% were getting 60 times more. And this comparison is based on the distribution between rich and poor countries. Adding the maldistribution within countries, the richest 20% of the world's people get at least 150 times more than the poorest 20%. (UNDP 1992, p. 1)

More recently, the 2015 HDR acknowledged that inequality in the world had increased yet again:

> In recent years rising incomes around the world have been accompanied by rising inequality in income, wealth and opportunities. […] With regard to global wealth, inequality is substantial. In fact, a small elite takes a large share of global wealth. The richest 1 percent held 48 percent of global wealth in 2014, a share projected to be more than 50 percent in 2016. Around 80 percent of the world's people have just 6 percent of global wealth […] Indeed, just 80 individuals together have as much wealth as the world's poorest 3.5 billion people. Such inequality has become a serious problem — both for economic efficiency and for social stability. (UNDP 2015, p. 65, attributed to Oxfam 2015)

Telleria (2017) synthesises that the current economic system is "benefiting 'a small elite' at the expense of others and that this tendency has been consolidated in recent decades." (p. 2153). Tellingly, although the UNDP emphatically acknowledges that inequality on such a massive scale "has become a *serious problem […] for social stability*" (UNDP 2015, p. 65, emphasis added), yet the entire report mentions 'social' terms only eight times (Fig. 5).

Acknowledging that the UNDP is espoused to "[c]ompetitive markets [as] the best guarantee for efficient production" (UNDP 1992, p. 1), Telleria (2017) concludes his analysis by underscoring that "the UNDP does not offer real alternatives: the proposed solutions strengthen the concrete political and economic system in which underdevelopment emerged and do not plan substantial, in depth changes." (p. 2154).

---

of fairness and social justice and of greater access to a better quality of life." (p. iv); (2) "Hence, our inability to promote the common interest in sustainable development is often a product of the relative neglect of economic and social justice within and amongst nations." (p. 14); (3) Timsina (2003). Promoting social justice and conserving mountain forest environments: A case study of Nepal's Community Forestry Programme." *Geographical Journal, 169*(3), 236–242 (p. 115); (4) "The 2011 Human Development Report offers important new contributions to the global dialogue on this challenge, showing how sustainability is inextricably linked to equity—to questions of fairness and social justice and of greater access to a better quality of life." (p. 185/back cover).

Therefore, UN ideology stands in marked contrast to other discourses that openly advocate for systemic changes. For instance, Woodward and Simms (2006) posit that "growth is failing the poor" because of the flagrantly "unbalanced distribution of the benefits and costs of global economic growth" (pp. 1–21). Further, Simms (2008) refutes what he calls the "[t]rickle-down myth":

> THE last line of defence for advocates of indefinite global economic growth is that it is needed to eradicate poverty. This argument is at best disingenuous. By any reasonable assessment it is claiming the impossible. Here's why. During the 1980s, for every $100 added to the value of the global economy, around $2.20 found its way to those living below the World Bank's absolute poverty line. During the 1990s, that share shrank to just 60 cents. This inequity in income distribution—more like a flood up than a trickle down—means that for the poor to get slightly less poor, the rich have to get very much richer. It would take around $166 worth of global growth to generate $1 extra for people living on below $1 a day. […] Perversely, under the current economic system, reducing poverty by a tiny amount will necessitate huge extra consumption by those who are already rich. To get the poorest onto an income of just $3 per day would require an impossible 15 planets' worth of biocapacity. In other words, we will have made Earth uninhabitable long before poverty is eradicated. *If we are serious about helping the poor rather than the rich, we need a new development model*. (p. 49, emphasis added)

Hickel (2015) has similarly calculated that

> Of all the income generated by global GDP growth between 1999 and 2008, the poorest 60 percent of humanity received only 5 percent of it. Given the existing ratio between GDP growth and the income growth of the poorest, it will take 207 years to eliminate poverty with this strategy, and to get there, we will have to grow the global economy by 175 times its present size. (para 13)

Overall, the research data and corresponding analyses presented in this paper (Figs. 5, 6, 7, 8, 9, 10 and 11) suggest that the UN exhibits no appetite to explore systemic change. On the contrary, according to discourse analysis conducted by Telleria (2017), UN HDRs have acknowledged the problem of growing inequality for many years, but at the same time "[t]he UNDP's discourse avoids reflecting on the political causes of the problem it supposedly aims to overcome." (p. 2154).

This concurs with analysis by Mouffe (1998), which Telleria (2017) cites as a substantiating point:

> Globalisation is the usual justification given for the 'there is no alternative' dogma. […] This kind of argument takes for granted the ideological terrain which has been established as a result of years of neo-liberal hegemony and transforms what is a conjunctural state of affairs into an historical necessity. Here, as in many other cases, the mantra of globalisation is invoked to justify the status quo and reinforce the power of big transnational corporations. (p. 18)

Overall, our data suggest that the UN remains doggedly espoused to 'economic' development as its sole panacea for poverty reduction, which it now advocates under the guise of 'sustainable' development (Figs. 5, 7, 9, 10 and 11). Even so, it needs to be asked whether an increase in so-called 'sustainable' development can indefinitely prolong 'economic growth' on a finite planet,[8] where resources are overwhelmingly finite and non-renewable? (WWF 2016; Salleh 2016).

---

[8]http://data.footprintnetwork.org/.

Scientists have recurrently warned that "[a] transition to sustainability cannot be achieved if our economic system is not radically changed, simply because limitless economic growth is impossible within a limited planet." (Pacheco et al. 2018, p. 238; cf. Hoekstra and Wiedmann 2014; Kendall 1997; Meadows et al. 1972; Ripple et al. 2017).

The Earth Overshoot Day (EOD) accounting concept sheds light on planetary constraints. According to the Global Footprint Network, "Earth Overshoot Day is the date when humanity's annual demand on nature exceeds what Earth can regenerate over the entire year." (EOD 2017, para. 2). Andrew Simms conceived and launched the EOD in 2006 in partnership with the Global Footprint Network (GFN). "At that time, Earth Overshoot Day fell in October."[9] Simms (2008) emphatically warned in 2008 against overshooting the Earth's biocapacity: "Humanity has been overshooting the biosphere's capacity to sustain our activities every year since the mid-1980s, and each year we do it sooner. In 2008, we had consumed the ration for the year by 23 September, five days earlier than the previous year." (p. 49). Incidentally, in 2017 the EOD fell on 2 August 2017, "the earliest date yet." (EOD 2017, para. 2).

Analysis by WWF (2016) concludes that the:

> GDP-growth-focused economic model has led to severe wealth inequality as well as culturally entrenched aspirations for material consumption. It has encouraged growth well beyond our basic needs and beyond what can be supported by the carrying capacity of a single Earth. (Hoekstra and Wiedmann 2014, cited in WWF 2016, p. 91)

Contrary to UN HDRs, which advocate "that the same competitive [global economic] system is the best way to overcome the problem" (Telleria 2017, p. 2153), there are now distinctive calls for systemic change and a reimagining of underlying hidden mental models that underpin the present global economic system:

> At the […] deepest level of thinking are the mental models of individuals and organizations that reflect the beliefs, values and assumptions that we personally hold. They are often hidden beneath a surface of rationalization for acting in a particular way (Maani and Cavana, 2007). Mental models—which can vary across cultures—are rarely taken into account in decision-making (Nguyen and Bosch, 2013). However, belief systems—'we need to get richer in order to be happier', 'people are poor because they don't try hard enough'—significantly affect all layers above. *They influence the design of system structures […]* (WWF 2016, p. 91, emphasis added)

It appears that the present global economic system accentuates inequalities, rather than redress or alleviate them. At the same time, UN HDRs continuingly advocate 'economic growth' as a solution instead of identifying it as a problem. Hence 'social' concerns remain side-lined in HDRs as a thinly veiled fringe concern (Figs. 5, 8, 9, 10 and 11), chiefly because the UN offers no meaningful corresponding remedial discourse. It appears that 'social' discourse is urgently needed to meaningfully engage the current systemic status quo. Even so, the UN seems to be an unlikely candidate to lead this charge, seeing that its silence on 'social' injustice and inequality is normatively determined. Notwithstanding, other actors are filling the discursive vacuum:

[9] https://www.overshootday.org/about-earth-overshoot-day/.

> Ultimately, addressing social inequality and environmental degradation will require a global paradigm shift toward living within safe Planetary Boundaries. *We must create a new economic system* that enhances and supports the natural capital upon which it relies. (WWF 2016, p. 122, emphasis added)

Hickel (2015) elaborates this necessity by calling attention to the UN's "failing economic model [as advocated via its SDGs] that requires urgent and deep structural changes" (para 31):

> What we need is to tackle the irrationality of endless growth head-on, pointing out that capitalist growth — as measured by GDP — is not the solution to poverty and ecological crisis, but the primary cause. And we need a saner measure of human progress — one that gears us not toward more extraction and consumption by the world's elite, but more fairness, more equality, more wellbeing, more sharing, to the benefit of the vast majority of humanity. (para. 32)

Simms (2008) concurs:

> [W]e have to overcome knee-jerk rejection of the "R" word—redistribution. With global growth constrained by the need to limit carbon emissions (remember that the poorest will be the first and worst victims of climate change), redistribution becomes the only viable route to poverty reduction. (p. 49)

In view of a critical reading of the data presented in Sect. 4, it may perhaps not seem quite so surprising that the UN's discourse on 'social development', 'social responsibility' and 'social justice' remains wilfully uninspiring over the course of almost three decades, and spanning 25 report publications (Fig. 11). And neither does it surprise much that "more than 70 years after the creation of the UN, the goal of promoting 'social progress and better standards of life in larger freedom' for every people (UN Charter, Preamble) is still an elusive aim." (Telleria 2017, p. 2143).

In summary, what our data and analyses make abundantly clear is that UN HDRs have not yet found a way to meaningfully acknowledge and include the 'social' dimension within human development discourse. Expressed in triple bottom line lingo, the UN's policy tripod to date leans overwhelmingly on 'economic' and 'sustainability' (growth) perspectives but painfully lacks any strength in its third 'social' leg. In this context, it appears unlikely that the UN can mount any credible development policy on this lopsided and teetering programmatic tripod, unless it can realign its normative narrative in favour of 'social' concerns (Hickel 2015; Henriques 2010). In short, the UN "can't solve the problem of poverty without challenging the pathologies of accumulation." (Hickel 2015, para. 17).

This is where the Triple Bottom Line (TBL) framework may provide an opportunity to reinforce the social and environmental pillars of sustainability within human development discourse. TBL originated as a more holistic framework of valuation that recognises the importance of the three pillars coexisting together. Hence, TBL evolved to address practical cross-sectional challenges the world faces (Elkington 1997). Thus, TBL may be more strategically placed to elevate the currency of social responsibility and environmental sustainability, alongside the traditional economic pillar (Sorrell 2010).

6 Conclusion

This study was motivated by a desire to better understand United Nations (UN) policy discourse at a meta-analytical level. The study is important as it offers unique perspectives on the UN's agenda for so-called 'human' development. More specifically, the study sought to investigate what prioritisation the UN ascribes to 'economic', 'sustainable' and 'social' development in its global flagship Human Development Reports (HDRs), published (almost) annually since 1990. HDRs were chosen as a data source as they constitute "the most influential [literature] in the field of development" (Telleria 2017, p. 2143, attributed to McNeill 2007). A systematic keyword search was conducted in all 25 HDRs published to date. Keyword analysis reveals that 'economic' perspectives dominate 'sustainability' and 'social' perspectives by a factor of 2 and 4.67 respectively. The findings show that 'sustainable development' and 'social development' are both conceptually subsumed in importance within 'economic development'. Further, 'social development' is accorded almost no prioritisation in UN HDR discourse. In synthesis, research data (Figs. 5, 6, 7, 8, 9, 10 and 11) and corresponding meta-analyses (Sects. 4 and 5) suggest that the UN's policy tripod to date leans overwhelmingly on 'economic' and 'sustainability' perspectives but painfully lacks any strength in its third 'social' leg. In short, the UN remains doggedly espoused to 'economic' development as its sole primordial panacea for poverty reduction, which it increasingly advocates under the guise of 'sustainable' development. Relatedly and importantly, UN HDRs continuingly advocate 'economic growth' as a solution instead of identifying it as a problem. Meanwhile inequalities all around the world continue to rise, remaining side-lined in HDRs as a thinly veiled fringe concern, chiefly because the UN offers no meaningful antidote and corresponding discourse to address this normative problem. In this context, UN HDRs exhibit little or no appetite to explore global economic systemic change, although this is increasingly called for by both science and society,[10] as exemplified by one sample:

> Ultimately, addressing social inequality and environmental degradation will require a global paradigm shift toward living within safe Planetary Boundaries. *We must create a new economic system* that enhances and supports the natural capital upon which it relies. (WWF 2016, p. 122, emphasis added)

To conclude, this meta-analytical review shows clearly that, despite its 'sustainability' asseverations, the UN is (subconsciously?) edging humanity towards a growth-based economic development model that is *de facto* 'unsustainable' and 'socially irresponsible' and poses great problems for humanity and the planet in the future. In response, this research proposes a Triple Bottom Line (TBL) approach as a possible bridging notion for a more holistic human development agenda. This may include integrating a measure of "social footprint" within the UN Human Development Index as advocated by Henriques (2010, pp. 169–171).

[10]See, e.g., EOD (2017), Hoekstra and Wiedmann (2014), Meadows et al. (1972), Oxfam (2015), Pacheco et al. (2018), Ripple et al. (2017), Simms (2008), among others.

7 Research Limitations and Opportunities for Further Research

The authors acknowledge the limitations of this research as follows: Following Walid and Luetz (2018), this systematic keyword review was limited in scope to UN HDRs, which were intentionally chosen as a data source "because of their global influence and appeal. Such focus could also be viewed as a limitation." (p. 805). Hence there exists an opportunity for further research that widens the keyword search scope to encapsulate additional data sources (e.g., World Bank World Development Reports (WDRs), published annually since 1978, among others). There are also opportunities to extend the scope of research internally within HDRs to other phrases of interest, e.g., 'corporate responsibility' or 'CSR', or to conduct comparative analyses between reports, e.g., HDRs and WDRs. Scrutinising ideological and/or ideational differences between the World Bank's WDRs (which overtly advocate economic indicators for measuring 'poverty'), or the United Nations Development Programme's HDRs (which advocate the capability approach and the use of composite indicators for measuring 'human poverty' more holistically), may also constitute fertile spaces for future research. Such studies may shed light on important differences that govern dissimilar institutional, ideological/normative or theoretical frameworks for development. Similarly, this meta-analytical review was limited in scope to HDRs published in the English language. Rothstein et al. (2006) caution that authors of systematic reviews "do not restrict their searches to those studies published in the English language, as evidence suggests that trials published in languages other than English may have different results than those published in English" (p. 51; attributed to Egger et al. 1997; Jüni et al. 2002). Hence there are opportunities to widen future searches to include HDRs published in languages other than English.

Furthermore, there is an element of interpretive subjective assessment in our analysis that may be seen by some readers as potentially limiting the robustness of quantitative findings presented in this paper. These authors suggest that quantitative findings are best comprehended as "roughly right [instead of] precisely wrong."[11] (Myers and Kent 1995, p. 33). Or expressed in the words of the ancient Greek philosopher: "It is the mark of an educated mind to rest satisfied with the degree of precision which the nature of the subject admits and not to seek exactness where only an approximation is possible." (Aristotle cited in Fripp et al. 2000, p. 55). Notwithstanding, comprehensive sampling ensured that the findings of this meta-analytical study were not obscured by 'publication bias' (Rothstein et al. 2006), multiple rounds of data collection and analysis maximised "test-retest reliability" (Punch 2014), and "construct validity" (pp. 238–240) was ensured by means of including three distinct keyword search strings for each TBL emphasis under investigation (Sect. 3.1).

Finally and importantly, this research raises opportunities for future engagement. In view of the limited currency of the 'social' dimension in UN HDRs highlighted

[11] As advocated by Myers and Kent (1995), "[in] a situation of uncertainty where not all factors can be quantified to conventional satisfaction, let us not become preoccupied with what can be precisely counted if that is to the detriment of what ultimately counts." (p. 33).

by this research, there are obvious opportunities to strengthen this emphasis. For example, Henriques (2010, pp. 167–177) has proposed integrating a measure of "social footprint" within the UN's composite Human Development Index (HDI):

> The point has often been made that social systems (and within them economic systems) are merely aspects of ecological systems; without a natural world there can be no society, whereas the reverse does not hold. […] So if social impact is an important component of sustainability, perhaps the concept of an ecological footprint can be extended to include social aspects […] Adding weight to this perspective, social and environmental issues are interconnected. Social issues relate to ecological issues in two key ways: (1) Social well-being is required if adverse environmental impacts are to be allocated sufficient resources and systematically addressed; (2) Adverse environmental impacts can form an important component of social issues. In both developed and less developed societies, environmental impacts such as pollution or the availability of natural resources are socially skewed. Possible candidates for integration into such an expanded footprint might include some sort of derivative of the UNDP's Human Development Index […] or some measure of social equity. Each of these measures appears to capture the extent to which a population has achieved important social outcomes. They are also calculable, and so satisfy the first condition for footprint indicators […] (pp. 169-170)

Promoting such 'social' emphasis may support more holistic human development perspectives. However, given that 'social' perspectives seem to be intentionally eclipsed in HDRs on normative/ ideological grounds makes it doubtful that the UN will rally to incorporate such 'social' measures within its HDI composite index. Even so, the opportunity remains compelling. Relatedly, Harper (2018) has called for a new politics of time, one which makes "free time a key measure of UK wellbeing—replacing GDP as a barometer of economic progress" (para 1), and gauging national wellbeing with a Free Time Index (cf. Walker 2018). In summary, opportunities for further research are plentiful, promising and highly pertinent.

Acknowledgements The authors of this meta-analytical review wish to thank Ms. Kirsty Andersen for her copy-editorial support, Kirk Huffman and Dr. Jens Unger for their constructive comments, and Ms. Karen du Plessis and Dr. Jer-Ming Chen for their assistance with data analysis.

References

Atu O (2013) Triple bottom line accounting: a conceptual expose. IOSR J Bus Manag (IOSR-JBM) 13(4):30–36. https://doi.org/10.9790/487x-1343036

Australian Government (1992) National strategy for ecologically sustainable development. Australian Government Publishing Service, Canberra

Carroll AB (2016) Carroll's pyramid of CSR: taking another look. Inter J Corp Soc Respon 1:3. https://doi.org/10.1186/s40991-016-0004-6

Basu K (2000) On the goals of development. In: Meier GM, Stiglitz JE (eds) Frontiers of development economics: the future in perspective. World Bank and Oxford University Press, Washington, DC, pp 61–86

Bowen HR (1953) Social responsibilities of the businessman. Harper, New York

Braat L (1991) The predictive meaning of sustainability indicators. In: Kuik O, Verbruggen H (eds) In search of indicators of sustainable development. Kluwer Academic, Netherlands, pp 57–70

Bryman A (2016) Social research methods, 5th edn. Oxford University Press, Oxford, UK

Carroll AB (1979) A three-dimension conceptual model of corporate performance. Acad Manag Rev 4(4):497–505
Cheeseman G (2015) The triple bottom line case for the UN's New sustainable development goals. Online article. https://www.triplepundit.com/special/climate-week-nyc-2015/the-triple-bottom-line-case-for-the-uns-new-sustainable-development-goals/. Accessed 1 Jan 2017
Choucri N (1997) Global system for sustainable development research TDP-MIT. Unpublished notes. MIT, Cambridge, MA
Commission for Social Justice (1994) Social justice: strategies for national renewal. Verso, London
Cranfield School of Management (2017) Cranfield knowledge interchange. Corporate responsibility and sustainability. Resource document. Author. http://www.som.cranfield.ac.uk/som/p16919/Knowledge-Interchange/Management-Themes/Corporate-Responsibility-and-Sustainability. Accessed 29 Dec 2017
Creative Cloud (2018) Adobe Acrobat PRO DC, 2018 release (continuous), Version 2018.011.20035. Adobe Systems Incorporated, San Jose, CA
Creswell JW (2014) Research design: qualitative, quantitative, and mixed methods approaches, 4th edn. Sage, Thousand Oaks, CA
Creswell JW, Plano Clark VL (2011) Designing and conducting mixed methods research, 2nd edn. Sage, Thousand Oaks, CA
Dahlsrud A (2005) A comparative study of CSR-strategies in the oil and gas industry. In: Paper presented at Navigating globalization: stability, fluidity, and friction, Trondheim, Norway
Dahlsrud A (2008) How corporate social responsibility is defined: An analysis of 37 definitions. Corp Soc Responsivity Environ Manag 15(1):1–13. https://doi.org/10.1002/csr.132
Dang G, Pheng L (2015) Infrastructure investments in development economies. Springer, Singapore
Douglas A, Doris J, Johnson B (2004) Corporate social reporting in Irish financial institutions. TQM Mag 16(6):387–395
Ebner D, Baumgartner R (2008) Queen's university conference papers. The relationship between sustainable development and corporate social responsibility. Resource document. Corporate Responsibility Research Conference. http://citeseerx.ist.psu.edu/viewdoc/download?doi=10.1.1.485.5912&rep=rep1&type=pdf. Accessed 5 June 2018
Egger M, Zellweger-Zahner T, Schneider M, Junker C, Lengeler C, Antes G (1997) Language bias in randomised controlled trials published in English and German. Lancet 350:326–329
Elkington J (1997) Cannibals with forks: the triple bottom line of 21st century business. Capstone, MN
EOD—Earth Overshoot Day (2017) On earth overshoot day (August 2), calculate your own Overshoot Day and #movethedate. Press Release. https://www.overshootday.org/newsroom/press-release-english-2017-calculator/. Accessed 4 March 2018
Fisher WF, Ponniah T (2003) Another world is possible: Popular alternatives to globalization at the World Social Forum. Zed Books, London
Friedman M (1962) Capitalism and freedom. University of Chicago Press, Chicago, USA
Friedman M (1970) The social responsibility of business is to increase its profits. New York Times Magazine, p 17
Friedman BM (2005) The moral consequences of economic growth. Knopf, New York
Friedman M (2007) The social responsibility of business is to increase its profits. In: Zimmerli WC, Holzinger M, Richter K (eds) Corporate ethics and corporate governance. Springer, Berlin, pp 173–178
Fripp J, Fripp M, Fripp D (2000) Speaking of science: notable quotes on science, engineering, and the environment. LLH Technology, Eagle Rock, VA
Griffin K (1989) Alternative strategies for economic development. Macmillan, London
Harper A (2018) Get rich or die trying: the case for a new politics of time — Is free time a better measure of economic success than GDP? New economics foundation. 15 October 2018. https://www.neweconomics.org/2018/10/get-rich-or-die-trying-the-case-for-a-new-politics-of-time? Accessed 28 Oct 2018

Henriques A (2010) Corporate impact: measuring and managing your social footprint. Earthscan, London

Hickel J (2015) The problem with saving the world: the UN's new sustainable development goals aim to save the world without transforming it. Jacobin Magazine. Accessed 19 June 2018, http://jacobinmag.com/2015/08/global-poverty-climate-change-sdgs/

Hoekstra AY, Wiedmann TO (2014) Humanity's unsustainable environmental footprint. Science 344(6188):1114–1117. https://doi.org/10.1126/science.1248365

Holmberg J (ed) (1992) Making development sustainable. Island Press, Washington, DC

Hopkins M (1999) Towards an index to measure the social responsibility of business and human development (No additional referencing information provided in UNDP 1999, p 115)

ISO—International Organization for Standardization (2010) Discovering ISO 26000: guidance on social responsibility. Genève, Switzerland: Author. https://www.iso.org/files/live/sites/isoorg/files/archive/pdf/en/discovering_iso_26000.pdf. Accessed 15 Apr 2018

Jüni P, Holenstein F, Sterne J, Bartlett C, Egger M (2002) Direction and impact of language bias in meta-analyses of controlled trials: empirical study. Int J Epidemiol 31(1):115–123

Kendall H (1997) World scientists' warning to humanity. Resource document. Union of Concerned Scientists. Cambridge, MA. https://www.ucsusa.org/sites/default/files/attach/2017/11/World%20Scientists%27%20Warning%20to%20Humanity%201992.pdf. Accessed 20 Nov 2017

Ledwith M (2005) Community development: a critical approach, 2nd edn. Policy Press, Bristol, UK

Levitt T (1958) The dangers of social responsibility. Harvard Bus Rev 36:41–50

Luetz JM, Bergsma C, Hills K (2019) The poor just might be the educators we need for global sustainability—a manifesto for consulting the unconsulted. In: Leal Filho W, Consorte-McCrea A (eds) Handbook of sustainability and humanities. Springer, Cham, Switzerland, pp 115–140. https://doi.org/10.1007/978-3-319-95336-6_7

Maani KE, Cavana RY (2007) Systems thinking, system dynamics: managing change and complexity, 2nd edn. Pearson Education, Canada

MacMillan (1988) Dictionary of the environment, 3rd edn. MacMillan Press, London

McGuire JW (1963) Business and society. McGraw-Hill, USA

McNeill D (2007) Human development: the power of the idea. J Human Dev 8(1):5–22

Meadows DH, Meadows DL, Randers J, Behrens WW III (1972) The limits to growth: a report for the club of Rome's project on the predicament of mankind. Universe Books, New York

Mouffe C (1998) The radical centre. Soundings 9:11–25. http://banmarchive.org.uk/collections/soundings/09_11.pdf. Accessed 4 Mar 2018

Mubarak F (2016) Sustainable development and its origin. Environ Cent Arab Towns-Dubai Municipality Pap 13:13–17. http://en.envirocitiesmag.com/articles/issue-13/3.pdf. Accessed 5 June 2018

Munasinghe M, Lutz E (1991) Environmental-economic evaluation of projects and policies for sustainable development. In: Environment Department working paper (no. ENV 42). World Bank, Washington, DC

Myers N, Kent J (1995) Environmental exodus: an emergent crisis in the global arena. Climate Institute, Washington, DC

Nguyen NC, Bosch OJH (2013) A systems thinking approach to identify leverage points for sustainability: A case study in the Cat Ba biosphere reserve, Vietnam. Syst Res Behav Sci 30(2):104–115. https://doi.org/10.1002/sres.2145

Nussbaum MC (2003) Capabilities as fundamental entitlements: Sen and social justice. Feminist Econ 9(2–3):33–59

O'Riordan T, Yaeger J (1994) Global environmental change and sustainable development, global change and sustainable development in Europe. In: Manuscript on file. Presented at the Wuppertal Institute, Nordrhein-Westfalen, Germany

Oxfam (2015) Wealth: having it all and wanting more. Issue Briefing. Oxford, UK. https://www.oxfam.org/sites/www.oxfam.org/files/file_attachments/ib-wealth-having-all-wanting-more-190115-en.pdf. Accessed 2 Mar 2018

Pacheco LF, Altrichter M, Beck H, Buchori D, Owusu EH (2018) Economic growth as a major cause of environmental crisis: comment to Ripple et al. Bioscience 68(4):238. https://doi.org/10.1093/biosci/biy006
Pearce DW (1988) Optimal prices for sustainable development. In: Collard D, Pearce DW, Ulph D (eds) Economics, growth and sustainable environments. MacMillan, London, pp 57–66
Phelps ES (1973) Economic justice. Penguin, Harmondsworth, UK
Punch KF (2014) Introduction to social research: quantitative & qualitative approaches, 3rd edn. Sage, London
Ripple WJ, Wolf C, Newsome TM, Galetti M, Alamgir M, Crist E, …Laurance WF (2017) World's scientists' warning to humanity: a second notice. BioScience 67(12):1026–1028. https://doi.org/10.1093/biosci/bix125
Rothstein HR, Sutton AJ, Borenstein M (eds) (2006) Publication bias in meta-analysis: prevention, assessment and adjustments. Wiley, New York, NY
Salleh A (2016) Climate, water, and livelihood skills: a post-development reading of the SDGs. Globalizations. https://doi.org/10.1080/14747731.2016.1173375
Savitz A, Weber K (2006) The triple bottom line. Jossey-Bass, San Francisco, CA
Sen A (2007) Unity and discord in social development. In: Keynote lecture delivered at the 15th symposium of the international consortium for social development at the Polytechnic University of Hong Kong, 16–20 July, Hong Kong, China (SAR)
Sidhu G, Burd-Sharps S, Kurukulasuriya S (n.d.) Ideas, innovation, impact: how human development reports influence change. Human Development Report Office, United Nations Development Programme. http://hdr.undp.org/sites/default/files/impact_publication.pdf. Accessed 3 June 2018
Sijbesma F (2015) UNEP, Our planet: triple-bottom-line. Online article. UN environment. http://web.unep.org/ourplanet/march-2015/articles/triple-bottom-line. Accessed 1 Jan 2017
Simms A (2008). Trickle-down myth. New Scientist. Special Report 2678:49
Simpson J, Speake J (2009) Oxford dictionary of proverbs. Oxford University Press, Oxford. http://www.oxfordreference.com/view/10.1093/acref/9780199539536.001.0001/acref-9780199539536-e-1221. Accessed 1 June 2018
Sorrell S (2010) Energy, economic growth and environmental sustainability: five propositions. Sustainability 2(12):1784–1809. https://doi.org/10.3390/su2061784
Stegenga J (2011) Is meta-analysis the platinum standard of evidence? Stud Hist Philos Biol Biomed Sci 42:497–507. https://doi.org/10.1016/j.shpsc.2011.07.003
SustainAbility (2004) Gearing up: from corporate responsibility to good governance and scaleable solutions. London: Author. http://www.rso-matrix.com/projectos/0017_RSO%20Matrix/docs/Gearing%20Up_From%20Corporate%20Responsibility%20to%20Good%20Governance%20and%20Scalable%20Solutions%20-%20en.pdf. Accessed 5 Mar 2018
Sze J, London JK (2008) Environmental justice at the crossroads. Sociol Compass 2(4):1331–1354
Telleria J (2017) Power relations? What power relations? The depoliticising conceptualisation of development of the UNDP. Third World Q 38(9):2143–2158. https://doi.org/10.1080/01436597.2017.1298437
Timsina NP (2003) Promoting social justice and conserving mountain forest environments: a case study of Nepal's community forestry programme. Geogr J 169(3):236–242
Todaro M, Smith S (2003) Economic development, 8th edn. Addison Wesley, Boston, MA
UNCTAD—United Nations Conference on Trade and Development (2014) Services: New frontier for sustainable development. Geneva, Switzerland: Author
UN-DESA—United Nations Department of Economic and Social Affairs (2015) Sustainable development knowledge platform. Sustainable development goals. Resource document. United Nations. https://sustainabledevelopment.un.org/sdgs. Accessed 12 Mar 2018
UNDP—United Nations Development Programme (1990) Human development report 1990: concept and measurement of human development. Oxford University Press, New York
UNDP—United Nations Development Programme (1991) Human development report 1991: financing human development. Oxford University Press, New York

UNDP—United Nations Development Programme (1992) Human development report 1992: global dimensions of human development. Oxford University Press, New York

UNDP—United Nations Development Programme (1993) Human development report 1993: people's participation. Oxford University Press, New York

UNDP—United Nations Development Programme (1996) Human development report 1996: economic growth and human development. Oxford University Press, New York

UNDP—United Nations Development Programme (1999) Human development report 1999: globalization with a human face. Oxford University Press, New York

UNDP—United Nations Development Programme (2006) Human development report 2006: beyond scarcity: Power, poverty and the global water crisis. Palgrave Macmillan, Hampshire, UK

UNDP—United Nations Development Programme (2007). Human development report 2007/8: fighting climate change: human solidarity in a divided world. Palgrave Macmillan, New York. http://hdr.undp.org/sites/default/files/reports/268/hdr_20072008_en_complete.pdf. Accessed 5 June 2018

UNDP—United Nations Development Programme (2009) Human development report 2009: overcoming barriers: human mobility and development. Palgrave Macmillan, Hampshire, UK

UNDP—United Nations Development Programme (2010) The real wealth of nations: pathways to human development. Palgrave Macmillan, Hampshire, UK

UNDP—United Nations Development Programme (2011) Human development report 2011: sustainability and equity: a better future for all. Palgrave Macmillan, Hampshire, UK

UNDP—United Nations Development Programme (2013) Human development report 2013: the rise of the south: human progress in a diverse world. UNDP, New York

UNDP—United Nations Development Programme (2015) Human development report 2015: work for human development. UNDP, New York

UNDP—United Nations Development Programme (2016) Human development report 2016: human development for everyone. UNDP, New York

Visser W (2008) Corporate social responsibility in developing countries. In Crane A, Matten D, McWilliams A, Moon J, Siegel DS (eds) The Oxford handbook of corporate social responsibility. Oxford University Press, Oxford, UK, pp 473–479

Walid M, Luetz JM (2018) From education for sustainable development to education for environmental sustainability: reconnecting the disconnected SDGs. In: Leal Filho W (ed) Handbook of sustainability science and research. World Sustainability Series. Springer, Cham, pp 803–826. https://doi.org/10.1007/978-3-319-63007-6_49

Walker P (2018) Never mind GDP: make free time the measure of UK wellbeing, say Greens. The guardian. 5 October 2018. https://www.gu.com/p/9hthf/sbl. Accessed 28 Oct 2018

Wartick S, Cochran P (1985) The evolution of the corporate social performance model. Acad Manag Rev 10(4):758–769

WCED—World Commission on Environment and Development (1987) Our common future. Brundtland Report. Oxford University Press, Oxford. http://www.un-documents.net/our-common-future.pdf. Accessed 5 Mar 2018

Winograd M (1995) Environmental indicators for Latin America and the Caribbean. In: Trzyna T (ed) A sustainable world: defining and measuring sustainable development. Published for IUCN by California Institute for Public Affairs, Sacramento, CA, pp 198–215

Wood J (1991) Corporate social performance revisited. Acad Manag J 16(4):691–718

Woodward D, Simms A (2006) Growth is failing the poor: the unbalanced distribution of the benefits and costs of global economic growth. In: DESA Working Paper No. 20. http://www.un.org/esa/desa/papers/2006/wp20_2006.pdf. Accessed 28 Feb 2018

WWF (2016) Living planet report 2016. Risk and resilience in a new era. WWF International, Gland, Switzerland. http://awsassets.panda.org/downloads/lpr_2016_full_report_low_res.pdf. Accessed 5 June 2018

Dr. Johannes M. Luetz is Senior Lecturer, Postgraduate Coordinator and Research Chair at CHC Higher Education in Brisbane, Australia, and Adjunct Lecturer at the University of New South Wales (UNSW) in Sydney in the School of Social Sciences, where he also earned his Ph.D. in Environmental Policy and Management with a thesis on forced human migration. Dr. Luetz has worked extensively with World Vision International on research projects raising awareness of the growing effects of climate change on poor and vulnerable communities in Asia, Africa and Latin America. He has travelled extensively and has a passion for holistic education and human development, which previously found expression in his former role as Lecturer in Development Studies at UNSW, Sydney.

Mohamed/Mometo Walid is International Behavioural & Economic Consultant with several organizations around the world. He has been posted in several stations with the United Nations, with his last in the Pacific Islands. He is currently serving as Behavioural Scientist to the UAE Government. He has six post-graduate master degrees from top universities including the London School of Economics from law to behavioural economics and policy-making to international development and security through to quantitative analytics. He has been identified as one of the top young leaders in Egypt and the MENA region before 2011.

An Inquiry to Consider CSR in Integrated Management Systems

Markus Will, Jana Brauweiler, Anke Zenker-Hoffmann and Bernd Delakowitz

Abstract Although there is an unprecedented need to make the activities of companies more sustainable, many companies still face difficulties in managing sustainability. The paper proposes an integrated management systems approach based on the assumption that is consideration of external requirements such as CSR better succeeds when linked to existing management systems. The main issues addressed in this paper are: the suitability of the high level structure for the integration of CSR aspects and a four-step model that may provide guidance how to integrated CSR. The paper is more theoretical and conceptual rather than an empirical study. It is targeted to an audience of practitioners and SME managers and builds on an extensive literature survey and own experience from hands-on integration projects and consulting to SME.

Keywords Integrated management systems · Integration · Management systems ISO 26000 · Corporate social responsibility

1 Introduction—Business Sustainability and Corporate Social Responsibility

Businesses of any kind, i.e., not only companies in the public domain or with a vivid public interest, such as utilities, mobility providers, or the banking sector, are expected to take responsibility for social and environmental impacts beyond fulfilling legal requirements. Key stakeholders, such as investors, analysts, and rating agencies, are integrating sustainability factors into their evaluations and investment decisions. Stakeholders such as consumers and non-governmental organizations increasingly require ethically correct behaviors and demand data regarding a company's environmental and social performance (Garcia et al. 2016; Milne and Gray 2013; Fonseca

M. Will (✉) · J. Brauweiler · A. Zenker-Hoffmann · B. Delakowitz
University of Applied Sciences Zittau/Görlitz, Theodor-Koerner Allee 16, 02763 Zittau, Germany
e-mail: M.Will@hszg.de

W. Leal Filho (ed.), *Social Responsibility and Sustainability*, World Sustainability Series, https://doi.org/10.1007/978-3-030-03562-4_17

2015; Souza and Alves 2018). Policymakers, regulators, and standard setters also put pressure on companies. In the European Union, for instance, reporting on sustainability, i.e., non-financial aspects are now mandatory for larger enterprises with public interests (Directive 2014/95/EU). Companies with more than 500 employees should disclose relevant and useful information regarding environmental matters, social and employee aspects, respect for human rights, anti-corruption and bribery issues, and diversity in their board of directors. Companies are therefore urged to operate in a more socially and environmentally sound way and to manage and improve their sustainability performance. Thus, sustainability has made its way to the boardroom (EY 2010); however, businesses often react to increasing external demands and expectations using rhetoric and means of public relations, i.e., external communication, reporting, and stakeholder participation. This is often more camouflage or "bluewashing" than *true corporate sustainability* (Dyllick and Muff 2016), which would focus on meeting corporate goals and needs of direct and indirect stakeholders without compromising this ability for future stakeholders (Dyllick and Hockerts 2002). The business model of an authentically sustainable company focuses on "*doing business, doing good*" including the creation of social and environmental value. However, overcoming unsustainable business models (Schaltegger 2012: 166; Willard 2009: 6; Boons and Lüdeke-Freund 2013; Del Baldo and Baldarelli 2017; Will 2017) is ambitious and needs a designed change and transformation process, which includes a necessity for appropriate management approaches and systems (Schaltegger et al. 2013). Numerous corporate scandals, namely VW-Dieselgate where economic and legal consequences are still not foreseeable (Schmid 2017) or the several catastrophic accidents in Asian textile factories[1] lead to the conclusion, that there is an unprecedented demand for the development of sustainability management models and, last but not least, their integration into other MS already in place, such as quality, environmental or energy MS. Corporate Social Responsibility (CSR) is widely seen as an approach to improve sustainability performance, i.e. to make the transactions and practices of corporations more socially responsible and transparent (Idowu and Leal Filho 2010; Banerjee 2001; ISO 26000 clause 3). Caroll's Pyramid, see the figure below, is one of the early CSR frameworks and compromises a fourfold CSR hierarchy, including economic, legal, ethical and discretionary responsibility (Carroll 1991).

Caroll's CSR principles are considered in the ISO 26000 that aims in providing guidance for organizations to practice social responsibility and thereby to improve their impacts on their workers, their natural environments and their communities.

Especially large and multinational companies are taking sustainability considerations into account, in order to gain competitive advantages by customer binding in the first place, but also by developing new competencies and constituent capabilities (Laszlo and Zhexembayeva 2011). The CSR concept and related frameworks looks to be tailored to larger companies (Baden et al. 2011; Fassin et al. 2011; Lynch-

[1]Such as Ali Enterprises in Pakistan 2012 with 300 fatalities, the fire in the factory Tazreen (112 fatalities) and the collapse of the Rana Plaza multistory factory in Bangladesh (1127 fatalities), not to speak of other corporate scandals like Enron, Parmalat, Ahold etc.

Wood et al. 2009; Perrini 2006; Murillo and Lozano 2006; Vázquez-Carrasco and López-Pérez 2013), as its realization involves complex transformations in order to connect to decision making and strategic decisions. On the one hand, the inherent complexity of sustainability makes it challenging for small and medium-sized enterprises (SME) to realize CSR beyond greenwashing. SME often have a rather limited scope of freedom and responsibility, especially in global supply chains with high competitive and price pressure (Müller-Christ 2011). On the other hand, SME can collectively make a huge positive sustainability impact top society (Von Weltzien and Shankar 2010; Morsing and Perrini 2009; Jenkins 2006; Senge et al. 2010). Especially owner-managed SME are considered to better be able to deal with CSR, because they are more flexible, can take rather quick decisions and they are less fixated on maximizing and earning rations and may be more willing to sacrifice some of their profits to serve for a greater good (Vives 2006; Jamali et al. 2009; Jenkins 2006). Several smaller companies also undertake sustainability initiatives at the operational level and sometimes unknowingly. For example, if a company creates positive impact when engaged in local neighborhood activities or other philanthropic initiatives, without labeling this as sustainable (i.e. "silent" or "sunken" CSR, see Jenkins 2004; Longo et al. 2005; Perrini 2006; Raynard and Forstater 2002; Roberts et al. 2006, Jamali et al. 2009). However, CSR activities seem to be on a rather operational level (Fisher and Bonn 2011; Souza and Alves 2018), instead of being integrated at all business levels. The transfer of CSR management frameworks and stakeholder engagement programs, tailored for larger organizations and MNC (such as UN Global Compact, GRI reporting frameworks and alike), is not straightforward. SME are no "little big companies" (Tilley 2000; Preuss and Perschke 2010), they are unique in a sense, being active in a quite specific context. There is also doubt that motivation for CSR is triggered much by external pressure. Non-customer stakeholders are rarely confronting single SME with claims and requests related to social aspects of sustainability. Indeed, if there is external pressure to demonstrate CSR credentials, it is rather weak and with focus on environmental issues mainly. Such external pressure then originates from B2B customers, who better should sweep at their own doorstep before asking their suppliers (interviewee response in Jenkins 2006). Often the results are "box-ticking exercises" in self-assessments or remote audits (Jenkins 2006). Thus, a transfer of generic assumptions and concepts to SME is often of little use. However, the implementation of management systems (MS) is a common standard in today's business contexts. Organisations implement MS to meet the needs and expectations of their customers and other stakeholders and to comply legal requirements. There is hardly a company, who is not required to keep an external certification of, most prominently, a quality management system such as ISO 9001, EN 9100 or ITAF 16949. This not only applies to large, multinational companies, but also to SMEs, which often act as their suppliers. There is the proposition that additional external requirements such as CSR may better succeeds when linked with (integrated) management systems that are already in place. This claim is supported by the fact that the High Level Structure (HLS) offers a structural basis and promotes integration of standards and requirements (see Chapter "Focusing Sustainable Human Resource Management—Framework for Sustainability Management in

Research Organizations"). Hence MS or IMS provide a unique gateway and are still flexible enough to allow adaptations of new issues such as CSR.

Although there is a plethora of studies dealing with business sustainability, literature on the integration of sustainability into business processes and in particular into management system standards is rather limited (Souza and Alves 2018; Asif et al. 2013; Jørgenssen 2008; Klute-Wenig and Refflinghaus 2015) and further research is required, especially with regards to the consideration of ISO 26000. This paper extends the body of literature by depicting the principal suitability of the high level structure (HLS) for the integration of CSR and by proposing a conceptual model and a how-to guideline.

2 Suitability of the High Level Structure (HLS) for the Integration of CSR Aspects

Since the revision of ISO 9001 and ISO 14001 in the year 2015, MS are structured according to the so-called High Level Structure (HLS). This is basically a structure that aims at emphasizing similarities and ranges of integration between different MS. The HLS is intended to ensure consistency of the ISO management standards. Upcoming ISO standardized MS are required to follow this structure. The similarities and the compatibilities of the standards are emphasized by ten clauses of the HLS. Thus, the range of possible integration is easy to be identified. Hence, there is the expectation that integration of MS is better supported. The structural analogy of MS is important, however, it's importance might be a bit overestimated, as the lack of compatibility in the structure of the standards has not prevented the merger of documentation (Wilkinson and Dale 2001) even in the years before the HLS.

The HLS and related compatibility and linkage tables, so called compatibility matrices, (see Fig. 2) can be used as an instrument to identify possible areas of integration. At the same time, areas where no integration is useful, for instance in case of function specific requirements, are uncovered. Figure 1 shows, for the example of a triple-integrated IMS (i.e. quality, environmental and OHS MS), that the HLS allows benefiting from existing synergies both at the structural level of the MS elements as well as the level of MS documentation and handbook.

ISO 26000 is different from other management standards. First of all, it is not structured by HLS and does not contain any specific requirements that need to be achieved for example in order to get certified by external bodies. Certification is not at all intended by ISO 26000, although there are agencies who may deliver such an external reward. Instead, ISO 26000 provides guidance, meaning that a learning process is associated and implied when an organization considers it. It should help to clarify the meaning of social responsibility and its contribution to sustainable development in order to translate principles into effective actions (see IWA 2017: 26 in; ISO 2017). It is therefore suggested, that companies recognize its social responsibility within its sphere of influence which implies a review of the principles and

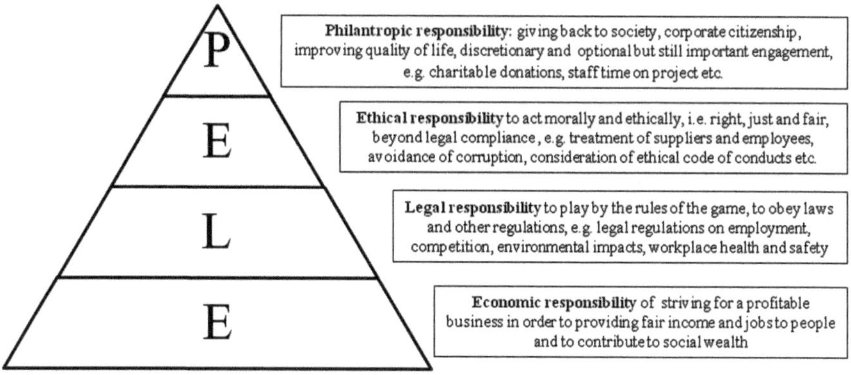

Fig. 1 Caroll's CSR Pyramid (Adapted from Carroll 1991)

core subjects of social responsibility and their integration in business policies, strategies, and operations. This involves also building competency and internal/external communication and a regular review according to the PDCA-model.

We assume that, taking a management system approach, i.e. providing a structural fundament across the core elements of MSS, probably enhances the ability of an organization to contribute to sustainable development through using the synergies. Some empirical studies suggest that, especially SME, integrate their MSs rather than keeping them separate (Labadová 2004; Zutshi and Sohal 2005; Rocha et al. 2007; Karapetrovic and Casadesús 2009; Karapetrovic and Willborn 1998; Zeng et al. 2007; Salomone 2008; Bernardo et al. 2009, 2017).

Based on our assumptions and consulting experience as well as in the light of empirical findings in the relevant literature, a conceptual model for the integration of ISO 26000 was developed and is described in this paper. We present in Chapter "Knowledge, Values and Attitudes Towards Marine Protected Areas in Gozo (Malta)" a Three-Level-Approach of Integrated Management Systems and provide in Chapter "Social Projects and the Internalization of Sustainability and Social Responsibility: Concepts for the Improvement of Quality of Life" a 4-step-strategy for integration of ISO 26000 guidance into a (Integrated) Management System.

3 Conceptual Remarks on the Integration of Management Systems

It seems to be challenging to implement broad sustainability aspects into corporate practice especially for SME. ISO 26000, although not being a management system itself, is principally in line with the *modus operandi* in other MS. Thus, it is reasonable that the work presented here starts from the assumption, that it would be a promising approach to start with sustainability activities in SME by the help of Integrated

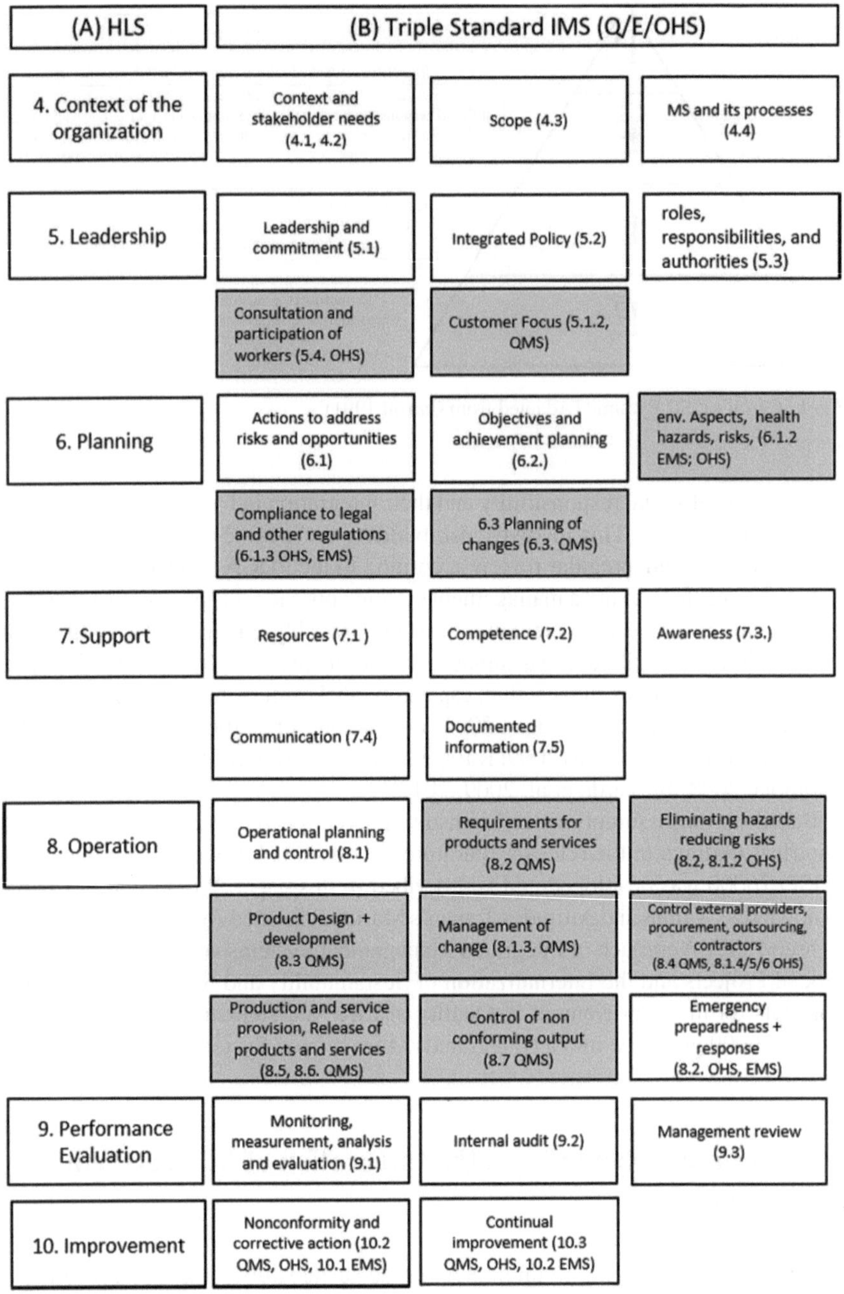

Fig. 2 Compatibility matrix: integration potentials in a triple standards IMS (column (B) on the basis of the HLS (column (A))—grey: function specific requirements

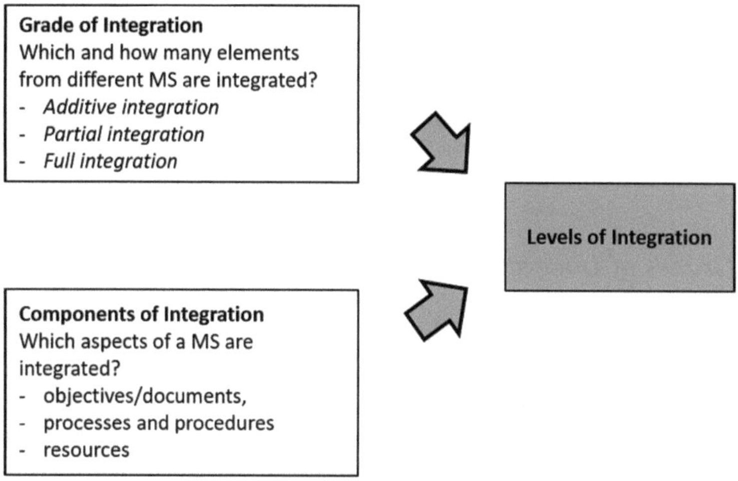

Fig. 3 Tree-level-approach of integrated management systems

Management System (IMS). A good definition of IMS was provided by Karapetrovic and Willborn: IMS are accordingly "*a set of interconnected processes that share a pool of human, information, material, infrastructure and financial resources in order to achieve a composite of goals*" (cit. according to Bernado et al. 2009 following Karapetrovic and Willborn 1998 and Karapetrovic 2003). Integration is then the technical, hands-on process of linking different management standards (Bernado et al. 2009) with the aim to improve the capability of a company to achieve its aims and to improve stakeholder's satisfaction. Hence, an IMS provides a management framework that allows to effectively merge different autonomous MS (MS), e.g. for quality management, environmental management, occupational safety, into one multi-purpose MS. Doing so is expected to have a number of benefits, resulting from the elimination of redundant regulations, documents and goal conflicts of individual MS, the exploitation of synergies and fewer effort and expenditures for (internal and external) audits (Abad et al. 2014; Zeng et al. 2011).

Integration simply means to put different function-specific MS together into one multi-purpose IMS (i.e. based on the High Level Structure). In addition, CSR aspects as described in ISO 26000, shall be considered and integrated into MS effectively in order to induce the translation of goals into concrete actions. However, the approaches taken in companies vary and different kinds of integration can be distinguished (Asif et al. 2013; Wilkinson and Dale 2001; Karapetrovic 2003; Beckmerhagen et al. 2003; Bernardo et al. 2009). For the sake of this paper, a Three-Level-Approach on Integration of MS is described (see also Fig. 3) that distinguishes:

- *Grades of Integration*, i.e. from additive integration, partial integration to full integration. Here it is considered if only selected or all components are integrated depending on the similarities across the standards—see Sect. 3.1.

- *Components of integration*, which relates to the range of elements or aspects of MS (i.e. objectives and documents, processes and procedures and resources) that are integrated—see Sect. 3.2.
- *Levels of Integrated Management Systems*, corresponds to a state of integration, which is shaped by the two perspectives "grade" and "components"—see Sect. 3.3.

3.1 Grades of Integration

The integration grade is a non-disjunct and non-directly measurable construct, representing an appraisal of the degree, to which the MS and their requirements are aligned and harmonized. It varies between the poles of '*low integration*' to '*complete integration*'. As with integration in general, each company has to decide which approach to choose, which range of integration is appropriate and which integration grade is to be achieved, based on its own situation and context. There seems to be no utterly accurate, generally applicable and transferable receipt for integration (Bernardo et al. 2009; Karapetrovic 2003). However, various theoretical and empirical studies have been carried out in order to construct ex-post different grades of integration (see Abad et al. 2014; Bernado et al. 2017; Pojasek 2006).

In our approach the following different degree of deployment and *inter alia* integration grades are distinguished:

- *Additional integration of system specific requirements*: Although considered as the first integration grade, it is rather an enlargement of the inventory of requirements, hence an addition to the MS. Specific components of the MS are kept separately. This is the case, when certain requirements are not required across all standards, but in one specific standard (for example: "8.6 Release of products and services" in ISO 9001:2015, "evaluation of occupational health only" in ISO 45001, "determination of environmental aspects" only in ISO 14001:2015). If there is a lack of similarities or compatibility regarding contents and/or procedures, let's say in a quality assurance audit of a certain production line with the goal to reduce failures, then integration is neither meaningful nor possible. If specific requirements are merely relevant for just one management system, it is obvious that those areas should be handled separately in documents, procedures and roles. The requirement is then simply added to the manual and documentation of the integrated management system, based on the HLS. This concerns particularly the so called core subjects 6.3–6.8 of ISO 26000.
- *Partial integration* means that not all components of the MS are integrated, for instance when certain activities/processes are required by only two standards but not across all in a multi-standard IMS. For instance, to manage the preparation and response to emergencies is only required in ISO 14001:2015 and ISO 45001:2018 (Clause 8.2) and thus the integration grade refers to partial integration. The same applies for the determination of environmental aspects (ISO 14001:2015), the identification of health hazards and risk (ISO 9001:2015) and the social responsi-

bility core subjects (ISO 26000). Those requirements imply activities, which have some methodological and procedural similarities. This is, congruously, considered as partly integrated, because there is no similar requirement in ISO 9001. However, there are requirements occurring in particular standards, but they could most appropriately be considered in all MS, such as emergency preparedness and compliance obligations. To integrate across all MS would imply *full integration* to be discussed next.
- *Full integration* refers to requirements that are similar across all the specific MS and can, therefore, be integrated into one multi-purpose IMS. Full integration seems feasible for the "core elements", such as context and stakeholder analysis (HLS 4.1, 4.2), scoping (HLS clause 4.3), policies (HLS clause 5.2), documented information (HLS clause 7.5), document and record control (HLS clause 7.5.3), internal auditing (HLS clause 9.2) or operational planning and control (HLS clause 8.1.). Requirements related to the HLS core elements can be fully integrated, meaning that goals, documented information, and procedures are in force across all MS and departments. There are quite a couple of items from ISO 26000, namely in relation to context (HLS 4), leadership (HLS 5) as well as performance evaluation and improvement (HLS 9, HLS 10), where also the integrated use of tools such as Balanced Scorecards may be appropriate.

3.2 Components of Integration

The range of the integration of MS relates to the integration of different management system components. We distinguish objectives and documents, processes and procedures and resources (Karapetrovic and Willborn 1998; Bernardo et al. 2009; Beckmerhagen et al. 2003). Examples are given in the Table 1.

3.2.1 Objectives and Documents

Objectives are, according to ISO 14001:2015, strategical, tactical or operational results to be strived for. This could be environmental or quality objectives that have to be specified for relevant functions and levels of an organization based on the organization's policies (e.g. integrated quality and environmental policy). Integration at this level means to harmonize the goals and objectives of the different MS to be integrated, in order to balance the goals and to deal with possible trade-offs. In an IMS those objectives that are common for all function specific MS need to be integrated first (Karapetrovic and Willborn 1998; Karapetrovic 2003; Karapetrovic and Jonker 2003; Bernardo et al. 2009). This finally implies that the objectives are described in one integrated document, namely the IMS policy and the IMS objectives.

The documentation of a management system is a set of documents, specifying the management system itself (i.e. the manual) or specific requirements for procedures (i.e. written procedures, work instructions) or documents that are used to state the

Table 1 Overview of management system components to be considered for integration

Objectives and documents	
• Commitment • Policy • Objectives • Scope of the MS	• Creating, updating and • Manual • Written procedures, instructions • Records
Procedures and processes	
• Context analysis • Stakeholder analysis • Risk and opportunity analysis • Determination of compliance obligations • Resource management • Determination of requirements • Internal and external communication • Training and competence development	• Operation planning and control • Monitoring, measurement, analysis and evaluation • Control of documented information • Internal audits • Management review • Improvements, nonconformity and corrective action
Resources	
• Organizational roles, responsibilities and authorities (e.g. Management System Manager, Management System Representative, Inspectors, Audit teams)	

results achieved and thus provide evidence for activities performed (e.g. records). It is the documentation that is the most visible, tangible characteristic of a management system which captures MS dimensions most (Abad et al. 2011; Blewett and O'Keeffe 2011; Abad et al. 2014). Some might say the documentation is the management system. For the integration of objectives and documents, the High Level Structure (HLS) of the recent revisions of ISO 9001 and ISO 14001, provides a structural basis. The similarities and the compatibilities of the standards are emphasized by the ten clauses of the HLS and thus easily be analysed.

3.2.2 Processes and Procedures

Processes and procedures refer to actual day-to-day activities at the operational level in an organization. A procedure represents the specific way to carry out a particular activity or process. Written procedures (or procedure documents) refer to instructions and records of activities undertaken to achieve conformity with specific requirements as well as policies and goals (ISO 9000:2005, § 3.4.3). Therefore, all the function specific requirements of the MS need to be considered at this operational level. However, the grade of procedural integration may vary. In the case of common activities and core elements of the MS, e.g. documented information, document and record control (HLS clause 7.5.3), internal auditing (HLS clause 9.2) or operational planning and control (HLS clause 8.1.), the procedural integration is rather high (Beckmerhagen et al. 2003). Function-specific requirements, i.e. determination of environmental aspects (ISO 14001:2015, clause 6.1.2) or control of externally provided processes, products, and services (ISO 9001:2015, clause 8.4) are managed within separate procedures. Also at the level of processes and procedures, the HLS provides a common basis for the identification of complementariness to figure out in a compatibility matrix (see Fig. 4).

3.2.3 Resources

The range of integration of management system may be further expanded to include also resources. Obviously, sufficient resources are needed for an effective management system and its continuous improvement. Resources include, generally speaking, human resources and specialized skills, as well as infrastructure, technology, financial resources and alike. It is especially the human resources that are of interest, because responsibility for different MS may fall into one person, namely the EHS-manager or management system representative. The integration of resources could also take place at top management level, where a strengthening of leadership is required as well as at the shop-floor-level, i.e. regarding environmental inspectors or quality engineers (organizational roles, responsibilities and authorities). However, empirical evidence regarding the integration of human resources is lacking (Bernardo et al. 2009).

3.3 Integration Levels

As sort of conclusion regarding the theory and practice of integration, the two perspectives "grade" and "components"[2] are now brought together. To put it simple:

$$\text{integration level} = \text{grade} + \text{components range (see Fig. 4)}$$

Finally, three levels of Integrated Management Systems can be considered:

- *Level 1*: refers to an additive integration, i.e. new requirements from stakeholders or MSS are added to the manual or documentation. The integration level is rather low, as it is simply an extension. The placement and subsumption, i.e. where to add in the manual index, is based on the HLS.
- *Level 2*: New requirements are merged and aligned with existing goals, documents, procedures, and resources, making use of common areas and similarities of the standards. This means, that MS are adopted by different business units and documentation and similar process are fused in order to exploit synergies and tackle trade-offs. However, this integration does neither cover all elements of a MS nor all existing MS in a company.
- *Level 3*: Goals, documents, processes are fully integrated so that the single constituting MS lose the unique identities resulting in a general management system.

[2]To be precise: it is the range to which the components are integrated.

Total integration is beyond the technocratic approach with a focus on documents and the structure of standards and manuals. This is a more business-excellence and holistic approach that follows the idea of TQM and EFQM that aims at embedding the IMS philosophy, balancing the requirements of the different function-specific MS in the organizational culture of a company (Jørgenssen et al. 2006; Jørgenssen 2008) (Fig. 4).

Referring the rare empirical studies (namely Bernardo et al. 2009; Abad et al. 2014), it can be summarized that a majority of companies integrate documents and objectives as well as procedures. Rather high grades of integration occurs with regard to organizational policies, objectives and the manual as well as procedures such as internal audits, management reviews, record control, internal communication and measurement, analysis and improvement. Procedures related to more operational issues and product realization seem to be least integrated. Thus, there is some empirical evidence that those elements with high compatibility and with higher strategical relevance are integrated first, while more function-specific and operational elements only partially integrated or kept separate. This makes sense from a pragmatic point of view, i.e. starting with obviously similar items before looking at specific, operational working procedures that are more difficult to standardize (Koubek and Pölz 2015: 19; Bernardo et al. 2009). There is a contradictory assumption with regard to the integration of resources, in particular, human resources and personnel. Based on experience and from more conceptual studies there is some belief in the integra-

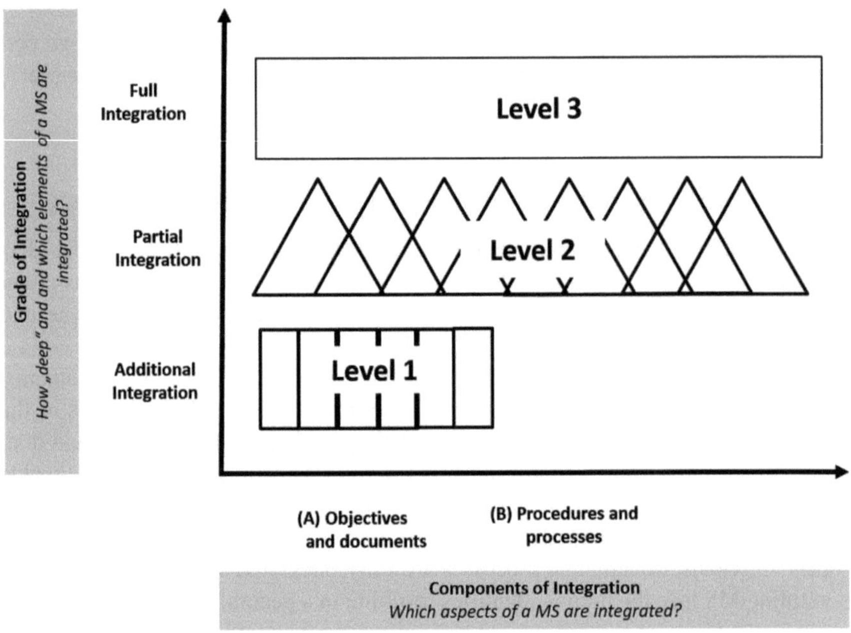

Fig. 4 Levels of integration based on range and integration grade

tion across different hierarchical levels, i.e. that responsibilities fall into one person or that interdisciplinary teams are established (Karapetrovic 2003; Beckmerhagen et al. 2003). However, this appraisal was not supported by other empirical studies (Bernardo et al. 2009).

Finally, the question arises how the presented 3-level-approach can be transferred to the integration of the requirements of ISO 26000 into an MS or into a single or double or triple standard MS. In Chapter "Social Projects and the Internalization of Sustainability and Social Responsibility: Concepts for the Improvement of Quality of Life" a corresponding procedure is shown.

4 Integration of ISO 26000 Guidance into a Triple-Standard Management System

In order to integrate CSR aspects based on the ISO 26000 guidelines into a HLS based triple-management system, a four-stage model is suggested. The model applies a combined project management and change management approach, where a distinction is made between project initiation, definition, planning, implementation, and improvement. The model is based on concepts published elsewhere (e.g. Karapetrovic 2003; Beckmerhagen et al. 2003), but also on long-term experience with SME in regard of the implementation of MS in general (see Brauweiler et al. 2015, 2018), as well on a multi case-project on the integration of ISO 50001 compliant energy management into a QMS (Brauweiler and Will 2017). Therefore, the model is considered as being generic. It is abstract and flexible enough to be adopted in the various contexts of companies. However it is not a cook-book recipe and may be adjusted to specific circumstances. An overview is provided in the Fig. 5.

4.1 Initiation of the Integration Project (Step 1)

Initiation and preparation aims at laying the foundations for CSR implementation. The integration project is initiated by an official top-management decision to foster the integration of ISO 26000 guidance issues and their consideration within the IMS-structures. It is required, that top and middle management understand and accept the necessity to deal with CSR in order to reduce socio-economic impacts of business activities. The integration approach needs to be carefully examined to determine whether or not it benefits the organization. During this phase, a decision making team will identify if the project can realistically be completed. This implies also, that no isolated approach to CSR is taken, but the existing MS framework and related management process should be used. The top-level decision needs, obviously, be closely connected to the approval of budget and resources. Therefore, responsible employees need to be appointed (i.e. sustainability managers with a specified qual-

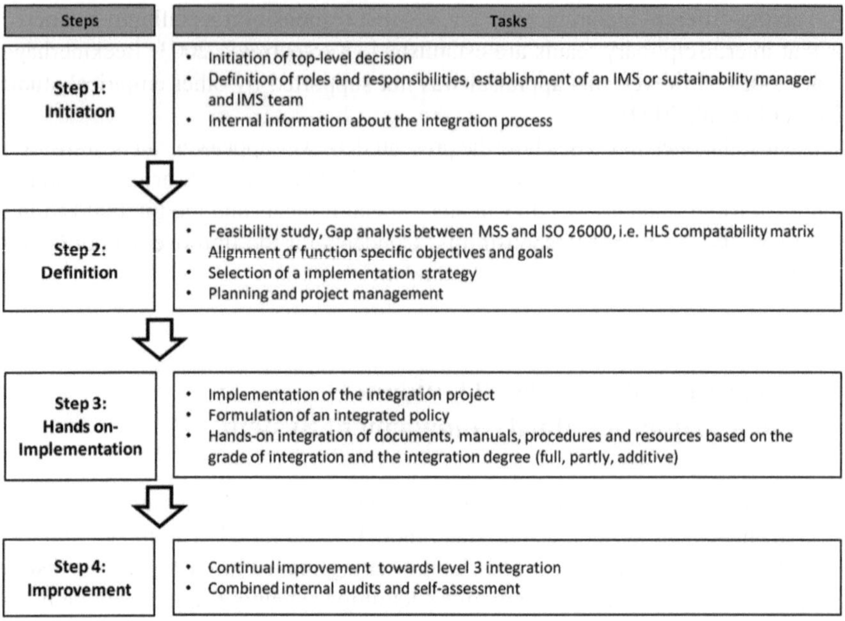

Fig. 5 Four-step model for integration of ISO 26000 into a triple standard IMS

ification profile, see Will et al. 2018) and an IMS teams is to be established, where roles and responsibilities are defined. Finally, the decision of CSR integration is communicated internally, i.e. announcement of the integration project.

4.2 *Definition of the Integration Project (Step 2)*

An appropriate integration strategy has to be chosen based on the analysis of compatibilities and similarities of ISO 26000 with existing MS. For the implementation of the integration strategy an appropriate project management needs to be set up. Step 2 is probably the most demanding one, as a number of important decisions have to be made and a collaboration of top and middle management is required to determine which integration level is desirable and feasible.

4.2.1 Feasibility and Gap Analysis Based on a Compatibility Matrix

This requires that a feasibility and gap-analysis as well as an alignment of general and functional specific objectives and goals and dealing with trade-offs. It is recommended to analyze structural and content related similarities of new requirements of ISO 26000 with the already implemented MS in a company. As an effective tool refers a *compatibility matrix*, which helps to identify significant overlaps and convergence in terms and topics, covered by standards and MS (i.e. perspective integration grade). It also allows finding cross-references with specific elements, which is important to adequately embed items in the applied HLS. The compatibility matrix serves to identify gaps and areas of non-compatibly and points to starting-points for further action. The grade of integration should be specified, i.e. it is fixed, which and how many elements from the different MS are to be integrated. Figure 6 provides a graphical representation of an extensive compatibility matrix for the integration of ISO 26000 into a triple-standard IMS (Q/E/OHS). Items with a high integration potential, i.e. where full integration can be achieved, are indicated (dark grey boxes).[3]

As shown in Fig. 6, the potentials for full integration are considerably lower at the operational level (i.e. "Do-Phase" of the PDCA, HSL Chapters "Enhancing Organizations' Social Responsibility by Workplace Health Promotion?" and "An Indigenous 'right way' Environmental, Social and Cultural Core-Benefits Verification Standard"), due to more function-specific requirements and tasks. The same is true for the integration of ISO 26000 items. Obviously, stakeholder and context analysis, as well as the determination of the scope of the MS with regard to the sphere of responsibility and influence and the monitoring and continuous improvement activities, offer large integration potentials. Competence building, awareness, raising and (internal and external) communication may differ in precise contents but can make use of already existing structures in staff development and public relations, hence it can be fully integrated. Decision making and triggering of voluntary initiatives for enhancing CSR may also be integrated into the overall operational planning (ISO 9001, 8.1.) anyway. The alignment of the ISO 26000 core subjects,[4] seems to be quite similar in the methodological sense with the determination of environmental impacts and the identification of health hazards, although there is a focus on external effects along the supply chain in ISO 26000.

Finally, there are quite a few aspects that are demanded within ISO MSS that should be considered for ISO 26000 integration. This considers all the items that make an MS an effective instrument to improve an organizations performance, namely repeatable steps that should be consciously implemented (see MS and its processes, HLS clause 4.4., operational planning and control, HLS clause 8.1)). The allocation of resources, responsibilities, and authorities is as well a precondition for the effective implementation of a CSR strategy which is not required but necessary.

[3]HLS 4 context of the organization, HLS 5 Leadership, HLS 7 Support particularly competence and awareness raising, HLS 9 and 10 performance evaluation, improvement.

[4]i.e. human rights (6.3), labour practices (6.4), environment (6.5), fair practices (6.6), consumer issues (6.7), community involvement development (6.8) i.e. human rights (6.3), labour practices (6.4), environment (6.5), fair practices (6.6), consumer issues (6.7), community involvement development (6.8).

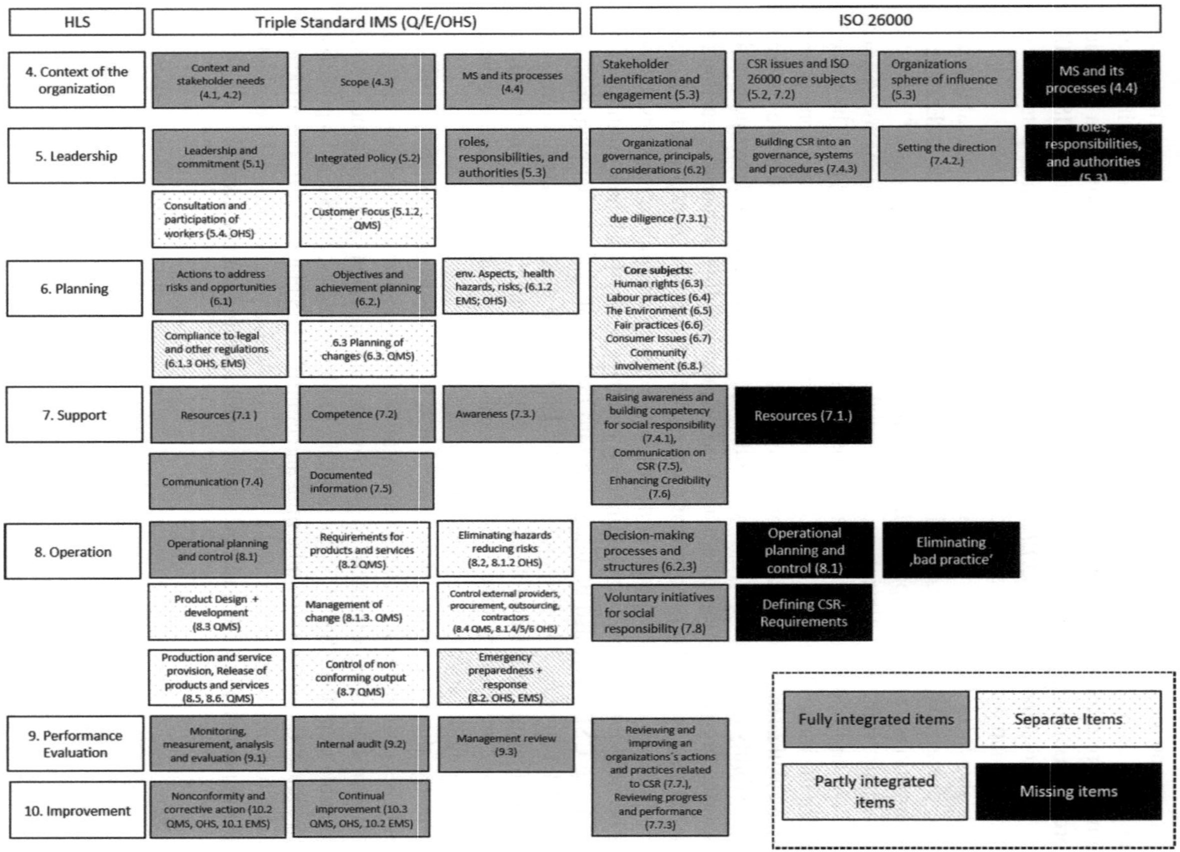

Fig. 6 Overview of integration potentials of ISO 26000 into a triple-standard management based on the HLS elements and integration grades (own elaboration, see also Souza and Alves (2018) and ISO (2017))

4.2.2 Selection of an Integration Strategy

In addition to the analysis of structural and content related similarities, the integration strategy has to be selected. Top management might decides that integration of goals and documents is sufficient for their company, without necessarily integrating process and procedures or combining resources. On the other hand, top management could also decide that a complete amalgamation ("total integration") with a full integration at all levels across goals, documents, processes, and resources should occur (Karapetrovic 2003; Beckmerhagen et al. 2003). It is also necessary to align function specific objectives and goals, including the identification of potential trade-offs and synergies.

The integration project itself is related to more or less profound organizational changes and therefore is likely that resistance in departments or divisions occurs like in most change processes. Additionally, as the previous analysis of the integration potentials may have already implied, there will be documents and procedures ("core elements", such as management review and internal audits as well as internal and external communication, document control etc.) which are easier to handle, while in other areas the situation might be more complicated and systemic and structural barriers appear. Therefore it might be a good strategy to start with a "low-hanging fruits" stepwise approach (Koubek and Pölz 2015: 199). This is, however, also a precarious strategy as this sort of "cherry-picking" only leads to a postponement, meaning, that with structural barriers and resistance has to be dealt with later on.

4.2.3 Project Plan

Whatever strategy is chosen, it is the responsibility of the top management to help middle managers and executors, also with the allocation of appropriated funds. This step includes a comprehensive planning and project management. During this phase, therefore, the IMS team should prioritize the project, and calculate a budget and schedule, and determine what resources are needed. A project plan, project charter and/or project scope may be put in writing, outlining the work to be performed.

4.3 Hands-on Integration of Documents, Procedures, and Resources (Step 3)

This step is, obviously, the most comprehensive one. It aims at the practical application of the integration, namely the technical, hands-on work of integration of documents, manuals, procedures and, if so, resources, as it is determined by the analysis and the compatibility matrix. Another issue here is the formulation and adherence to an integrated policy of the organization, including definition of roles, responsibilities, competencies, and authorities.

4.4 Continual Improvement (Step 4)

The continual improvement process according to the ubiquitous PDCA-model is deeply anchored in the MS. Therefore, the effectiveness of the integration process itself, as well as suitability and adequacy of the IMS, is the object of continual improvement. Internal IMS audits, self-assessment and a linkage of indicators across the function-specific performance measures allow monitoring of progress. Continual improvement means, on the one hand, the further promotion of the integration towards full integration of goals, documents, procedures and resources where feasible and practicable (or "amalgamation" following Beckmerhagen et al. 2003), which is considered as level 3 integration. On the other hand, improvement would mean to go beyond the technocratic approach of merging documents towards a more holistic approach striving for and business-excellence and development of an IMS-appropriate organizational culture. Finally, further, improvement may also lead to the opportunity and a state of organization, that allows integrating management standards of the future, that are certain to come in a world of "mushrooming management standards" (cit. Karapetrovic 2003).

5 Conclusion

Only a few studies exist on the integration of sustainability issues into standardized management systems (Souza and Alves 2018; Asif et al. 2013; Jørgenssen 2008; Klute-Wenig and Refflinghaus 2015). This conceptual paper provides a theoretical analysis of integration of management systems in general and the potentials for the integration of ISO 26000 guidelines in particular. Furthermore the existing body of academic and applied research literature is extended by the first attempt to integrated ISO 26000 into a triple-standard management systems. It is the first study of its kind that refers to the most recent international MSS, namely ISO 45001:2018.

For instance, a three-level approach to integration is described, which distinguishes grades of integration (e.g. additive integration, partial integration to full integration) and the range of components (i.e. objectives/documents, procedures and resources).

Full integration is appropriate whenever the requirements are similar across all MS and a multi-purpose IMS can be established. If not all components of MS can be integrated, because they are not required in the standards or it seems to be not appropriate, the grade is partial integration. Addition of system specific requirements occurs when the requirements are particularly function specific and therefore another kind of integration is not possible. With the help of a compatibility matrix, the grade of integration can be determined. An example is provided in Fig. 6. However, the identification and attribution to the HLS can also differ depending in the context of particular organizations. One interesting finding of this paper is that there are a several requirements in the IMS standards and the ISO 26000 that can be fully integrated, which is also in line with previous publications (i.e. Souza and Alves 2018; ISO 2017). Another important finding is that only few aspects, namely the definition of roles and responsibilities, the allocation of resources and the establishment of MS-processes, are missing in ISO 26000.

The range of components describes the deployment of the integration, for example if the integration covers merely objectives and document or of processes and procedures as well as resources are considered. The highest level of integration is accomplished when full integration is achieved and all components are integrated right up to resources.

Finally, a five-Step model for integration of ISO 26000 into a triple standard IMS was presented. The model provides a guideline for practitioners in companies. The model is rather abstract and can therefore be adapted to the specific needs.

The integration of management systems will continue to be a relevant area both for academic and applied research as well as the practice in organisations. Future research is needed to conceptualize the levels of integration and to apply this conceptualisation in empirical studies. As a part of this effort, the development of supporting instruments for each requirement of an IMS, including self-assessments, maturity models and internal auditing is required. In addition to conceptual work focussing on management system standards and their application, it is necessary to further study another perspective from business administration, namely the business process modelling approach.

References

Abad J, Dalmau I, Vilajosana J (2014) Taxonomic proposal for integration levels of management systems based on empirical evidence and derived corporate benefits. J Clean Prod 78:164–173

Abad J, Mondelo P, Sanchez-Toledo A (2011) Management systems integration e-Spanish survey reveals key benefits. ISO Focus 2(1):42–45

Asif M, Searcy C, Zutshi A, Fisscher OAM (2013) An integrated management systems approach to corporate social responsibility. J Clean Prod 56:7–17

Baden D, Harwood I, Woodward D (2011) The effects of procurement policies on downstream CSR activity: content analytic insights into the views and actions of SME owner–manager. Int Small Bus J 29:259–277

Banerjee SB (2001) Corporate environmental strategies and actions. Manag Decis 39(1):36–44

Beckmerhagen I, Berg H, Karapetrovic S, Willborn W (2003) Integration of management systems: focus on safety in the nuclear industry. Int J Qual Reliab Manag 20(2):210–228

Bernado M, Gianni M, Gotzamani K, Simon A (2017) Is there a common pattern to integrate multiple management systems? A comparative analysis between organizations in Greece and Spain. J Clean Prod 151(2017):121–133

Bernardo M, Casadesus M, Karapetrovic S, Heras I (2009) How integrated are environmental, quality and other standardized management systems? An empirical study. J Clean Prod 17(8):742–750

Blewett V, O'Keeffe V (2011) Weighing the pig never made it heavier: auditing OHS, social auditing as verification of process in Australia. Saf Sci 49(7):1014–1021

Boons F, Lüdeke-Freund F (2013) Business models for sustainable innovation: state-of-the-art and steps towards a research agenda. J Clean Prod 45:9–19

Brauweiler J, ZenkerHoffmann A, Will M (2018) Umweltmanagementsysteme nach ISO 14001 Grundwissen für Praktiker. Springer, Wiesbaden

Brauweiler J, Will M (2017) Handlungshilfe Integriertes Energiemanagement - Einführung eines Energiemanagementsystems in ein bestehendes Qualitätsmanagementsystem. SAENA - Sächsische Energieagentur GmbH. http://www.saena.de/projekte/integrierte-managementsysteme.html

Brauweiler J, Will M, Zenker-Hoffmann A (2015) Auditierung und Zertifizierung von Managementsystemen. Grundwissen für Praktiker. Springer Essentials. Springer, Wiesbaden

Carroll AB (1991) The pyramid of corporate social responsibility: toward the moral management of organizational stakeholders. Bus Horiz 34:39–48. https://doi.org/10.1016/0007-6813(91)90005-g

Del Baldo M, Baldarelli MG (2017). Renewing and improving the business model toward sustainability in theory and practice. Int J Corporate Soc Responsib 2:3. https://doi.org/10.1186/s40991-017-0014-z

Directive 2014/95/EU of the European Parliament and of the Council of 22 October 2014 amending Directive 2013/34/EU as regards disclosure of non-financial and diversity information by certain large undertakings and groups Text with EEA relevance

Dyllick T, Hockerts K (2002) Beyond the business case for corporate sustainability. Bus Strategy Environ 11:130–141. https://doi.org/10.1002/bse.323

Dyllick T, Muff K (2016) Clarifying the meaning of sustainable business—Introducing a typology from business-as-usual to true business sustainability. Organ Environ 29(2):156–174

EY (2010) The sustainability journey: from compliance, to opportunity, to integration. Insights for North American Committee Members. Ernst & Young Assurance, Tax, Transactions, Advisory. EYGM Limited

Fassin Y, Van Rossem A, Buelens M (2011) Small-business owner–managers' perceptions of business ethics and CSR-related concepts. J Bus Ethics 98:425–453

Fisher J, Bonn I (2011). Business sustainability and undergraduate management education: an Australian study. High Educ 62:563–571. https://doi.org/10.1007/s10734-010-9405-8

Fonseca LM (2015) Strategic drivers for implementing sustainability programs in Portuguese organizations e let's listen to aristotle: from triple to quadruple bottom line. Sustainability 8:136–142. https://doi.org/10.1089/SUS.2015.29004

Garcia S, Cintra Y, Torres R, Rita de Cássia SR, Lima FG (2016) Corporate sustainability management: a proposed multi-criteria model to support balanced decision-making. J Clean Prod 136:181–196. https://doi.org/10.1016/j.jclepro.2016.01.110

Idowu SO, Leal Filho W (eds) (2010) Professionals' perspectives of corporate social responsibility. Springer, Heidelberg, Dordrecht, London, New York. ISBN 978-3-642-02629-4. https://doi.org/10.1007/978-3-642-02630-0

ISO (2017) IWA 26:2017—Using ISO 26000: 2010 in management systems. ISO/TMBG Technical Management

Jamali D, Zanhour M, Keshishian T (2009) Peculiar strengths and relational attributes of SMEs in the context of CSR. J Bus Ethics 2009(87):355–377

Jenkins H (2006) Small business champions for corporate social responsibility. J Bus Ethics 2006(67):241–256

Jenkins H (2004) A critique of conventional CSR theory: an SME perspective. J Gen Manage 29(4):55–75

Jørgenssen TH (2008) Towards more sustainable management systems: through life cycle management and integration. J Clean Prod 16(2008):1071–1080

Jørgenssen TH, Remmen A, Mellado MD (2006) Integrated management systems—Three different levels of integration. J Clean Prod 14(2006):713–722

Karapetrovic S, Jonker J (2003) Integration of standardized management systems: searching for a recipe and ingredients. Total Qual Manag 14(4):451–459

Karapetrovic S (2003) Musings on integrated management systems. Meas Bus Excell 7(1):4–13

Karapetrovic S, Casadesús M (2009) Implementing environmental with other standardized management systems: scope, sequence, time and integration. J Clean Prod 17(5):533–540

Karapetrovic S, Willborn W (1998) Integration of quality and environmental management systems. TQM Mag 10(3):204–213

Klute-Wenig S, Refflinghaus R (2015) Integrating sustainability aspects into an integrated management system. TQM J 27(3):303–315. https://doi.org/10.1108/TQM-12-2013-0128

Koubek A, Pölz W (2015) Integrierte Managementsysteme: Von komplexen Anforderungen zu zielgerichteten Lösungen. Carl Hanser Verlag GmbH & Co. KG, Munich

Labadová A (2004) Implementing integrated management systems using a risk analysis based approach. J Clean Prod 12(2004):571–580

Laszlo C, Zhexembayeva N (2011) Embedded sustainability: the next big competitive advantage. stanford business books

Longo M, Mura M, Bonoli A (2005) Corporate social responsibility and corporate performance: the case of Italian SMEs. Corp Gov 5(4):28–42

Lynch-Wood G, Williamson D, Jenkins W (2009) The over-reliance on self-regulation in CSR policy. Bus Ethics 18:52–65

Milne MJ, Gray R (2013) W(h)ither ecology? The triple bottom line, the global reporting initiative, and corporate sustainability reporting. J Bus Ethics 118:13–29. https://doi.org/10.1007/s10551-012-1543-8

Morsing M, Perrini F (2009) CSR in SMEs: do SMEs matter for the CSR agenda. Bus Ethics: Eur Rev 18(1):1–6

Müller-Christ G (2011) Sustainable management coping with the dilemmas of resource-oriented management. Springer, Heidelberg, Dordrecht, London, New York

Murillo D, Lozano J (2006) SMEs and CSR: an approach to CSR in their own words. J Bus Ethics 67:227–240. https://doi.org/10.1007/s10551-006-9181-7

Perrini F (2006) SMEs and CSR theory: evidence and implications from an Italian perspective. J Bus Ethics 67:305–316. https://doi.org/10.1007/s10551-006-9186-2

Pojasek R (2006) Is your integrated management system really integrated? Environ Qual Manage 16(2):89–97

Preuss L, Perschke J (2010) Slipstreaming the larger boats: social responsibility in medium-sized businesses. J Bus Ethics 92(4):531–551

Raynard P, Forstater M (2002) Corporate social responsibility: implications for small and medium enterprises in developing countries. United Nations Industrial Development Organization, Vienna, Austria

Roberts S, Lawson R, Nicholls J (2006) Generating regional-scale improvements in SME corporate responsibility performance: lessons from responsibility northwest. J Bus Ethics 67:275–286. https://doi.org/10.1007/s10551-006-9184-4

Rocha M, Searcy C, Karapetrovic S (2007) Integrating sustainable development into existing management systems. Total Qual Manag Bus Excell 18(1):83–92

Salomone R (2008) Integrated management systems: experiences in Italian organizations. J Clean Prod 16(16):1786–1806

Schaltegger S (2012) Die Beziehung zwischen CSR und Corporate Sustainability. In: Schneider A, Schmidpeter R (eds) Corporate social responsibility. Springer, Berlin, Heidelberg

Schaltegger S, Beckmann M, Hansen EG (2013) Corporate sustainability meets transdisciplinarity. Bus Strategy Environ 22:217–218. https://doi.org/10.1002/bse.1770

Schmid CU (2017) "Germany". In: Frigessi di Rattalma M (ed) The dieselgate a legal perspective. Springer, Switzerland

Senge P, Smith B, Kruschwitz N, Laur J, Schley S (2010) The Necessary revolution: how individuals and organizations are working together to create a sustainable world. Crown Business, New York

Souza JPE, Alves JM (2018) Lean-integrated management system: a model for sustainability improvement. J Clean Prod 172(2018):2667–2682

Tilley F (2000) Small firm environmental ethics: how deep do they go? Bus Ethics Eur Rev 9:31–41

Vázquez-Carrasco R, López-Pérez ME (2013) Small and medium-sized enterprises and corporate social responsibility: a systematic review of the literature. Qual Quant 2013(47):3205–3218. https://doi.org/10.1007/s11135-012-9713-4

Vives A (2006) Social and environmental responsibility in small and medium enterprises in Latin America. J Corp Citizsh 21:39–50

von Weltzien H, Shankar D (2010) How can SMEs in a cluster respond to global demands for corporate responsibility. J Bus Ethics 2011(101):175–195. https://doi.org/10.1007/s10551-010-0708-6

Wilkinson G, Dale BG (2001) Integrated management systems: a model based on a total quality approach. Manag Serv Qual: Int J 11(5):318–330. https://doi.org/10.1108/09604520110404040

Will M, Brauweiler J, Neumann C (2018) Professional, methodical and didactical facets of ESD in a masters course curriculum—A case study from Germany. In: Leal W et al (eds) Implementing sustainability in the curriculum of universities. Springer, pp 307–322. https://doi.org/10.1007/978-3-319-70281-0_19

Will M (2017) A sustainable company is possible—Some cases studies and a maturity model. In: Velazquez L (ed) International sustainability stories: enhancing good practices. Universidad de Sonora. ISBN 978-607-518-250-6

Willard B (2009) The sustainability champion's guidebook—How to transform your company. New Society Publishers, Gabriola Island

Zeng S, Shi J, Lou G (2007) A synergetic model for implementing an integrated management system: an empirical study in China. J Clean Prod 15(18):1760–1767

Zeng S, Xie X, Tam C, Shen L (2011) An empirical examination of benefits from implementing integrated management systems (IMS). Total Qual Manag Bus Excell 22(2):173–186

Zutshi A, Sohal A (2005) Integrated management system: the experiences of three Australian organizations. J Manuf Technol Manag 16(2):211–232

Socioenvironmental Policies in Brazil and England in a Cosmetics Industry—A Comparative Study

Amanda Espósito Borges, Lucas Ferrari Pinto dos Santos and João Luiz de Moraes Hoefel

Abstract Currently, in the business context, due to the relevance and necessity of policies aiming socioenvironmental preservation, many organizations have being pressured to invest in new ideals and actions that generate less social and environmental impacts, besides the adoption of sustainable practices in their management. This study seeks to analyze and identify sustainable practices carried out by a Cosmetics Company in two countries: Brazil, where its subsidiary is located and in the United Kingdom, where its head office is based, with more established sustainable practices. In order to carry out the study, examples of sustainable socioenvironmental actions were analyzed and collected in the literature and in websites of recognized institutions as examples of sustainable good practices, to compare later them with practices used by the Cosmetics Company. Thus, this paper presents a comparative analysis between existing environmental policies used in Brazil and England, and how they affect the practice and management of the same company in these two countries. In both cases, socioenvironmental sustainability actions were identified as an effective part of each company organizational culture, but there are also differentiated positions in some areas of action, as well as significant differences in socioenvironmental and business policies of both countries.

Keywords Environmental policies · Corporate sustainable · Sustainable practices
Socioenvironmental practices · Social responsibility

A. E. Borges (✉) · L. F. P. dos Santos
Curso de Administração - Faculdades Atibaia, Estrada Municipal Juca Sanches
1050, Atibaia, SP 12954-070, Brazil
e-mail: amanda_eb4@hotmail.com

L. F. P. dos Santos
e-mail: lucasferrari2@hotmail.com

J. L. de M. Hoefel
Curso Tecnológico em Gestão Ambiental e Núcleo de Estudos em Sustentabilidade e Cultura - Faculdades Atibaia, Estrada Municipal Juca Sanches 1050, Atibaia, SP 12954-070, Brazil
e-mail: jlhoeffel@gmail.com

W. Leal Filho (ed.), *Social Responsibility and Sustainability*, World Sustainability Series, https://doi.org/10.1007/978-3-030-03562-4_18

1 Introduction

This paper deals with a comparative study between social responsibility and environmental policies in Brazil and in the United Kingdom. For that reason cosmetics companies were used as examples to emphasize the importance of these policies for the growth of the sector, in a way that does not harm employees and the environment. During the process of industrialization, all over the world, natural resources were exploited in a disorderly way, causing negative effects on the environment and humans. Initially, there were major environmental problems and national and international events to discuss these aspects (Nascimento 2008), which highlights the importance and growth that these environmental concerns have generated for the development of many companies and especially the economic impact that has been observed. Later, the ISO 26000 comes up as the first norm of Social Responsibility that integrates socioenvironmental policies into the organizations. Thus, according to the Associação Brasileira de Normas Técnicas (ABNT 2010), the ISO 26000 encourages the organizations to go beyond legal compliance, recognizing this is an essential part for their internal actions of social responsibility, because it takes into consideration societal, environmental, legal, cultural, political and organizational diversity.

Hence, examples of socioenvironmental norms and practices were used in order to be exercised by companies that contribute to improvements to the environment and that also bring positive social and economic results.

Consequently, the main objective of this work is to compare the practices of companies in the cosmetics sector, to study the reasons why a company uses certain sustainable practices and to carry out a case study to understand the internal and external difficulties of this sector.

In addition, definitions are presented on Sustainability in Organizations, Environmental Management System, Social Responsibility and how to adhere to these practices within Organizations.

2 Sustainability and Corporate Social Responsability

The historical context of sustainability has an important landmark which began in 1972 with international concern for the environment related to technological and economic development, which promoted the United Nations Conference on the Human Environment in Stockholm/Sweden, and which sought to stimulate the balance between human beings and minimize environmental degradation.

Therefore, it is necessary to understand that this concern arose from the expansion of the economic productivity that, increasingly, used natural resources for benefits the public and private sectors.

Until then, pollution and unconcern were seen as gain and in the same decade (1970) environmental laws were created and are characterized by the flexibility to meet the specific needs of each region.

According to the World Commission on Environment and Development (Santos and Wagner 2008, p. 1), defining sustainability is "meeting the needs of the present without compromising the ability of future generations to meet their needs".

In that case, it is possible to relate Social Responsibility to the Sustainability of organizations, because it is based on the economic, environmental and social aspects of a company, thus seeking profit and capital remuneration, together with the rational use of natural aspects with the minimization of impacts environmental practices carried out within the corporate environment, providing interaction between these parties.

That is the reason why so many companies have adopted positions and behaviors that promote the well-being of its internal and external public, contributing to a better society.

If sustainable development seeks to meet current needs without compromising the future and social responsibility it has to stimulate behaviors that contribute to a fair society, and it can be said that both have the same purpose, regarding their concerns with others, so we can insert, as a consequence of this relationship, the 17 goals of sustainable development defined by the United Nations that describe the relationship between sustainability and social responsibility that should be implemented in all countries by 2030.

Thus, the 17 objectives defined by the UN are related to different aspects of human society being possible to highlight reduction of poverty and hunger, adequate health education and water access, clean energy, decent work with good infrastructure and innovation in the industrial sector, less inequalities in sustainable cities and communities, responsible consumption and production, climate action and general partnerships to accomplish the goals.

However, little is seen in the theoretical researches as a link between corporate sustainability with corporate social responsibility and environmental management, because it still has as main purpose, actions that add value to the company, avoiding the main objective, which is the contribution that generates value for society in general, including of course the environment without initially aiming for profit, which should be a consequence and not a goal.

As a result, social responsibility must promote improvements to the community in which it is inserted in general, adding value in all aspects that involve a final product or service.

2.1 Concept

Sustainability has always been seen as human actions that do not seek to affect other generations, that is, it is directly related to economic issues without aggression to the environment. In addition, it searches for correct uses of natural resources without compromising its future. The intelligent use of these natural resources ensures that resources are constantly maintained so that there is no shortage of natural goods for future generations.

For Gallon et al. (2007), the commitment to the environmental issue has increased and acquired enormous proportions, no longer being a discussion only of environmentalists and technicians in the area.

This concern has grown more and more, because in the business world there are several requirements that involve its competitiveness. In this way it can be noticed when this concern arises in fact from the intention and concern on the part of the organizations, or simply by the fact that this concern is a necessity for its consumers.

According to Andrade and Tachizawa (2008, p. 1), sustainability is growing and its economic results depend on business decisions and point out that:

> There is no conflict between profitability and the socio-environmental issue; the sustainability movement is growing worldwide; clients and the community in general are increasingly valuing the adoption of socio-environmental practices by organizations; corporate billing is under pressure from consumer behavior that increasingly emphasizes its preference for environmentally sound products and organizations.

Sustainability has become essential, precisely because society itself has given preference to natural resources, increasingly taking into account the environmental issue in different situations and this is a tendency already experienced by some companies. Thus, "the consumer of the future, including in Brazil, will privilege not only price and quality of products, but mainly, the social behavior of companies that manufacture these products" (Andrade and Tachizawa 2008).

Another issue that organizations have worried about is certainly the economic consequences that may arise along environmental degradation and future environmental impacts.

Nowadays, some companies have already noticed some impacts, and as Smith (2017, p. 5) say:

> (…) several companies and governments have already verified that the environmental crisis exists and is indefinite, because it is not clear what their consequences will be in the long term. What we can say is that it brings us new uncertainties and new challenges, intrinsically linked to our development and survival.

As it is well known, this question is often unused, there is no basis for what influences they are dealing with when we talk about the natural environment and its direct consequences to the economy. Therefore, Barbieri (1997), analyzes the emergence in the agents, governments and three other stages in Environmental Perception, mentioned below:

- 1st Step: Localized Environmental Problems—Attributed to Ignorance and Neglect of Producers and Consumers.
- 2nd Step: Problems at the level of National States.
- 3rd Step: Perceived as a planetary problem.

The main difficulties related to environmental problems since the Industrial Revolution and how to solve them tend to be more visible. Since the rise of the capitalist economy, the indiscriminate and insufficient use of natural resources has grown abundantly, even though it recognizes the possibility of scarcity of these resources, and with this, the process of environmental destruction has not been altered and even the acceleration of consumption.

In addition, according to Barbieri (1997), the perception for solving global problems seeks not only to reduce degradation in the physical and biological environment, but also to solve social, political and cultural issues such as poverty and social exclusion and so reaching sustainable development.

Regarding Barbieri's (1997) citation, it is essential to social development that awareness is more present in human lives, especially the consumers and producers, as these are the main causes of environmental degradation, justified by the law of supply and demand. The abundant use of natural resources should not be part of companies if consumers do not agree with it, if there is conscious consumption, companies will automatically have to adapt to the requirements of their customers. For that reason, sustainable development must evolve.

Consequently, there may not initially be a perception of profitability for companies that have adapted to a green world, but they will certainly make more money because their customers trust them.

According to Castro (1996), companies that make a commitment to the environment, demonstrate confidence and bet on the future. In other words, they seek a new perspective in which environmental care ceases to be obstacles to the company's activity and becomes the guarantee that it will establish itself in the market with greater business opportunities.

According to Morandi and Gil (1999), the process of environmental management implies a continuous process of analysis formed of decision, organization and control of development activities, as well as an evaluation of the results to improve the formulation of policies and their implementation for the future.

This is precisely what sustainable development seeks to achieve. It is clear that it is necessary to raise awareness among all, including consumers, suppliers, companies, people in general, but it is the small attitudes that initiate green consciousness process and it is through this that environmental impacts such as global warming, deforestation and ecosystems destruction should be reduced and mostly avoided. This is the view that currently only the minority of companies have, but they can already feel the results in this new perspective.

2.2 *How the Organizations Are Incorporating the Concept of Environmental Management System (EMS)*

Environmental Management System (EMS) can be defined as a set of procedures to manage or administer an organization, in order to obtain the best relationship with environment (Nascimento 2008). Thus, it can be understood by an organizational structure that seeks to introduce balance between productive, economic and environ-

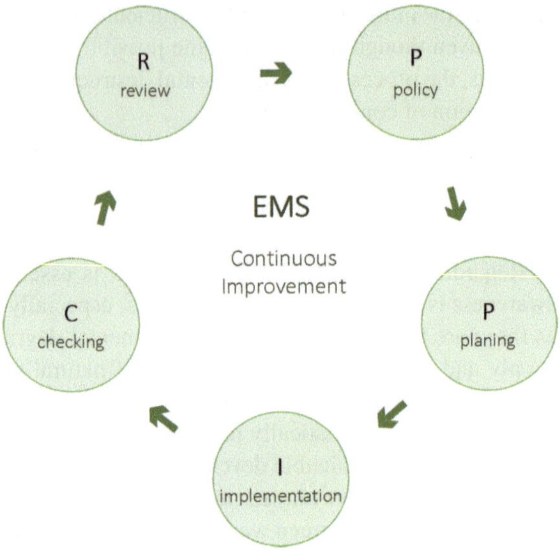

Fig. 1 EMS implementation cycle. *Source* The authors, 2018

mental activities, and is supported by the International Standard Organization (ISO), corresponding to a set of interrelated organizational policies, practices and procedures for an overall performance. Therefore, its main objective is to optimize the use of natural resources and minimize environmental impacts related to economic activities of each company. This system has been fundamental in the issue of environmental management, since it somehow obliges the implementing company to comply with environmental standards. Among the advantages, we can highlight compliance with environmental legislation, reduction of environmental impacts, reduction of costs, improvement of the company's image, increase of productivity, ease of reaching and entering foreign markets, among others. In accordance with Regulation (EC) 761/2001 on the EMS, there is a continuous improvement model that involves the accomplishment of five stages (Fig. 1):

- Environmental Policy—Ensure compliance with environmental standards, such as whether it is appropriate to nature, whether it has a commitment to the process of continuous improvement, compliance with legislation, etc.
- Planning—Identify the most significant environmental problems to start the project.
- Implementation and Functioning—The resources should be well defined, like any other quality management, develop a hierarchical structure, such as all well-trained professionals with their responsibilities directed and ready for emergency situations to put the plan into practice.
- Verification—Evaluate the results and always promote improvements as preventive and corrective measures (if necessary).
- Review of the Management—Analysis by the Management of the Organization to ensure its compliance with the EMS.

Once all the steps mentioned above have been completed, annual audits are required to ensure that the process is maintained and contains new improvements linked to it.

2.3 Sustainable Practices

For the implementation of an Environmental Management System, certain behaviors of the company towards the environment, that classify it or not as sustainable, are necessary. Therefore, environmental practices should not only indicate good manners, but also physical results for the environment and for society as a whole and not only for the economic growth of a company without aggression to the environment. However, one can cite some examples of sustainable practices, according to Laville (2009):

- Use of water treatment and reuse systems.
- Rational use of water and electricity.
- Recycling of solid waste.
- Reuse of leftover raw materials.
- Creation of educational projects aimed at preserving the environment.
- Adoption of projects that aim at the education and cultural development of the community in which the company is inserted.
- Use of recyclable materials for the packaging of products.
- Use of biodegradable bags (Supermarkets, for example).
- Use of filters that retain pollutants emitted at certain stages of industrial production.
- Do not dispose of sewage or chemical residues of any kind in rivers, streams or lakes.
- Do not pollute the soil with chemicals or any other polluting material.
- No use, under any circumstances, of child labor, forced or slave labor.
- Full compliance with the country's environmental laws.
- Do not adopt practices aimed at taking advantage of public competitions. The sustainable company should not, under any circumstances, adhere to corruption schemes. It is worth remembering that public resources diverted by corrupt people mean less investment in essential areas for the population (health, education, transport, leisure, etc.).
- When possible, use clean and renewable energy sources in the production processes.
- No use of discrimination (race, color, religion, sexual orientation, etc.) in employee selection processes. Use of fair forms, respecting the principles of equal rights in the selection process.
- Respect to the labor laws of the country, making the payment in a fair way and guaranteeing all the rights of the workers.
- Use of production practices that ensure the total safety of employees in the work environment.

- Production of goods and services that do not endanger the health and physical or psychological safety of consumers.
- Use of contracts with consumers and other companies that are clear, objective and fair.
- Providing an efficient customer service system (CSS).
- Inform consumers adequately about the characteristics of the products they sell or the services they provide. In addition, it is important that the company advises its consumers about the discard of packaging, products with expired validity or that will no longer be used for any other reason.
- Adoption of reverse environmental Logistics. This is to prevent certain products from being disposed of in the environment. Manufacturers of tires, batteries, medicines and other products that may pollute the environment should use this process.

Reflecting on the sustainable practices mentioned above, we can understand the great importance of its application to an improvement in environmental, social and economic quality.

It is also noted that the ultimate impact of all these practices is positive in relation to all internal and external organizational issues. The reduction and efficient use of natural resources, social commitment to employees, and adoption of ecological systems guarantee not only physical results in short term, but mainly in the long term, and the positive consequences are evident in the results for a company with all the benefits sustainable practices entail.

A significant aspect of the whole analysis is the changing consumer mindset that influences companies to adapt to new environmental standards and their concerns about the outcome of these practices across the social, economic, environmental and even health context.

2.3.1 Sector of Cosmetics and Environmental Policies in Brazil

The evaluation practices at diverse contexts of a local place allow to analyze the development of the actions practiced, which in brief to Dye (2008) means learning about the consequences of this policy. However, such evaluations are based on results, using them in real situations through comparisons of actual practices.

In addition, it is important to consider this same process for Environmental Policy decisions, so that all sustainable practices, whether social, environmental, or even economic, are analyzed.

The most important in the role of environmental policies is the need to induce economic actors to introduce environmental practices to reduce pollutants released into the environment and to avoid scarcity of non-renewable resources.

This introduction of sustainable practices was required to force organizations to engage in activities that do not harm the environment, such as their non-use, because only consumer interest did not induce all companies to adhere to sustainable practices, and it was necessary to use a greater power, such as environmental legislation, to introduce sustainable activity as a certain obligation.

Thus, Paavola (2001) believes in some aspects, such as, for example, that public interventions are more effective than individual opinions. Such public actions carry out environmental changes with lower costs than individual or small group actions and also generate faster results, since the guarantee of human well-being is emphasized. In this way, environmental policies help not only the environment and the economy, but also ecosystems in general, avoiding certain diseases caused by environmental degradation itself.

Thus, it is possible to relate several aspects and practices mentioned above in several productive sectors and service providers, and in this work specifically in large cosmetic industries existing in Brazil.

An example of good environmental practice is Natura, which was one of the United Nations (UN) award winners as a pioneer in sustainable production that created the option to sell its products in refill, reducing by 54% the use of regular packaging, in addition to making measurements on the intelligent use of the natural resources used in the manufacturing process.

Another example would be the company O Boticário that in its physical stores carries out besides the sale, the reverse logistics of its packaging that are donated to recycling cooperatives for the reutilization in several other productive processes. One innovation developed by the company is 3D skin technology, which simulates a human organ to perform tests of effectiveness and quality of its products.

In addition to environmental practices, it is possible to mention LUSH, a company that has stood out in the Brazilian market with its social practices that collaborate financially with diverse groups and NGOs that are aligned with its values. For that, it uses campaigns to market its products and at the end of it, the value is destined to the groups or NGOs that motivated the campaign. An example of such projects is Charity Pot, which conducts several campaigns throughout the year to offer assistance to various institutions worldwide that support social, environmental and animal causes.

All the companies mentioned above must comply with the Agência Nacional de Vigilância Sanitária (ANVISA), the Brazilian National Health Surveillance Agency, standards and regulations for commercialization, such as microbiological control parameters, registration procedures, list of substances that should not be used in manufacturing, etc.

2.3.2 Sector of Cosmetics and Environmental Policies in United Kingdom

It is clear that any country located in the European Union and operating in the European market must comply with the required environmental policies and standards. Besides these policies, there are also individual policies of each country that must be inserted in the daily life of these companies.

An existing standard in the United Kingdom is the annual obligation to disclose environmental practices. It was presented by the British Companies Act, three new requirements from 2006, such as respect for the environment by the managemen-t—Article 172; disclosure of environmental performance in the business review of each company—Article 417, item 5; use of environmental indicators to justify the environmental performance of each company—Article 417, item 6.

Such requirements have stimulated concern with environmental practices as part of the business and this has been evidenced by the environmental performance of several economic sectors, not only in England, but across Europe.

Prior to that, in the 1970s, the English Parliament passed the Equal Pay Act in 1970 and the Health and Safety Act at work in 1974. This has generated great repercussions as it has influenced activist groups and even the public on environmental issues that companies to treat the environment responsibly and show their actions through corporate reporting (Idowu 1989).

In order to strengthen certain practices mentioned above, two large green companies in the UK, which are examples of environmental practices, can be used as an example.

The first is the company THE BODY SHOP, founded by Anita Roddick, known for her real social concern. It pioneered the handling of natural beauty products and offered campaigns against animal testing and through a petition calling for a ban on the marketing of products tested on animals, it was the first international company to sign and be recognized by the Humane Cosmetic Standard.

LUSH, meanwhile, has been performing the same practices since its opening, since it was funded by THE BODY SHOP itself. It is known worldwide for the use of fruits, vegetables, among other organic and natural materials for the manipulation of its cosmetics, which is why it is 100% vegetarian, in addition to inserting naked products (free of packaging). It also offers products that are manufactured with their own plastic packaging as a way of disposing of their waste and through their fair market projects.

Both companies have great concern for socioenvironmental causes and annually carry out several campaigns to alleviate certain problems existing around the world. These are examples of what type of concern exists in the UK, and some of these practices are mandatory for both domestic and external market action.

3 A Case Study on Environmental Policies and Practices in a Cosmetics Industry in Brazil and in England

The case study presented involves an analysis of existing environmental policies in two countries: Brazil and England and which of them are duly practiced by the Cosmetics Company under analysis and also which practices could be implemented. In this way it will be possible to relate the difficulties encountered in maintaining the customs of a company in different countries knowing the need to prioritize their individual policies. The case study will be presented in three parts, they are: (1)

Presentation of the company, (2) Methodology and (3) Presentation and Analysis of Results.

3.1 Industry Presentation

Sustainable policies and practices are being increasingly adopted by companies that aim at competitive advantage and good economic performance.

The implementation process is not easy and not even fast. Planning is the main factor in properly adapting environmental practices and in order to avoid any dysfunction in this process. The company that is the object of this analysis is a Cosmetics Company, which has been in the market since the 1990s and has since been distinguished by its socioenvironmental practices throughout the world.

When it began its manufacture, it had in mind the production of cosmetics that could be handled in kitchens, with fresh materials that everyone knew of its origins, and since that time, the concern with the origin of the materials, was already part of the daily life of consumers. From then on, it began the development of differentiated products that increasingly came to be sought all over the world.

3.2 Methodology

Environmental experts say that in a short time companies tend not to survive without the sustainable use of the environment, since they relate to aggregate preservation to profit (Gestão 2014). That is why the importance of sustainable management even to collaborate in a strategic vision of the companies.

To compare the sustainable management of the Cosmetics Company in Brazil and in England, this study was realized from August 2015 to December 2017. Examples of sustainable practices cited by Laville (2009), mentioned before, were used as a research base.

In addition to comparing these sustainable practices, we analyzed the possible practices that could be used without compromising the Cosmetics Company in Brazil's environmental policies.

3.3 Presentation and Analysis of Results

After analyzing the Cosmetics Company, it can be said that several sustainable policies are practiced, involving social, economic and also environmental aspects.

In addition, it is possible to exclude two items from the sustainable practices mentioned by Laville (2009), since there is no need for their adoption, such as filters to contain the emission of pollutants emitted during the production phase—since the process is not industrialized, and therefore, there is no emission of pollutants. Another example is the practices that aim at competitive advantage in public competitions, considering that the Cosmetics Company does not participate in any hypothesis of the provision of public services.

In this way, we can cite the following practices adopted by the Company of Cosmetics:

- Rational use of water—By reducing the use of water in its products and consequently throughout the process and by the introduction of renewable energy.
- Use of recyclable materials in their packaging, including another item that involves the use of reverse logistics to reuse packaging.
- Use of recyclable bags—Bags and Packaging in general certificate by Forest *Stewardship Council* (FSC).
- Do not pollute the soil with chemical pollutants—the production leftovers and wastes are both incinerated.
- Non-use of slave or unfair labor—Main adoption, as it develops small producers of certain materials used in the process and manufacturing to insert them into the market with fair payments.
- Total respect for Brazil's environmental laws—Main factor in which the Cosmetics Company adopts, not making use of all brand policies.
- No use of discrimination—It can be noted by the diversification of manufacturing and retail employees mainly.
- Full respect of labor laws.
- Handler Safety.
- Manipulation of cosmetics that do not pose risks to the health and physical or psychological safety of the consumers.
- Use of contracts with consumers and other companies that are clear, objective and fair. These are performed annually.
- Providing an efficient customer service system (CSS).
- Inform consumers about the characteristics of the products that are sold and advice on the disposal of the packaging.
- Adoption of reverse environmental logistics.
- Creation of educational projects aimed at the preservation of the environment—The most concrete example of this practice is the search for producers using permaculture—an agricultural methodology that studies various functions of species and how one complements the other, in the sense that what one discards or does, can be used by another for its own evolution, without any soil degradation.

These practices are related to the main permaculture goals analyzed by Holmegren (2013) that include three main actions like earth care, fair share and people care. These three actions can be inserted with observe and interact, catch and store energy, obtain a yield, apply self-regulation and accept feedback, use and value renewable resources and services, produce no waste, design from patterns to details, integrate rather than segregate, use small and slow solutions, use and value diversity, use edges and value the marginal and creatively use and respond to change.

In addition to these practices already adopted by the Company, it is possible to develop several other practices that often do not generate any cost, such as:

- Adoption of projects that aim at the education and cultural development of the community in which the company is inserted.
- Use of water treatment and reuse systems.
- Reuse of raw material leftovers.

The remaining practices could be adopted, by means of the disposition and necessity that a company has to insert them in its day to day, and in view of the nonexistence of these services in the city in which the Cosmetic Company is inserted:

- Recycling of solid waste.
- Do not dispose of sewage or chemical residues of any kind in rivers, streams or lakes.
- When possible, use clean and renewable energy sources in the production processes.

This analysis will vary from country to country. It is necessary that each subsidiary of the Cosmetics Company first meets the laws of the country installed and subsequently go to suit the policies of the brand.

Therefore, it is possible to cite a very simple example of difficulty that other units of the Cosmetics Company have, but which indicate exactly the difficulty existing to apply the same existing norms of its Head office. This is the case with packaging. In the head office it is possible to identify that in its stores, all naked products are exposed on the shelves without any packaging and after the purchase of the products, the customer can choose to take them inside bags or not, which is precisely to justify the type of product. In Brazil, the situation is different, however many products without packaging are exposed throughout the store, no customer can choose to take it with or without packaging, because the Cosmetics Company in Brazil, is obliged, following ANVISA regulations, to issue labels which contains lot information, manufacturing, etc. Thus, in Brazil there was a need to assign such practice to meet the country's environmental and health policies that are not the policies of the company itself.

Another example is the existence of a sector specialized in socioenvironmental projects in England that does not exist in Brazil, which takes care exactly of the destination of some materials used, for example—plastic bottles that are used to make handkerchiefs; reuse of resources that are unusable by other companies, such as demolition timbers—which are used to manufacture the furniture of the Cosmetics Company; in addition to projects that seek to help environmental causes around the world.

In addition to these, we can cite another sustainable practice that is carried out by the Cosmetics Company, which deals with the annual revaluation of the consumption of natural resources so that there is no lack or end of these resources, and a very important example of this reevaluation is the search for the reduction of use of water throughout the process. Only with the development of products with low concentration of water and in solid versions, the use of 450 thousand liters of water is avoided.

All the practices mentioned above are of fundamental importance for the existence of the company, in this way, there must be interaction and positive impacts with the environment, therefore, ecological issues, are the heart of the company and through these, many actions are carried out by the Cosmetics Company daily.

Given this, it is possible to identify practices that are performed in the Cosmetics Company in England and that for some specific reason, are not yet performed in Brazil, such as:

- Use 100% of organic and non-tested animal cleaning products in Retail

This practice is used throughout the retail of England, but not yet in Brazil. This has been under development and adaptation in Brazil, including in the manufacturing units. There are use of organic cleaning products, but there are still exceptions.

- Own power generation

This practice is still under development in England, with the main objective of reducing the use of electric energy. In Brazil, there is still no green project to reduce and generate electricity for the company own consumption.

- Recycling of marketed pots

In England, there is an internal recycling center of the Cosmetics Company, which is responsible for the reverse logistics process of marketed pots, after the sale, that are returned to the stores and factory (through the promotion, that every 4 pots returned, you get a fresh mask—facial cleanser) and then they are reshaped and reprocessed into new pots. Unfortunately in Brazil there is only the existence of the promotion as a way of not discarding recyclable products in the common trash, but not yet carried out the reverse logistics of pots, considering that they are still not produced in Brazil, but imported from England.

- Recycling Service for retail and office

The Cosmetics Company in England seeks to use the best recycling service for retail and office waste. In addition, it carries out trainings with the employees so that the separation is done correctly. This situation is different in Brazil, where the recycling service is public and there is a great deficiency in the process.

4 Final Considerations

Increasingly in recent years, companies have been pressured to take actions that mitigate the environmental impacts it has caused in their overall processes. Each has a motivational factor that will imply, in practice or not, in alternative and sustainable issues in its processes. In addition, it is noticed that quality certifications have been increasingly sought by companies that seek to adapt to the environment and have already noticed the importance of this issue in several aspects, in addition to the environmental. Not only for obtaining certification, but also for offering everyone involved in the process a better quality of life through the good practices acquired by companies.

For companies, adopting Socioenvironmental Management has become paramount, because society gives preference and requires the adoption of sustainable resources and actions. "The consumer of the future, including in Brazil, will privilege not only price and quality of products, but mainly, the social behavior of the companies that manufacture these products" (Andrade and Tachizawa 2008, p. 5).

In this way, the case study carried out at Cosmetics Company highlights the differences in environmental policies in England and Brazil, with a comparison between two subsidiaries of the Company.

For this, we analyzed the manufacturing process and existing sales in the Cosmetics Company in both countries. After the analysis, it was possible to identify sustainable practices that exist in England and that have not yet been inserted in the process of Brazil and with that, it is indicated what activities Brazil should put in practice to equal the English management.

This paper intends to present and to contribute to future researches, with a focus on the evolution and use of environmental policies in Brazil, using a comparison between Brazil and England and how the internal policies of each country influence the actions of the organizations. So, it is possible to identify certain rules that exist in Brazil that initially stand against the image that the Cosmetics Company intends to offer, such as use of after sale packaging and that contain information regarding validity and manufacturing, which is not carried out in England, only if it is at the customer's request and rules that don't exist in Brazil but are used in England.

There is a clear concern about the environment in both countries, for adopting various global strategies to achieve this goal. But it is possible to identify several improvements in the process so that there is further environmental preservation.

Among them, it is possible to mention sustainable practices that could be inserted in the Brazilian process and that would not violate legal environmental policies, such as:

- Reuse of natural resources
- Solid waste recycling
- Adequate disposal of waste
- Use of clean energy
- Use 100% of indirect organic materials.

Such practices are difficult to execute, since they require a great investment by the company and also training for correct use, but its adoption would be a great differential.

Thus, the essence of the brand, which is related with its environmental, ethical and moral values, will prevail in a good relationship management between employees, consumers and suppliers.

References

Andrade ROB, Tachizawa T (2008) Gestão Socioambiental: estratégias na nova era da sustentabilidade. Campus, Rio de Janeiro

Associação Brasileira de Normas Técnicas—ABNT (2010) NBR ISO 26000: guidance on social responsibility. Retrieved from http://www.abntcatalogo.com.br/norma.aspx?ID=80716. Accessed on 13 March 2018

Barbieri JC (1997) Desenvolvimento e meio ambiente: as estratégias de mudança da Agenda 21. Vozes, Petrópolis

Castro (Coord) N (1996) A questão ambiental: o que todo empresário precisa saber. SEBRAE, Brasília

Dye T (2008) Understanding public policy, 12th edn. Person Education, Upper Saddle River

Gallon A et al (2007) Produção científica e perspectivas teóricas da área ambiental: um levantamento a partir dos artigos publicados em congressos e periódicos nacionais da área de contabilidade e administração. In: Congresso Usp de Contabilidade e Controladoria, vol 7. USP, São Paulo

Gestão ambiental mais consolidada (2014) Revista Análise Gestão Ambiental. São Paulo, p 51

Holmegren D (2013) Permacultura: princípios e caminhos além da sustentabilidade. Via Sapiens, Porto Alegre, p 416

Idowu SO (1989) Accountable to society. Pass Magazine, November: 14–15

Laville E (2009) A Empresa Verde, 1st edn. Õte, São Paulo

Lush Fresh Handmade Cosmetics Web Page (2017) Retrieved from https://br.lush.com/artigo/um-pouco-sobre-o-nosso-charity-pot, https://uk.lush.com/tag/our-policies. Accessed on 3 May 2017

Morandi S, Gil IC (1999) Tecnologia e meio ambiente. Copidart Editora, São Paulo

Nascimento LF (2008) Gestão Ambiental e Sustentabilidade. Sistema Universidade Aberta do Brasil, Florianópolis

Paavola J (2001) Economics, ethics and green consumerism. In: Cohen M, Murphy J (eds) Exploring sustainable consumption: environmental policy ant the social sciences. Elsevier, Oxford, pp 79–96

Santos LP, Wagner R (2008) Gestão estratégica de pessoas no contexto de demanda por Sustentabilidade. ENANPAD, Rio de Janeiro

Smith SBM (2017) O Consumidor Ambientalmente Responsável. Retrieved from www.communita.com.br/pdf/consumidorambientalmenteresponsavel.pdf. Accessed on 3 April 2017

Amanda Espósito Borges is concluding a Business Administration Graduation Course at Atibaia College (FAAT), Atibaia, São Paulo, Brazil. Currently is Material Planning Analyst in a chemical Industry.

Lucas Ferrari Pinto dos Santos is concluding a Business Administration Graduation Course at Atibaia College (FAAT), Atibaia, São Paulo, Brazil. Currently is Tax Assistant in a construction company.

João Luiz de Moraes Hoefel has a Ph.D. in Social Sciences at Campinas State University (IFCH/UNICAMP), São Paulo, Brazil, with a focus on environmental issues and developed a Post-Doctoral research at the Center for Environmental Education and Public Policies at São Paulo University (ESALQ/USP). He teaches at Atibaia College (FAAT), Atibaia, São Paulo, where he also coordinates the Center for Sustainability and Cultural Studies (NESC/FAAT) developing research projects on Natural Resources Conservation, Environmental Education and Environmental Planning, and at University of São Francisco, Bragança Paulista. He also develops research activities at NEPAM/UNICAMP on Global Environmental Changes and Quality of Life as a Collaborator Researcher.

The Legacy Café—A Trial of Intergenerational and Sustainable Learning in an Early Childhood Centre in Liverpool

Diane Boyd

Abstract Early childhood is a transformative period where attitudes and foundations for life are laid (Siraj-Blatchford et al. Education for sustainable development in the early years, 2010). The principles of education for sustainability reflect a holistic and interconnected approach, similar to the ecological context of early childhood. This interconnectedness is further highlighted with the three pillars of sustainability (economic, environmental and socio/cultural), that they are all interrelated (Brundtland, Our common future: the world commission on environment and development. Oxford University Press, Oxford, 1987). The ecological context centres the child in their own contextual and cultural environment. Bronfenbrenner (The ecology of human development: Experiments by nature and design. Harvard University Press, Cambridge, 1979) noted a key element of this environment or community was its "a dynamic entity which is constantly changing" (Keenan and Evans, An introduction to child development. Sage, London, 2010: 35) reflecting flexibility and the bi-relational aspect when the child not only interacts with their environment, but influences it too. The Legacy Intergenerational sustainable skill café is a socially cultural integrated model, bringing generations within communities together, building a more sustainable society, a "community of practice" (Lave and Wenger, Communities of practices: learning, meaning and identity. Cambridge University Press, Cambridge, 1998) researching through a "place of possibilities" (Dahlberg and Moss, Dialogue with Reggio Emilia. Listening, researching and learning. Routledge, London, 2006, p. 12). The elderly willingly share cultural traditions with families and children in disappearing or lost skills that are being 'divorced' (Langlands, Craft: an inquiry into the origins and true meaning of traditional crafts. W.W Norton and Company, New York, 2018) from our identity or cultural heritage, reflecting a "collective responsibility" (Dahlberg and Moss, Dialogue with Reggio Emilia. Listening, researching and learning. Routledge, London, 2006: 10) validating the position of the family as a socialising agent' (Mbebeb 2009, p. 25).

D. Boyd (✉)
School of Education, Liverpool John Moores University, Liverpool, UK
e-mail: d.j.boyd@ljmu.ac.uk

W. Leal Filho (ed.), *Social Responsibility and Sustainability*, World Sustainability Series, https://doi.org/10.1007/978-3-030-03562-4_19

1 Introduction

Within the Brundtland Report (1987, p. 9) there is an urgent collective call to "all citizen groups" to work towards "furthering the common understanding and common spirit of responsibility so clearly needed in a divided world." This message of social responsibility is aimed at amongst others, educational institutes and early childhood has a responsibility to act.

Seedsman (2013, p. 355) highlights that "major economic, social, cultural, political and demographic changes have altered and transformed traditional family, community and social structures". Society is seemingly more fractured with less emphasis on civic participation, community cohesion and the retention of culturally specific skills and traditions. Additionally, with a rapidly increasing elderly population whilst the birth rate is decreasing, Weckström et al. (2017) also notes the deterioration of communication between generations in nuclear families. The Cambridge Primary Review (2014) noted "family life and community are breaking down" and as a consequence "that respect and empathy both within and between generations" was also in decline. Early childhood is a time when fundamental attitudes are formed, emphasising the importance of good role models and sustainable behaviours. Children learn through observation, by direct experiences of collaboration, participation and through books and the media (Bandura 1977). Research, shows that children are "capable of sophisticated thinking in relation to socio-environmental issues and that the earlier ESD ideas are introduced the greater their impact and influence can be" (Siraj-Blatchford et al. 2010, p. 6). The Intergenerational legacy café demonstrates a socially cohesive community that endeavours to address this social imbalance and help communities to become more sustainable, whilst honouring their traditions and cultures. It is as Dahlberg and Moss (2006, p. 20) note "it offers a sense of belonging" within a "relationship of hope" to move communities down a more sustainable and socially responsible path.

2 Early Childhood

Within early childhood, children learn through socially constructing and reconstructing their worlds through experience and observation. Vygotsky (1978) stressed the importance of both cultural and historical tools that enable children to master and participate freely in their world. These tools reflect the culture and traditions of the history and community into which the child is embedded. Lave and Wenger (1998) highlighted the importance of developing 'communities of practice' and early childhood provides such contexts for opportunities that bind children, their families and communities together. Smidt (2009, p. 29) states this binding could be "their values, their religion, their feelings, their concerns", any principles and behaviours that have been developed through interactions. A 'community of practice' must reflect diversity and difference recognising that each child (and family) retains their identity, but bringing their own individual uniqueness, history, needs, hopes and cultural habitus. Within Early Childhood key pioneering giants have influenced aspects of current practice but elements of their original thinking and pedagogy needs to be retained and highlighted, as Froebel, Steiner and Montessori all stressed the importance of community, inclusion, family and peace (Boyd 2018). This 'community of practice' should utilise a "guided participation" (Rogoff 1996) approach where "individuals transform their understanding of and responsibility for activities through their own participation in activities and in the process they become prepared to engage in similar subsequent activities." The contributors to this process of meaning making, reflect the principles of Reggio Emilia when early childhood children are in "continuous encounters with others and the world, and the child and the teacher are understood as co-constructors of knowledge and culture" (Rinaldi 2006, p. 6). Within Reggio Emilia there is an assumption that the child, adult and environment are interconnected and an equal co-constructor. Rogoff (2014, p. 124) also stressed a mutual interconnection stating that when the child is in the process of thinking that it is not a separate entity from "cultural functioning".

Theoretically the community and locality surrounding the child and their families provides a social/cultural environment where knowledge is seen as a "process of meaning making" (Rinaldi 2006, p. 6). Frederick Froebel is universally recognised as the originator of the 'kindergarten' where children are active participants in their own learning, an environment that is "an agent for learning" (Liebschner 1993, p. 43). Froebel highlighting agency was innovative at this time, moving away from children just playing as a purely passive occupation. He developed a pedagogical system where children understood moral values, sustainable living and a deep rooted awareness of the four seasons, believing that children "benefitted from being exposed to a wider community" (Joyce 2012, p. 52). He stressed that the environment must be a place where "social and moral values could be experienced by children" (Liebschner 1993, p. 49) resonating with Loris Malaguzzi's view of the environment as a 'third teacher'. A key principle of Steiner Waldorf curriculum also recognises and encourages all to "enable and value the *contribution* of individuals, groups and communities to the improvement of our common human culture" (Avison and Rawson 2016, p. 16). This is also true of Montessori's prepared environment in which the child "absorbs knowledge … simply by living." (Standing 1998, p. 263).

3 Education for Sustainability

Davis (2015, p. 9) suggests that a "popularised description" of education for sustainability is meeting "the needs of the present generation without compromising the ability of future generations to meet their own needs," (Brundtland Report 1987, p. 8). In 2015 following the Millennium goals (2000–2015) in Paris countries from around the world signed a declaration to try to work towards a more sustainable future. There were seventeen sustainable development goals (SDG) highlighted that over the next fifteen years Governments and non-Governments organisations (NGO) must strive to achieve. One of these goals (SDG 4) specifically focuses upon education, emphasising equality of opportunity and quality practice. This goal highlights the importance of "all learners" acquiring the necessary knowledge and skills to campaign, for human rights, peace, gender equality and an "appreciation of cultural diversity and of culture's contribution to sustainable development" (UNDES 2015, p. 17). Bourn et al (2016) highlighted that although children are aware of 'global issues' it is mainly as a consequence of social media and digital images, and rather than engaging in actual experiential learning about global issues, whilst Hunt (2012) questions exactly how globally aware or literate they actually are.

In 1992 at the Earth Summit in Rio the terminology of three interconnecting and influential pillars—Social/cultural, economic and environmental was introduced in regard to education for sustainability. It is important to acknowledge that these three pillars are "mutually reinforcing" (Siraj-Blatchford et al. 2010, p. 5) and practice and policies developed to support ESD are weakened if they are considered in isolation. The Environmental pillar focuses upon aspects of current lifestyles and how they are affecting the environment, such as climate change, pollution and overflowing land fill sites. Early Childhood has always been closely associated with an understanding of and utilising the environment pedagogically, Davis (2009) however identified a "research hole" around early childhood education for sustainability. Her research noted that although there was evidence of both practice and research focusing on children 'in' and 'about' the environment there was very little emphasis on children as agents of change (education *for* the environment). She concluded that early childhood has something to contribute in the "development of human capacities that underpin" learning how to be sustainable (2009, p. 239) emphasising the need for more sustainable projects in early childhood communities.

The Social cultural pillar emphasises an ethos of diversity, compassion, inclusion, equity amongst genders and a peaceful just world, which are also the cornerstones of the pioneering ideas of early childhood. In early childhood behaviours, attitudes and an awareness of the world are formed and therefore a greater emphasis must be placed upon quality education through role models and advocating an anti-bias approach to the world. Vygotsky (1978) emphasised the importance of learning through socialisation and that children became part of the culture through their interaction with the cultural tools associated with it. When considering the influence of Paulo Freire, Smidt (2014) encourages practitioners to follow his lead and ensure children learn to 'read their world' with all the cultures, histories and stories attached to it, making sense of it and everything in it. The Social/cultural pillar also underpins all aspects of the Universal Declaration on Cultural Diversity, (UNESCO 2002) and Article 7 highlights a key aspect relevant to the context of the legacy cafes, stating;

> Creation draws on the roots of cultural tradition, but flourishes in contact with other cultures. For this reason, heritage in all its forms must be preserved, enhanced and handed on to future generations as a record of human experience and aspirations, so as to foster creativity in all its diversity and to inspire genuine dialogue among cultures.

This focus on heritage and culture seems absent from the English EYFS (DfE 2017: 9) with a more generic definitions of awareness of "similarities and differences between themselves and others, and amongst families, communities and traditions," which is open to interpretation. This is in comparison to the other three curriculum of the home countries of the United Kingdom, which all have clear and specific geographic and cultural references to a positive identity (Boyd et al. 2018). Langlands (2018) also highlights the need to recognise and celebrate our culture and likens what he perceives as a loss of traditional knowledge, to a 'divorce' from our identity or cultural heritage. These shared crafts or skills include knowing how to cook traditional recipes, how to sew a button on a shirt and simple knitting. Kuttner (2015, p. 75) asks what individuals in our society are "taught to value" culturally, reflecting that the EYFS (DfE 2017: 4) has no explicit mention of traditional crafts or skills rather a focus on ensuring they are "ready for school" with emphasis on more formal methods of learning (Ofsted 2017). Mbebeb (2009, p. 25) argues that education should be "part and parcel of culture", validating "the position of the family as a socialising agent" and the legacy café offers an opportunity to empower families and communities together.

The Economic pillar is seen as the least understood or recognised in practice (Siraj-Blatchford et al. 2010, p. 25) describing practitioners awareness of it as "extremely weak". The three pillars are interconnected and Robins and Roberts (1998) argue that all practice should reflect an awareness of these connections. Mbebeb (2009, p. 26) further stresses the importance of developing children with creativity and entrepreneurial mind-sets whilst highlighting that when schooling attitudes changed towards a "school-like cognitive competence" there came a disconnection from family values and cultural values. Loris Malaguzzi notes the same in his 100 languages of children poem noting that schools steal ninety-nine of the ways of being, seeing and doing, with the emphasis solely on the head (Malagguzzi n.d) with Moss (2017) arguing that this style of education makes teachers "technicians." Dewey (1916, para 6) highlights instead the necessity of educational "aims and habits of the social group have to be rendered cognizant of them," as without them "the group will cease its characteristic life." Nsamenang (2007) suggests this leads to self-regulated and responsible individuals, resonating with Froebel's original idea of agency of children, key components of ESD.

4 Intergenerational Learning

The Brundtland report (1987) also stressed the need to "look across cultural and historical barriers" and "reach the minds and hearts of people young and old" (1987, p. 8) which reflects aspects of the New Zealand early childhood curriculum (1996), likened to a mat with all elements of generations of family, community and society woven together, ensuring a "reciprocity of benefit" to all Nicholls (2004, p. 27). Dewey (1916) suggests that "communities and social groups sustains itself through continuous renewal, and that this renewal takes place by means of the educational growth of the immature members of the group". Seedsman (2013, p. 350) suggests that mixed intergenerational learning allows for "the coordination and engagement of diverse community groups... through linking, joining, interacting and sharing".

Wyness (2015, p. 278) explores this community of practice (Lave and Wenger 1998) further with attention focused on the communication and dialogic skills necessary to build such a community, arguing for "*active* citizens" with an emphasis on "authentic participation". Arnstein (1969, p. 2) critiques this terminology however, stating that "there is a critical difference between going through the empty ritual of participation and having the real power needed to affect the outcome of the process," resonating with Shier's (2001) model of participation used in education. There is much discussion regarding the rights of the child (United Nations 1991) but less so on the rights of the Elderly. In 1991 the General Assembly recognised the "contribution that older persons make to their societies" with particular emphasis placed on opportunities "for willing and capable older persons to participate in and contribute to the ongoing activities of society". The Article highlights significantly that the elderly must "share their knowledge and skills with younger generations" resonating with Ballantyne et al. (2006) who suggested a 'unidirectional' approach where the elderly mentor and teach traditional skills and traditions to the younger children. However, Davis (2015) states children are capable of making choices and participation, and influencing their family and communities too, reflecting a more bi-relational (Vygotsky 1978) approach.

There can be tensions in terms of intergenerational communication with Giles and Coupland (1991, p. 159) suggesting it involves "different internally differentiated cultural groups, who possess different values and beliefs about talk, different social and existential agendas, and different language codes". Their research around 'communication accommodation theory' (CAT) noted that the elderly tended to not make any accommodations for the young, whilst the young tended to over accommodate. Intergenerational community learning, such as the legacy café has a bi-relational approach, which encourages mutual respect and informality, with a focus on dialogue through meaningful work.

Intergenerational learning is a key component of the Te Whariki (1996) curriculum where the family and community is the central tenet of the relational practice. According to Nicholls (2004, p. 27) this learning environment enables opportunities "for each new generation to be nurtured and to grow in knowledge of the past" of traditions, skills and their culture, developing "a strong sense of identity." Whilst the Te Whariki (1996) has its roots firmly in New Zealand culture and history, the legacy café model offers potential opportunities for the same type of intergenerational learning. The elderly are demonstrating key sustainable traditions and practices which have meaning and purpose to them, their history and culture, as they are specific to each and every cultural context. Graue and Walsh (1995, p. 148) call this learning 'mediated learning' as it is "located within specific cultural and historical practices and time." Additionally, this intergenerational learning draws upon the principles of Reggio Emilia which "offers a sense of belonging to people, longing for other values, relationships and ways of living" (Dahlberg and Moss 2006, p. 19).

Relationships are an integral part of intergenerational learning which develop organically through sharing of activities and skills, extended meaningful conversations and the feeling of belonging, through the cultural connections. Generation United (2007) evaluated intergenerational learning and noted that it had many values socially. It highlighted the enhancement of productive and engaged elders, with increased health and functional outcomes, with an emphasis on improved health for all participants. As a consequence of informal intergenerational learning, the Legacy café recognises and builds responsive and reciprocal relationships which provide opportunities for connections between generations. Bostrom (2003, p. 5) also reminds practitioners that learning skills such as sewing, knitting, mending punctures, can be facilitated by "people not trained, paid or acknowledged as teachers" reflecting a cohesion within a community of learners. This resonates with the Sustainable development goal (4.7) (UNESCO 2015–30) which advocates for "all learners acquire knowledge and skills needed to promote sustainable development".

5 Context

The trial was for a six month period, starting in January 2018 until June 2018 running once a month. It was held in the public area of the children's centre to ensure public liability insurance covered all aspects of the trial. Full ethical approval was sought and granted by the university before commencement and all participants consented and agreement was gained at the onset.

The legacy Intergenerational Sustainable Skill café trial was situated in Everton a socio-economically deprived area of Liverpool, in the North West of the United Kingdom. The children consisted of two groups,—three–four year old children from the nursery at Everton with their key worker Julie and the second group consisted of two–three year olds that came from a private day nursery that rented space within Everton. The same children attended the café over the 6 months to ensure consistently. The elderly participants came from within the Liverpool 6 district of Everton and were brought and fro from to the centre on a mini bus.

The activities included cooking and baking from scratch using local produce, traditional meals (for example, scouse or short bread), sewing activities—threading, weaving, leading to using either binca or felt pieces, knitting with children's needles, mending punctures, wheels on bicycles and scooters. All these activities could be deemed to potentially be a lost art or skill. Each child participated in both a cooking and sewing type activity at each café session visited.

6 Methodology

The theoretical underpinning of the research is socio-constructivism with a strong Reggio Emilian approach. The trial being located in a children's centre resonates with the idea of their concept of school, as "a 'participatory' public event, as an ongoing process." (Rinaldi 2006: 146). Within this idea of co-constructing knowledge from within the children's centre itself, was the understanding that within this 'place' "culture is not only transmitted but produced." Robinson and Vaealiki (2015,

p. 104) articulates the importance of ethical considerations within the (research) group, are based upon "collective understandings and/or cultural wisdoms." There was no attempt to privilege one culture over another in the trial, with ongoing dialogues around suggestions of activities or content.

During the trial there was a variety of participants that also reflected the theoretical aspect of Reggio Emilia. There was a genuine co-operation between all layers of staff participating within the trial resonating with their principles of "working *together*, indeed, *being* together, is deeply rooted in everything that is Reggio," (Abbott and Nutbrown 2007, p. 6) and Rinaldi (2006) reminds us that Malaguzzi recognised that the "teachers professional development is a cultural education."

The trial was situated within the community public space likened to the '*piazza*' familiar within both the city and school of Reggio Emilia with access to the outside communal area reflecting "a living pattern" idea of interaction, as described by Alexander (1979, p. 10). Over the course of the trial the 'piazza' reflects time, place, days of the week, history and reinforce through these interactions of both animate and non- animate, the sense of identity and community.

The research reflects a socio constructivist approach with knowledge being "co-constructed, in relationships with others" (Moss 2007, p. 128) but crucially recognising that the child is an equal, and a "researcher actively seeking to make meaning of the world," (2007, p. 129). This 'equality' reflects Vygotsky's view of an "interactive reciprocal nature" or bi-relational approach to learning as all participants have potentially a part to play within this process making (Gray and MacBlain 2015, p. 98). Within the process, however, there were opportunities for the elderly to share and model their skills, with an 'enabling' approach with the young child being supported through the process by a 'knowledgeable other'. This was apparent with the differentiation of task and support offered during the trial.

The research had an ethnographic flavour with the researcher and participants mingling freely and openly during the trial. Brewer (2000, p. 6) notes that this type of research will "capture social meanings and ordinary activities" in an attempt to understand how the culture works. The actual methods chosen for this trial were participant observation through notes and photographic evidence. Lutz (1986, p. 108) likens this approach as "participant observation of a society or culture through a complete cycle of events that regularly occur as that society interacts with its environment." Limitations of this are that there is potential for missed opportunities during the café sessions or a more biased focus from the one researcher recording. This café is also only a small representation of one children's centre within Liverpool, and therefore reflects this cultural context only.

At the completion of the trial an informal group chat (interview) with the participants was included to ascertain their feelings about the trial. The group element maintained a "community of practice" feel and it stressed a 'group' characteristic, rather than individual. Denscombe (2014, p. 188) suggests this is "more illuminating" as it offers participants a chance to "listen to alternative points of view" trading on the group dynamics. As a consequence of informal conversations, observations and the group chat, several themes emerged as a result which will be examined in the findings and discussion.

7 Findings and Discussion

The action research cycle consists of both action and analysis (reflection) and looking for particular patterns or codes. Paige-Smith and Craft (2008, p. 15) remind practitioners of the "ongoing complexity" of reflection embedded into early childhood contexts. Schön (1983) highlights both reflection 'in' the action, (what happened during the trial) as a consequence of observations and informal conversations, but also reflection 'on' the process, identifying through dialogue and consultation, different themes or codes. MacNaughton and Hughes (2009, p. 174) suggest that codes are just "simply a label that you apply to your data" and that this can "simplify and standardise." But through analysing the codes and in reflection and with subsequent discussion, several codes or themes emerged as a result of the data analysis.

8 Intergenerational Learning

O'Brien (2013, p. 8) highlights the need for a new educational model that is not just preparation for "conventional employment" but instead arguing for a "vision that contributes to well- being for all, forever" (2013, p. 9), coining the term "sustainable happiness" (2013, p. 14). This 'sustainable happiness' was apparent in how all those participating had only positive comments to make.

> I was just saying how fab this is – this little boy is not usually confident, but working with this old lady Mary, he's really confident and come out of his shell. (Practitioner)
>
> I loved being with the children. I loved helping, talking to them and getting to know their names and learning how to knit. (Elderly lady)

Intergenerational learning bridges generations and by being bi-relational in practice, offers opportunities for all participants to contribute. Duvall and Zint (2007, p. 15) recognised that children can promote effective intergenerational learning, as well as "consumer choices made by parents". Observing the children enthusiastically engaged in preparation, cooking and eating 'new' recipes impacted upon some parents.

> It's laziness on my part. I suppose that I can give it a go now – cook some chicken, buy the veg with him, the whole thing, rather than just give him hot dogs, noodles and nuggets. (Dad)

Kenner et al. (2007, p. 220) highlighted the "transmission of knowledge or prolepsis between generations" as a consequence of intergenerational learning, recognising that "culture is seen as understandings shared between members of a community jointly created through shared practice." (2007, p 22).

> Because I want to make my own clothes, design them and I have come here to learn those skills (Mum)
>
> It's been really good. I have learnt how to thread the needle using a needle threader (Mum)

The community feeling was apparent in the atmosphere created together, bringing different generations and cultures together with one agreed specific aim—to be more sustainable together.

> It's nice to see them mixed together- talking to them and learning about their experiences (Children centre practitioner)
>
> It's important for all this- it's a life skill to be able to mix (Practitioner)

9 Sustainability

O'Brien (2013) challenges the idea that happiness is about material wealth or material consumerism. Economically there is a need for an awareness of how consumerism and the throwaway society it creates, impacts on the environment often being entangled with an idea of being happy. Kasser (2006, p. 200) cautions that society may confuse the "path to the 'good life' as the 'goods' life". This is a time of fast fashion and dangerous levels of plastic rubbish polluting the oceans and threatening our planet. This was apparent in comments noted during the trial.

> They can't be bothered to sew. They just throw away stuff. (Elderly lady)
>
> I sew buttons on but I don't think people do that anymore do they? (Mum)
>
> Parents could not cut a real pineapple and had to be shown, as usually they pay more for a tinned one which is expensive (Student).
>
> What an amazing idea! It's really difficult to get past school admin officers. Sent seventy seven emails and had two responses. Here we feel involved. (Uniform exchange)

Vaillant (2002) highlighted the true meaning of how intergenerational learning can support and ensure traditional skills and values are passed down through generations and therefore contribute to preserving the planets ecological systems, describing them as 'keepers of the meaning'. Resonating with the principles of the UN rights for the Elderly (1991) Vanderven (2004, p. 85) suggests this approach "preserves what is meaningful from ones past life".

> I want Jamie to grow up doing these old fashioned things. (Mum)
>
> Look! Look at them! They are fixing our teddy. She is sewing but I am threading (4 year old child)
>
> They cook in the nursery but they have never seen anyone sewing or knitting before (practitioner)
>
> A necessary art (Elderly lady)

Another dimension of the social pillar is an awareness of equality, diversity and challenging gender stereotypes which was needed to challenge role assumptions of gender- mums only sew, cook and bake, and dads build and mend bicycles. This became a crucial element of the trial from the onset.

> Look at all the Grandma's sewing! Grandads don't sew but Grandma's do! (4 year old child)

The little girl was later challenged cognitively when she saw a dad knitting with a furrowed brow evident of her perplexity, with even the dads recognising it was a gender challenge.

> Men think it's a women's role to sew and knit. (Dad knitting)

10 Health and Well-Being Through Reciprocal Relationships

Vanderven (2004, p. 78) proposes that Intergenerational learning "can help build and sustain a true caring community among generations who are in danger of being harmfully separated." An Age UK Report (2015, p. 11) highlighted that the elderly "valued the ability to talk, listen and share information with another human being… provid(ing) a sense of belonging". This resonates with Helliwell et al. (2018) which recognised that happiness has its roots firmly in social foundations and family connections. This was clearly demonstrated with the wide ranging comments around personal feelings and worth as a consequence of attending the legacy cafes.

> I feel like I have a purpose being here (Dad)
>
> I came back to see the cheeky monkeys (Elderly lady)
>
> It gets people together –it creates a community (Elderly lady)
>
> Coming here – you don't feel so alone (Elderly lady)
>
> As you get older and your grandchildren grow up – you feel not needed- out on a limb. We have more time now we are older to pass on these skills (Elderly lady)
>
> This keeps me grounded – I like working with the community. (Dad)
>
> It's a good atmosphere. The smell of the soup. A nice feeling and I leave here with good energy. I like to be involved. (Dad)

There is evidence of how sewing and knitting are beneficial to the health of the participants. Knit for Peace (n.d, p. 3) researched the benefits of knitting on the elderly saying it has "positive health benefits, physical and mental." Brayshaw (2017) reported on how sewing and embroidery referred to as "meditative, transformative work" was offered to recovering servicemen after World War One, challenging the gender construct of that time. This was reflected in the trial.

> It (knitting) passes the time. It takes your mind off things – worrying. (Dad)
>
> It keeps your hands busy (Dad)

The importance of relationships within communities is highlighted when children became aware of and friendly with older generations as a consequence of the trial. One mum said it was nice for her child "as *we don't have grandparents here.*"

Children noted differences and the café presented an opportunity to learn about 'others' in a relaxed and informal way.

> You have grey hair! (4 year old)

And

Why do you have grey hair? (4 year old)

The elderly lady replied saying "*because I am old*" to which the young girl replied "*I have never seen anyone with grey hair before!*" This little girl has no grandparents in Liverpool and within her community it is rare for people to go grey demonstrating the need for children to see and observe differences in natural environments within their community and that surely is the social responsibility of early childhood.

11 Conclusion

This research trial of the legacy café supports the interconnectedness of three pillars of sustainability from a community perspective sharing social responsibility, with the evidence recognising it has positive benefits for all participants. As Dewey (1916) noted "where learning is the accompaniment of continuous activities or occupations which have a social aim and utilize the materials of typical social situations, from under these conditions, the school becomes itself a form of social life, a miniature community." This was evident in the intergenerational legacy café with a wide range of participants that attended voluntarily, a crucial component of the trial. Families that were normally seen as being 'marginalised' or even 'troubled' came willingly, dads found a 'purpose' and a 'more meaningful' way of connecting with their children and the elderly participants felt there was a reason to leave their flat and enter the outside world. Gender stereotyping was a noted theme and it was apparent the need to ensure all young children had an opportunity to observe a wide range of perceived role reversals. Dads cooking in the kitchen, mums mending scooters and punctures, dads knitting or sewing a button on and the elderly talking with the children and their families. This must be a continued element of all intergenerational projects, illuminating anti-bias and opening up equal opportunities for all, breaking down stereotyping organically and authentically in the early years, when fundamental values and attitudes are developed.

These findings also highlighted the need for further research into the health benefits of sustainable community groups, with a particular emphasis on mental health. At present in England it is noted there is an ever increasing and constant rise in mental health issues across all age divides, and the legacy café if established on a national scale would offer opportunities for all generations to support each other through conversation, companionship and sharing of historic and cultural tasks. There is neuroscientific research that supports the need to keep both the head and hands busy, and this again needs to be a focus of research. These elements will be a crucial aspect of the next phase of trials in Liverpool. Public health in England struggle under the cost of the growing elderly population and accessing universal health benefits. There is a growing awareness of the need for more support for both the elderly in terms of social isolation and being cognitively active, but also mental health issues in younger and younger children, and this intergenerational community approach might be a potential solution. This café project is unique in that it opens up the potential for generational research and development rather than focusing on one particular age. There are already several developments leading on from this legacy café, with Liverpool City Council trailing the model in five children and family centres (three with dementia hubs attached) over the next year, particularly because it offers an authentic not tokenistic approach of cohesion and social responsibility. There are also four other culturally different trials starting in Finland, the USA, and Tasmania, and through a premier league football club in Liverpool. The café is a simple but unique research project, which represents a positive move towards "our common future" (Brundtland 1987).

References

Abbott L, Nutbrown C (2007) Experiencing Reggio Emilia; implications for pre-school provision. Open University Press, Maidenhead

Age UK (2015) Evidence review: loneliness in later life. Retrieved from https://www.ageuk.org.uk/globalassets/age-uk/documents/reports-and-publications/reports-and-briefings/health–wellbeing/rb_june15_lonelines_in_later_life_evidence_review.pdf

Alexander C (1979) A timeless way of building. Oxford University Press, Oxford

Arnstein SR (1969) A ladder of citizen participation. J Am Plan Assoc 35(4):216–224

Avison K, Rawson M (2016) The tasks and content of the Steiner -Waldorf curriculum. Floris Books, Edinburgh

Ballantyne R, Connell S, Fien J (2006) Students as catalysts of environmental change: a framework for researching intergenerational influence through environmental education. Environ Edu Res 12(3–4):413–427

Bandura A (1977) Social learning theory. Prentice Hall, Englewood Cliff, NJ

Bostrom A (2003) Lifelong learning, intergenerational learning and social capital. From theory to practice. Institute of International Education, Stockholm

Bourn D, Hunt F, Blum N, Lawson H (2016) Primary education for global learning and sustainability. Research report for the Cambridge Primary Review Trust. Retrieved from www.cprtrust.org.uk

Brayshaw E (2017) How embroidery therapy helped first world war veterans find a common ground. The Guardian. Retrieved from https://www.theguardian.com/commentisfree/2017/apr/25/how-embroidery-therapy-helped-first-world-war-veterans-find-a-common-thread. Accessed from 25th April 2017

Brewer JD (2000) Ethnography. Open University Press, Buckingham

Bronfenbrenner U (1979) The ecology of human development: experiments by nature and design. Harvard University Press, Cambridge

Brundtland G (1987) Our common future: the world commission on environment and development. Oxford University Press, Oxford

Boyd D (2018) Early childhood education for sustainability and the legacies of two pioneering giants. Early Years 38(2):227–239

Boyd D, Hirst N, Siraj-Blatchford J (eds) (2018) Understanding sustainability in early childhood education. Case studies and approaches from across the UK. Routledge, London

Cambridge Primary Review (2014) Community soundings: the primary review regional witness sessions. Retrieved from http://cprtrust.org.uk/wp-content/uploads/2014/06/Comm-soundings-briefing-REVISED-2014.pdf

Dahlberg G, Moss P (2006) Introduction. Our Reggio Emilia. In: Rinaldi C (ed) Dialogue with Reggio Emilia. Listening, researching and learning. Routledge, London, pp 1–23

Davis J (2015) Young children and the environment. Early education for sustainability. Cambridge University Press, Melbourne
Davis J (2009) Revealing the research 'hole' of early childhood education for sustainability: a preliminary survey of the literature. Environ Educ Res 15(2):227–241
Denscombe M (2014) The good research guide for small scale research projects. Open University Press, Maidenhead
Department for Education (2017) Statutory framework for the early years foundation stage: setting the standards for learning, development and care for children from birth to five. Department for Education, Runcorn
Dewey J (1916) Democracy and education: an introduction to the philosophy of education. Project Gutenberg, Salt Lake City. Retrieved from http://www.gutenberg.org/ebooks/852?msg=welcome_stranger
Duvall J, Zint M (2007) A review of the effectiveness of environmental education in promoting intergenerational learning. Journal Environ Educ 38(4):14–24
Generation United (2007) The benefits of intergenerational programs. Retrieved from https://www.gu.org/
Giles H, Coupland N (1991) Language: contexts and consequences. Brooks/Cole, Pacific Grove
Graue E, Walsh D (1995) Children in context: interpreting the here and now of children's lives. In: Hatch JA (ed) Qualitative research in early childhood settings. Praeger, Westport, Connecticut, pp 135–54
Gray C, MacBlain S (2015) Learning theories in childhood. Sage, London
Helliwell J, Layard R, Sachs J (eds) (2018) World happiness report. Retrieved from http://worldhappiness.report/ed/2018/
Hunt F (2012) Global learning in primary schools in England: practices and impacts. DERC Research paper no. 9. UCL Institute of Education, London
Joyce R (2012) Outdoor learning. Past and present. Sage, London
Kasser T (2006) Materialism and its alternatives. In: Csikszentmihalyi M, Csikszentmihalyi I (eds) A life worth living: contributions to positive psychology. Oxford University Press, Toronto, pp 200–214
Kenner C, Ruby M, Jessel J, Gregory E, Arju T (2007) Intergenerational learning between children and grandparents in east London. J Early Child Res 5(3):219–243
Keenan T, Evans S (2010) An introduction to child development. Sage, London
Knit for Peace (n.d). The health benefits of knitting. Retrieved from http://www.knitforpeace.org.uk/wp-content/uploads/2017/05/The-Health-Benefits-of-Knitting-Preview.pdf
Kuttner P (2015) Educating for cultural citizenship: reframing the goals of arts education. J Crit Inq 45:1
Langlands A (2018) Craft: an inquiry into the origins and true meaning of traditional crafts. W.W Norton and Company, New York
Lave J, Wenger E (1998) Communities of practices: learning, meaning and identity. Cambridge University Press, Cambridge
Liebschner J (1993) A child's work. Freedom and guidance in froebel's educational theory and practice. Lutterworth Press, Cambridge
Lutz F (1986) Ethnography; the holistic approach to understanding schooling. In: Hammersley M (ed) Controversies in classroom research. Open University Press, Milton Keynes, pp 107–120
MacNaughton G, Hughes P (2009) Doing action research in early childhood studies. Open University Press, Maidenhead
Malagguzi L (n.d.) One hundred languages poem. Retrieved from https://reggioemilia2015.weebly.com/the-100-languages.html
Mbebeb F (2009) Developing life skills in children. Priming entrepreneurial mindsets through socialisation in family occupations. Inter J Early Child 41(2):23–34
Ministry of Education (1996). Te Whariki: early childhood education curriculum. Ministry of Education, Wellington, New Zealand

Moss P (2007) The 'Otherness' of Reggio. In: Abbott L, Nutbrown C (eds) Experiencing Reggio Emilia. Implications for pre-school provision. Open University Press, Maidenhead, pp 125–138
Moss P (2017) Lecture to LJMU November 2017. Retrieved from http://www.ecsdn.org/FileUploads/Peter%20Moss%20Presentation%20Workforce_1465598743_5.pdf
Nsamenang BA (2007) Cultures in early childhood care and education. Education for all global monitoring report. UNESCO, 2007/ED/EFA/MRT/PI/3
Nicholls M (2004) Cultural perspectives from Aotearoa/New Zealand. J Intergenerational Relat 1(4):25–34
O'Brien C (2013) Education for sustainable happiness and well-being. Routledge, London
Ofsted (2017) Bold beginnings: the reception curriculum in a sample of good and outstanding primary schools. Retrieved from https://assets.publishing.service.gov.uk/government/uploads/system/uploads/attachment_data/file/663560/28933_Ofsted_-_Early_Years_Curriculum_Report_-_Accessible.pdf
Paige-Smith A, Craft A (2008) Developing reflective practice in the early years. Open University Press, Maidenhead
Rinaldi C (2006) In dialogue with Reggio Emilia. Listening, researching and learning. Routledge, London
Robins N, Roberts S (1998) Consumption in a sustainable world accelerating international action. In: Background report for the workshop on consumption in a sustainable world. Kabelvaag, Norway, 2–4 June
Robinson L, Vaealiki V (2015) Ethics and pedagogy at the heart of early childhood education for sustainability. In: Davis J (ed) Young children and the environment. Education for sustainability. Cambridge University Press, Melbourne, pp 103–123
Rogoff B (1996) Observing sociocultural activity on three planes: participatory appropriation, guided participation, and apprenticeship. Retrieved from http://methodenpool.uni-koeln.de/situierteslernen/Teaching%20As%20Learning.htm
Rogoff B (2014) Children's guided participation and participatory appropriation in sociocultural activity. In: Wozniak R, Fischer K (eds) Development in context. Acting and thinking in specific contexts. Psychology Press, New York, pp 121–155
Schön D (1983) The reflective practitioner: how professionals think in action. Temple Smith, London
Seedsman T (2013) A mosaic of contemporary issues, insights and perspectives surrounding intergenerational practice. J Intergenerational Relat 11(4):347–355
Shier H (2001) Pathways to participation, openings, opportunities and obligations. A new model for enhancing children's participation in decision making in line with the article 12.1 of the United Nations convention on the rights of the child. Child Soc 15:107–117
Siraj-Blatchford J, Smith KC, Pramling Samuelsson I (2010) Education for sustainable development in the early years. Available online: https://www327matters.org/Docs/ESD%20Book%20Master.pdf
Smidt S (2009) Introducing Vygotsky: a guide for practitioners and students in early years education. Routledge, London
Smidt S (2014) Introducing Freire. A guide for students, teachers and practitioners. Routledge, London
Standing EM (1998) Maria Montessori. Her life and work. Penguin, London
UNDES (2015) Transforming our world: the 2030 agenda for sustainable development. United Nations Department for Economics and Social Affairs, New York. Retrieved from http://unesdoc.unesco.org/images/0024/002463/246300E.pdf
UNESCO (2002) Cultural declaration on cultural diversity (Culture diversity series no.1). UNESCO, Paris
United Nations (1991) United Nations principles for older persons. Resolution 46/91. OHCHR, Geneva
United Nations, Human Rights, Office of the High Commissioner. Retrieved from https://www.ohchr.org/EN/ProfessionalInterest/Pages/OlderPersons.aspx
Vaillant G (2002) Aging well. Little Brown, Boston

Vanderven K (2004) Intergenerational theory in society. J Intergenerational Relat 2:3–4
Vygotsky L (1978) Mind in society. the development of higher mental psychological processes. Harvard University Press, Cambridge
Weckström E, Jääskeläinen B, Ruokonen I, Karlsson L, Ruismäki H (2017) Steps together: children's experiences of participation in club activities with the elderly. J Intergenerational Relat 15(3):273–289
Wyness L (2015) "Talking Citizenship…" exploring the contribution an intergenerational, participatory learning project can make to the promotion of active citizenship in sustainable communities. Local Environ 20(3):277–297

The Important Role of Universities in Enhancing Sustainability: The Case of the University of Milano-Bicocca

Alessandro Capocchi, Paola Orlandini, Cinzia Vallone, Murad Harasheh and Stefano Amelio

Abstract During the last decade, an increasing number of universities have started to play an active role in creating a more sustainable future, with many implementing sustainability initiatives on campus. Following recent studies of other significant initiatives, this paper aims to analyse recent sustainability policies introduced by the University of Milano-Bicocca to examine how these policies have worked and how improvements have been measured. The results of this analysis, based on an evaluation of the sustainability initiatives, indicate that sustainability reporting in Italian universities is still in its infancy but, at the same time, suggests (i) that certain areas of progress and positive results can be identified, with several benefits for the stakeholders and for the local community, and (ii) that the example offered by the university provides a potential pathway to the creation of specific modules in the degree programs. Based on the experience of the University of Milano-Bicocca, we suggest that standardization in sustainability reporting will play a central role in the near future. Standardization is very important, not only in terms of the creation of sustainability reports but also in terms of their interpretation. This paper explains the current work in progress and the efforts of an Italian university to better incorporate and institutionalize sustainability and to become a template for the Italian University System in facilitating the standardization and the implementation of a unique assessment and reporting model capable of adoption across the National University System.

A. Capocchi (✉) · P. Orlandini · C. Vallone · M. Harasheh · S. Amelio
Department of Business Administration, Finance and Law (DiSEADE), University of Milano-Bicocca, Via Bicocca degli Arcimboldi n. 8, 20126 Milan, Italy
e-mail: alessandro.capocchi@unimib.it

P. Orlandini
e-mail: paola.orlandini@unimib.it

C. Vallone
e-mail: cinzia.vallone@unimib.it

M. Harasheh
e-mail: murad.harasheh@unimib.it

S. Amelio
e-mail: stefano.amelio@unimib.it

W. Leal Filho (ed.), *Social Responsibility and Sustainability*, World Sustainability Series, https://doi.org/10.1007/978-3-030-03562-4_20

Keywords Sustainable development · Reporting · Universities · Education
Voluntary disclosure · CSR

1 Introduction

In high-pollution economies, concern over sustainability is greater than ever. In addition to facing high-pressure competition, various economic sectors must increasingly pay attention to resource usage, waste treatment, air emissions, water pollution, employee welfare, and so on. Failing to manage these sustainability issues can substantially damage the image of a company and thus affect its performance. In 1987, the Brundtland Report first established the concept of sustainable development, describing it as "development that meets the needs of the present without compromising the ability of future generations to meet their own needs" (WCED 1987).

In general, social legitimacy is a conventional view as to why economic entities perform sustainability reporting. In this era of alarming global warming, economic actors communicate their environmental performance to their varied stakeholders to renew their 'social license' as a responsible social and economic entity. However, sustainable practices were—at earlier stages—viewed by different economic actors as an annoyance. The changing stakeholder concerns currently highlight environmental conservation through reduced greenhouse gas emissions and pollution prevention. These caused the construct of CSR to evolve further into environmental performance measures—which together create the sustainability concept—in different variables of environmental management and life cycle assessment.

Sustainability in the last decades has started to be considered an increasingly important issue in several scientific disciplines. Sustainability had been introduced in several programs for at least 20 years in the government and private sector, contributing to an increase in the awareness of negative environmental effects (Jabbour 2010). Any effort to create a more sustainable world must consider the important role played by the education system. At the Stockholm Conference in 1972, education was formally recognized internationally as playing an important role in fostering environmental protection and conservation. This was the first step in a long process over the subsequent decades, which led to many academic declarations, charters and partnerships designed to foster environmental education to encourage sustainable development. The idea of sustainability evolved in three distinct strands: society, education, and higher education.

The education sector is called to express its own role in two different ways: (i) by encouraging and implementing sustainable development in university programs and (ii) by reducing the negative impacts on the environment, society, and health, generated from its use of resources (Cole 2003).

In the past, universities have played an important role in transforming societies and in serving the public good. Today, universities are seen as key players in the promotion of sustainable development (Cortese 2003; Lozano 2006). Universities have a fundamental moral responsibility to contribute to sustainable development and to guide societies on their path towards a sustainable future (Waas et al. 2010). Universities also provide a significant contribution to the development of our societies, and they therefore have a special societal responsibility, particularly with regard to youth training and public awareness about sustainability (Alshuwaikhat and Abubakar 2008). Since the 1990s, universities worldwide have increasingly embraced the sustainable development movement, and more than 1000 academic institutions have signed international declarations towards implementing sustainability through environmental literacy initiatives; curriculum development; research; partnering with government, non-governmental organizations and industry in developing sustainability initiatives; and "greening" physical operations. These actions are very important, but they represent merely the early stages of a more complex series of processes since much remains to be done for sustainable development to become genuinely and comprehensively implemented at universities and for universities to become true leaders in sustainable development (Waas et al. 2010). An increasing number of higher education institutions and universities have started to enrich their educational offerings by incorporating and institutionalizing sustainability into their curricula, research, operations, outreach, and assessment and reporting (Calder and Clugston 2003; Cortese 2003; Lozano-Ros 2003). The enhancement of educational offerings is an important step, but the question remains whether, considering the pivotal role that universities can play in the education of new generations, this should be considered enough. Education programs more or less focused on sustainability issues are not sufficient. The focus on sustainability by universities has also become important in managerial and organizational modules (Wright 2007; Corcoran et al. 2004; Calder and Clugston 2003; Social Sciences and Humanities Research Council of Canada 2002). Universities have started to create a link between education, on the one hand, and organizational and managerial models, on the other hand: both hands are oriented towards creating a new sustainable culture for new generations of students. Following Lozano (2011), it remains the case that the number of universities engaged in sustainable development is still small compared to the total number of universities in the world or to the number of companies that have published sustainability reports. For this reason, this paper, following recent studies concerning other significant experiences and considering the Graphical Assessment of Sustainability in Universities tool, aims to analyse the case of the University of Milano-Bicocca with specific regard to the recent sustainability policies implemented by the university itself. These policies are the foundation of sustainability reporting as voluntary activity oriented (i) to assess the current state of the organization's economic, environmental and social dimensions and (ii) to communicate the organization's efforts and sustainability progress to its stakeholders. The experience of the University of Milano-Bicocca aims to be a benchmark for Italian universities in order to contribute to the creation of a more sustainable culture in new generations. The goal of this paper is to examine the most current practices related to sustainability reporting, designs and measures adopted by academic institutions; present the case of the University of Milan-Bicocca according to the acceptable adopted measures and standards for sustainability reporting—which is considered a pioneering activity in the Italian universities—and finally to provide novel recommendations and related policy implications to render university sustainable reporting a significant matter for policy and decision making. *This paper seeks to bring together the recent advances in sustainability evaluation in reporting and design and, acknowledging measures adopted elsewhere on standards for sustainability, produce a coherent policy for sustainability reporting across all Italian universities.*

2 Literature Review

The first concept to establish in the literature review is the meaning of a "sustainable" university. According to Amaral et al. (2013), the concept of a sustainable university has a dual meaning: in one sense, it is related to the sustainable use of university infrastructures, which include all the elements that comprise the university campus. Operating a highly complex entity such as a university necessarily involves a large amount of energy and resources and generates vast quantities of waste. In another sense, the meaning is related to the capacity of the university to introduce the sustainability concept into the curriculum. Linking the two aspects, demonstrating sustainable practices and promoting the sustainability concept generally—the sustainable university can be defined as one that preaches excellent behaviour and then practices what it preaches by minimizing the negative environmental, economic, societal and health effects generated from its use of resources (Velasquez et al. 2006). The sustainable university is directly and demonstrably concerned with (i) protecting the "*health and well-being of humans and ecosystems*"; (ii) using the knowledge produced to "*address the ecological and social challenges*" (Cole 2003); and (iii) promoting within society the effort of energy and resource conservation, waste reduction, social justice and equity (Alshuwaikhat and Abubakar 2008). In the literature, Cortese (2003) defined the university as a four-dimensional system based on *education, research, campus operations* and *community outreach*. Following this setting Lozano-Ros (2003) added a fifth dimension represented by assessment and reporting activities.

Concerning environmentally sustainable practices, universities are not new to this concept (Stephen et al. 2013), as are other public and private actors. Many studies seek to identify energy consumption in universities in three distinct categories: green building initiatives (Filippín 2000; Alshuwaikhat and Abubakar 2008), management initiatives (Clarke and Kouri 2009; Bero et al. 2012; Chang 2013; Fisher 2003; Noeke 2000; Price 2005; Spellerberg et al. 2004), and management systems (Richards and Gladwin 1999; Shriberg 2002; Posner and Stuart 2013; Velasquez et al. 2006).

The transition from green building initiatives to sustainable management systems is necessary given the complexity of universities when compared to other institutional actors: universities must combine the three strands of sustainability shared by all institutions (economic, environmental and social) with the five requirements of their mission (education, research, operations, community outreach and reporting), only some of which are shared by other types of operations. In universities, the sustainability issue needs to be integrated within the organization, following Richards and Gladwin (1999). Sustainable management requires organizations to go beyond eco-efficiency by integrating environmental and social goals into all aspects of decision making. Due to this complexity, the sustainability issue in the university requires a management system able to create models.

Velasquez et al. (2006) proposed a model to achieve this based on the Deming‘s method: Plan, Do, Control and Action (PDCA). The model describes four phases, organized from strategic to more operational, and based on four strategies. The structure of a sustainable university should also include the following: networks with other universities; sustainability audits to monitor, analyse and control the performance of sustainable initiatives; and adherence to the continuous improvement principle.

Another model was proposed by Alshuwaikhat and Abubakar (2008) based on a framework for a more suitable approach in achieving campus sustainability that could outperform the limitations of traditional environmental management practices. This model aims to integrate three different strategies: EMS, public participation, social responsibility, and sustainability in teaching and research. Following this model, the implementation of an EMS should have a positive impact on the three keystones of sustainability: economy (conserving energy and resources), environment (minimizing negative impacts and waste generation), and social (giving equal opportunities to all in the areas of teaching and research).

In this theoretical context, assessment and reporting combine have an important role in the context of the complexity, which distinguishes the case of universities.

Following Dalal-Clayton and Bass (2002), there are three main approaches to measuring sustainability: (i) accounts, (ii) narrative assessment and (iii) indicator-based assessment.

In short, the accounts approach involves the use of data that must be converted to a multiple common minimum; narrative assessments are more oriented to the story-telling method using text, maps, graphics and statistical data; indicator-based assessment requires the identification of variables and involves the use of different data types, organized around one or more indicators.

There are several tools devised to represent and measure sustainability, although some are not specifically just for universities. For this reason, Cole (2003) analysed and compared 13 of the most important tools used to measure sustainability, including the Global Reporting Initiative (GRI), ISO 14000 series and ecological footprint. Cole sought to adapt these tools to the specific needs of a higher educational establishment. At the same time, there have been several attempts to develop university-oriented tools specifically designed to assess sustainability in higher educational environments (Roorda 2001; AASHE 2012; Lozano 2009).

There are other tools already deployed in certain areas to assess and report the sustainability of universities, such as (i) the National Wildlife Federation’s State of the Campus Environment (McIntosh et al. 2001, 2008), (ii) the Princeton Review’s Guide to Green Colleges (The Princeton Review 2012), (iii) the Graphical Assessment of Sustainability in Universities (GASU) (Lozano 2006, 2011), the GRI G3 as a set based on 63 indicators (Lopatta and Jaeschke 2014; Alonso-Almeida et al. 2015), and a set of four assessment tools used by Kamal and Asmuss (2013) in their particular case.

In all of these tools, the balance between environmental, economic and educational requirements demands particular care in the education field, which is still the least measured.

Sustainability reports measure the current state of the economic, environmental and social objectives and practices of an organization and aim to communicate sustainability performance and related progress to stakeholders (Lozano 2006). The most used standard guidelines are represented by the Global Reporting Initiative (GRI) (KPMG 2011, 2013). However, there is neither a sector-specific GRI supplement nor any other accepted sustainability reporting standard for universities (Adams 2013).

According to Sassen and Azizi (2018), sustainability reporting by universities is still in its infancy (Lopatta and Jaeschke 2014; Adams 2013), and the same is true for research on sustainability reports disclosed by universities (Ceulemans et al. 2015; Vagnoni and Cavicchi 2015).

3 Method and Design

The research design of this paper is consistent with existing studies (Lozano 2011; Lopatta and Jaeschke 2014; Sassen et al. 2014).

In this paper, the research focuses specifically on the case of the University of Milano-Bicocca, which may be considered a template for the Italian education system. For this university, the following two research questions were considered fundamental: Which sustainability dimensions are reported by the University of Milan-Bicocca? What exactly is reported? To answer these questions, the case study approach has been adopted, according to the GRI benchmark, in explorative and descriptive ways. A review of the recent related literature has been conducted to ensure that the latest "state of the art" findings and recommendations for sustainable practice and reporting by academic institutions were evaluated. The study was performed in 2018 using data, academic information, and financial reports from the last two years.

To ensure the most up-to-date accuracy, the information provided in the report was evaluated using the GRI's environmental, economic and social performance indicators in accordance with studies developed by Sassen et al. (2014) and the surveys conducted by Lozano (2011) and Lopatta and Jaeschke (2014). The case study research method has been adopted for the following main reasons: (i) as Eisehnardt observed, the case study is a research strategy that focuses on understanding the dynamics present within single settings and typically combines data collection methods such as examining archives, interviews, questionnaires, and observations (Eisenhardt 1989); (ii) similarly, Yin stated that the case study is a research method that allows the observation of several variables related to the same phenomenon (Yin 1994); (iii) Yin observed also that the case study allows the deep comprehension of a complex phenomenon through the use of multiple direct sources, such as the analysis of evidence, internal documents, surveys, interviews and any direct observations (Yin

1994); (iv) the process of implementation of a new organizational-managerial model involves those persons actually engaged in the organization and the work that they daily carry out; (v) the necessity of a direct observation of the phenomenon has made the use of purely statistical quantitative methods objectively not always adequate: with the case study method, as Gobo wrote, the research can reduce the possible errors of measurement and their incidence on the relationship between probability, representativeness and generalizability of the results obtained (Gobo 2004); and, finally, (vi) the case study research method is a very good way to start the empirical analysis of new or specific phenomena that can promote re-evaluation in the literature and new perspectives that may inform the future creation of quantitative models. This is particularly true for the social sciences, where we can agree with Eisehnardt that "building theory from case study research is most appropriate in the early stages of research on a topic or to provide freshness in perspective to an already researched topic" (Eisenhardt 1989). This is true for the analysis of the sustainability report in Italian universities, where we are still far from a standardization of tools and methods.

4 Results and Analysis

4.1 The University of Milan-Bicocca

The University of Milan-Bicocca has chosen to actively engage in sustainability as fully as possible, ensuring that its facilities, activities and services offered are sustainable from an environmental, social and economic point of view. To achieve this goal, BASE (*Bicocca, Ambiente, Società, Economia*) was born as an initiative that aims to create a network within the university between research and application experiences in the field of environmental, economic and social sustainability. BASE intends to actively involve staff and students in the subject of sustainability in order to allow the university to achieve significant results, both internally and in the surrounding area.

Universities also have the formative role in the social promotion of and guidance towards a path to sustainability for the local and global landscape through the study, implementation and dissemination of strategies and activities that tend to make the economic and social reality more sustainable. This is the mission of the Italian Network of Universities for Sustainable Development (RUS), of which the University of Milan-Bicocca is among the promoting institutions. Bicocca is also part of the International Sustainable Campus Network (ISCN), the worldwide network of Sustainable Universities: the principles and guidelines required in this international association are the concrete management objectives that the university will pursue through BASE. Bicocca Sustainability (BASE) is promoted by the Rectorate and by the General Management and led by a Scientific Committee for Sustainability composed of 9 professors belonging to a variety of university departments. BASE is carried out by the Sustainability Office, which is based in the Department of Environmental and Earth Sciences, in collaboration with the Infrastructure and Procurement Area. The above structure illustrates the necessity to link the organizational imperatives to management practices in order to enhance the sustainability practices of the university and make the university a leader in sustainability with respect to its stakeholders. This structure governing the sustainable actions of the University of Milano-Bicocca is represented in Table 1.

Energy management

Improvements should be both structural (plant management) and behavioural (promotion of good behaviour towards staff and students). The monitoring of energy consumption for the years 2011–2015 showed similar trends with a tendency towards reduction due to the optimization of some consumption and the implementation of some structural actions aimed at energy efficiency (e.g., installing LED lights in all the public areas of the university). In energy consumption, the university's carbon footprint was calculated, which is approximately 15,600 tons of CO_2eq/year. The carbon footprint for energy consumption was calculated from 2011 to 2014. In this period, the average consumption of electricity was 2578 MWh/month. The trend of monthly consumption in the years 2011–2014 is similar, and the lowest values occurred in the months of April and May, in which heating and cooling systems are not used or only partially used. The calculation of the equivalent emissions was carried out according to the emission factor 0.524 kgCO_2eq/kWh in reference to the production of electricity according to the Italian national electricity generation mix. The equivalent emissions related to energy consumption for the year 2014 are 15,662 tons of CO_2eq. To evaluate the actual practical impact, the CO_2eq value of the individual buildings was calculated in relation to their surface area. There are clear differences between the actual surface of buildings and the emissions generated, as these are related to the intended use: the buildings of a scientific nature, such as the buildings U1–U4 and U5, have greater consumption and emissions due to the presence of laboratories and research facilities. The management buildings have smaller footprints, as seen for U6, the largest building by area, which is home to administrative offices. The buildings U9 and U14, home of the IT facilities (with university servers), have very high level of energy consumption. The U17 building, which hosts the student secretariats, has a high emission/surface ratio since its lighting costs include the lighting of the square in front; at the moment, it is not possible to distinguish these contributions.

Mobility management

The latest survey on home-to-work transfers of university staff, carried out in 2014, provided indications of the personal choices staff made to reach the workplace (private mobility 38%, TPL 51%, slow mobility 10%). The emissions produced are approximately 2203 T. of CO_2eq.

The significant interventions already implemented are as follows: (a) Realization of three surveys of home-to-work transfers of university staff (2002, 2009, 2014); (b) Availability of 70 bicycles for the university staff, thanks to an agreement with the

Table 1 Sustainable actions in the environmental dimension

Action	Scope	Indicators
Energy management	Monitoring the progress, identifying inefficiencies, optimizing management, promoting energy savings are all fundamental actions for the sustainability of the University	Monitoring of energy consumption (electrical and thermal) on an annual basis
		Calculation of the Carbon Footprint related to energy consumption in order to reduce emissions of greenhouse gases
		Implementation of a sustainable energy management plan
		Creation of a database of energy consumption for each space (classrooms, offices, common areas) with relative mapping
		Implementation of structural measures to improve energy efficiency
Mobility management	Promoting sustainable mobility policies and at improving the environmental, economic and social performance of the University and indirectly of the city area in which the Milano-Bicocca University is located	Drafting of a home-to-work transfer plan for the University of Milan-Bicocca
		Extension of the student mobility survey
		Implementation of interventions to promote the improvement of modal split and recourse to a more effective inter-modality
		Experimentation of a more careful policy of parking in the car parks in the University structure
		Reduction of CO_2 emissions associated with home-to-work travel and to University personnel missions
		Promotion of a culture of sustainable mobility in the University and in the surrounding neighborhood
Waste management	Improving the environmental, economic and social performance of the University of Milan-Bicocca and to reduce CO_2 emissions. Waste management goes through the principles of reduction, reuse and recycling	Effective application of the waste management model throughout the University

(continued)

Table 1 (continued)

Action	Scope	Indicators
		Sample monitoring of waste production over time and quantity/quality of RD
		Consolidation of PolApp
		Design of separate waste collection in outdoor spaces
		Promotion of policies to reduce, reuse and recycle
		More incisive and coordinated involvement of the population of the University, main actor of these participatory processes of sustainability
Water and Food	The water resource is often used improperly and it is necessary to make everyone aware of how to use it correctly. The food production and consumption chain is responsible for significant environmental impacts, while everyone's food choices can have a substantial impact on the environment and health	6000 steel flasks purchased from Bicocca: they are distributed to all staff and during events or initiatives to students
		13 installed regulators, located in the main campus buildings (click to see where they are located—users reached: 90%)
		18,100 L the average water distribution expected in a year for a single dispenser
		253,000 L of total annual distribution (for 14 suppliers)
		506,000 plastic bottles that can be avoided for use every year
		18,200 kg of CO_2 in the atmosphere avoided in one year thanks to the suppliers (1516 per month, 1300 ksCO_2/year for each provider)
		5 the number of plastic bottles that create the same impact as a single steel bottle: this means that using the water bottle for just one week makes it possible to cancel the impact on the environment (in terms of production, transport, distribution) compared to the use of plastic

municipality of Milan; (c) Activation of a free Ecobus shuttle service for trips within the campus; (d) Ability to access an annual subscription to the TPL for university staff; (e) Installation of a mobile Ciclofficina in Piazza della Scienza and Piazza dell'Ateneo Nuovo; and (f) Availability of a car sharing service. The university provides Guidami with a parking space on the Bicocca campus at the U5 building in Viale dell'Innovazione. The Guidami car is booked by students at discounted rates.

Waste management

Monitoring carried out at the university (years 2014–2015) highlighted the separate collection of 25% of a total amount of waste produced equal to approximately 530 tons. A new waste management system has therefore been developed in an agreement with Amsa (a waste collection manager in Milan) and Comieco, a consortium for the separate collection of paper.

The system, called "Bicocca makes the difference", allowed the organization to bring the separate collection to about 70% (data related to the experimentation in U1–U4 buildings). An integral part of the new system is PolApp, an application developed by the POLARIS Research Center and Geomatic Lab, which allows quantitative and qualitative monitoring of waste by users of the university.

Water and Food

Water and food are fundamental resources to be safeguarded for sustainability. To limit the production of plastic waste and to encourage users to drink tap water, the university has decided to install a smooth and carbonated water dispenser in each building. Currently, there are 13 dispensers operating, mostly installed at the refreshment areas. The dispensers are equipped with activated carbon filters and UV lamps mounted on the delivery nozzles to guarantee sterility. Water analysis is performed on a monthly basis in accordance with prevailing legislation. The quality of water intended for human consumption is governed by Legislative Decree No. 31 of 2001, which incorporates the Directive 98/83/EC and applies to all water intended for drinking.

The numbers indicated in Table 1 were calculated starting from a one-year experimentation with 6 dispensers. In a year, an average of 2600 L of water were consumed per week (approximately 530 litres per day) for a total of 114 thousand litres of water, corresponding to a hypothetical saving of approximately 228 thousand half-litre plastic bottles. A plastic bottle is responsible for the emission of 36 grams of CO_2 equivalent over its life cycle (Data Plastics Europe). In a year, the presence of 6 water dispensers allowed the avoidance of the hypothetical introduction into the atmosphere of 8150 kg CO_2eq, corresponding to the amount of emissions produced by a car that travels approximately 12 thousand km or 10 round trip flights from Milan to New York.

For each week, it means a savings of approximately 160 kg CO_2eq, similar to the emissions produced by a PC running for 150 days or a class A +++ refrigerator in two years or a methane boiler used for 400 showers. For every day, it means a savings of 32 kg CO_2eq, like the emissions of an 80 W light bulb turned on uninterruptedly for a month. These emission values, even if only indicative, make clear how much the choice of drinking dispenser water instead of that in a plastic bottle can be very significant in the production of climate-altering emissions.

An integrated Bicocca sustainable system

The four actions are integrated and together they produce the Bicocca integrated sustainable system. With this system, it is clear how the University of Milano-Bicocca is playing its role of leader from two perspectives: the internal dimension involving all stakeholders such as academics, students, staff etc. and the external dimension involving the local community and the whole vicinity in which the University of Milano-Bicocca is located.

In the University of Milano-Bicocca, the path of sustainability began to take on greater significance after 2012, the year in which the first surveys on the university's environmental footprint were produced by the organization. These led directly to the identification of critical aspects within certain activities for which operational interventions were devised and implemented.

Over the years, the university has developed numerous projects concerning the dimensions of sustainability, both with the aim of making its facilities compliant with the principles and objectives of sustainability and thus reducing costs and the waste of resources in its infrastructure while promoting social wellbeing and with the educational purpose of spreading a sustainable culture for its stakeholders, both internal and external. This commitment has led Bicocca to establish the Sustainability Office, details of which will be presented later in greater detail. In 2017, a Sustainability Action Plan was drawn up at the university, which anticipated implementation over a three-year period.

The University Environmental Sustainability Plan for the three-year period 2017–2020 envisages three types of actions: structural (AS) support in collaboration with the Infrastructures and Procurement Area (InAp), management (AG), and educational-behavioural (AC).

In implementing the strategy, with which the Bicocca University aims to achieve its objectives, sustainability also plays a major role in social engagement, and, to achieve any objective in this direction, the starting point resides in listening to and analysing the needs of the university's stakeholders. This aspect is of fundamental importance for the stakeholders to be actively involved in the activities of the university. In this regard, the university is very attuned to the need for the creation of a sense of social well-being with regard to the people involved in carrying out its activities: the students, teachers, researchers, employees and collaborators, as well as all the other stakeholders.

To contribute to the general wellbeing of the company and the planet, it is essential, in fact, to direct attention and commitment not only to the internal stakeholders of its structure but also to external ones, representing the local community, the institution and the company.

Attention to the social aspect in the university is expressed through different methods, as represented in Table 2.

As shown in Table 2, concerning the social aspects, it is important to mention:

Table 2 The social sustainable dimension in the University of Milano-Bicocca

Commitment of the University of Milano Bicocca on the social dimension	
Input	Output
Analysis of the needs of its stakeholders	Stakeholders must be involved in the activities of the University
Action 1—The presence of committees that ensure the welfare of their clients, within the university structure (such as the Technical-Administrative Staff Committee); it means in this regard the presence of the single guarantee committee for equal opportunities	
Action 2—The presence of: code of ethics and code for the protection of dignity, well-being of people and organization	
Action 3—The creation and dissemination of services for the promotion of activities with artistic, cultural, educational and sports content; of services for the benefit of students, staff, teachers and communities; services designed to support the weakest categories and services to promote cultural integration within the University	
Human resources	Assessment composition and trend
Action 1—Analysis of the composition of academic staff and administrative and technical staff members	
Action 2—Analysis of the distribution of the academic staff and administrative and technical staff members	
Action 3—The guarantee committee has promoted, as every year, a report for the well-being of the technical-administrative staff during the performance of their duties	
Students	Assessment composition and trend/performances
Social value generated	Actions
Action 1—Children's nursery Bicocca	
Action 2—Service for students with disability and with DSA	
Action 3—University sport committee	
Action 4—Between	
Action 5—Fellowship and right to study	
Action 6—Committee for equal opportunities	

- The presence of committees that ensure the welfare of their clients within the university structure (such as the Technical-Administrative Staff Committee), i.e., the presence of the Single Guarantee Committee for equal opportunities.
- The presence of a Code of Ethics and code for the protection of the dignity and well-being of the people and the organization itself. The first is aimed at ensuring the university's development in research, education and training, together with the dissemination of a culture based on the principles of environmental protection, human rights and international solidarity; the second is aimed at integrating into the university structure principles on the protection of human rights and health, making reference, among others, to the principles of the Italian Constitution and of International Law.

– The creation and dissemination of services for the promotion of activities with artistic, cultural, educational and sports content; services for the benefit of students, staff, teachers and communities; services designed to support the weakest categories and services to promote cultural integration within the university.

4.2 Actions, Assessment and Reporting

The results, based on matching sustainability actions, show that sustainability reporting in Italian universities is still in its early stages when compared to sustainability reporting in corporations or in international academic institutions. But the analysis does show (i) that many positives can be identified, with several benefits for the stakeholders and the local community, and (ii) that the example offered by the university organization is an essential premise to the creation of specific subjects in the degree programs.

In Italy, few universities have started to create sustainability reports; particularly significant are the experiences of University of Torino and University Ca'Foscari of Venice. Concerning the University of Milano-Bicocca, there is not a sustainability report yet, but one is in the process of construction, with three major strands: environmental, which is focused on the actions described; economic and financial, which is under construction following the GRI standard dataset series 200; and social, which is already being compiled.

The analysis in this paper aims to propose a possible integrated model for the sustainability report of the University of Milano-Bicocca according to the more significant international experiences (Sassen et al. 2014; Lozano 2011; Lopatta and Jaeschke 2014), involving three dimensions (environmental, economic and finance, and social) and following the implementation of the GRI standards. For the economic and finance dimension, it is useful to understand the implications of the GRI Standards Series 200. The economic and finance dimension of sustainability concerns an organization's impacts on the economic conditions of its stakeholders and on economic systems at local, national, and global levels. The Standards in the Economic series (200) address the flow of capital among different stakeholders and the main economic impacts of an organization throughout society. (GRI 201, D). These standards are designed to be used by organizations to report their impacts on the economy, the environment, and society (GRI 201, A). The series 200 is composed of 6 main standards:

- GRI 201 (Economic performance 2016): "addresses the issue of economic performance. This includes the economic value generated and distributed (EVG&D) by an organization; its defined benefit plan obligations; the financial assistance it receives from any government; and the financial implications of climate change" (GRI 201, D).
- GRI 202 (Market presence 2016): "addresses the topic of an organization's market presence, covering its contribution to economic development in the local areas or

communities where it operates. For example, this can include the organization's approaches to remuneration or local hiring" (GRI 202, D).

- GRI 203 (Indirect economic impacts 2016): "addresses indirect economic impacts, which are the additional consequences of the direct impact of financial transactions and the flow of money between an organization and its stakeholders. GRI 203 also addresses the impacts of an organization's infrastructure investments and services supported" (GRI 203, D).
- GRI 204 (Procurement practices 2016): "addresses the topic of procurement practices. This covers an organization's support for local suppliers, or those owned by women or members of vulnerable groups. It also covers how the organization's procurement practices (such as the lead times it gives to suppliers, or the purchasing prices it negotiates) cause or contribute to negative impacts in the supply chain" (GRI 204, D).
- GRI 205 (Anti-corruption 2016): "addresses the topic of anti-corruption. In this Standard, corruption is understood to include practices such as bribery, facilitation payments, fraud, extortion, collusion, and money laundering; the offer or receipt of gifts, loans, fees, rewards, or other advantages as an inducement to do something that is dishonest, illegal, or represents a breach of trust. It can also include practices such as embezzlement, trading in influence, abuse of function, illicit enrichment, concealment, and obstructing justice" (GRI 205, D).
- GRI 206 (Anti-competitive behavior 2016): "addresses the topic of anti-competitive behavior, including anti-trust and monopoly practices" (GRI 206, D).

Following the new set of GRI standards (valid from 1 July 2018), the indicators that can be part of the proposed sustainability report of the University of Milano-Bicocca in accordance with the economic and finance dimensions are as follows: (i) G4 EC1 that corresponds to GRI 201-1: Direct economic value generated and distributed; (ii) G4 EC4 that corresponds to GRI 201-4: Financial assistance received from government; and (iii) G4 EC7 that corresponds to GRI 203-1: Infrastructure investments and services supported. The implementation of these standard datasets related to the economic and finance dimension of the University of Milano-Bicocca is represented in the following Tables 3 and 4:

In GRI 201, "Information on the creation and distribution of economic value provides a basic indication of how an organization has created wealth for stakeholders." This evidence shows the fundamental role played by the funding of the Ministry of Education (MIUR) in supporting the university's activity, equal to 60% of the total funds, while funds coming from didactics represents only 19%. The direct economic value generated is distributed to the stakeholders in an amount equal to 87%, while the remaining 13% is held by the university. In particular, most of the value generated is attributed to human resources at 51%: 38% to academics and researchers and 13% to technical-administrative staff. The rest of the value is transferred to operating expenses (36%) based on the definition of "operating costs" provided by GRI 201 in the section entitled "Guidance" and to providers of capital and government (less than 1%) (Table 5).

Table 3 Direct economic value generated and distributed (EVG&D)

Direct economic value generated: sources of funds	€ 241,614,046.24	
MIUR and other central administrations contributions	€ 144,416,124.79	59.77%
Revenues from students	€ 46,610,950.92	19.29%
Revenues from research with competitive funding	€ 17,712,683.47	7.33%
Public entities contributions	€ 16,411,249.44	6.79%
Other revenues and income	€ 10,043,173.36	4.16%
Revenues from commissioned research and technology transfer	€ 3,681,575.28	1.52%
From others (private) contributions	€ 2,671,658.98	1.11%
European union and other international organisms contributions	€ 66,630.00	0.03%
Total	€ 241,614,046.24	100.00%

Table 4 Economic value distributed and retained

Economic value distributed (Uses of funds)		
Employee wages and benefits		
Teachers and researchers	€ 91,625,683.58	37.92%
Technical-administrative staff	€ 31,841,011.00	13.18%
	€ 123,466,694.58	51.10%
Operating costs	€ 87,212,790.82	36.10%
Payments to providers of capital	€ 44,885.62	0.02%
Payments to government by country	€ 340,410.00	0.14%
Community investments		
Total	€ 211,064,781.02	87.36%
Total direct economic value generated	€ 241,614,046.24	100.00%
Total economic value distributed	€ 211,064,781.02	87.36%
Total economic value retained	€ 30,549,265.22	12.64%
Depreciation	€ 9,534,910.20	3.95%
Provisions for risks and charges	€ 9,672,699.18	4.00%
Other costs	€ 2,167,244.79	0.90%
Extraordinary result	€ 263,790.42	0.11%
Profit	€ 9,438,201.47	3.91%

The University of Milano-Bicocca receives substantial contributions from the Italian government and from the European Union at 67% of total revenues.

It is important to analyse this dimension, as "the significant financial assistance received from a government, in comparison with taxes paid, can be useful for developing a balanced picture of the transactions between the organization and government" (GRI 201, Guidance).

In the specific case, these data, compared with taxes paid, show a deficit between public funding received (€ 160,894,004.23) and taxes paid (€ 340,410.00) equal

Table 5 Financial assistance from the central government

Financial assistance received from government		
MIUR and other central administrations contributions	€ 144,416,124.79	59.77%
Public entities contributions	€ 16,411,249.44	6.79%
(European Union and other International Organisms contributions)	€ 66,630.00	0.03%
Total financial assistance received from government	€ 160,894,004.23	66.59%
Total funds and contributions	€ 241,614,046.24	100.00%

Table 6 Reclassified balance statement

Reclassified balance sheet		
Intangibles	€ 2,539,162.53	0.52%
Material	€ 287,632,835.27	58.71%
Financial	€ 7,119,134.73	1.45%
Long-term assets	€ 297,291,132.53	60.68%
Others	€ 192,648,298.73	39.32%
Invested capital	€ 489,939,431.26	100.00%
Equity	€ 186,691,450.97	38.11%
Liabilities	€ 303,247,980.29	61.89%
Acquired capital	€ 489,939,431.26	100.00%

to € 160,553,594.23. Transactions between the government and the university are unbalanced in favour of the university. However, this situation can be explained in terms of the type of organization analysed, that is, this is not a private company but a public university with a social role for the Italian state community (Table 6).

In particular, the statement of financial position shows an important consistency of investments in fixed assets; in fact, this constitutes 61% of invested capital. Most of the investments are long-term material assets. Land and buildings play the main role. In particular, Bicocca used loans to acquire the majority of its campus buildings. Other buildings are rented (U6, U22, U72), while others are granted a concession for use, and for the latter a fee is paid that takes into account the services that the university offers to the surrounding area. The U16 building is home to a kindergarten and a nursery school, which can benefit families in the neighbourhood, as well as offer training courses.

In this asset analysis, therefore, the work of valuing the contribution made by the university to the local area has offered a useful insight into previously undervalued elements of the university's real estate investments.

These investments were financed by the allocation of state funds and capital grants, charged to the balance sheet using the deferred income method; these represented equity reserves and will gradually be recorded in the income statement in proportion to the depreciation rates of the assets. The composition of assets shows an incidence of 61% of fixed assets while loans and liquid assets amount to 39%. The liability structure shows that shareholders' equity amounts to 38% of the capital acquired, while the sources referring to loans and provisions for risks and accruals is equal to 62%. The comparison between the asset and liability structure highlights the sustainability of the University organization given a sustainable balance between the structure of loans and that of investments.

5 Conclusion and Discussion

This study addressed the sustainability issue of the University of Milano-Bicocca by analysing the several activities in which the university is engaged over a period of years and the assessment and reporting process, which is still under construction with particular regard to economic, finance and social dimensions. For these reasons, the paper proposes an integrated model of assessment and reporting which is founded on GRI standard datasets integrating the environmental, economic and financial, and social dimensions.

The experience of the University of Milano-Bicocca shows just how important the sustainability report is, not merely as a document, but as a complex process that involves inside and outside players and stakeholders. Organization, accounting and administrative procedures are part of the process. Environmental, economic and finance, and social dimensions are key points of the report and of the reporting process.

The experience shows also how the University of Milano-Bicocca is aware that sustainability issues need to be integrated with the organization, as described by Richards and Gladwin (1999), since sustainability management requires organizations to go beyond eco-efficiency by integrating environmental and social goals into all aspects of decision making. Because of this complexity, the sustainability issue in the case of Milano-Bicocca is still a work in progress concerning the assessment and reporting processes, as they require a management system able to create an integrated model. In this integrated model, the environmental dimension is strongly integrated with the information provided by the university website. The economic and finance dimension, allied with the social, is still under construction, but in the near future, the university will complete the assessment. The sustainability report will be the main output of an integrated process oriented towards both the internal and external stakeholders. The conducted analysis confirms that universities could learn from the experiences of corporate sustainability reporting efforts and incorporate them into their efforts as learning organizations to better align their systems with sustainability. There are also important implications for education issues.

In short, this study is important and contributes to the international literature debate for two main reasons: first, it enabled an outline of the major integrated areas in which universities—even in Italy—can contribute to the achievement of sustainability goals. The evaluation is unique and contributes several findings in some less-researched areas along with several practical, social, and research implications. Second, in following a business/economics approach, it reaffirms universities' commitment to sustainable research and to sensitizing and involving stakeholders through the promotion and organization of behaviours supportive of sustainable development. If the role of a university is to promote and provide education, this objective could be more coherently reached by integrating new subjects in the didactic programs with concrete actions involving all the stakeholders. With the integration of education and concrete actions, universities can induce a spillover of sustainability in their specific local areas.

The process in the University of Milano-Bicocca is still in progress and will also take into consideration the evidence coming from other experiences at the international level (Lopatta and Jaeschke 2014; Sassen and Azizi 2018). This is also a way to ensure comparability with the results of previous studies. For these reasons, we suggest the implementation of a fixed set of indicators (Sassen et al. 2014) embracing the GRI G4 guidelines, which do not address university-specific indicators, and integrating this standard with more specific indicators in accordance with other international experiences (Lozano 2011; Lopatta and Jaeschke 2014; Sassen et al. 2014). As an objective, the work in progress aims to implement the use of a standardized indicator system that may have the advantage of ensuring greater objectivity, transparency and comparability among the results. This is important also for the Italian legal system, where transparency issues are becoming more important, and for accountability generally.

For the future efficacy of ongoing projects, the impact of sustainability on performance measurement and reporting systems, the standardization in sustainability reporting, and the implication for the financial system require in-depth studies. Standardization is very important not only in terms of the creation of sustainability reports but also in terms of their interpretation as a unified method of reporting, which could reduce the amount of effort both for universities and for their stakeholders. This is an important issue leading to savings in cost, time and manpower as well as to a better allocation of resources. A unified reporting system would necessarily consider stakeholder expectations, as the information disclosed would naturally be of interest to them. Furthermore, the set of indicators for the university dimension should be developed further and be better adapted to the needs of universities.

The originality of the paper lies in the fact that (i) the business economics approach can be used as a basic framework for those who wish to determine the state of sustainability in their national or local university systems, (ii) it describes the activation of mechanisms to improve sustainability performance, (iii) it establishes the need for an academic debate on the responsibilities universities have towards society in terms of promoting models of sustainable behaviour, and (iv) it is able to help Italian universities make a change and increase their commitment to becoming strong examples of green behaviour in educating students and communities for sustainability and in promoting, at an international level, research to improve the quality of life without compromising the environmental requirements of future generations.

References

AASHE (2012) Association for the advancement of sustainability in higher education (AASHE). 2012. STARS. Available at: https://stars.aashe.org/. Last Accessed 20 June 2018

Adams CA (2013) Sustainability reporting and performance management in universities: challenges and benefits. Sustain Account Manag Policy J 4(3):384–392

Alonso-Almeida M, de Navarrete CF, Rodriguez-Pomeda J (2015) Corporate social responsibility perception in business students as future managers: a multifactorial analysis. Bus Ethics Eur Rev 24(1):1–17

Alshuwaikhat HM, Abubakar I (2008) An integrated approach to achieving campus sustainability: assessment of the current campus environmental management practices. J Clean Prod 16(16):1777–1785

Amaral LP, Martins N, Gouveia JB (2013) Quest for a sustainable university: a review. Int J Sustain High Educ 16(2):155–172

Bero BN, Doerry E, Middleton R, Meinhardt C (2012) Challenges in the development of environmental management systems on the modern university campus. Int J Sustain High Educ 13(2):133–149

Calder W, Clugston RM (2003) International efforts to promote higher education for sustainable development. Plan High Educ 31(3):30–44

Ceulemans K, Lozano R, Alondo-Almeida M (2015) Sustainability reporting in higher education: interconnecting the reporting process and organisational change management for sustainability. Sustainability 7(7):8881–8903

Chang HC (2013) Environmental management accounting in the Taiwanese higher education sector: issues and opportunities. Int J Sustain High Educ 14(2):133–145

Clarke A, Kouri R (2009) Choosing an appropriate university or college environmental management system. J Clean Prod 17(11):971–984

Cole L (2003) Assessing sustainability on Canadian University campuses: development of a campus sustainability assessment framework. Royal Roads University, Victoria, Canada

Corcoran SP, Evans WN, Schwab RM (2004) Changing labor-market opportunities for women and the quality of teachers, 1957–2000. Am Econ Rev 94(2):230–235

Cortese AD (2003) The critical role of higher education in creating a sustainable future. Plan High Educ 31(3):15–22

Dalal-Clayton B, Bass S (2002) Sustainable development strategies, 1st edn. Earthscan Publications Ltd, London

Eisenhardt KM (1989) Building theories from case study research. Acad Manag Rev 14(4):532–550

Filippín C (2000) Benchmarking the energy efficiency and greenhouse gases emissions of school buildings in central Argentina. Build Environ 35(5):407–414

Fisher RM (2003) Applying ISO 14001 as a business tool for campus sustainability: a case study from New Zealand. Int J Sustain High Educ 4(2):138–150

Gobo G (2004) Sampling, representativeness and generalizability. In: Seale C, Gobo G, Gubrium JF, Silverman D (eds) Qualitative research practice. Sage, London, p 435

Jabbour CJC (2010) Greening of business schools: a systemic view. Int J Sustain High Educ 11(1):49–60

Kamal ASM, Asmuss M (2013) Benchmarking tools for assessing and tracking sustainability in higher educational institutions: identifying an effective tool for the University of Saskatchewan. Int J Sustain High Educ 14(1):449–465

KPMG (2013) International survey of corporate responsibility reporting. Amsterdam, The Netherlands, KPMG. https://assets.kpmg.com/content/dam/kpmg/pdf/2015/08/kpmg-survey-of-corporate-responsibility-reporting-2013.pdf

KPMG International Cooperative (2011) KPMG International survey of corporate responsibility reporting 2011. KPMG International Cooperative, Bern, pp 4–7

Lopatta K, Jaeschke R (2014) Sustainability reporting at German and Austrian universities. Int J Educ Econ Dev (Inderscience Enterprises Ltd) 5(1):66–90

Lozano R (2006) Incorporation and institutionalization of SD into universities—breaking through barriers to change. J Clean Prod 14:787–796
Lozano R (2009) Auditing the contributions of Cardiff university curricula to sustainable development. In: GIN2009 Conference—Joint Actions on Climate Change, Aalborg
Lozano R (2011) The state of sustainability reporting in universities. Int J Sustain High Educ 12(1):67–78
Lozano-Ros R (2003) Sustainable development in higher education. Incorporation, assessment and reporting of sustainable development in higher education institutions. IIIEE, Lund University, Master thesis. http://lup.lub.lu.se/student-papers/record/1325193
McIntosh M, National Wildlife Federation & Princeton Survey Research Associates (2001) State of the campus environment: a national report card on environmental performance and sustainability in higher education: a national wildlife federation report
McIntosh M, Gaalswyk K, Keniry LJ, Eagan DJ (2008) Campus environment 2008—A national report card on sustainability in higher education. National Wildlife Federation, Reston, VA
Noeke J (2000) Environmental management systems for universities—a case study. Int J Sustain High Educ 1(1):237–251
Posner SM, Stuart R (2013) Understanding and advancing campus sustainability using a systems framework. Int J Sustain High Educ 14(3):264–277
Price TJ (2005) Preaching what we practice: experiences from implementing ISO 14001 at the University of Glamorgan. Int J Sustain High Educ 6(2):161–178
Richards DJ, Gladwin TN (1999) Sustainability metrics for the business enterprise. Environ Qual Manage 8(3):11–21
Roorda N (2001) AISHE: Auditing instrument for sustainable higher education. Dutch Committee for sustainable higher education. Available at: www.dho.nl/documents/AISHE-Book1.5.pdf. Last Accessed 20 June 2018
Sassen R, Azizi, L (2018) Assessing sustainability reports of US universities. Int J Sustain High Educ
Sassen R, Dienes D, Beth C (2014) Nachhaltigkeitsberichterstattung deutscher Hochschulen (sustainability reporting of German higher education institutions). Zeitschrift Für Umweltpolitik & Umweltrecht 37(3):258–277
Shriberg M (2002) Institutional assessment tools for sustainability in higher education. Int J Sustain High Educ 3(3):254–270
Social Sciences and Humanities Research Council of Canada (2002) Annual report. http://publications.gc.ca/collections/Collection/CR1-2002E.pdf. Last Accessed 23 June 2018
Spellerberg IF, Buchan GD, Englefield R (2004) Need a university adopt a formal environmental management system? Progress without an EMS at a small university. Int J Sustain High Educ 5(1):125–132
Stephen S, Maxey L, Heather L (2013) The sustainable university. Routledge studies in sustainable development. Routledge, New York
The Princeton Review (2012) College hopes and worries survey report. https://www.prnewswire.com/news-releases/princeton-reviews-2012-college-hopes–worries-survey-reports-on-10650-students–parents-top-10-dream-colleges-and-application-perspectives-44338495.html
Vagnoni E, Cavicchi C (2015) An exploratory study of sustainable development at Italian universities. Int J Sustain High Educ 16(2):217–236
Velasquez L, Munguia N, Platt A, Taddei J (2006) Sustainable university: what can be the matter? J Clean Prod 14:810–819
Waas T, Verbruggen A, Wright T (2010) University research for sustainable development: definition and characteristics explored. J Clean Prod 18:629–636
WCED (1987) Our common future. Report of the World Commission on Environment and Development. In: Brundtland GH (ed). Oxford, Oxford University Press, p 45
Wright T (2007) Developing research priorities with a cohort of higher education for sustainability experts. Int J Sustain High Educ 8(1):34–43

Yin RK (1994) Case study research: design and methods. Applied social research methods series, 5. Biography, Sage Publications, London

Professor Alessandro Capocchi is Associate Professor of Business Economics at the University of Milano-Bicocca. He has a Ph.D. in Business Economics in Public Administrations, and he has strong experience in research and academic activities.

Professor Paola Orlandini has a degree in Economics. She is currently Full Professor in the University of Milan-Bicocca. Her current research interests include sustainability, business administration and accounting history.

Cinzia Vallone is assistant Professor of Accounting, at the university of Milan-Bicocca. She has a Ph.D. in Business Administration and Strategy. Her research interests include Tourism management, sustainability and Family business studies.

Murad Harasheh received his Ph.D. in Economics, Law, and Institutions from the IUSS Institute for Advanced Studies (Pavia, Italy) in 2014 and master's and bachelor's degrees (with distinction) in Business and Finance from Birzeit University in Palestine from 2008–2011. He worked as a full-time Lecturer in Finance at Birzeit University from 2007–2011 and as a financial consultant and analyst from 2015–2017 serving as an economist and energy market analyst at the Italian Authority of Energy Regulations. Currently, he is a post-doctoral research fellow at the University of Milan-Bicocca, with primary research interests related to firm valuation, energy finance, and sustainability economics. He is the author of various publications in international journals in finance, business valuation, and energy.

Stefano Amelio received a master's and bachelor's degree in Business Administration (with distinction) from the University of Insubria (Varese) and a doctorate in Business Administration and Marketing (Ph.D.) from the University of Bergamo. In 2014–2015, he worked as an Associated Chartered Accountant and Statutory Advisor. From 2011 to 2015, he was a Voluntary Teaching Assistant in Business Administration; from 2015 to 2017, he was a Lecturer in Business Administration and Accounting at the University of Insubria; and since 2017, he has been a Lecturer in Business Administration at the University of Milan-Bicocca. His current research interests include sustainability, business administration and accounting history.

Social Responsibility and Sustainability: Initiatives of an Energy Company

Issa Ibrahim Berchin, Icaro Roberto Azevedo Picolli, Gabriela Almeida Marcon, João Marcelo Pereira Ribeiro, Jacir Leonir Casagrande and José Baltazar Salgueirinho Osório de Andrade Guerra

Abstract There is an increasing demand from society for more responsibility, transparence, accountability and sustainability from organizations. Organizations are responding to these demands by providing social services and empowering communities. Therefore, this paper aims to understand how energy justice, as a source of social responsibility, is approached in practice. In order to do so, a literature review on energy justice and social responsibility was developed, followed by a review of the case of a community park developed and maintained by a power generation company. According to the literature, studies on energy justice are growing in quantity and relevance, showing the increasing importance of this subject. Thus, case studies may be interesting ways of understanding its application in organizations' activities and projects.

I. I. Berchin (✉) · J. M. P. Ribeiro · J. B. S. O. de Andrade Guerra
Center for Sustainable Development (Greens), University of Southern Santa Catarina (Unisul), 219 Trajano Street, Florianopolis, Brazil
e-mail: issaberchim@gmail.com

J. M. P. Ribeiro
e-mail: jacir.unisul@gmail.com

J. B. S. O. de Andrade Guerra
e-mail: baltazar.guerra@unisul.br

I. R. A. Picolli · G. A. Marcon · J. L. Casagrande
Unisul, Florianopolis, Brazil
e-mail: icaropicolli@gmail.com

G. A. Marcon
e-mail: gabriela@almeidamarcon.com

J. L. Casagrande
e-mail: joaomarceloprdk@gmail.com

J. B. S. O. de Andrade Guerra
Department of Land Economy, Cambridge Centre for Environment, Energy and Natural Resource Governance (C-EENRG), University of Cambridge, Cambridge, UK

W. Leal Filho (ed.), *Social Responsibility and Sustainability*, World Sustainability Series, https://doi.org/10.1007/978-3-030-03562-4_21

Keywords Energy justice · Social justice · Renewable energies
Social responsibility · Sustainable development

1 Introduction

Increasing global economic, social and environmental challenges require innovative thinking and behavior to support sustainable development and promote quality of life for all. Social responsibility of organizations represents the voluntary commitment of the institution to perform its economic responsibilities, whilst complying with ethical behaviors sustainability (Sarkar and Searcy 2016; Palihawadana et al. 2016). Thus, corporate social responsibility works as a tool to promote justice and sustainability transformations of the society, which helps to improve social wellbeing, environmental preservation and economic development, avoiding injustices—such as those related to energy and the environment (Sovacool and Dworkin 2015).

Energy justice is key to social wellbeing, seeking to emphasize the social dimension of energy, going beyond technologies to improve energy efficiency, productivity, storage and transmission. Energy justice includes the ways which energy is produced, their external costs and who have access to it; it also integrates ethical, social and environmental concerns (Yenneti and Day 2016). In this context, energy justice as a practice for social responsibility of organizations comply with social, economic and environmental dimensions of sustainability. Sustainability of organizations imply a balance between its economic development and growth with its social responsibilities towards its employees, customers, suppliers and the community as a whole, respecting working conditions, human rights and promoting wellbeing; whilst promoting environmental preservation/conservation and avoiding environmental degradation and liability.

Regarding the emerging field of energy justice studies, further research is required to understand the relations between corporate social responsibility, sustainability and energy justice. Therefore, this study aims to understand how energy justice, as a source of social responsibility, is approached in an energy generation company.

2 Towards an Integrative Approach to Energy Justice: A Literature Review

Considered as a relatively new research area, energy justice emerged from the studies on environmental justice (Yenneti and Day 2015; Sovacool and Dworkin 2015; Heffron et al. 2015; Chatterton et al. 2016). In the same perspective, Jenkins et al. (2016) allege that energy justice arose from a new agenda of social sciences to incorporate justice principles in energy policy.

The concept of energy justice is defined by Sovacool and Dworkin (2015, p. 436) as "a global energy system that fairly disseminates both the benefits and costs of energy services, and one that has representative and impartial energy decision-making". In this respect, energy justice involves three key elements: distribution of costs and externalities through the society, distribution of benefits with the society, and procedures, ensuring that decision-making process are representative and respect due processes (Sovacool et al. 2016). Accordingly, Heffron and McCauley (2014, p. 437) allege, "energy justice aims to provide all individuals, across all areas, with safe, affordable, sustainable and secure energy sources".

Goldthau and Sovacool (2012), consider energy injustice as the lack of access to modern sources of energy (i.e. billions of people currently living on energy poverty). Therefore, energy justice can be achieved when all people can have access to modern sources of energy. There are three main challenges for global energy systems: energy security, energy justice and low carbon energy transition (Goldthau and Sovacool 2012).

According to Jenkins et al. (2016, p. 175), energy justice "evaluates where injustices emerge, which affected sections of society are ignored, which processes exist for their remediation in order to reveal, and reduce such injustices". Energy justice enables the identification of injustices and the development of processes to avoid and remediate it, it also "provides a new framework for bridging existing and future research on energy production and consumption, both of which can subscribe to the common goal of achieving just energy-based processes and outcomes" (Jenkins et al. 2016, p. 175).

Yenneti and Day (2015) also indicate some matters of interest to energy justice: fuel poverty, energy access, ethical energy consumption, renewable energy development and global coverage. Energy and fuel poverty can be considered major concerns for energy justice and policy-making (Sovacool 2015; Walker et al. 2016). According to Walker et al. (2016), energy poverty is the lack of access to affordable energy. Walker and Day (2012, p. 69), state that "addressing fuel poverty has to involve seeking justice in terms of the cultural and political recognition of vulnerable and marginalised social groups and pursuing procedural justice through opening up involvement and influence in decision-making processes".

Walker and Day (2012), Heffron and McCauley (2014), Heffron et al. (2015) identify three key elements of energy justice: distributional justice, procedural justice and justice as recognition. Distributional justice is related to a more inclusive and broad energy system. Procedural justice refers to the use of equitable procedures that inform and allow all stakeholders to participate in the decision-making. Justice as recognition means, "individuals must be fairly represented", having equal political rights. In this regard, Walker et al. (2016) also highlight the importance of participatory processes to define the necessities of a society and engaging people in policy-making.

By presenting the case of a solar energy park, Yenneti and Day (2015, p. 672) indicate that "procedural justice principles-providing detailed information, valuing local knowledge, listening to the communities through responding to their concerns, and securing the involvement of all affected communities" are important for both acceptances of the project and mitigation of externalities, and to promote a distributive justice. Thus, failures in procedural justice can increase energy injustices, increasing inequality, and reducing the acceptability of a project (Yenneti and Day 2016).

McCauley et al. (2016, p. 141) consider energy justice as "a new framework for assessing the justice implications—or simply the injustices—of current policy decisions as well as making practical recommendations". By developing an energy justice metrics, Heffron et al. (2015, p. 169) allege that energy justice can achieve a "just and equitable balance between the three dimensions of the Energy Trilemma" (i.e. the energy trilemma is the relations among energy finance, energy security, and climate change mitigation), diminishing injustices through fair policies.

Considering that people living below the poverty line pay proportionally more for energy, it imped them to accumulate wealth and run out of poverty (Sovacool and Brown 2010; Goldthau and Sovacool 2012), emphasizing energy injustice. In this respect, Hall (2013) promotes the debates on the importance of ethical consumption to promote energy justice.

The concept and implications of energy justice are intrinsically connected to ethical debates and, consequently, good governance. According to Sovacool and Dworkin (2015, p. 436), energy justice goes beyond philosophical debates and "can directly impact community livelihoods and the bottom line of energy corporations".

Still according to Sovacool and Dworkin (2015), energy justice can be achieved when three elements (social costs, benefits and procedures) are considered and respected, promoting benefits to producers and consumers, also enabling the fair share of costs among all stakeholders of a project. Demski et al. (2015, p. 66) indicate that among key public perspectives on their acceptance of sustainable energy transitions, there are "social justice and fairness", meaning that energy systems are developed in respect to "people's abilities to live healthy lives", resulting from a fair, inclusive and transparent process to all stakeholders.

The complexities of social responsibility practices have evolved from the attendance to legal/normative determinations to will of organizations to promote sustainable development. Accordingly, social responsibility of organizations can be defined as the voluntary will to perform its economic responsibilities, going beyond legal obligations, being ethical in their behavior and observing the impact of their actions on the stakeholders, further contributing to global sustainability (Sarkar and Searcy 2016; Palihawadana et al. 2016; Maas and Reniers 2014; Benites-Lazaro and Mello-Théry 2017).

Justice is a central element to the well-functioning of societies, so consultation processes with the community influenced by a project are essential to increase its acceptability and fairness. By analyzing the outcomes of a wind farm pilot study, Gross (2007, p. 2727) concluded, "perceptions of fairness do influence how people perceive the legitimacy of the outcome, and that a fairer process will increase acceptance of the outcome". In the same way, Swofford and Slattery (2010, p. 2517) also recognize that, especially for energy projects, "increasing levels of public participation during the early stages of a project will increase the likelihood of a project being accepted by the public". Still presenting the case of wind energy projects, Cowell et al. (2011) allege that the benefits of such projects to the communities directly influenced can only be achieved by involving the community in the decision-making process, also empowering them to veto the project if needed.

Toke et al. (2008, p. 1136) also allege that "when investors come from outside the community or when higher tiers of government try to site wind power facilities without involving local communities, they may easily create mistrust and the process can be perceived as unfair. It has been argued that developers will have greater success if they involve the local community (the public, as well as stakeholders) in decisions". Swofford and Slattery (2010) also indicate that this participation in early stages of the project "can consist of a variety of formats including survey questionnaires, public meetings, focus groups, and semi-structured interviews".

3 Methods

This study is a qualitative analysis of the social responsibility practices of a energy generation company through its environmental park initiative. The object of study is an international energy company based in Brazil.

The data was collected mainly from in-depth interviews with senior managers of the company though semi-structured interviews: the first interviewee was the manager of the Environmental Park (I1) and the second was the sustainability advisor of the company (I2). The interviewees were chosen due to their higher knowledge and their direct responsibility for the company's sustainability plan and the environmental park. The interviews were made between February 2017 and May 2017.

Accordingly, the main procedures of this research were: Literature review to understand energy justice and its main principles, selection of the case study, development and application of interviews, and analysis of the data collected from the interviews related to the principles of energy justice gathered from the scientific literature.

4 Case Study: An Energy Generation Company in Brazil

Earlier named Tractebel Energia, the company was bought by Engie in 2016, and the core business of the company is the energy generation. The company have the installed capacity to produce 7010 MW, equivalent to 6.2% of total Brazilian needs per year and nearly 5% of the total installed capacity in Brazil—i.e. which was equivalent to 140.858 MW in 2015 (Brazilian Energy Balance 2016; Engie 2017).

The company have 29 power plants in Brazil, nine of them are hydroelectric power plants, four thermoelectric power plants, and sixteen complementary plants based on three biomass plants, nine windfarms, three small hydropower plants and a solar park (Engie 2017). The French-Belgian group Engie controls Engie Brasil, which is the largest independent power producer in the world with an installed capacity of

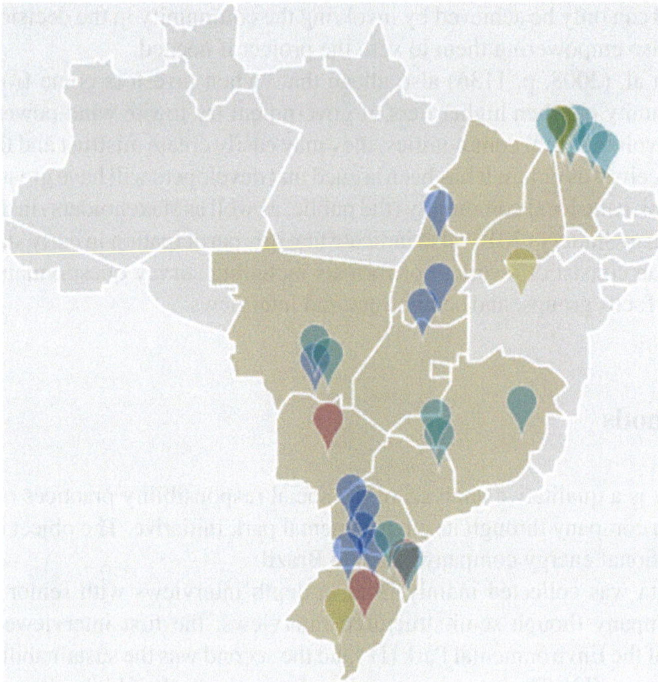

Fig. 1 Engie power plants in Brazil. *Source* Engie (2017)

117 GW. In 2016, Engie Brasil valued nearly 5 billion dollars, employing more than a 1000 people (Engie 2017).

Engie is the largest private energy generation company in Brazil. The company has power plants in many parts of the country, as illustrated on Fig. 1 (Engie 2017).

The main installed electricity generation capacity of the company in Brazil is illustrated on Table 1.

The Environmental Park was inaugurated in 2014 in a city located in the South of Brazil. The park was designed in a collaboration between five institutions from different sectors: Engie Brasil (former Tractebel), which was the founder and financer of the project; the University of Southern Santa Catarina (Unisul); the University of Capivari de Baixo (FUCAP); the Association of Municipalities of the Laguna region (AMUREL), and the Association of Tractebel Energia and Eletrosul Employees (ASES) (Engie 2017).

The idea of building the Environmental Park emerged from the will to recover the degraded area occupied by the Thermoelectric Complex of the company (which mainly used coal for energy generation) and improve the quality of life in the surrounding communities. The exploitation of coal was the main activity responsible for the economic development of the southern cities in Brazil between 1940 and 1980. In

Table 1 Installed electricity generation capacity of the company in Brazil

Hydroelectricity power plants		Thermoelectricity power plants		Additional sources (Eolic, solar, small hydropower plants)	
Power plant	Installed capacity (MW)	Power plant	Installed capacity (MW)	Power plant	Installed capacity (MW)
Cana Brava	450	Jorge Lacerda A	0	Areia Branca	20
Estreito	436	Jorge Lacerda A1—2 × 50 MW	100	Beberibe	26
Itá	1.127	Jorge Lacerda A2—2 × 66 MW	132	Cacimbas	19
Machadinho	404	Jorge Lacerda B	262	Estrela	19
Passo Fundo	226	Jorge Lacerda C	363	Ferrari	81
Ponte de Pedra	176	William Arjona	190	Fleixeiras I	30
Salto Osório	1.078			Fotovoltaica Cidade Azul	3
Salto Santiago	1.420			Guajiru	30
São Salvador	243			Ibitiúva	23
				José Gelazio	24
				Lages	28
				Mundaú	30
				Pedra do Sal	18
				Rondonópolis	27
				Santa Mônica	19
				Trairí	25
				Tubarão	2

Source Engie (2017)

1960, with the possibility to use metallurgical coal for thermoelectricity generation, the Thermoelectric Complex was built (Engie 2014).

The wastes produced by the burning of coal in Brazil is the largest source of resources for thermoelectricity generation in the country. The environmental impacts produced by these power plants are mainly characterized by the incorrect waste disposal, by the particles (heavy and light ashes) emitted as pollutants in the air, soil and water, and by the high greenhouse gas emissions (Barros 2013; Alves et al. 2011). Accordingly, power generation companies might develop and adopt responsible and sustainable technologies and practices to avoid negative social and environmental impacts, reducing the amount of pollutants produced (i.e. reducing greenhouse gas emissions, sustainably managing its wastes disposal, and attending to the main regulation policies and civil responsibilities) (García-Ubaque et al. 2013; Barros 2013; Alves et al. 2011).

Due to the tons of residues emitted by these practices, the production of metallurgical coal and its use for thermoelectricity generation creates several externalities for both the environment and the communities in the surroundings. The tons of wastes generated by the burning of metallurgical coal from the Thermoelectric Complex was stored/disposed on the ground for many year, without any environmental control or monitoring of its impacts for neither human population, nor the environment. Only by the end of 2008, after removing 2.1 million tons of coal disposed on the ground, with an investment of nearly 2 million dollars, the company accomplished the recovery of the area, currently used as the Environmental Park (Engie 2014).

Currently the park is a space for community leisure, disposing of a Center for Culture and Sustainability, a Theater with 370 seats, an Exhibition space with 730 m^2, Outdoor Plants parts Museum, Multipurpose area for recreation, Acoustic shell for shows for up to 30 thousand people, Space for gymnastics, 3.5 thousand meters of bike paths, 4 thousand meters of hiking trails, Photovoltaic module that can supply the needs of up to 30% of the park and lake with 14,500 m^2. Therefore, this entire infrastructure resulted in the improvement of air quality and the better utilization of the area (Engie 2014).

5 Results and Discussion

Transparency and ethics are at the core of energy justice and the social and environmental responsibility of organizations, demanding them to seek the wellbeing of the communities affected by their actions (Sovacool and Dworkin 2015; Yenneti and Day 2015; Hall 2013). Despite the broadness of the term, energy justice is related to the promotion of clean, renewable, accessible and affordable energy to all, since energy is produced in safe ways to both society and the environment, attending to three dimensions: procedural justice, distributional justice and justice as recognition (Sovacool et al. 2016; Yenneti and Day 2015; Walker and Day 2012; Heffron and McCauley 2014; Heffron et al. 2015; Fuller and McCauley 2016; Newell et al. 2011; Sovacool and Dworkin 2015).

Thus, organizations might avoid externalities that affects local communities, but they must compensate these externalities whenever it happens, also promoting tools to improve the wellbeing of these communities and enabling them to benefit from the organizations' activities (Sovacool and Dworkin 2015; Walker et al. 2016; Swofford and Slattery 2010). Energy justice require more transparency and engagement of civil society in decision making processes, raising awareness about the organization's plans and activities, increasing its acceptability and legitimacy (Sovacool et al. 2016; Walker et al. 2016; Newell et al. 2011; Gross 2007; Cowell et al. 2011; Toke et al. 2008; Swofford and Slattery 2010) (Table 2).

The development of interviews demonstrated the applicability of energy justice principles in a power generation company, specifically regarding the development and maintenance of a community park to increase the community's wellbeing and reduce the negative impacts of the organizations' activities in the past.

Table 2 Main dimensions of energy justice

Dimension	Definition	Authors
Procedures	It is related to the development of transparent, fair and equitable procedures "that inform and allow all stakeholders to participate in the decision-making"	Sovacool et al. (2016), Yenneti and Day (2015), Walker and Day (2012), Heffron and McCauley (2014), Heffron et al. (2015), Fuller and McCauley (2016), Newell et al. (2011), Yenneti and Day (2016), Sovacool and Dworkin (2015)
Distribution and inclusion	It is related to the access of all people to modern energy, broadening energy systems, promoting inclusion and development	Sovacool et al. (2016), Yenneti and Day (2015), Walker and Day (2012), Heffron and McCauley (2014), Heffron et al. (2015), Fuller and McCauley (2016), Sovacool and Dworkin (2015)
Justice as recognition	It is related to the acceptability of an activity of project by the community affected/influenced by it. Therefore, it requires the engagement of all possible stakeholders, their awareness and their empowerment in decision-making	Sovacool et al. (2016), Walker and Day (2012), Heffron and McCauley (2014), Heffron et al. (2015), Walker et al. (2016), Newell et al. (2011), Sovacool and Dworkin (2015)

The understanding of the interviewers regarding energy justice are: I1 "*My knowledge about energy justice is very shallow, I understand that should be a very new topic. I think energy justice is linked to not allowing electricity shortages, for it to be used in actions of* […] *subsistence of the population and what is necessary for production, to meet the primary elements of production of a given community*" accordingly I2 understand the term as "*My conception today is more on the basis of distributed generation than in the centralized one. The Brazilian project "Light for All"* [Luz para Todos], *has its merits. Now, solar energy is becoming cheaper, this has already happened and it is in my view a great solution for a country like Brazil, with so many areas still isolated, with so much social need, is the prospect of exchanging energy with local concessionaires, and this, with a distribution network*".

When asked if and how the company embraces energy justice, the interviewees answered that I1 "*Within the environmental park we predict that in the future each unit will be autonomous in its energy production, i.e. … There is inside the environmental park a pilot production of energy through solar panels. We see that the conversion capacity in solar and electric energy in the country, and in that region, is very large*", thus, I2 stated that "*Engie and energy justice: This is one of Engie's mottos of the parent company in Paris. Energy is an essential good for life, that's the idea. Even Engie has a very interesting initiative called 'le rationnement de l'énergie'. The group now focuses on three axes: decentralization, digitization and decarbonization*".

Both interviewees stated that the organization developed the park to diminish its environmental impacts by recovering the quality of the soil (i.e. which served as a disposure of wastes of the power plant), to improve the wellbeing and the quality of life of local population that suffered from low infrastructure for leisure in the region and from bad air quality in the cities due to the gases emitted by the power plant. According to E1, the private investments made by the company was based on (1) soil recovery over the sedimentary basins and coal deposit, (2) addition of layers of quality/fertile soil, (3) reforestation with native/endemic species, (4) development of installations for maintenance of the park and facilities for community's leisure and entertainment that "*has changed all the dynamics of the surrounding cities*".

Both interviewees highlighted the initiative of the company supported by the engagement of the university, the civil society and the local government was essential to develop, implement and maintain the park and ensure its acceptability, legitimacy and success.

When asked if/how the community was involved in the project, I2 answered that "*the community participated in the idealization of the park, mainly through the association of the residents of the region*" and still "*this association has a voice in the administration of the park*". I2 emphasized that the park raises awareness about sustainability issues by enabling the community to have contact with solar panels and wind power generators, and with other mechanisms in the park such as lectures, rainwater storage and recycling. Thus, considering that the region lack of resources and infrastructure "the park was a considerable increment in both leisure and culture" for the surrounding cities.

According to I1 and I2, the park receives nearly 100 thousand visitors per year and it is free of charges for visitors. They also stated that the park is seeking for sustainability in all its dimensions including economic sustainability and financial autonomy from the company, by defining an annual cultural agenda supported by governmental financial programs.

Further explaining the benefits of the park, I1 stated that comparing to the total profits earned by the company, the development and maintenance of the park is very small and contributes not only to increase the quality of life of the population, but also you improve the company's social balance and its brand, and increase the willingness of the employees that in majority live in the surrounding cities, enhancing their psychological bound with the company.

Regarding the three dimensions of energy justice presented (i.e. Procedures, Distribution and Inclusion, and Justice as Recognition); it is possible to interpret the discourse of the interviewees as if the company, particularly regarding the development and maintenance of the Environmental Park, attended to these principles, as it is illustrated in Table 3.

6 Conclusions

By reviewing the scientific literature, it is noticeable the increase of publications on energy justice, however, it is necessary to further understand the scope of this

Table 3 Implementation of energy justice principles and responsibilities by Engie through the Environmental Park

Dimension	I1	I2
Procedures	The local community was engaged in the development and implementation of the park, being aware and being consulted about all its actions	The local community, local higher education institutions and the local government were aware of the project for developing the park and were engaged with it. Thus, the monitoring and follow up of the project is made by the number of people visiting the park and attending to the events and courses promoted by the park
Distribution and inclusion	The development of the park attending to sustainability criteria (i.e. storing rainwater and saving and reusing water, and producing renewable energy) people had access to these technologies, increasing their awareness	The implementation of solar panels and an Eolic generator in the park allow people to have access to modern energy production, raising awareness and knowledge in the community
Justice as recognition	To reduce the externalities of the Thermoelectric Power Plant and engage the community in the development, implementation and use of the park, the company consulted them about their preferences. To avoid the development of a park that wouldn't be used by the community, the company developed socioenvironmental activities, sociocultural activities and activities related to culture, leisure, education and entertainment, attracting people to the park and its activities	The stakeholders, especially the local community influenced by the project, were engaged sensitized since the beginning of the project in a collaboration between higher education institutions, the local government, the civil society and the company. Thus, the community still have a seat in the management of the park

emerging area. In this regard, by analyzing practical/specific realities, case studies could support the consolidation of this emerging area in the literature.

The increasing societal demands for more responsibility, transparency, accountability, ethics and sustainability of public and private organizations require these institutions to transform themselves into drivers of social change towards more equitable, fair and sustainable development. Among the different paths to meet these demands, social services and community engagement and empowerment appear to be a reasonable solution. Therefore, organizations as energy generation companies are seeking to promote initiatives that reflect their responsibility and commitment to social development and environmental protection when planning, developing and implementing their projects. Also, creating mechanisms and environments to improve the community's wellbeing.

According to the main dimensions of energy justice (Table 2), the company regarded the concept of energy justice, attending to its social responsibilities within the community, recovering the degraded land and returning it to the community as a source of leisure, culture and education, involving them in the planning, development and maintenance of the park (Procedural justice). Thus, by empowering the community, including them in the processes and promoting renewable energy generation that meet the park's needs and allow the community to use this energy freely (Distributional justice and Inclusion), the company increased its image/brand, increasing its acceptability and the legitimacy of its actions, also enhancing the psychological bond of the community to the company (Justice as Recognition).

This paper contributes to the literature on social responsibility by adding the concepts of energy justice into its scope. Thus, reviewing the initiative of a large energy company helps to understand the responsibility of companies towards its operations in the production and disposal of wastes and regarding the wellbeing of the communities in its surroundings. Future studies could investigate social responsibility and sustainability practices in companies from other sectors, comparing their practices and sustainability initiatives.

References

Alves MR, Rodríguez M, Roseta-Palma C (2011) Sectoral and regional impacts of the European carbon market in Portugal. Energy Policy 39:2528–2541. https://doi.org/10.1016/j.enpol.2011.02.018

Barros J (2013) Performance measurement of thermoelectric generating plants with undesirable outputs and random parameters. Electr Power Energy Syst 46:228–233. https://doi.org/10.1016/j.ijepes.2012.10.019

Benites-Lazaro L, Mello-Théry N (2017) CSR as a legitimatizing tool in carbon market: evidence from Latin America's clean development mechanism. J Clean Prod 149:218–226. https://doi.org/10.1016/j.jclepro.2017.02.095

Brazilian Energy Balance (2016) Brazilian energy balance: final report 2016. https://ben.epe.gov.br/downloads/Relatorio_Final_BEN_2016.pdf. Accessed on 11 Nov 2017

Chatterton TJ, Anable J, Barnes J, Yeboah G (2016) Mapping household direct energy consumption in the United Kingdom to provide a new perspective on energy justice. Energy Res Soc Sci 18:71–87. https://doi.org/10.1016/j.erss.2016.04.013

Cowell R, Bristow G, Munday M (2011) Acceptance, acceptability and environmental justice: the role of community benefits in wind energy development. J Environ Planning Manage 54:539–557. https://doi.org/10.1080/09640568.2010.521047

Demski C, Butler C, Parkhill KA, Spence A (2015) Public values for energy system change. Glob Environ Change 34:59–69. https://doi.org/10.1016/j.gloenvcha.2015.06.014

Engie (2014) Parque Ambiental Tractebel completa um ano. http://www.engieenergia.com.br/wps/wcm/connect/087ce848-b745-4ae3-b123-b37b41607b87/WEB_tcb_revista_boasnovas_n51.pdf?MOD=AJPERES&CACHEID=ROOTWORKSPACE087ce848-b745-4ae3-b123-b37b41607b87. Accessed on 11 Nov 2017

Engie (2017) The company. http://www.engieenergia.com.br/wps/portal/internet/a-companhia. Accessed on 11 Nov 2017

Fuller S, McCauley D (2016) Framing energy justice: perspectives from activism and advocacy. Energy Res Soc Sci 11:1–8. https://doi.org/10.1016/j.erss.2015.08.004

García-Ubaque CA, González-Hässig A, Vaca-Bohórquez ML (2013) Ceramic bricks made from municipal solid waste incineration-derived clay and ashes: a quality study. Ingeniería Investigación 33(2):36–41

Goldthau A, Sovacool BK (2012) The uniqueness of the energy security, justice, and governance problem. Energy Policy 41:232–240. https://doi.org/10.1016/j.enpol.2011.10.042

Gross C (2007) Community perspectives of wind energy in Australia: the application of a justice and community fairness framework to increase social acceptance. Energy Policy 35:2727–2736. https://doi.org/10.1016/j.enpol.2006.12.013

Hall SM (2013) Energy justice and ethical consumption: comparison, synthesis and lesson drawing. Local Environ Int J Justice Sustain 18:422–437. https://doi.org/10.1080/13549839.2012.748730

Heffron RJ, McCauley D (2014) Achieving sustainable supply chains through energy justice. Appl Energy 123:435–437. https://doi.org/10.1016/j.apenergy.2013.12.034

Heffron RJ, McCauley D, Sovacool BK (2015) Resolving society's energy trilemma through the Energy Justice Metric. Energy Policy 87:168–176. https://doi.org/10.1016/j.enpol.2015.08.033

Jenkins K, McCauley D, Heffron R, Stephan H, Rehner R (2016) Energy justice: a conceptual review. Energy Res Soc Sci 11:174–182. https://doi.org/10.1016/j.erss.2015.10.004

Maas S, Reniers G (2014) Development of a CSR model for practice: connecting five inherent areas of sustainable business. J Clean Prod 64:104–114. https://doi.org/10.1016/j.jclepro.2013.07.039

McCauley D, Heffron R, Pavlenko M, Rehner R, Holmes R (2016) Energy justice in the Arctic: Implications for energy infrastructural development in the Arctic. Energy Res Soc Sci 16:141–146. https://doi.org/10.1016/j.erss.2016.03.019

Newell P, Phillips J, Mulvaney D (2011) Pursuing clean energy equitably. United Nations Development Programme, New York. http://hdr.undp.org/en/content/pursuing-clean-energy-equitably. Accessed on 11 Nov 2017

Palihawadana D, Oghazi P, Liu Y (2016) Effects of ethical ideologies and perceptions of CSR on consumer behavior. J Bus Res 69(11):4964–4969. https://doi.org/10.1016/j.jbusres.2016.04.060

Sarkar S, Searcy C (2016) Zeitgeist or chameleon? A quantitative analysis of CSR definitions. J Clean Prod 135:1423–1435. https://doi.org/10.1016/j.jclepro.2016.06.157

Sovacool BK, Brown MA (2010) Competing dimensions of energy security: an international perspective. Annu Rev Environ Resour 35:77–108. https://doi.org/10.1146/annurev-environ-042509-143035

Sovacool BK, Dworkin MH (2015) Energy justice: conceptual insights and practical applications. Appl Energy 142:435–444. https://doi.org/10.1016/j.apenergy.2015.01.002

Sovacool BK (2015) Fuel poverty, affordability, and energy justice in England: policy insights from the Warm Front Program. Energy 93:361–371. https://doi.org/10.1016/j.energy.2015.09.016

Sovacool BK, Heffron RJ, McCauley D, Goldthau A (2016) Energy decisions reframed as justice and ethical concerns. Nat Energy 1:1–6. https://doi.org/10.1038/NENERGY.2016.24

Swofford J, Slattery M (2010) Public attitudes of wind energy in Texas: local communities in close proximity to wind farms and their effect on decision-making. Energy Policy 38:2508–2519. https://doi.org/10.1016/j.enpol.2009.12.046

Toke D, Breukers S, Wolsink M (2008) Wind power deployment outcomes: how can we account for the differences? Renew Sustain Energy Rev 12:1129–1147. https://doi.org/10.1016/j.rser.2006.10.021

Walker G, Day R (2012) Fuel poverty as injustice: integrating distribution, recognition and procedure in the struggle for affordable warmth. Energy Policy 49:69–75. https://doi.org/10.1016/j.enpol.2012.01.044

Walker G, Simcock N, Day R (2016) Necessary energy uses and a minimum standard of living in the United Kingdom: energy justice or escalating expectations? Energy Res Soc Sci 18:129–138. https://doi.org/10.1016/j.erss.2016.02.007

Yenneti K, Day R (2015) Procedural (in)justice in the implementation of solar energy: the case of Charanaka solar park, Gujarat, India. Energy Policy 86:664–673. https://doi.org/10.1016/j.enpol.2015.08.019

Yenneti K, Day R (2016) Distributional justice in solar energy implementation in India: the case of Charanka solar park. J Rural Stud 46:35–46. https://doi.org/10.1016/j.jrurstud.2016.05.009

Incorporation of Materials that Improve Durability, Resistance to Water and Anti-seismic Behaviour for Sustainability Earthen Construction Techniques Used in Rural Housing in Ráquira, Colombia

Franz Calderon

Abstract In Latin-American countries materials such as concrete are still a preferred option in the construction of housing, even though, their exploitation has caused environmental damages, and they are expensive. This situation is affecting the most vulnerable population, especially in developing and not industrialized countries. This paper proposes an ecological and rural housing building innovation: a panel made from natural and low cost materials for the construction of housing in rural areas of developing countries. The study was developed in Ráquira, a rural region located 120 km from Bogota, Colombia. For the purpose of the project, we were studying, about 10 samples and 3 prototypes of the panels were made. Tests were performed by mixing different materials such as sisal, hay, recycled rubber, wood in very small sections, and wire mesh of steel. In addition, the testing provided additional data about the materials that would give the earth more stability and strength.

Keywords Earthen construction · Innovation · Sustainable housing Sustainable materials

1 Introduction

In this research, the CO_2 emissions of each of the evaluated materials were quantified and in the final prototype, the emissions were well below the values emitted by other traditionally used materials (steel and cement), guaranteeing the sustainability of the resulting materials. Additionally, the project is socially responsible, since the final prototype was built in family housing with few economic income, formed two women as family heads and a child of 5 years. Taking into account the aspects contemplated

F. Calderon (✉)
Universitaria Agustiniana, Avenida Ciudad de Cali 11B-95, Bogotá, Colombia
e-mail: franz.calderon@uniagustiniana.edu.co

W. Leal Filho (ed.), *Social Responsibility and Sustainability*, World Sustainability Series, https://doi.org/10.1007/978-3-030-03562-4_22

by the ISO 26000 standard, responsibility to the environment, responsibility to the community and society.

The Frugal Innovation is a manufacturing concept has focused on industrial processes that has three basic factors, low cost in materials, low cost in manufacturing and minimal features with basic functions. The Frugal Innovation is being used by large multinationals in emerging countries, mainly China and India, to achieve product development in these countries companies not only apply the three factors, but also establish local organizational structures that allow the development Are tailored to specific needs and are developed with the technology available in each country, this translates into boosting local R&D, with local knowledge about the product to be developed.

Companies that want to commit and start in their subsidiaries Frugal Innovation should understand that the main characteristic of Frugal Innovation is high benefits for the customer with very low production costs. The road to its implementation involves the creation of local R&D teams and the construction of a local infrastructure to facilitate the work of these R&D teams.

In architectural design, this concept is new although without being in the initial objectives of the research, the final results incorporated the basic concepts of frugal innovation, incorporated in the sustainable development. Since the formal definition of sustainable development established in the UN document known as the Brundtland report as "meeting the needs of the present without compromising the ability of future generations to meet their own needs" (Brundtland report WCED 1987). It has been adapted to the needs of each field of study. Consequently, the International Union of Architects (UIA) and the American Institute of Architects (AIA) in Chicago in 1993, defined the principle of sustainability in its, Declaration of Interdependence for a Sustainable Future, with the following commitment: 'to place environmental and social sustainability at the core of our practices and professional responsibilities'.

Currently, the Building Services Research and Information Association (Association for Information and Research Facility Buildings, BSRIA) have defined sustainable construction as 'the creation and management of healthy buildings based on ecological principles and the efficient use of resources' (Kibert 2009). Moreover, recognized practitioners as Norman Foster + Partners, have defined sustainable architecture as "the creation of buildings that are efficient in terms of energy consumption, health, comfort and flexibility in use and are designed to have a long life".

However, the concept of sustainable development has numerous variables, but a common denominator, which is based on growth and social welfare; it must ensure the conservation of environmental resources by the present generation, for the benefit of future generations. The materials used in the construction exert environmental impact caused by the extraction, processing, transport, use and disposal. This impact occurs on the global and regional level affecting the climate, biodiversity and health of people. The material earth used by pre-Colombian cultures for making Bahareque and which was built in Colombia was gradually replaced by cement in the twenties with other materials that emerged from the industrial revolution Colombia, such as steel and cement. Then the earth was as material was associated with poverty and replaced by modern alternatives (Riveros 2007).

Traditionally, this material has four ways to be worked as a construction material are adobe, rammed earth, the mud and the BTC (Block compact earth), the latter in Colombia was named in the sixties CINVA-RAM thanks to research in CINVA ("Centro Interamericano de Vivienda") conducted by the National University of Colombia and OEA, had as main objective to facilitate the construction of housing with economic and sustainable materials. Investigations are focused on adding sustainable earth mix that will allow it to withstand compressive loads and improve their physical quality materials (Peña and Martha 2008).

Currently the earth used as building material in Colombia has been rediscovered thanks to their sustainable characteristics and its history. Moreover, the earth as a building material is economical and easy accessibility for users. Designers should envisage "simple" products that can be synthesized with a minimal number of components and/or features for minimizing resource consumption (Balkrishna 2017). The urge to design products for all round sustainable development has led to the advent of methodologies such as design for circularity. These approaches, which are rooted in sustainability, conserve materials by reusing, recycling and reduction mechanisms; controlling material throughput through each stage of product life cycle; and offering a mix of products and services that cater sustainable to the needs of society; respectively.

According to the growing interest of architecture in the development of sustainable materials and construction techniques, the aim of this paper is to prove a new construction technique based on earth as traditional material, natural fibres and a low quantity of expensive materials. Moreover, this research can contribute to develop frugal innovations in this topic and enrich the sustainable architecture as field of study.

This paper is divided into five sections. After this introduction, Sect. 2 describes the research method used to find answers to the research question. Section 3 analyses the outcomes of our research and prototyping process. Section 4 presents the discussion of the results. Finally, Sect. 5 concludes with a summary of the main research results.

2 Methods

This project was developed in Candelaria's Desert Ráquira in 2016, had a duration of 10 months. It developed following two research methods. In the first stage, qualitative data gathered through semi-structured interviews to the rural population. In this stage, the main objective was to determine and identify materials typology, used in the zone, for the construction of houses.

The house selected for the case study, was chosen for its antiquity and the materials under which it was built (Earthen bricks, adobe), in this paper we will call it the Fanny's house. The analysis started by making an architectural drawing and photographic reconnaissance.

After this initial analysis some adobes were taken and mixing earth with lime to be used as a coating, see Fig. 1. Based showing the adobe Fanny's house, with

Fig. 1 The Fanny's house, old earthen bricks used in the first test

the first prototype, gaveras were created (wood moulds for making adobe) to prove that sustainable materials could improve the resistance of the earth. Recycled rubber, sisal, horse droppings and cement among others were used.

The methodology was considered a chronological experimental quantitative case study. The response provided by them is that: (a) co-variation between the independent variable (materials) and the dependent (s), (existing housing) (b) the application of the independent must precede the measurement of the Dependent variable and (c) other variables, other than the independent one, have to be discarded as possible explanations of the changes observed in the dependent ones.

Three panels were built in wood with a different core material, with the aim to simulate different types of Bahareque (construction technique of Latin-Americans indigenous) (de León 2012). In each of these panels were exposed to compression. The compression test is performed on a material to determine the Strength of the material to the compressive stress, the material is subjected to the stress until deformed and to fail by the forces applied to the ISO 604. The tests with a mixture of earth and cement, determine the resistance to horizontal movement and the degree of adhesion of this mixture to the material that served as support, we finally managed to get a stable mixture and the core material will not deform under the weight of the earth.

Based on these tests, a panel was chosen in the construction of a prototype timber in the Fanny's house and the patent registration was initiated in the official agency.

3 Results

In the first stage, these were the results of mixing materials and earthen Adobe bricks, the objective were to find a material with better resistance than Adobe bricks. Each mixture was subjected to a compression test, only because the objective of this research was not to find a seismic-resistant material, but a sustainable material that worked as a structural enclosure. See Table 1.

The results were clear, the mixture improved the resistance to compression, 90% of earth, 6% grey cement, 4% hay; these percentages was used in the manufacturing of the panels in Bahareque. In the research, process took four mixtures of more outstanding materials and taking data BEDEC the ITEC (Catalan Institute of Construction) calculate the CO_2 emissions of these adobe bricks from aspect mix 3, it generated less emission than other tests. See Table 2.

In the manufacturing process of the adobes, it was found that adobe bricks have a very high weight, almost 10 kg per unit. Due to the weight of the bricks and based on the fact that the construction of houses would be carried out by women of the region this material was discarded. This information was used for this phase to the fabrication of a stable mixture.

After doing the tests with adobe bricks, the tests were started with panels of bahareque, in these tests the objective was to find sustainable materials that following the premises frugal innovation will improve the traditional manufacturing technique.

Table 1 Testing different materials in Adobe bricks fabrication

	% Earthen	% Cement	% Recycled rubber tire	% Horse droppings	% Hay	% Dramix	% Ceramic ships	Resistance to compression kg/cm^2
1	82	4	–	–	4	10	–	44.5
2	88	2	10	–	–	–	–	34
3	90	6	–	–	4	–	–	74
4	80	4	–	–	4	–	10	15.5

Table 2 Kg/m^2, CO_2 materials emissions

ID	Earthen	Cement	Recycled rubber tire	Horse droppings	Hay	Dramix	Ceramic ships	Total emissions CO_2 kg/m^2
1	0.001	2.57	–	–	0.013	11.43	–	14,014
2	0.001	2.57	4.5	–	–	–	–	7071
3	0.001	2.57	–	–	0.013	–	–	2584
4	0.001	2.57	–	–	0.013	–	0.5	3084

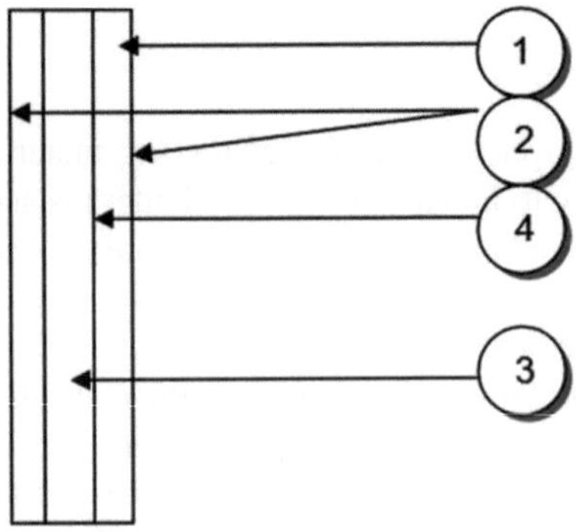

Fig. 2 First panel, materials description section

Three panels were made, from which the pros and cons were analysed to finally create a fourth panel that was patented (Fig. 2).

Panel 1

On the next stage, the construction of Bahareque's models was implemented. The panel 1, wood, fabric sisal and earth mixture 90% and cement 10%, resulted in a very unstable panel, in this frame, it was determined that the fabric of pine has little resistance in addition to raising manufacturing costs. Weaving Fique was not well anchored to the wooden frame and also show much instability when putting on the mixing, this instability resulted in an effect of runoff of the mixture, sisal fibres adhere very well to the earth, which does not favour their use in rural households. See Figs. 3 and 4.
CO_2 emissions, distribution of materials:

1. Earth Thermal Inertia: 1000 J/kg °C, CO_2 emissions: 0.001 kg/m^2, Percentage used in the mix 94%
2. Wood frame, CO_2 emissions: 0.99 kg/m^2
3. Fique[1]'s tissue CO_2 emissions: 0.1 kg/m^2
4. Cement, CO_2 emissions: 2.57 kg/m^2, Percentage used in the mix 6%.
   Total CO_2, emissions: 2.25 m^2 × 3341 kg/m^2 = 7517 kg.

In this first test it was observed that the earth did not adhere to the support material and also we did not use sufficient structural elements, besides the size used caused the panel to deform. See Fig. 2.

Panel 2

It was made from a mixture of wood, hay, galvanized steel net, earth and cement. This module worked quite well, but the hay is a very soft material and did not attach

[1]Fique: Plant with long, triangular, fleshy leaves, ending in a strong sting, and yellowish flowers in bouquet on a central bohordo; It is originally from Mexico; It is used in the manufacture of textile fibres and in the production of pulque, mezcal and tequila. Synonyms: agave, maguey, pita.

Fig. 3 Panel 1, Fique's tissue model

it to the wooden structure, circles were printed on this panel to reduce cracks by drying earth. In this frame, it was found that the hay that works as thermal insulation and rubber improved crushing strength with earth mix. See Fig. 5.
Dimensions: 60 cm × 60 cm × 15 cm.
CO_2 emissions, distribution of materials:

1. Earth, Thermal Inertia: 1000 J/kg °C, CO_2 emissions: 0.001 kg/m^2, Percentage used in the mix 95%
2. Wood Frame, CO_2 emissions: 0.99 kg/m^2
3. Hay, CO_2 emissions: 0.013 kg/m^2
4. Rubber crushed, CO_2 emissions: 1.3 kg/m^2, Percentage used in the mixture 5%.
5. Henhouse Mesh type, CO_2 Emissions: 10.5 kg/m^2.
   CO_2 Total emissions: 0.36 m^2 × 12,804 kg/m^2 = 4609 kg (Fig. 6).

In this second panel the steel mesh used did not have enough rigidity because it had a very little thickness this caused the panel to deform on its outer faces, it was observed that the hay, worked very well as thermal insulation and the size used made that the framework was more structurally rigid.

Fig. 4 Panel 1, Fique's tissue with earthen cement mixture

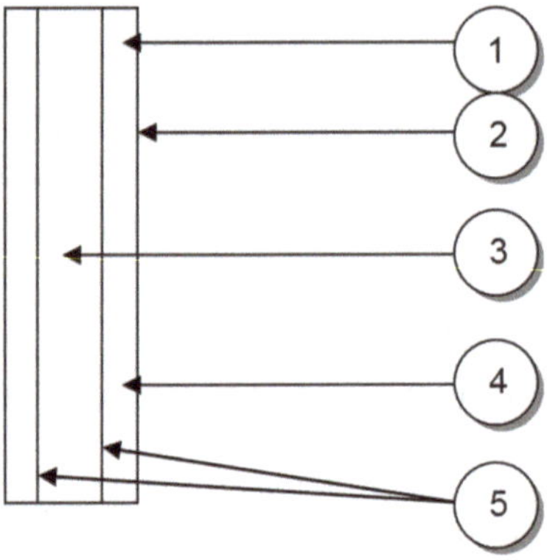

Fig. 5 Second panel, materials description section

Panel 3

It observed that the steel mesh type vein adheres well to earth, however, needs some reinforcements to prevent buckling, the use of steel mesh and cement increased CO_2 emissions of this prototype. This panel was used in Mrs. Fanny's house and it was based on this test. See Fig. 7.
Dimensions: 1.50 m × 1.50 m × 14 cm.

Fig. 6 Panel 2, Hay between two steel meshes

CO_2 emissions, distribution of materials:

1. Earth, Thermal Inertia: 1000 J/kg °C, CO_2 emissions: 0.001 kg/m^2, Percentage used in the mix 94%
2. Wood Frame ("Pino Sabanero"), CO_2 emissions: 0.99 kg/m^2
3. Steel wire, CO_2 emissions 61.43 kg/m^2
4. Cement, CO_2 emissions: 2.57 kg/m^2, Percentage used in the mix 6%.
   CO_2 Total emissions: 1.5 m × 1.5 m = 2.25 m^2 × 64,991 kg/m^2 = 146 kg.

4 Patented Panel

Finally, we make a panel with materials used in testing panels 1, 2 and 3. This panel was patented by SIC (Colombian Commerce Industry Review Office). This panel is unique in its type, responds to environmental needs, has a cost very low and is

Fig. 7 Panel 3, steel mesh core with two earthen mixtures layers

sustainable by incorporating materials with low CO_2 emissions in its manufacture, the panel is an assembly of made with products already processed and ready for use, the only element manufactured is the mixture of earth and cement.

It is manufactured with four basic materials, wood, earth, steel mesh and cement, for its assembly it is required of threaded rods and screws for wood. It is based on the construction with traditional Bahareque (de León 2012), we have incorporated a new production methodology in addition to incorporating a material such as a mesh of steel. It is mainly intended for rural housing construction thanks to the ease that they would have to obtain the material, nevertheless their use in environments is not ruled out rural.

The patent was published the 4th of June of 2017 in The Industrial Property Gazette with the identification number is NC2016/0001687. Figures 8 and 9 are part of the drawings of this document.

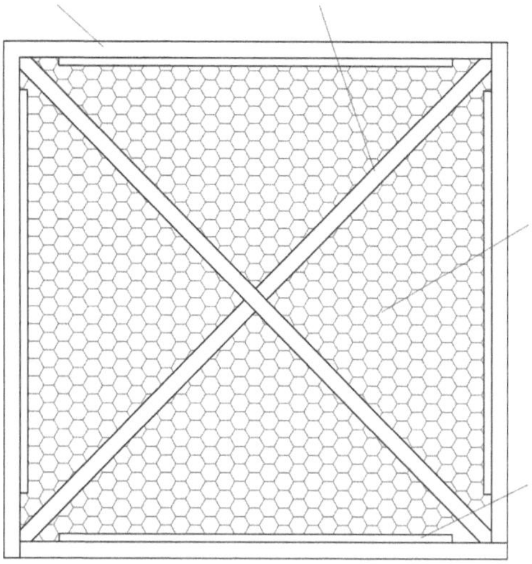

Fig. 8 Frontal view of patented panel

5 Discussion

The frugal innovation traditionally is used in industrial process, but in this project we use sustainable materials to produce a low cost panels. In architectural housing the frugal innovation, introduce new premises that could change some materials specially the cement and concrete, the most contaminant material in the building.

The reintroduction of traditional architectural techniques with sustainable materials for rural housing in developing countries as Colombia is necessary. Latin American countries have a very restrictive earthquake resistant regulations and this has been the main excuse not to investigate, or use the earth as a building material. However, it is possible to reach a compromise, in which the wood in the structure and earth elements of the facade that are not structural in Chile and is being built in this way and we believe that in Colombia it is possible to reintroduce these construction techniques used.

6 Conclusions

This research project is important because I provide a patent of invention that can be used as a point of reference for future research, since thanks to its new features allowed residents of the area to improve the construction of walls built on land,

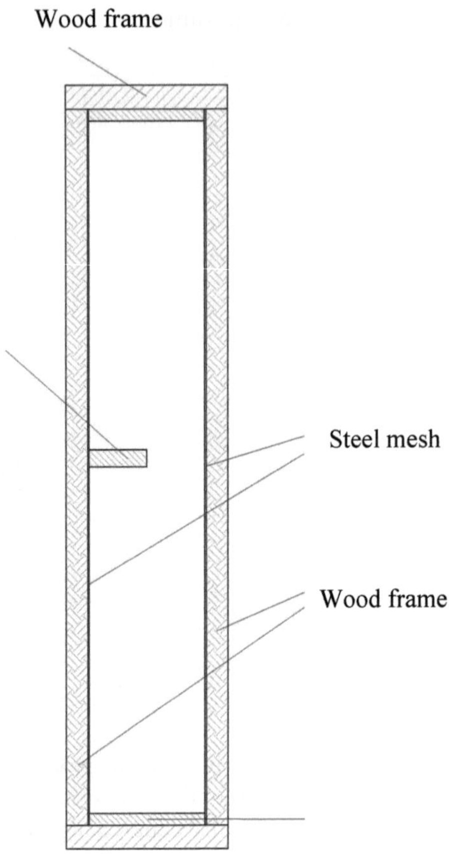

Fig. 9 Section of patented panel

thanks to the incorporation of new materials in the process of building their homes, in addition to reducing CO_2 emissions considerably.

The construction of rural homes with prefabricated panels on the earth is totally unprecedented in Colombia and this project opens the doors to the implementation and research of this new process implemented in a real prototype, the possibility of retaking abandoned construction techniques makes this project initiate the road to new research on the sustainability of rural housing in Colombia in addition to its social and economic implications necessary for rural development in Colombia. See Fig. 10.

Fig. 10 Women residents working in the patented Bahareque's panel

References

Balkrishna CR (2017) Revisiting classical design in engineering from a perspective of frugality. Heliyon 3(5):e00299. https://doi.org/10.1016/j.heliyon.e00299. http://www.sciencedirect.com/science/article/pii/S2405844016318643, ISSN 2405-8440

Brundtland WCED (1987) Report of the world commission on environment and development: our common future acronyms and note on terminology chairman's foreword. Oxford, Brundtland; Oxford University Press, New York

de León AH (2012) Aproximación a un estudio sobre las lesiones del bahareque en el estado Zulia, Venezuela. Algunas recomendaciones para su intervención. Informes de La Construcción 64(525):63–74. https://doi.org/10.3989/ic.08.049

Kibert R (2009) Sustainable built environment—volume I, Resource conscious building design methods. Eolss Unesco, pp 186–205

Peña R, Martha L (2008) El Programa CINVA y la acción comunal, Revista Bitácora Urbano Territorial [en linea], 12 (Enero-Junio): [Fecha de consulta: 30 de junio de 2017] Disponible en: http://www.redalyc.org/articulo.oa?id=74811916012 ISSN: 0124-7913

Riveros S (2007) El uso masivo de la Tierra como material de construcción. Revista Apuntes 354–363

Franz Calderon is an architect graduated from the National University of Colombia (1999), he did doctoral studies in Visual Communication in architecture and design at the Polytechnic University of Catalonia, in Barcelona, Spain. He is currently working as a full-time researcher at the Universidad Agustiniana in Bogotá Colombia and develops research on land architecture and sustainability.

Strengths and Weaknesses of Community-Based Systems in Municipal Services Delivery: The Case of a Community-Based Water Supply Scheme in Dar es Salaam, Tanzania

Yohannes Edwin Kachenje

Abstract Towards the end of the 20th century, the world experienced serious deterioration of infrastructure services, thus raising concerns to the global community. Because of that, municipal services delivery, particularly water supply in most cities of the world witnessed institutional reforms and increased participation of non-public actors, including community-based actors. In Tanzanian city of Dar es Salaam, community-based water supply schemes have become an important institution through which a number of urban communities get their potable water supply. Although these schemes have established themselves as provider institutions, more knowledge of the extent to which their operations are sustainable is still needed. This paper therefore, discuses strengths and weaknesses of the community-based schemes, in fulfilling their social responsibility of service delivery to the community. The paper is mainly based on primary data together with some document analysis. The data was collected through key informant interviews, focus group discussions and some household interviews. This paper points out that community-based schemes derive their strength for sustainable operations from inter-alia their good organization structures, adherence to the rules, and application of innovative monitoring approaches. Weaknesses are related to outdated system of data storage and processing, and insufficient creativity in competition with other schemes.

Keywords Coordination · Institutions · Sustainable service delivery
Water supply · Dar es Salaam

Y. E. Kachenje (✉)
Institute of Human Settlement Studies (IHSS), Ardhi University, Dar es Salaam, Tanzania
e-mail: ykachenje2001@yahoo.co.uk

W. Leal Filho (ed.), *Social Responsibility and Sustainability*, World Sustainability Series, https://doi.org/10.1007/978-3-030-03562-4_23

1 Introduction

Towards the end of the 20th century, municipal services at global level experienced a significant push and move from pure government or public provision to including/accommodating non-public actors in the social responsibility of service provision (Bel and Fageda 2008; Philippe 2009). The move was due to the failure of many governments or public institutions to meet the demand for services, particularly in urban areas. As such, the need for improving municipal services delivery has been among the global key issues from the 1990s, following concerns on underperformance (IDRC 1997; Makurira and Mugumo 2003; WSP and Consumer International 2004; Bel and Fageda 2008; Philippe 2009). During this period the global policy directions turned into favouring markets over government provisioning, and decentralized over centralized governance (UN-Habitat 2003). Eventually, many countries embarked on ambitious institutional reforms that allowed inter alia participation of community-based institutions along with other non-public actors in service delivery, (Philippe 2009). Solid waste collection, water supply and sanitation are among the municipal services that were reformed and allowed participation of non-public institutions (Bel and Fageda 2008).

In Tanzania, just like in many other Sub-Saharan countries, the government embarked on Local Government Reform Policy with the overall objective of improving the delivery of services to the public (URT 2005). The main strategy in fulfilling that objective was decentralisation by devolution, which entails the transfer of powers, functional responsibilities and resources from central government to local government authorities (ibid.). The government and local authorities engaged in institutionalization of an enabling framework for permitting community-based organizations (CBOs), non-governmental organizations (NGOs), individual households and the private sector to contribute towards this social responsibility of provision and maintenance of urban services (Kombe and Kyessi 2008; Kombe and Ndezi 2015).

For the case of water supply, the National Water Sector Development Strategy (NWSDS) of 2006 has given Water Supply and Sanitation Authorities (WSSAs) and Community-Owned Water and Sanitation Organisations (COWSOs) the power to enter into contract with service providers including public, private and community-based, so that they can continue supplying water within such areas in accordance with the signed contracts (Triche 2012).

Community-based water supply systems are known to fill in two important gaps that public and private water and sanitation suppliers may not be able to adequately provide. First, in participatory point of view, community-based systems have a better potential for the active involvement of local communities, as compared to the public and private water supply systems. And, secondly, these community-based systems mainly operate where no public system of water supply exists—whether it is yet to be provided or it existed and failed. In other words, the community-based water supply system was also meant to cover the gap of water supply left by the public system (Kyessi 2002; WaterAid 2008; Kachenje 2017).

Since the community-based systems of water supply attempt filling the two aforementioned gaps, it may be easy to forget analyzing weaknesses and strengths related to their operations, and instead concentrate on the service provision. As such many researchers have underlined the importance of the community-based systems of water supply, and gave less importance to discussion on sustainability of such systems. In that way little is known about their sustainability, as far as the fulfillment of their social responsibility of service provision to the community is concerned.

This paper therefore, discusses sustainability of community-based systems of water supply in terms of their strengths and weaknesses in service delivery. The discussion of strengths and weaknesses is based on one selected community-based water supply scheme, and is pursued around some influential factors to such schemes' sustainability, including institutional factors, rules, collaboration, resource mobilization, capacity building, efficiency, ICT, and political will.

The study contributes to the body of knowledge on the delivery and management of municipal services to urban communities. Delivery and management of services such as water supply, highly requires analysis of sustainability for improved services. Thus, through the analysis of strengths and weaknesses, this study contributes some important aspects for discussing sustainability of community-based schemes of contexts similar to the studied case.

2 Community-Based Water Supply System in Dar es Salaam

Community-based as one of the non-public systems of water supply, does occur in separate areas to the ones under the public system of water supply (Figs. 1 and 2). The institutional framework for water supply in the city recognizes community-based schemes of water supply as formal. However, where the public system happens to co-exist with the community-based system there also occurs unregulated competition, in which the latter tends to be the loser (Kachenje 2017). It is very rare for community-based schemes to peacefully co-exist with private or public scheme of water supply. Kyessi (2002) observed such peaceful co-existence in Tabata Dar es Salaam, whereby one community-based water supply scheme harmoniously shared service hours with a private service provider who supplied water when the community-based water system did not supply.

A total of 130 community-based water supply schemes existed in the municipality by the year 2015 (Kachenje 2017). All of these had water committees formed under the facilitation of the Municipal Council, which also monitors their operations through the Municipal Water Engineer (MWE) office. Below this level are the Ward Executive Officer (WEO) and finally the *Mtaa* level. These schemes vary in the number and size of administrative areas they serve, although most of them serve more than one *Mtaa* area.

Fig. 1 Water kiosks in Mamboleo 'B' community-based water scheme. *Source* Kachenje (2017)

Although there are different sources of funding through which the community-based schemes are initiated, most of them are funded through the municipal council. This local authority uses its Water Engineer office to facilitate or supervise construction tasks. Among development partners who funded such projects in the municipalities are WaterAid (Tanzania) and Belgium Technical Cooperation (BTC). Also through the World Bank some funds were solicited, involving the Dar es Salaam Water Supply and Sewerage Authority (DAWASA).

Despite their good value in this social responsibility of water supply, community-based schemes are regarded by many as temporary, in anticipation that some day they will be replaced by the public water supply system.

3 Methodology

This study is mainly qualitative, thus understanding of the study phenomenon mattered more, rather than frequencies of data and generalized statements (see Mason 2010). The study applied a case study approach, whereas one community-based water supply system in Dar es Salaam was studied, after being selected through

Fig. 2 Water kiosks in Mamboleo 'B' community-based water scheme. *Source* Kachenje (2017)

criteria sampling, from among 130 community-based water supply systems in one municipality, named Temeke (see Fig. 3). Research methods used included document analysis, key informant interviews, household interviews and focus group discussion. The focus group discussion involved 14 members and leaders of community-based water supply systems in the ward. The study was performed in the period 2013–2015, whereas the main phase of data collection started in September 2014 and ended in February 2015. Table 1 lists the interviewed key informants.

Although this study applied criteria sampling, its results are not statistically representative. However, the findings provide an understanding of the strengths and weaknesses of community-based water supply systems, as a measure of sustainability.

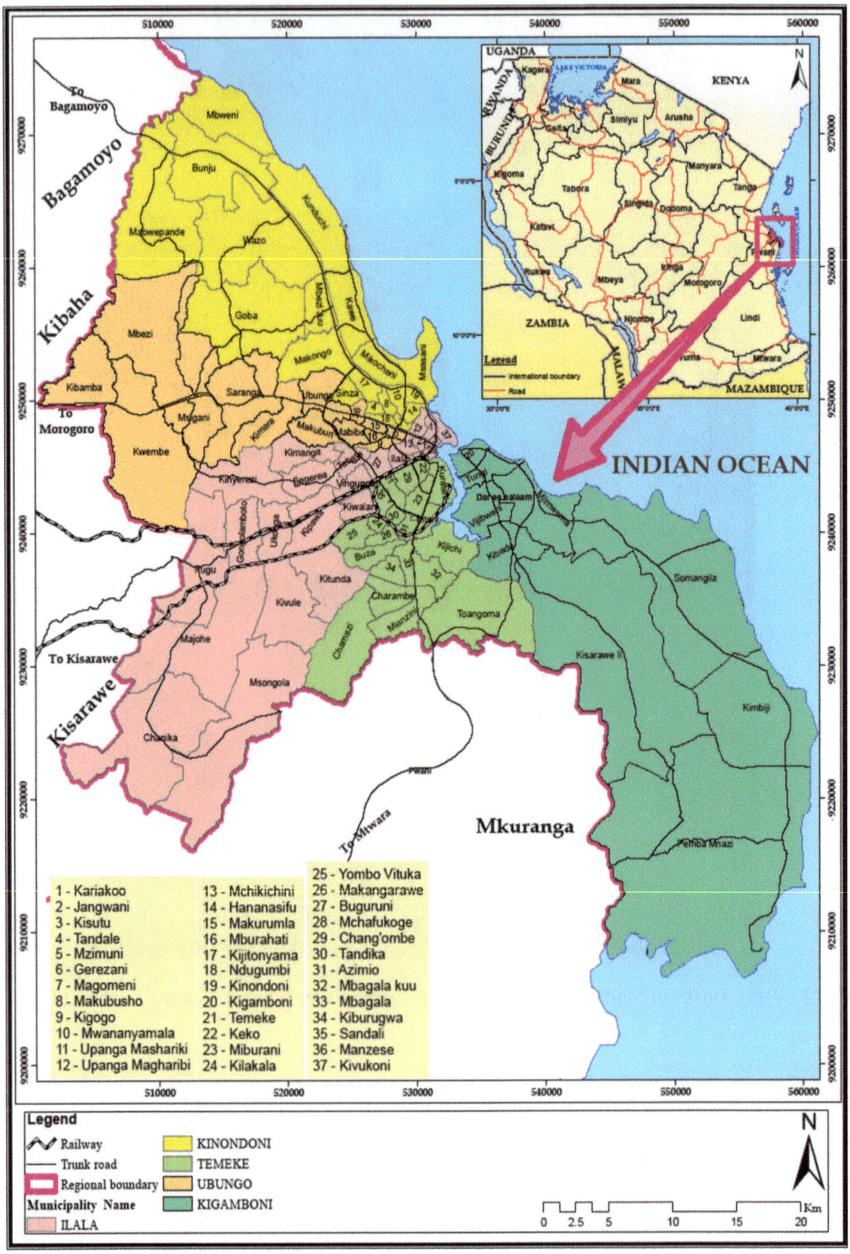

Fig. 3 Location of Dar es Salaam and its municipalities. *Source* Ministry of Lands Housing and Human Settlements Development (2016)

Table 1 Key informants interviewed

S. No.	Key informant	Number	Remarks/description
1	*Mtaa* (sub-ward) leaders	5	2 chairmen; 2 members, and 1 executive officer
2	Water committee leaders	3	Chairman, secretary and a member
3	Water sales attendants	3	In 3 scattered water supply points
4	Municipal officials	4	Water engineers; 1 community development officer
5	Utility services officials	4	Public service providers (DAWASA and DAWASCO)
6	Wateraid (Tz)	1	A representative
7	Ward executive officer	1	Sandali ward
8	Ewarema consult	1	Private consultant
9	Basin water officials	3	Wami Ruvu basin water office delegates

Source The author

In pursuing strengths and weaknesses of the community-based water supply scheme, this study applied mainly three theories, namely coordination, institutional economics and network theories, on the basis of what they contribute in explaining social responsibility and sustainability in relation to municipal services delivery. Thus, the three theories also were the main source of the variables developed in this study. The coordination theory explains about actors, roles and relationships as central in analysing coordination (Malone and Crowston 1994; Crowston et al. 2004). Thus, the theory is very important even for understanding sustainability related to organisations in water supply services in this context, since there are many actors, with many inter-depending roles in the community-based scheme. However, the definition of the individual actors and their respective roles is based on rules, a situation that compelled application of the institutional economics theory, as the theory puts forward that formal and informal rules empower organisations to pursue what they pursue (Hodgson 2006; North 1991). Since there are some actors with equivalent status and roles among the water supply actors, this study applied also the network theory so as to explain such kind of relationships among actors (see Kadushin 2004; Hudalah et al. 2010).

The findings were derived from applying seven variables, around which sustainability of the community-based water supply system was analyzed. The variables included collaboration; regulatory mechanism; institutional factors; motivation; resource mobilisation; capacity building; and information and communication. The next section presents the findings and analysis.

Table 2 Description of the service area of the scheme

S. No.	Parameters	Distribution in the *Mtaa* Areas		Totals
		Mamboleo 'B'	Kisiwani	
1	Population	4421	3804	8225
2	Number of households	1200	1028	2228
3	Number of houses	302	262	564
4	House connections (community scheme)	23	22	45
5	Water supply points (community scheme)	8	7	15
6	Water supply points (private scheme)	5	1	6
7	House connections (private scheme)	7	18	25

Source The author

4 Results and Discussion

4.1 The Community-Based Water System Case

The studied community-based case is known as Mamboleo 'B' Community-based water scheme. It is a scheme that serves two neighbouring *Mtaa* (sub-ward) areas, namely Mamboleo 'B' and Kisiwani, with a total population of 8225. In this service area which is located about seven kilometres from the CBD, water supply through the public system has never been available. Even the public piped water supply network is yet to be in place there. Since the community-based scheme is for two *Mtaa* areas, it is operated by a local Water Committee (WC) composed of members from both of the areas. Some details of the service area are summarized in Table 2.

As such, at *Mtaa* level, the WC is accountable to both *Mtaa* areas, particularly the residents as led by the two *Mtaa* Committees, which work jointly in water supply issues (see Fig. 4). Acting on behalf of the Temeke Municipal Director, the Municipal Water Engineer (MWE) office is the top responsible actor of the scheme, working in collaboration with the Sandali Ward Development Committee (WDC), the *Mtaa* Committee and the WC itself.

The WC of the scheme has also engaged 15 people as water sales attendants, one for each communal water supply point. Apart from managing water supply house

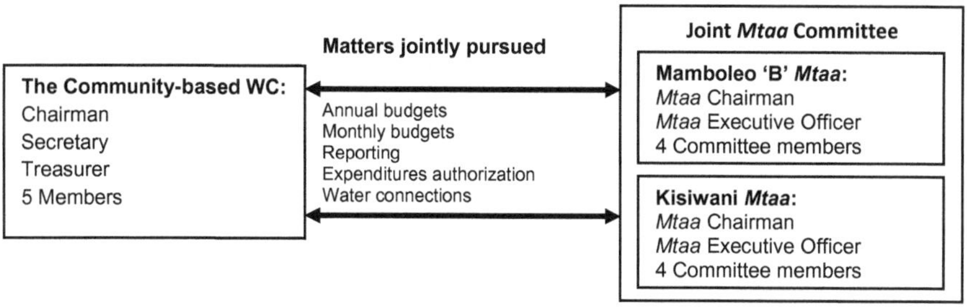

Fig. 4 Settlement level organization of the community-based water scheme. *Source* The author

connections, the committee is responsible for all other tasks related to operation and maintenance. It is the 8-members WC that does the managing at this level.

4.2 Strengths of the Community-Based Scheme

4.2.1 Favourable Organisation Structure

Among the strengths of the community-based water scheme was its organization structure. The structure consists of defined institutional levels (*Mtaa*, Ward and Municipality) at which very important coordination roles are pursued, including monitoring and supervision through the accountability chain. In association with the efficient reporting mechanism, that leads to a reasonably effective institutional coordination in the scheme. The institutional characteristic of the WC for the Community-based scheme gives it the opportunity to plan and implement activities without much bureaucracy.

The Community-based water scheme is organized into three main institutional levels: the settlement, the ward and the municipal levels. These levels form a hierarchy that constitutes a chain of accountability based on the Constitution of Water Committees, 2003. The settlement level organization is made up of the WC of the community-based water scheme and the joint *Mtaa* committee for the service area of the scheme. Beyond the settlement level are the ward level and the municipal level, as presented in Fig. 5.

4.2.2 Financial Sustainability of the Scheme

The Water Committee as a local organization for managing water supply does not rely on external financing for implementing its planned activities. Although its plans and implementation have to be approved from the *Mtaa* level to the municipality, it is not too bureaucratic to undermine efficiency. The committee is widely accepted in the community as an institution for managing water supply because of its consis-

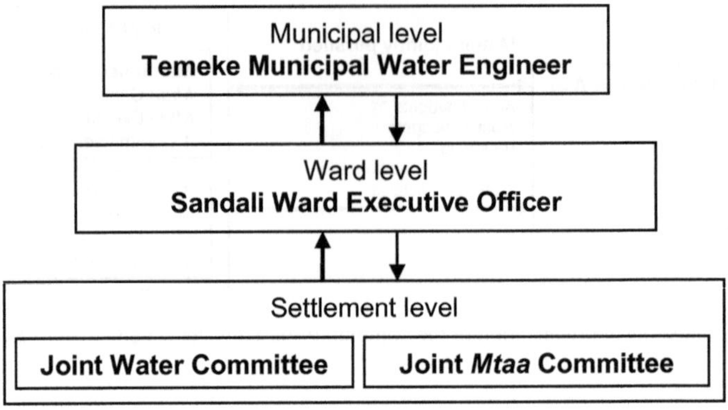

Fig. 5 Institutional levels for the community-based water scheme. *Source* Author

Table 3 The scheme's revenue and expenditure for December 2014[a]

S. No.	Item	Amount (TZS)
1	Water sales at communal supply points	2,192,000.00
2	Water sales at house connections	978,000.00
3	Cash in hand (brought forward)	78,540.00
4	**Total revenue**	**3,248,540.00**
5	**Total expenditure**	**3,180,200.00**
	Cash balance at 31/12/2014	68,340.00

[a]As at 07th August 2018 1 USD was equivalent to TZS 2245
Source The author

tent service in the area. Its financial performance makes the committee manage all its operation and maintenance activities properly and achieve cost recovery. Some expenditure like maintenance, transport, communication and cleanliness vary from month to month. Therefore, there are some months in which very limited or no amount is banked, while in some other months much is deposited. Table 3 presents a comparison of revenue collection and expenditure in the scheme in one of the months.

Based on Table 3, the scheme's revenue collection for December, 2014 covered its all costs related to operation and maintenance, although the amount of cash balance was relatively small.

Despite a somehow limited cash balance, the situation presents a successful cost recovery for operation and maintenance. As this was considered among the months with high maintenance costs, the financial performance has been good. The accountability chain pursued through the reporting process and the monitoring contributes greatly to the good financial performance. This is very important since it ensures financial sustainability to the WC for successfully managing the service. This is in line with Konteh's (2000) findings on the importance of internal ability to manage financial resources available, to become fully operational and achieve objectives.

4.2.3 Acceptance to the Served Community

The reality that the objectives of the WC are a direct contribution to the well-being of the community has given strength to that committee as a local institution, because achieving such objectives also implies achieving accountability to the Mamboleo 'B' and Kisiwani community. This is strength of its own, with synergetic relationship embedded, in the sense that, ensuring good and continued service provision determines acceptance of the WC to the residents, and residents' acceptance of the WC tends to make the institution attempt to maintain it, so that it keeps being accepted. Water supply has been consistent in the settlement for years now, after a crisis ended in 2012. Even in case of emergencies, the WC reacts quickly and ensures that the supply is back as soon as possible, depending on the extent of a particular breakdown.

4.2.4 Adherence to the Established Rules

Actors and institutional levels depend on rules and procedures for their existence and performance of their interdependent roles (Kyessi 2002; Hodgson 2006; Kachenje 2017). The Water Committees Constitution of 2003 as prepared by the Temeke Municipal Council is the key regulatory instrument for the Community-based Water Scheme. This is applied in association with directives from the municipal level authorities issued from time to time. Based on the Constitution and directives, the Water Committee is required to produce a report on revenue and expenditure on monthly basis. Also, the WCs Constitution requires the joint *Mtaa* Committee and the Water Committee to meet every month, to discuss the monthly reports submitted by the WC. Both of these requirements are well kept by these two institutions, something that strengthens their position, acceptability and sustainability at large. An important aspect regarding the application of the constitution and directives is the use of the local government administration structures of the *Mtaa*, ward and municipal council in facilitating enforcement. The municipal council through the municipal water engineer is the top level institution in managing the community-based water schemes. It also ensures through the accountability chain that the ward and *Mtaa* levels supervise the working of water schemes.

4.2.5 Settlement Level Linkages and Capacity Building

Strong linkages among actors and simplified capacity building process are among important strengths for the community-based water supply scheme. A combination

Table 4 Linkages and capacity building at the local level

	Capacity building component	Participants/collaborators	Recipients
1	Meter reading	15 Water sales attendants; 3 Water committee leaders; 5 Water committee members	Water sales attendants
2	Pump operation	15 Water sales attendants; 3 Water committee leaders	Water sales attendants
3	Reports preparations	Water committee treasurer and other leaders and members; *Mtaa* committee leaders; water sales attendants	Water committee members and water sales attendants

Source The author

of the linkages and knowledge exchange ensures sustainability of the water scheme and its operations.

In the studied community-based water scheme, settlement[1] level linkages are horizontal, and basically, involve three sets of actors: firstly, the *Mtaa* Committee with the WC; secondly, the WC members and leaders of the WC; and thirdly, the WC and the water sales attendants. At all these levels there are evidences of linkages and capacity building occurring successfully. Table 4 summarizes on components of the capacity building processes and the participants.

Linkages between the WC and attendants in the Mamboleo 'B' Water Scheme basically involve water sales attendants, WC members and the Treasurer of the committee as the key actors.

Each of the actors has specified work content in routines (specified activities) which he/she has to follow, meaning that he/she knows in advance what he has to produce and the procedure, as a water pump attendant witnesses[2]:

> My work is to sell water and operate the water pump. I have been doing this job for four years now, with my knowledge obtained through training by some experienced and trained members of the WC. Every evening I hand over the cash from daily water sales to the Treasurer of the WC. I record meter readings every morning and evening along with their respective sales. I submit the records to the Treasurer after every three days for reconciling them with the revenue collected and the readings recorded. In case of any breakdown or power cut-off, I normally report to the treasurer for action.

The quote above provides a picture of how successful the capacity building process has been in the studied scheme, through collaboration of the actors. In that way sustainability of the scheme is enhanced.

[1]Settlement here implies the service area of the scheme, that is, the two *Mtaa* areas.

[2]Interview held in the study area in October, 2014.

4.2.6 Application of Innovative Monitoring Approaches

Water flows, water sales, infrastructure working condition and the overall running of the water supply services are all well monitored in this scheme, and in that way it ensures sustainable water supply services to the community. Monitoring is mainly applied by the WC members on the water sales attendants, through observing and recording the daily meter readings in the water supply points. This is done every evening for the purpose of establishing the quantity of water sold and matching it with the relative revenue collected. The closing unit of the water meter in the evening becomes the opening unit for the next morning. This practice is very good and effective, as it controls sales, ensures accurate recording and promotes transparency in the system.

Monitoring within the WC involves the members and leaders. The chairman who has the leading role in the committee calls monthly internal meetings which he uses as a forum for integrating the work processes of the committee members and leaders. In such meetings, the various information and contributions from the individual actors are recorded and integrated into one report. Compilation from the actors may include information on the infrastructure condition obtained through the daily round-ups; the daily and monthly readings of water meters; the daily and monthly collections from water sales and other sources; the number of customers and new applications; the main activities done in the month; problems and challenges; achievements and ideas on the way forward.

Depending on the individual roles and tasks, each participant has the obligation to report in the meeting, a contribution that forms part of the overall report. The committee has a schedule for monthly distribution of tasks (particularly meter reading) to the members. This schedule helps in monitoring accountability of the members to the committee. As such, the reporting process becomes an important mechanism to coordinate the intra-committee actors. The report which is an output from the committee, becomes an important input for the monthly meeting with the *Mtaa* Committee. The structure of the report is uniform for all the WCs in the municipality. Once the whole WC is satisfied with its own report, copies are made and distributed to the *Mtaa* Chairmen in the scheme, the WEO and the Municipal Director on whose behalf is the MWE. In this way the monitoring ensures sustainability of the scheme.

4.2.7 Efficiency in Processing Water Supply Connections

High efficiency in the processing of water connections is among the qualities that are highly commended by the residents served by the scheme. The process from applying to getting a water supply connection to a house requires three to six days in the studied community-based scheme as summarized in Fig. 6.

It implies that, the whole process from application to connection is pursued within a week in this scheme, as opposed to four weeks in the public water supply system. The duration in this community-based scheme is considered short enough in the

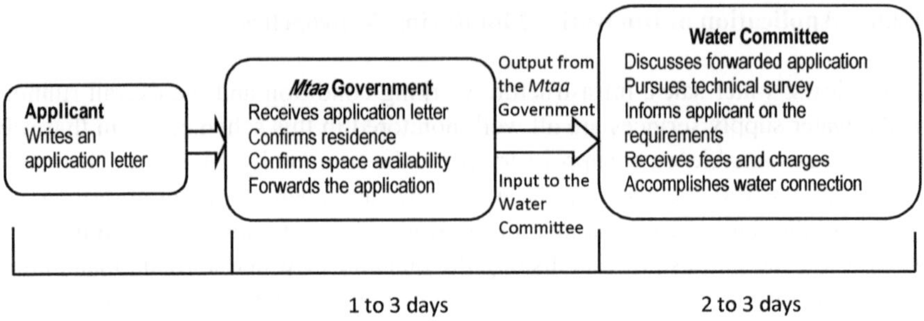

Fig. 6 Efficient water supply connection process in the scheme. *Source* The author

absence of an established standard duration. It shows, however, good responsiveness of the actors in the processing of connections, a quality that needs to be maintained.

4.3 *Weaknesses of the Community-Based Scheme*

4.3.1 Outdated System of ICT Facilities

One of the key weaknesses of the community-based scheme is the outdated system of facilities for data/information storage and processing at the WC level. This is risky since it undermines effectiveness in terms of storage, access/retrieval, analysis, display and other uses. Weakness in this area is likely to make the scheme collapse soon, since it will fail to manage its data properly, and subsequently weaken the monitoring process. The WC is still solely relying on storing and processing of its data in hard copies in this world of advanced information and communication technology. For the monthly reporting, the WC repeatedly types its outputs in secretarial bureaus.

The office of the WC lacks ICT facilities such as computers and printers. For the type and extent of information/data that the committee collects and processes, together with the monthly reports that it prepares, it may be undesirable and limiting to rely on external secretarial and ICT services to that extent.

4.3.2 Limited Creativity of the Water Committee

Limited creativity on the part of the WC is another weakness in the community-based scheme. The committee and the served community are all aware that the water supplied by the community-based scheme is somehow salty. At the same time, there is a private service provider whose water is not salty. But the WC has been silent about it, while the customers are talking about it. Some few are even not using the water from the community-based scheme for drinking but other purposes only.[3] The WC appears concerned about the competition from external private service providers,[4] but it does not have a clear strategy for addressing the issue.

4.3.3 Unknown Future Status of Community-Based Schemes

Community based schemes are recognised in the institutional framework for water supply in Tanzania and in the city of Dar es Salaam as well. But the superiority of the public system of water supply sheds uncertainty on the future of the community-based schemes and other non-public schemes, as the public system is regaining and extending its services towards the areas that have been exclusively under the community-based scheme for many years. As such, the competition from the public system may be too much for the community-based scheme to cope with. It could be better for the policy to put it clear on the fate of the non-public schemes, particularly community-based schemes. It is also an issue for further research, as to how the future situation will be.

5 Conclusion

Sustainability of systems for community-based municipal services is an issue that is influenced by some inter-linked factors within and beyond individual schemes. The studied case indicates that, community-based schemes derive their strengths for sustainable operations from factors related to inter-alia, their organization structures, adherence to established rules, application of innovative monitoring approaches; financial sustainability; responsiveness and creativity of actors; as well as the use of ICT facilities. A number of such strengths can be replicated to many other developing countries' urban settlements for improving management of their municipal services, particularly in community-based systems.

Despite the current status of sustainability in this and some other community-based water supply schemes, future research is needed for establishing the fate of community-based schemes in the wake of the on-going improvement and extension of the public system of water supply in the city. The question is, will the community-based schemes automatically die out, or will they be accommodated into the extension and improvement of the public system, and how?

[3]Household interviews held in September, 2014.

[4]Minutes of meetings (October and November, 2014).

References

Bel G, Fageda H (2008) Reforming the local public sector: economics and politics in the privatisation of water and solid waste. J Econ Policy Reform 11(1):45–65

Crowston K, Rubleske J, Howison J (2004) Coordination theory: a ten year retrospective. Syracuse University, School of Information Studies, New York, p 38

Hodgson GM (2006) What are institutions. J Econ Issues XL(1):1–25

Hudalah D, Winarso H, Woltjer J (2010) Policy networking as capacity building: an analysis of regional road development conflict in Indonesia. SAGE online version http://plt.sagepub.com/content/9/4/315. Last Accessed 25th June 2012

IDRC (1997) Urban water management in developing countries: research priorities, water supply and sanitation collaborative council, Oslo, Norway, p 15

Kachenje YE (2017) Institutional coordination in municipal services delivery: the case of water supply in Dar es Salaam, Tanzania. Unpublished Ph.D. thesis, Dar es Salaam, p 408

Kadushin C (2004) Introduction to social network theory, Chapter 2: some basic network concepts and propositions, http://www.communityanalytics.com/Portals/0/Resource_Library/Social%20Network%20Theory_Kadushin.pdf. Last Accessed 14th June 2012

Kombe WJ, Kyessi AG (2008) Measuring service delivery in Southern Africa project: Socio-political dynamics of service delivery report. Tanzania, Dar es Salaam, p 114

Kombe WJ, Ndezi T (2015) Water justice city profile: Dar es Salaam, Tanzania, DPU, UCL, http://www.bartlett.ucl.ac.uk/dpu/water-justice. Last Accessed 25th Mar 2016

Konteh R (2000) Enhancing community development in Sierra Leone. The need for coordination. Doctoral thesis, SPRING Centre, University of Dortmund, p 146

Kyessi AG (2002) Community participation in urban infrastructure provision, servicing informal settlement in Dar es Salaam. Doctoral thesis, SPRING Centre, University of Dortmund, p 405

Makurira H, Mugumo M (2003) Water sector reforms in Zimbabwe: the importance of policy and institutional coordination. A paper in the proceedings of the African regional workshop, Nairobi, Kenya, 8–10 Oct 2003, FAO, Rome, pp 158–167

Malone T, Crowston K (1994) The interdisciplinary study of coordination. ACM Comput Surv 26(1):87–119

Mason M (2010) Sample size and saturation in PhD studies using qualitative interviews (63 paragraphs). Forum: Qual Soc Res 11(3), Art. 8, http://nbn-resolving.de/urn:nbn:de:0114-fqs100387. Last Accessed 17th Nov 2015

North DC (1991) Institutions. J Econ Perspect 5(1):97–112

Philippe M (2009) Public-private partnerships for urban water utilities: a review of experiences in developing Countries. The World Bank, Washington, p 191

Triche T (2012) A case study of public-private and public-public partnerships in water supply and sewerage services in Dar Es Salaam. World Bank, Washington, p 74

United Nations Human Settlements Programme (UN-HABITAT) (2003) Water and sanitation in the world's cities, ch.5, changing perspectives and roles in urban water and sanitation provision: privatization and beyond. Routledge, London, pp 158–192

URT (2005) Decentralization for service delivery in Tanzania. A paper presented by D.M.S. Mmari, permanent secretary, PO- RALG, URT at the conference on Building capacity for the education sector in Africa, Oslo, Norway, 12th–14th Oct 2005, p 30

WSP and Consumers International (2004) Moving from protest to proposal: Building the capacity of consumer organizations to engage in urban water supply and sanitation sector reform in Africa. Consumers International and WSP, Harare, pp 1–23

WaterAid (2008) Why did city water fail? The rise and fall of private sector participation in Dar es Salaam's water supply, online version http://www.wateraid.org/. Last Accessed 22nd Nov 2015

Doctor Yohannes Edwin Kachenje has an honours degree in Urban and Regional Planning (2000), Masters degree in Built Environment Analysis at KTH, Stockholm (2006), and Doctoral degree in Built Environment (2017) at Ardhi University, Dar es Salaam. Dr. Kachenje is currently employed as Research Fellow at the Institute of Human Settlement Studies, Ardhi University, Dar es Salaam. His research interests include physical infrastructure management, institutional coordination in municipal services delivery, housing and human settlements development.

Managing Energy Efficient Refurbishment Within Homeowners Associations (HOA): A Qualitative Study

Antje Bretzmann, Stephanie Huber and Thomas Bäumer

Abstract As far as climate protection is concerned, the German government has set itself the target of a virtually climate-neutral building stock by the year 2050. Currently, a refurbishment backlog exists especially for homeowner associations (HOA) in Germany: Nearly 70% of condominiums are considered unrefurbished. At the same time, HOA share 22% of the total building stock in Germany. Therefore, these stakeholders are a key success factor for the ongoing implementation of climate protection targets. The decision to implement energetic refurbishment measures is basically the responsibility of the HOA. Within this process, the property manager is eminently important: He plans, controls and monitors tasks and acts as an interface between the owners and other stakeholders (e.g. craftsmen). In order to identify possible strategies for the implementation, and to determine influencing factors of property managers, we conducted eight qualitative expert interviews with selected property managers. As a result of this study, we identified several barriers for the implementation of energy efficient refurbishment as seen from the perspective of property managers (e.g. a lack of remuneration). Thus, recommendations which are bound to enhance the rate of energetic refurbishment within homeowners associations could be formulated.

Keywords Energy efficient refurbishment
Low and zero carbon technologies (LZC) · Sustainability · Property management

A. Bretzmann (✉) · S. Huber · T. Bäumer
Department of Business Psychology,
Hochschule für Technik Stuttgart (HFT), University of Applied Sciences Stuttgart,
Schellingstraße 24, 70174 Stuttgart, Germany
e-mail: antje.bretzmann@hft-stuttgart.de

S. Huber
e-mail: stephanie.huber@hft-stuttgart.de

T. Bäumer
e-mail: thomas.baumer@hft-stuttgart.de

W. Leal Filho (ed.), *Social Responsibility and Sustainability*, World Sustainability Series, https://doi.org/10.1007/978-3-030-03562-4_24

1 Introduction

The EU has set itself ambitious targets for energy policy: The current Energy Efficiency Directive aims the reduction of greenhouse gas emissions by 20% by the year 2020 compared to 2008 (Energy Efficiency Directive 2012/27/EU 2012). The Federal Republic of Germany will miss this low-threshold objective to a great extent: The National Energy Efficiency Action Plan (NAPE) foresees 'a reduction of 7.2–10.1% by 2020 compared to the base year 2008 depending on three different scenarios' (BMWi 2014, p. 10). This means the reduction will only be about half of the target envisaged. Great efforts are needed in the building sector in particular: this sector accounts for nearly 40% of energy consumption in Germany. Energy efficient refurbishments are a suitable measure for reducing CO_2 emissions for generating and consuming of heating and heat in existing buildings.

As far as climate protection is concerned, the efforts of the Federal republic of Germany go beyond the EU requirements: The German government has set itself the target of a virtually climate-neutral building stock by the year 2050. Therefore it is necessary to reduce the primary energy demand in that sector by 80%. On the one hand, this can be achieved by means of energy refurbishment, which reduce the energy requirements of buildings and increase their energy efficiency. On the other hand, the use of renewable energies in electricity and heat generation is likely to effectively reduce CO_2 emissions as well.

Currently, especially the building stock of condominium communities has a renovation backlog. Almost 70% of condominiums in Germany are considered unrefurbished and the current refurbishment rate is below 0.6% per year compared to the average of 0.8 among all building types (Diefenbach et al. 2010). The share of HOA in the total building stock is about 22%. There are over 9 million condominiums in Germany—with an upward trend (Kaßler 2017, p. 43). Thus, this building type with its target group of homeowners is highly relevant for the achievement of climate objectives and, particularly in Germany, the implementation of the energy transition.

Refurbishment management from the perspective of the property manager
The decision for energetic refurbishment measures within a HOA is basically the responsibility of the owner community. This decision is usually made with respect to specific occasions (e.g. the exchange of a central boiler) and certain motives (e.g. environmental protection). The property manager of the HOA plays an essential role hereby: He takes over central planning, controlling and monitoring tasks and acts as an interface between the owners and other stakeholders (e.g. craftsmen, architects). This is mainly due to his/her legal obligations under the German residential property law ('WEG-Gesetz'): Property managers are responsible for maintaining and setting up, thus preserving the value of the entrusted real estate.[1] Deriving from these major tasks, property managers have pertinent duties and obligations in the course of their activities: Property management includes, in particular, the duty to build up an appropriate maintenance reserve and to prepare the annual economic plan as well

[1]c.f. §§ 27 f. WEG Wohnungseigentumsgesetz (2014).

as the statement of accounts (Deutsche Energie-Agentur GmbH (dena)/Institut für Energie- und Umweltforschung (ifeu)/Passivhausinstitut 2017, p. 53). Furthermore, the property manager legally represents the HOA externally and manages the annual owners' meetings (Kaßler and Benusch 2013, p. 18).

In order to describe the topic of energy efficient refurbishment in the building sector, it seems promising to look at the ecological as well as the economic and social aspects. The concept of the 'triple bottom line' as a basic concept of sustainable development assumes that sustainability is carried by the three pillars of ecology, economy and social responsibility. While this paper has a clear focus on environmental sustainability in the building sector, within the subject of "Social Responsibility and Sustainability" it would also be possible to look at the economic dimension, which is briefly outlined in the following.

Thus for the involved stakeholders (owners, property management companies) the implementation of an energetic refurbishment would be an essential part of their sustainable entrepreneurial activity: 'In contrast to sustainable economy, sustainability management refers to the individual economic level and means that all entrepreneurial activities are systematically designed so that […] the external effects triggered by the respective core business on an ecological, economic level are handled responsibly' (Popović 2018, p. 82). Property management companies thus act sustainably if they consider the consequences of their entrepreneurial activity whilst acting economically and/or ecologically sensible. In addition, the individual property manager will also do justice to his role towards his customers by focusing on preserving the value of the property.

From a social responsible point of view, energetic refurbishments in homeowner associations (HOA) would include considering aspects of participation and fairness within the decision-making process (Bretzmann et al. 2017) or increasing energy poverty (cf. Bouzarovski and Petrova 2015).

In this paper an attempt is made to shed light on the influence of property managers on homeowners within a HOA. Are they capable, in addition to their core tasks, of promoting energy recovery and thereby contributing to the achievement of climate protection goals? Which conditions would have to be fulfilled?

Energetic refurbishment measures in homeowner associations require versatile competencies, commercial and technical as well as legal and communicative. In order to plan the process accordingly, a comprehensive refurbishment management with predefined tasks and goals is ideally required after each implementation phase. Taken together, the implementation of energetic refurbishment measures causes a comprehensive task complex for the property manager.

Barriers for the implementation

Following the current state of research, significant barriers to the implementation of energy efficient refurbishment in HOAs from a property manager's point of view are above all:

1. the (perceived) complexity of the processes and the upcoming effort,
2. missing (financial) incentives and
3. a lack of know-how (DDIV 2014).

In addition to these barriers, there is also a legal dimension which apparently also affects the immediate working environment of property managers: high liability risks and a lack of decision-making power have a negative effect on the implementation of energy refurbishment measures as well (Zentrum für Europäische Wirtschaftsforschung 2015, p. 28).

2 Research Questions and Methodology

In order to evaluate the process of energetic refurbishment within a HOA and to identify possible key factors to increase the likelihood of implementation, as well as to investigate the influence of property managers on energetic refurbishment in general, the following research questions were formulated:

1. What are the predominant occasions for refurbishment concerning HOAs from the point of view of property managers and how is a need for refurbishment detected?
2. What challenges do property managers face when planning an energetic refurbishment?
3. What measures would optimize the planning process of energetic refurbishment for the property manager?

These questions were pursued by means of guided expert interviews. All questions focused on the property manager's view of the process. The order of questions asked corresponds to the (ideal) process of initiation, implementation and evaluation of energy efficient refurbishment. Therefore, at the end of the survey possible solutions for difficulties encountered were requested and a possible solution suggested.

The aim of this survey was to answer the mentioned research questions with the help of the practical knowledge of the experts and to be able to make statements about the role of energetic renovations in the daily work of a property manager.

The surveyed managers have been working for many years in a large property management company with a focus on homeowner associations. Thus, the managers had gained great experience in the field of energy efficient refurbishment. It seemed to be an advantage to interview a representative of a branch association in addition to the respondents who work directly in a real estate management company. For this reason, the content of the interview guidelines have been adapted specifically for this interview partner in order to meet its more strategic approach. For the interview guideline, the research questions were translated into single topics. The questions were formulated as openly as possible in order to obtain unaffected statements (Gläser and Laudel 2010, p. 115; Friebertshäuser et al. 2013, p. 37).

A total of eight interview partners were selected. In addition to a representative of the German Association of Property Managers (DDIV) (=IP 1), the owner of a property management company in the greater Stuttgart area (=IP 2), the divisional manager of the business field HOA in a property management company (=IP 3), the managing director of a property management company (=IP 4) as well as the board of a building cooperative with the business field residential property management (=IP 5) were asked.[2] In order to be able to gain insights into the operative business, three property managers of a large company in the metropolitan area of Stuttgart (=IP 6-8) were also included into the sample (see also Appendix A).

The survey took place from 08 September to 29 September 2017 using both telephone and face-to-face interviews depending on the availability of the interviewees. The respondents received the interview guideline one week before the interview in order to get the possibility to prepare. On average, the interviews lasted about 40 min. After all the interviewer's questions were answered, the experts had the opportunity to address aspects that were not considered in the interviews.

3 Results

3.1 *Starting Point: Occasions and Initiation of Energy-Efficient Refurbishment*

'*What are renovation occasions for condominiums from the perspective of property management?*'

When interviewing homeowners directly, there are usually two occasions for refurbishment: the transfer of ownership or upcoming repairs of existing defects (Stieß et al. 2009, pp. 18; 62, and Deffner et al. 2012, p. 22). The present survey of property managers revealed some similarities to home owners' reasons: the majority of the interviewees mentioned damage or defects in the substance of the building or the heating system as a reason for a refurbishment (IP 2, 3, 4, 5, 6, 8). This means that for property managers upcoming repair measures are a major refurbishment occasion. In contrast, the transfer of ownership was not mentioned by the surveyed managers.

In the course of the interviews, further occasions could be identified:

1. With low interest rates, investing financial resources in the capital market is less lucrative and makes real estate investments more attractive (IP 3). Therefore, a low level of interest rates in the capital market enables a (higher) investment in one's own property and, thus, in energy-efficient refurbishment. Low interest rates are bound to increase energetic refurbishment measures.
2. Another reason mentioned is a high level of energy costs (IP 4, 6). The higher the energy costs, the higher the expected financial savings through the reduction of energy consumption.
3. Legal requirements, for example those resulting from the German Energy Saving Ordinance (EnEV, Verordnung über energiesparenden Wärmeschutz und energiesparende Anlagentechnik bei Gebäuden 2013) or the Renewable Heat Act (EWärmeG, Erneuerbare-Energien-Wärmegesetz 2015), are also causes for

[2]Since the gender of the interviewees had no effect on the answers, the male form is consistently used in the following presentation of the results.

energetic measures (IP 3, 6, 7). As an example, owners' associations are required to exchange their boilers after 30 years according to applicable law (EnEV 2013). Furthermore, according to German regulations (EWärmeG), after the replacement of the boiler, 15% of the energy consumption must be provided by renewable energy.

Apparently, for the interviewed experts there are two independent dimensions, which are bound to trigger decisions for owners: an economic and a legal dimension. The economic dimension also covers the preservation and increase in value of the building. According to property managers, in addition to considerations of economic efficiency and legal compliance, the visual appearance sometimes seems to be important, too: 'Especially for the owners who live in them, or when they sell their apartments. Of course they want to have a certain visual appearance.' (IP 8) Apparently, this has not necessarily to be the case with owners renting their apartments.

In negative terms, a visual change of the building can be the reason for the rejection of an energetic refurbishment as well:

> What I have seen from my experience lately is that the full thermal insulation here is not so popular anymore, and that is because the owners see the danger of the windows becoming loopholes, as they say. Meaning the renovation is exaggerated. And second the photovoltaic system, which the owners do not want to have due to visual impairments. (IP 6)

'*How is a required rehabilitation discovered?*'
In practice, rehabilitation needs are often identified during regular site inspections or other on-site appointments by the property manager (IP 2, 3, 4, 5, 8). On this occasion damage to the building, such as cracks in the facade, is noticed frequently. Instead of considering the insulation, the focus can also be on energy production. Therefore, a need for repair can be based on the installation engineering. If necessary repair measures are piling up, it is usually an indication that a replacement of this system must be made in the near future. In property management, there are internal guidelines, according to which the individual components should be rehabilitated, for instance a roof after 40 years approximately (IP 7).

In addition to the property management, craftspeople who work in the building occasionally point out a rehabilitation requirement (IP 2, 3). Rarely, the notification is made by individual owners (IP 2, 3, 5, 8).[3]
'*Who initiates the implementation of energetic redevelopment?*'
In principle, energetic redevelopment can be initiated by three parties: the property manager, an administrative advisory board or individual owners. In most cases, however, the property manager himself acts as a 'spark' (IP 2, 3, 4, 5, 8) and usually draws up the advisory board in a timely manner.

Furthermore, a dedicated advisory board can initiate an energetic refurbishment: 'Often, however, the advisory board comes up to us and says [...] we could do that, or get an expert here. How does it look like for you?' (IP 8).

[3]Frequently implemented measures are the replacement of the windows (IP 2, 3, 4, 5, 6, 7, 8) or the heating system (IP 2, 4, 5, 6, 8) as well as the insulation of the top floor or basement ceiling (IP 2, 3, 6, 7, 8).

In rare cases, individual owners also initiate an energetic refurbishment (IP 2, 3, 4). Most of them are owner-occupants who benefit directly from this measure e.g. due to an improved heating system. This means, an insulation of the basement ceiling is rather triggered by a self-occupying owner of a ground floor apartment than an owner of a top floor apartment (IP 3). The different views of owner-occupiers and investors are a phenomenon existing especially in HOAs and becoming particularly obvious in the implementation of an energetic refurbishment: 'The one who owns the apartment only for investment, that is, who has rented it somehow, would like in principle almost not to renovate. There is always a big discussion in the owners' community if there is a need or there is no need.' (IP 5).[4]

3.2 *Challenges in Implementation*

3.2.1 Challenges in Dealing with the Owners

During the process of energy efficient refurbishment 'the owners have a significantly [...] increased need for information' (IP 4) and it lies within the responsibility of the property manager to provide the community with the necessary information: 'So the ongoing information, "let's keep that up to date for the owners", that's the way to say it.' (IP 8). Overall, the managers of energetic refurbishment measures also have social responsibilities: 'During the refurbishment, the trustee is almost a pastor.' (IP 4).

Furthermore, some on-site appointments with owners or craftsmen are required (IP 3) and advisory board meetings (IP 5) as well as extraordinary owners' meetings must be held (IP 3, 5): 'Naturally, you will need additional owners' meetings. We have already started with that, because to push such measures in an ordinary owners' meeting does not make sense. The owners have an extremely high demand for information.' Besides, the property manager must constantly coordinate the involved experts (IP 4).

The considerable (additional) effort can only be provided by property management companies with an appropriate remuneration. Even if such measures usually have a special remuneration in the management contracts, they can hardly cover the additional expenses of the property managers, especially for extensive measures, e.g. the organization of additional owner meetings or expert workshops (IP 1).

If refurbishment is required, it is often difficult to convince all owners of the necessity for a measure: 'When it comes to the owners own purse, the borders are always set very fast.' (IP 5)

'Exactly, so the heading is money. And that is also the killer argument by the way. If you have 70-year-olds in there, firstly they cry 'I will not live that long anymore'. That's it.' (IP 7)

[4]Nevertheless, in Germany there are means for landlords to transfer the costs of energy-efficient refurbishment to the tenants later.

In general, it is difficult for the property manager to coordinate the many heterogeneous views and needs of owners: ‘The desires of the individual apartment owners are sometimes extremely hard to reconcile.’ (IP 4)

‘Otherwise, difficulties are that you have to somehow reconcile the community and have to see that all is running […]’ (IP 3) ‘Troublemakers in the Community’ (IP 2) further complicate the work of the property manager (IP 1).

As a rule, the property manager is expected to have expertise of many kinds (e.g. legal, technical, communicative …), which cannot be met by a single manager who often has a commercial background. However, he frequently finds himself forced to justify their work and meet with owners’ misunderstandings (IP 2, 4):

‘For many there is no understanding that you have […] no X-ray vision.’ (IP 4)

‘We are no engineers and we are not lawyers either. We are not tax advisors. It takes a high level, but I do not have to be an expert of all kinds as a property manager.’ (IP 2)

‘As a property manager, you have to have so many jobs. You have to be a construction worker, you have to be a lawyer, you have to be an energy consultant, you have to be a finance expert. All these tasks.’ (IP 6)

3.2.2 Difficulties in Dealing with Banks

In addition to this high level of support and coordination, other skills are required. If a loan or the use of subsidies is decided upon, applications must be submitted by the property manager and the necessary documents have to be compiled (IP 2, 3, 4, 6, 8). There is an ‘increased paper and monitoring effort’ (IP 3). In particular, the implementation of contracting models using cogeneration units often poses major difficulties for property managers:

> I’ve had one so far. Then this just boils down to the fact that the owner community is doing business. With the whole sales tax registration etc. And somebody has to do that and we are not able to do that. Neither in terms of time nor knowledge. (IP 6)

A particular difficulty in dealing with energy efficient refurbishment in Germany is the use of subsidies, for instance with the German Kreditanstalt für Wiederaufbau (KfW), the world’s largest national development bank.[5] The application for subsidies sometimes involves a great effort: ‘If we […] deal with subsidies or with KfW loans […], it is a very incalculable expenditure of time.’ (IP 4) ‘You have just a huge package of documents.’ (IP 3) ‘So this administrative effort is crazy.’ (IP 7)

> KfW, for example, is always complex. You have to meet higher standards than normally required by the Energy Saving Ordinance. Then there’s a lot more paperwork, documents you have to fill out, and everything must be documented. (IP 8)

In addition, KfW’s schedule requirements regarding the start of the measure are in many cases difficult to meet due to the often complex and long-lasting decision-making process of the HOA (IP 3). Reducing the complexity of the application could

[5] A specialized bank that forwards public subsidies through loan programs.

facilitate the work of the property managers and therefore the implementation of energetic refurbishment.

Furthermore, there are also difficulties due to legal requirements: If a measure comes to unforeseen events, often a timely decision of the owner community is required, which can only be made within a meeting. Under German law, this takes a 14-day loading period and should preferably take place in the evening. This inhibits or at least complicates a flexible and timely response to the event (IP 2, 4).

3.3 *Possible Solutions: Optimizing the Process*

In order to evaluate the process of energy efficient refurbishment, property managers made two suggestions for optimizing the process: optimizing the cooperation with the owners and constantly improving their own qualifications.

With respect to the owners, property managers try to ensure a high transparency within the whole process. (IP 2, 4, 5). 'It's like being in a relationship: talking, talking, talking.' (IP 2)

> Create as much transparency as possible. This is the most important thing for the owners. Involve them, give them as much information as possible, and give them the opportunity to interact with one another. Transparent information is most important. If they feel like something is decided past them and they were not able to talk about it, it's always difficult. (IP 4)

This can be done through regular meetings with owners and experts (IP 5) as well as through postings in the building or construction journals (IP 4). Another possibility is the founding of a building committee (IP 5).

It is also proposed by the experts to use an (online) modular system to provide the necessary information on the possibilities of energetic renovations:

> And if you could say, sweepingly on the facade you have several options, at the roof you have to consider these things, or the following possibilities are available for heating to order a special package here, you could propose a viable way to the owners. So you can suggest to proceed for example by first, second, third, fourth, fifth. It would all make sense. (IP 6)

In addition, an adaptation of the legal regulations regarding the response to unexpected events could simplify the process for property managers: 'There should be a way […] to evade this monster of the owner's meeting with its 14-day period of notice.' (IP 4)

Another way to professionalize the process is to maintain and promote the management through continuous training and the exchange of colleagues (IP 1, 2, 3, 4, 5). However, there is still a need for training within the professional field that needs to be pursued in the future. 'Legally, technically, commercially—these are the three pillars.' (IP 1) Special attention is required especially for liability issues and compliance (IP 2, 3).

The curriculum of the apprenticeship for property managers in Germany is characterized thoroughly commercially and does not involve the transfer of technical knowledge, especially regarding energetic refurbishment measures. As a result, after completing their training, property managers frequently lack a technical understanding (IP 4, 6, 8). A possible solution to this problem could be to increase the qualification by including technical topics into the curriculum. Up to now, there is no protected professional title for German property managers and no special certification process for this job profile.

In the course of the interviews, the experts were asked to comment on the use of a 'carer' within the company. This person could be responsible for energetic remediation and has the necessary competencies to provide fundamental advice to the owner community. However, he (or she) should not replace the work of an architect or engineer.

The experts interviewed have largely rejected this idea. The reasons for the rejection were, in particular, the serious or impossible financial viability of this position and the liability, which the company is subject to (IP 2, 4, 7): 'Yes, it is the case, the liability risk compels us to employ specialists here because we cannot accept this liability risk. There are insanely high monetary risks that we would be about to take, that are, compensation claims.' (IP 6) However, it seems possible to set up the task externally by energy consultants or within the administration of cities and municipalities.

4 Discussion and Conclusion

Recent studies on the subject of energetic refurbishment focus mainly on single-family and semi-detached house owners (cf. Albrecht and Zundel 2010; Stieß et al. 2009). HOAs are scarcely in the focus of research examination, although they take a considerable share of existing real estate throughout the EU and have to be energetically renovated in the long term if EU energy efficiency projects and action plans for climate protection are to be successful. While investigations among single-family and semi-detached house owners concentrate, for example, on aspects of user behavior (Bundesinstitut für Bau-, Stadt- und Raumforschung 2017) or technology acceptance (Schäfer and Keppler 2013), further aspects become important for the special case of HOAs: in addition to the design of a participative decision-making process, other stakeholders besides owners can be examined. This exploratory study tried to highlight one of them, the property manager's role in the initiation and implementation of an energy efficient refurbishment measure. A basic assumption of the study was that the property manager plays a vital role within the process. In the course of events, he has to cope with a high level of care as an interface for all stakeholders.

As this study has shown, difficulties which property managers encounter in the course of an energetic renovation within a homeowners association are, above all, the coordination and integration of all owners. Solutions are proposed by the experts themselves: the greatest possible transparency in dealing with the owners can defuse conflicts at the beginning. If the owners participate in all phases of the process of an energy efficient refurbishment, the number of inquiries by owners can usually be reduced and the understanding be promoted (cf. Bretzmann et al. 2017).

Furthermore, the activity of the property manager is hindered by an elaborate application process for subsidies and an inflexible handling of unforeseen events due to legal provisions. This applies especially to HOAs in Germany on account of extensive legal provisions and a market regulated by state and municipal subsidies. Solutions could be provided via suitable political instruments: In the context of legal requirements (the German residential property law, 'WEG-Gesetz'), property managers could be enabled to react more flexibly to decisions that are to be taken in a timely manner. Besides, the complexity of applying for subsidies could be reduced.

In order to increase the refurbishment rate in homeowner associations in the long term, it is essential to endow the property manager as an important driver of the refurbishment process with all necessary resources. This includes a fair remuneration of the extra effort, but also an extensive knowledge and further training. As far as the field of communication is concerned, the supply with information material seems promising and expedient. By strengthening the role of the property manager, the implementation of energy efficient measures can be facilitated in the long term.

5 Outlook

As demanded by the surveyed experts, an improved education and training of property managers in Germany is foreseeable: from August 2018, a professional admission regulation for commercial property managers will come into force, which includes a general obligation to continue training. Within three years, 20 h of training must be proven. This could be a first step in overcoming the knowledge gap described in the present article. But the effect will take years and it is debatable whether the training obligation is sufficient.

As another example, an amendment to the legal regulations of the homeowner associations ('WEG-Recht'), as called for by the experts in this paper, is currently on the agenda of the new government in Germany: 'We will reform the regulations of the residential property law and harmonize them with the tenancy law in order to prepare and conduct homeowners' decisions on building measures, particularly in the areas of accessibility, energy efficiency, promotion of electromobility and burglary protection (Koalitionsvertrag zwischen CDU, CSU und SPD 2017 line 5193)'. This corresponds to the demands of the surveyed property managers to fundamentally simplify the process in the field of energy renovation. This declaration of intent is certainly also due to the fact that it is necessary to advance energy efficient refurbishment, especially in existing buildings, if the EU's climate protection objectives are to be achieved.

In the future, the requirements for energy efficiency of new and existing buildings will drastically tighten at the whole EU level: The amendment to the EU Directive 2010/31/EU (Energy Performance of Buildings Directive 2010) stipulates that from 2021 within the EU only low energy houses may be built. For existing buildings requirements for energy efficient refurbishment (insulation) will apply. Accordingly, the need to convince especially owners within HOAs and to make them participate in the energetic refurbishment will be much higher due to the legal situation. Again, property managers hereby can play a crucial role. It will be a necessary step if the EU's climate objectives are to be achieved in the long run, and if both economic and social sustainability should not be lost sight of.

Acknowledgements The authors gratefully acknowledge financial support from the Federal Ministry for Economic Affairs and Energy on the basis of a decision by the German Bundestag (Grant No. 03ET4017B).

Appendix A: Overview of the Interview Partners

1. Representative of the German Association of Property Managers (DDIV) (=IP 1)
2. Owner of a property management company in the greater Stuttgart area (=IP 2)
3. Divisional manager of the business field HOA in a property management company (=IP 3)
4. Managing director of a property management company (=IP 4)
5. Management Board of a building cooperative with the business field residential property (=IP 5)
6. Property manager of a large real estate management company in the greater Stuttgart area (=IP 6)
7. Property manager of a large real estate management company in the greater Stuttgart area (=IP 7)
8. Property manager of a large real estate management company in the greater Stuttgart area (=IP 8).

References

Albrecht T, Zundel S (2010) Gefühlte Wirtschaftlichkeit – Wie Eigenheimbesitzer energetische Sanierungsmaßnahmen ökonomisch beurteilen. Senftenberg

Bouzarovski S, Petrova S (2015) The EU energy poverty and vulnerability agenda: an emergent domain of transnational action. In: Tosun J, Biesenbender S, Schulze K (eds) Energy policy making in the EU: building the agenda. London, pp 129–144

Bretzmann A, Bäumer T, Huber S (2017) The role of participation and communication for energy efficient refurbishment. Econ Environ Stud 17(4):675–691

Bundesinstitut für Bau- Stadt- und Raumforschung (2017) Quantifizierung von Rebound-Effekten bei der energetischen Sanierung von Nichtwohngebäuden/ Bundesliegenschaften: Energiekennwerte, Prebound-Effekt und Verhalten der Nutzer/innen vor einer energetischen Sanierung. Bonn

Bundesministerium für Wirtschaft und Energie (BMWi) (2014) Ein gutes Stück Arbeit. Mehr aus Energie machen: Nationaler Aktionsplan Energieeffizienz. Berlin

Dachverband Deutscher Immobilienverwalter e. V (DDIV) (2014) Effektivere Anreize zur energetischen Sanierung von Wohnungseigentümergemeinschaften notwendig: Ergebnisse der Umfrage von DDIV und KfW zum Thema energetische Sanierung in WEG. Berlin

Deffner J, Stieß I, Reisenauer S (2012) Neue Wege in der Kommunikation energetischer Sanierungen für Eigenheimbesitzer/innen: Konzeptbausteine für eine integrierte Kommunikations- und Marketingstrategie mit Schwerpunkt auf dialogischen Maßnahmen. Frankfurt am Main

Deutsche Energie-Agentur GmbH (dena)/Institut für Energie- und Umweltforschung (ifeu)/Passivhausinstitut (2017) Handbuch für Energieberater: Anleitung mit Tipps und Tricks zur Umsetzung. Berlin

Diefenbach N et al (2010) Datenbasis Gebäudebestand. Darmstadt

Energy Efficiency Directive 2012/27/EU 2012, EED

Energy Performance of Buildings Directive 2010, EPBD

Erneuerbare-Energien-Wärmegesetz 2015, EWärmeG

Friebertshäuser B, Langer A, Prengel A (2013) Handbuch qualitative Forschungsmethoden in der Erziehungswissenschaft. Weinheim

Gläser J, Laudel G (2010) Experteninterviews und qualitative Inhaltsanalyse als Instrumente rekonstruierender Untersuchungen, 4th edn. Wiesbaden

Kaßler M (2017) Scheitert die Energiewende an den Wohnungseigentümergemeinschaften? In: Pöschk J (ed) Energieeffizienz in Gebäuden: Jahrbuch 2017. Berlin, pp 43–49

Kaßler M, Benusch S (2013) Zugangsvoraussetzungen für Immobilienverwalter schaffen: Ein Plädoyer für mehr Qualität und Verbraucherschutz beim Wohneigentum. Berlin

Koalitionsvertrag zwischen CDU, CSU und SPD: Ein neuer Aufbruch für Europa. Eine neue Dynamik für Deutschland. Ein neuer Zusammenhalt für unser Land (2017). Berlin

Popovic´ T (2018) Sustainable Finance als Katalysator für die Zukunft des nachhaltigen Wirtschaftens? In: Rogall H et al (eds) Jahrbuch Nachhaltige Ökonomie 2017/2018. Marburg, pp 78–92

Schäfer M, Keppler D (2013) Modelle der technikorientierten Akzeptanzforschung: Überblick und Reflexion am Beispiel eines Forschungsprojekts zur Implementierung innovativer technischer Energieeffizienz-Maßnahmen. Discussion paper. Eigenverlag, Berlin

Stieß I, Birzle-Harder B, Deffner J (2009) So ein Haus ist auch die Sparkasse von einem: Motive und Barrieren von Eigenheimbesitzerinnen und -besitzern gegenüber einer energieeffizienten Sanierung. Ergebnisse einer qualitativen Untersuchung. Frankfurt am Main

Verordnung über energiesparenden Wärmeschutz und energiesparende Anlagentechnik bei Gebäuden 2013, EnEV

Wohnungseigentumsgesetz 2014, WEG

Zentrum für Europäische Wirtschaftsforschung (2015) Finanzierungsbedingungen der energetischen Gebäudesanierung durch Wohnungseigentümergemeinschaften in Baden-Württemberg: Studie im Auftrag der L-Bank. Mannheim

Antje Bretzmann M.A. studied German literature and Communication at Freie Universität Berlin and worked for several years in the field of corporate communication for large transportation and energy companies. Her research interests focus on organizational communication tools and participation concepts. She is currently working on a project which aims to address and motivate energy efficient refurbishment within a HOA.

Stephanie Huber received her Ph.D. in the field of business administration and marketing at the Johannes-Gutenberg University of Mainz in 2003. Her research focus was on brand management and brand choice. Between 2000 and 2013 she worked for a strategy and marketing consulting company in Mainz. During this time, she primarily worked on projects in the fields of strategy, product and brand management and pricing. Since 2013 she is professor of marketing and marketing research at the HFT Stuttgart.

Thomas Bäumer received his Ph.D. in the field of social psychology at the University of Heidelberg. During this time his research focused on judgement and decision making. After his Ph.D. he was working for 8 years as a market researcher at the GIM (Gesellschaft fuer Innovative Marktforschung) and developed his skills in conducting applied surveys. Today he is professor for business psychology at the HFT Stuttgart with a focus on (psychological) market research and survey design.

The Relationship Between Corporate Social Responsibility and Financial Performance (A Case Study from Finland)

Mari Kooskora, Miia Juottonen and Katlin Cundiff

Abstract The impact of Corporate Social Responsibility (CSR) on the company's performance has become an increasingly important issue among investors, companies and company's management. Despite the fact, that many studies have been conducted on this topic, the relationship between CSR and financial performance is still unclear regarding the causality and different categories of CSR. Therefore, the aim of this paper is to study if corporate social responsibility (CSR) has an impact on financial performance (FP) and to find out, what the nature of the impact is. This study uses correlation and multiple linear regression models in order to examine the relationship between CSR and the financial variables. The sample consists of 30 publicly listed companies in Finland whose financial data and CSR activities during the years 2013–2016 are analyzed. The accounting based model of Return on Assets (ROA) and the market-based model of Earnings per share (EPS) are selected to measure financial performance and CSRHub rates to estimate the corporate social responsibility (CSR). The control variables: capital structure, risk, size and industry were chosen for this research, because of their tendency to have association with the financial performance.

Keywords Corporate social responsibility · Financial performance · Finland

M. Kooskora (✉) · M. Juottonen
Estonian Business School, Tallinn, Estonia
e-mail: mari.kooskora@ebs.ee

M. Juottonen
e-mail: miia.juottonen@gmail.com

K. Cundiff
Drury University, Springfield, USA
e-mail: katlincundiff@hotmail.com

W. Leal Filho (ed.), *Social Responsibility and Sustainability*, World Sustainability Series, https://doi.org/10.1007/978-3-030-03562-4_25

1 Introduction

The role of business in the society has been discussed and debated over heavily for decades and the overwhelming consensus leans towards companies not only earning profit but adding value to the community, or at least taking care that its activities do not negatively affect the surrounding community. For that purpose, many companies have begun to report different CSR activities to show their interest and investments into social welfare. Furthermore, companies are interested to better monitor and measure their social responsibility activities. These investments in social welfare have also raised the question whether CSR activities may lead to better financial benefits or is it only a cost item.

The relationship between CSR and financial performance has been studied in different ways and many researchers have found that CSR and financial performance are positively connected (McGuire et al. 1988; Karagiorgos 2010; Simpson and Kohers 2002). Others have found negative relationship between these two variables (Vance 1975; Lopez et al. 2007; Makni et al. 2009) while some researchers argue that there is no connection at all (Peng and Yang 2014; Nelling and Webb 2009). Despite the fact, that many studies have been conducted on this topic, the relationship is still unclear regarding the causality and different categories of CSR. Therefore, the aim of this paper is to study if, and what kind of impact, corporate social responsibility (CSR) has on financial performance (FP) in Finland.

2 Theoretical Framework

There is an impressive and long history behind the concept of corporate social responsibility (CSR) but a universal definition of what CSR exactly is still missing (van Marrewijk 2013). Dahlsrud (2008) recognized 37 different definitions of CSR in his study, which shows the wide variation in these definitions. One of the most widely accepted definition comes from the European Commission (2011) who has defined CSR as "the responsibility of enterprises for their impacts on society".

According to Visser (2006) Carroll's CSR Pyramid is probably the most well-known model of CSR. Its four interrelated layers encompass the economic, legal, ethical and philanthropic expectations that the society has of organizations at the given point in time (Carroll 1991). Several authors have gone beyond Carroll's model. Rahman (2011) derived ten distinguishing characteristics from numerous CSR definitions: obligation to society, stakeholder involvement, improving the quality of life, economic development, ethical business practice, law-abiding, voluntariness, human rights, protection of the environment, as well as transparency and accountability. Kooskora and Vau (2011) for example proposed that CSR is a process which is defined by six separate characteristic: (1) accountability; (2) sustainability; (3) economy; (4) integrity; (5) transparency; and (6) trust. McWilliams and Siegel (2001) defined CSR as actions that appear to further some social good, beyond the interest of the firm and that which is required by law. The lack of one-size-fit all definition offers benefits to apply the most fitting definition of CSR, yet at the same time open doors for opportunism to apply CSR in the loosest sense.

According to McWilliams et al. (2006) there is no common certainty, why companies engage in CSR activities. Sprinkle and Maines (2010) have proposed that companies either see CSR as a "right thing to do" in order to be a good global citizen, or seek stakeholder approval and try to avoid negative publicity. Dechant and Altman (1994) has also pointed out that stakeholders may have strong impact on the company's choice of engaging in CSR activities. Beside the pressure from stakeholders, the other motives behind CSR engagement include beneficial impacts, such as company image and reputation; employee motivation, retention and recruitment, cost savings and increased revenue from sales and market share; and CSR-related risk reduction or management (Weber 2008).

2.1 CSR Frameworks

Thomson (2007) identified 33 different groups of theories used as theoretical frameworks in CSR studies. According to Fernando and Lawrence (2014) the CSR theoretical perspectives can be classified into two categories, which are "Economic Theories" such as and "Social and Political Theories". The economic theories, such as usefulness theory, agency theory and positive accounting theory consider more an economic aspect of CSR practices. The three mainstream social and political theories such as stakeholder theory, legitimacy theory and institutional theory, provide more insightful perspective than economic theories (Gray et al. 1996).

The "stakeholder perspective" suggests that besides the shareholders, groups and constituents such as employees and the local community are affected by the company's activities (Freeman 2001) and should therefore be considered in managers' decisions as equally as possible with shareholders. According to this view, it might be beneficial for the company to engage in some CSR activities which are perceived to be important by non-financial stakeholders as those groups might otherwise lose their support for the company (Freeman 2001).

Legitimacy theory is the most widely used theory in the literature that discusses the CSR disclosures of organizations (Campbell et al. 2003) and relies on the idea that there is a "social contract" between a company and the society in which it operates (Deegan 2006). Lindblom (1994) explains that an organization's value system is congruent with the value system of the larger social system of which the entity is a part, and when a disparity, actual or potential, exists between the two value systems, there is a threat to the entity's legitimacy.

Institutional theory has not been often used in CSR literature (Gray et al. 2009). According to Oliver (1991), "*institutional theory views organisations as operating within a social framework of norms, values, and taken-for-granted assumptions about what constitutes appropriate or acceptable economic behavior*". The institutional theory includes two dimensions: isomorphism and decoupling (Dillard et al.

2004). Moll et al. (2006) refer to isomorphism as "*how competitive forces drive organizations towards adopting least-cost, efficient structures, and practices*". The other dimension: decoupling is related to the distinction between the external image of an organization and its actual practices (Moll et al. 2006). Deegan (2009) has claimed that the voluntary engagement in CSR activities should be considered as part of an institutional theory as it ties organizational practices for example, including CSR practices, to the values of a society within an organization operates.

Usefulness theory, also known as decision usefulness theory, assumes that investors find the social information exposed by company useful in their decision making (Spicer 1978). In this theory, the primary purpose of accounting, including CSR reporting is to provide information to permit informed decisions and it does not focus mainly on the needs of financial stakeholders as agency theory does (FASB 2010). Usefulness theory mostly focuses on user's perspective instead of reporting entity, as legitimacy theory does (IASB 1989; FASB 2010).

Agency theory describes the self-interested relationship between two parties: "the agent" who acts on behalf of another party called "the principal" (Ross 1973). Deegan and Unerman (2006) have argued that this theory assumes the existence of transactions costs and information costs. If the relationship between self-serving agents and principals is in balance, then both parties will benefit. However, if the relationship is not in balance, then "agency costs" will arise Benn and Bolton (2011).

Positive accounting theory focuses on how accounting is managed between two groups—individuals, the providers of resources, and the organization, the receiver of the resources (Deegan and Unerman 2006). Watts and Zimmerman (1978, 1986) argue that the main objective of Positive accounting theory is to explain the accounting practices chosen by managers and proposed three hypothesis of Positive accounting theory—bonus plan hypothesis, debt hypothesis, and political cost hypothesis. Bonus plan hypothesis note that the bonus receiving managers are more likely to use accounting methods that increase the reported income of current period. If publishing CSR information can increase the income of the company, managers through bonuses are more willing to participate (Scott 2014). Debt hypothesis assumes that the managers use the accounting method, which offers the highest debt/equity ratio of the company. If applying CSR activities offers higher debt/equity ratio, those activities would more likely be applied (Scott 2014). The cost hypothesis state that large companies are more likely to use those accounting methods that reduce reported profit. The more wealth the company has, the more it is subject to potential wealth transfers in the political processes, and therefore managers are likely to use accounting methods that reduce such a transfer (Setyorini and Ishak 2012). In perspective of CSR, this would mean that managers will not apply CSR, because this can increase the income and therefore affect their political examination (Scott 2014).

2.2 CSR Reporting

CSR reporting is a rather new phenomenon compared to financial reporting and regulation on reporting is diversified (KPMG 2011). Organizations are often developing their own standards for social accounting, reporting and auditing. Nevertheless, the purpose of CSR reporting is to measure, exposure and show how the company is liable towards the sustainable development and provide this information to its internal and external stakeholders (Simpson and Taylor 2013). Although there are no universally agreed common guidelines for CSR reporting, United Nations Environment Program's (UNEP) Global Reporting Initiative (GRI) is the most widely accepted reporting guideline. The main objective of GRI is to improve the comparability of CSR reports and their mission is "developing globally applicable guidelines for reporting on economic, environmental and social performance" (GRI 2016). The other commonly known sustainability reporting frameworks are AA1000 by ISEA (Institute of Social and Ethical Accountability), ISO 14000 series by International Organization for Standardization, EMAS by Eco-Management and Audit Scheme and SA 8000 labor standard by Social Accountability International.

According to Shnayder et al. (2015) one common practice to reporting is to apply Triple Bottom Line framework. Triple Bottom Line framework includes three dimensions of performance: social, environmental and financial (Slaper and Hall 2011). TBL also refers to three entities: people, planet and profit, more commonly known as the 3P's.

In Finland, as in most countries, CSR reporting is voluntary. Finnish Government has however stated that companies that are wholly or partially owned by Finnish government should either publish separate reports on corporate responsibility or include CSR information in annual financial reports (Valtioneuvosto 2011: 5). KPMG's Corporate Responsibility Reporting survey among 100 large Finnish companies in 2013, showed that up to 85% of surveyed companies report their CSR activities (KPMG 2014). According to PwC Corporate Responsibility Barometer, more than 65% of the companies report CSR activities in their annual reports, 38% publish separate CSR report, and 60% of companies use GRI as a reporting standard (PwC 2013). In 2016, Finnish government passed an Accounting Act amendment, which based on EU directive, obliges large publicly listed companies with a turnover greater than 40 million euros and with an average of more than 500 employees, to report their policies concerning the environment, employees, social issues, human rights and tackling corruption, bribery, CSR and Diversity (Ministry of Economic Affairs and Employment 2017).

2.3 Association Between CSR and FP

The association between CSR and financial performance has been examined in many different studies and the results and areas studied vary in these cases. There has been no consensus whether CSR has an impact on company's financial performance. The study results have fallen into three types of conclusions which are positive relationship, negative relationship or no association.

Most of the studies have found a positive relationship between CSR and financial performance, e.g. Karagiorgos (2010), Simpson and Kohers (2002), Allouche and Laroche (2005). McGuire et al. (1988) also added that low CSR performance could lead the company to larger risks than high performing firms. On the other hand, Vance (1975) used ROE as a performance measurement and concluded that CSR and ROE have a negative relationship. Lopez et al. (2007) looked at the Dow Jones Sustainability Index as a measurement of CSR concluded that engagement in CSR causes extra costs and therefore lower on financial performance. Makni et al. (2009) used both accounting and market-based measurements and concluded that socially responsible companies had lower profit than the companies less socially responsible. There are also studies that have found no correlation between CSR and financial performance (Soloman and Hansen 1985; Peng and Yang 2014). Nelling and Webb (2009) used both accounting- and market-based measurements for financial performance; McWilliams et al. (2006) used a regression model as a measure of FP and social performance, industry and expenditure for research and development as CSR measures. Neither studies found a relationship between CSR and financial performance.

2.4 CSR Measurements

There are a wide variety of different ways to measure the connection between corporate social responsibility and financial performance. Turker (2009) has presented four different approaches: reputation indices and databases; single or multiple issue indicators; content analysis of corporate publications; and scales measuring CSR performance of individuals. Reputation indices and databases include methods as KLD database, Fortune reputation index or CSRHub. Single or multiple issue indicator, such as pollution control rate, has many limitations since it does not consider the whole structure of CSR (Maignan and Ferrel 2000). The content analysis of corporate publications can provide a lot of information, however, reported information may not reflect the actual activities performed (McGuire et al. 1988). The scales measuring CSR performance of individuals, focus on different interest groups (Clarkson 1995), however it's limitation is the difficulty to provide scale measurement for CSR performance (Turker 2009).

The three most commonly used financial performance measures are accounting based, market based, or the combination of the two for. The accounting based measures are often return on equity (ROE) and Return on Assets (ROA) (Davidson and Worrell 1990; Tang et al. 2012). Wu (2006) applied other accounting measures such as Return on sales (ROS), Return on investment (ROI) and Profit margin. Market based measures are market return, price to earnings ratio, and market value to book value (Pava and Krausz 1996). Brammer et al. (2006) used stock market performance as a market based measure, while Martinez-Ferrero and Frías-Aceituno (2015) applied market value as a measurement. McGuire et al. (1988) and Pava and Krausz (1996) used the combination of accounting and market based measures and concluded that the use of the combination method may predict the relationship better and result the positive association between CSR and financial performance.

3 Research Methods

In order to answer the research question whether corporate social responsibility has an impact on company's financial performance, in 2017 and early 2018 quantitative data was collected from CSRHub, Nasdaq, Kauppalehti (Finnish newspaper) and The Wall Street Journal from the years 2013–2016. The time period was limited to four years, since CSR reporting is still a relatively new concept in Finland and there was not enough data available for a longer time period.

Accounting based measure (ROA) and Market based measure (EPS) were used to measure financial performance of the 30 Finnish companies listed on the Helsinki Stock Exchange. The data concerning financial performance (ROA) was collected from Wall Street Journal and EPS data was collected from Kauppalehti. CSR data has been gathered from CSRHub. The data concerning control variables size and risk, was collected from Wall Street Journal, capital structure data from Kauppalehti, and industry data from Nasdaq. Wall Street Journal and Kauppalehti provides yearly data from all listed companies.

3.1 Dependent and Independent Variables

Return on Assets (ROA) is being used in this study as an accounting-based measure for financial performance and Earnings per share (EPS) is being used as a market-based measure. The dependent variable ROA measures the profitability of a company relatively to its total assets and separates financial activities of the company from operational and investment activities (Nelling and Webb 2009). The formula used for calculating ROA is following according to Hackston and Milne (1996):

$$\text{ROA} = \text{Net Profit}/\text{Total Assets}$$

Earnings per share (EPS) is often been used when determining the value of stock and it is defined as the amount of company's income attributed to each share of common stock (Das and Zhang 2003). The formula used for calculating EPS is:

$$\text{EPS} = \text{Net earnings}/\text{number of outstanding shares}$$

The independent variable of this study is CSR. The data concerning CSR was collected from CSRHub, which aggregates ESG datasets from leading analysts: ASSET4 (Thomson Reuters), CDP (Carbon Disclosure Project), IW Financial, MSCI (ESG Intangible Value Assessment, ESG Impact Monitor, Governance Metrics, and Carbon Tracker), RepRisk, Trucost and Vigeo EIRIS. CSRHub scores company's social responsibility performance on a scale of 1–100 points. The total points of the company are calculated as the weighted average of four sub-categories. The sub-categories are Community, Employees, Environment and Governance.

3.2 Control Variables

There are four control variables used in this research: capital structure, firm size, risk and industry. The capital structure is measured by equity ratio. Equity ratio describes the relative solvency of the company's operations, in which the amount of equity is measured in relation to the total capital of the company (Leppiniemi and Leppiniemi 2006). The formula used for calculating capital structure is following:

$$\text{Capital Structure} = \text{Equity}/\text{Total Capital}$$

Firm size is selected as one of the control variables because large companies might be more capable and active to implement CSR activities than small companies. There are different measures used for this variable in previous literature, such as asset value, sales and number of employees. (McWilliams and Siegel 1997; Waddock and Graves 1997). This study utilizes ln of total assets as a measurement of firm size.

Risk is commonly used control variable in the previous studies and the risk tolerance of the company's management affects its attitude toward activities that have the potential to either obtain savings, incur present or future costs, or build or destroy markets (Waddock and Graves 1997). In this research, the risk was measured by the total debt to total assets ratio. The formula is following:

$$\text{Risk} = \text{Total Debt}/\text{Total Assets}$$

Industry was selected as a fourth control variable because of its tendency to influence the association between CSR and financial performance. Furthermore, some industries may perform financially better than others (Waddock and Graves 1997; McWilliams and Siegel 2000). The US Standard Industrial Classification (SIC) codes were used in this research to measure industry. SIC codes are four-digit numerical codes assigned to companies to identify the primary business of the establishment.

3.3 Methods of Data Analysis

This study utilizes a multiple linear regression model to determine the impact of the independent variable CSR on dependent variables ROA and EPS. The first regression model uses ROA as an accounting-based measure, and the second regression model uses EPS as a market-based measure.

The regression model created to dependent variable ROA is following:

$$Yi,t = \alpha + \beta 1CSRi,t + \beta 2CAPSTRUCi,t + \beta 3SIZEi,t + \beta 4RISKi,t$$
$$\beta 5INDi,t + \varepsilon i,t$$

where *Yi, t*—Dependent variable (ROA) of company i, (Net Profit/Total Assets)

α	regression constant
CSRi	CSR score of a company i
CAPSTRUCi	a proxy for capital structure of company i, (Equity/Total Capital)
SIZEi	a proxy for the size of company i, (ln of total assets)
RISKi	a proxy for the risk of company i, (Total debt/Total Assets)
INDi	industry of company i, (4 digit SIC code)
i	residual, part of the observed FPi which is not explained by the model
t	year index (year 2013–2016)
i	company index

The regression model created to dependent variable EPS is following:

$$Yi,t = \alpha + \beta 1EPSi,t + \beta 2CAPSTRUCi,t + \beta 3SIZEi,t + \beta 4RISKi,t$$
$$\beta 5INDi,t + \varepsilon i,t$$

Yi, t = α + β1*EPSi, t* + Where *Yi, t*—Dependent variable (EPS) of company i, (Net earnings/Number of outstanding shares)

α	regression constant
CSRi	CSR score of a company i
CAPSTRUCi	a proxy for capital structure of company i, (Equity/Total Capital)
SIZEi	a proxy for the size of company i, (ln of total assets)
RISKi	a proxy for the risk of company i, (Total debt/Total Assets)
INDi	industry of company i, (4 digit SIC code)
i	residual, part of the observed FPi which is not explained by the model
t	year index (year 2013–2016)
i	company index

3.4 Hypothesis

There are four hypotheses in this research:

H0a: There is no significant connection between CSR and company's financial performance in Finnish listed companies using accounting based measure.
H1a: There is a significant connection between CSR and company's financial performance in Finnish listed companies using accounting-based measure.
H0b: There is no significant connection between CSR and company's financial performance in Finnish listed companies using market-based measure.
H1b: There is a significant connection between CSR and company's financial performance in Finnish listed companies using market-based measure.

The null hypothesis can be rejected if corporate social responsibility is found to be statistically significant meaning that *P*-value is less than 0.05. Then the connection between CSR and company's financial performance can be stated. In the event that corporate social responsibility is found to be statistically insignificant meaning that *P*-value is larger than 0.05, it may be stated that there is no connection between CSR and company's financial performance and the null hypothesis will be failed to reject. The above mentioned hypothesis will be tested and explained in the next chapter.

4 Results and Discussion

4.1 Descriptive Analysis

Table 1 shows the descriptive statistics and provides an overview of the dependent, independent, and control variables that are used in this study. ROA has a mean of 8.9233, while EPS has a mean of 1.0213, which indicates that the majority of the Finnish listed companies in this research have a ROA of 8.92% and EPS of €1.02. The mean of CSR is 59.1417 showing that majority of the Finnish listed companies in this research have scored close to 59 points in CSRHub. Capital Structure shows that the equity ratio of most of the companies is 44.14% and size indicates that companies have approximately €61.87 million worth of total assets. Risk has a mean of 58.1 which is generally high. Industry has a mean of 3960, this explains that most of the companies in this research are manufacturing companies.

4.2 Correlation Analysis

Table 2 shows the Pearson correlations between ROA and CSR, including the variables related to the regression models. The Pearson correlation between ROA and CSR is weak −0.095. The value indicates that higher value of CSR decreases the ROA of the companies. This correlation is anyhow insignificant as *P* value is 0.302. Also Risk and Industry are negatively correlated with ROA. Risk appears to have stronger correlation with ROA −0.489 than Industry, as this correlation is only −0.028. The correlation between ROA and Risk is significant as *P* value is less than 0.05 but

Table 1 Descriptive statistics

	N	Minimum	Maximum	Mean	Std. deviation
ROA	120	−4.40	30.00	8.9233	6.51262
EPS	120	−2.95	3.09	1.0213	0.89740
CSR	120	45.00	69.00	59.1417	5.80828
CAPSTRUC	120	21.50	73.80	44.1375	10.20929
SIZE	120	679,000.00	767,500,000.00	61,875,916.6583	157,082,161.10232
RISK	120	26.75	78.57	58.1000	11.41069
IND	120	500.00	9500.00	3960.0000	2435.67325
Valid N (listwise)	120				

Source Authors' calculations

the correlation between Industry and ROA is insignificant. Capital structure and size have a positive correlation with ROA. Both are significant as *P* value is less than 0.05. Capital structure has a moderate correlation 0.528 with ROA, while the correlation between size and ROA is weaker 0.217. It should be also noted that the situation, where independent variables are highly correlated, called collinearity, may cause problems in interrelationship of the resulting multiple regression equations (Elliot and Woodward 2014). In this correlation results, there is a high negative correlation 0.951 between capital structure and risk.

Table 3 presents the Pearson correlations between EPS and CSR, including the variables related to regression models. The Pearson correlation between EPS and CSR is also negative −0.113, which indicates that higher value of EPS decreases the EPS of the companies. The correlation is anyhow insignificant as *P* value is more than 0.05. Capital structure and industry has positive correlation with EPS. The correlation between Capital structure and EPS is significant but only 0.280. The correlation between Industry and EPS very weak as only 0.001, this is anyhow insignificant. Risk is negatively correlated with EPS. The correlation is significant but weak as −0.292. Size is also negatively correlated with EPS −0.104 but the correlation is insignificant as *P* value is larger than 0.05. In this correlation results also, capital structure and risk variables are highly negatively correlated.

4.3 *Multiple Regression*

In this section the multiple regression analysis conducted for two financial performance measures will be discussed. The first equitation contains independent variable (CSR) and four control variables: capital structure, risk, size and industry, which explains the first measured dependent variable ROA. The value of R (0.586) presents the multiple correlation between ROA, CSR and control variables. The value of

Table 2 Correlation table (ROA)

		ROA	CSR	CAPSTRUC	SIZE	RISK	IND
ROA	Pearson correlation	1	−0.095	0.526**	0.217*	−0.489**	−0.028
	Sig. (2-tailed)		0.302	0.000	0.017	0.000	0.760
	N	120	120	120	120	120	120
CSR	Pearson correlation	−0.095	1	0.186*	−0.058	−0.115	−0.006
	Sig. (2-tailed)	0.305		0.042	0.527	0.212	0.948
	N	120	120	120	120	120	120
CAPSTRUC	Pearson correlation	0.526**	0.186*	1	0.131	−0.951**	−0.046
	Sig. (2-tailed)	0.000	0.042		0.155	0.000	0.615
	N	120	120	120	120	120	120
SIZE	Pearson correlation	0.217*	−0.058	0.131	1	−0.177	−0.100
	Sig. (2-tailed)	0.017	0.527	0.155		0.053	0.278
	N	120	120	120	120	120	120
RISK	Pearson correlation	−0.489**	−0.115	−0.951**	−0.177	1	−0.003
	Sig. (2-tailed)	0.000	0.212	0.000	0.053		0.973
	N	120	120	120	120	120	120
IND	Pearson correlation	−0.028	−0.006	−0.046	−0.100	−0.003	1
	Sig. (2-tailed)	0.760	0.948	0.615	0.278	0.973	
	N	120	120	120	120	120	120

**Correlation is significant at the 0.01 level (2-tailed)
*Correlation is significant at the 0.05 level (2-tailed)
Source Authors' calculations

Table 3 Correlation table (EPS)

		EPS	CSR	CAPSTRUC	SIZE	RISK	IND
EPS	Pearson correlation	1	−0.113	0.280**	−0.104	−0.292**	0.001
	Sig. (2-tailed)		0.219	0.002	0.257	0.001	0.994
	N	120	120	120	120	120	120
CSR	Pearson correlation	−0.113	1	0.186*	−0.058	−0.115	−0.006
	Sig. (2-tailed)	0.219		0.042	0.527	0.212	0.948
	N	120	120	120	120	120	120
CAPSTRUC	Pearson correlation	0.280**	0.186*	1	0.131	−0.951**	−0.046
	Sig. (2-tailed)	0.002	0.042		0.155	0.000	0.615
	N	120	120	120	120	120	120
SIZE	Pearson correlation	−0.104	−0.058	0.131	1	−0.177	−0.100
	Sig. (2-tailed)	0.257	0.527	0.155		0.053	0.278
	N	120	120	120	120	120	120
RISK	Pearson correlation	−0.292**	−0.115	−0.951**	−0.177	1	−0.003
	Sig. (2-tailed)	0.001	0.212	0.000	0.053		0.973
	N	120	120	120	120	120	120
IND	Pearson correlation	0.001	−0.006	−0.046	−0.100	−0.003	1
	Sig. (2-tailed)	0.994	0.948	0.615	0.278	0.973	
	N	120	120	120	120	120	120

**Correlation is significant at the 0.01 level (2-tailed)
*Correlation is significant at the 0.05 level (2-tailed)
Source Authors' calculations

$R2$ (0.343) shows that independent variables explain 34.3% of dependent variable ROA. ANOVA table shows that for the model the F-ratio is 11.915. This model is significant, because ($P < 0.05$) and it indicates that overall the regression model statistically significantly predicts the outcome variable. 0 for this model is −10.759 and the first regression coefficient is negative (−0.194) which indicates that the association between ROA and CSR is negative using this model. Increasing the independent variable by one point, results 0.194 decrease on dependent variable. This result shows that an increase in CSR results a decrease of the companies ROA. The association is significant ($P < 0.05$), which shows that this variable has statistically significant impact on the outcome variable. This means that the null hypothesis may be rejected and the connection between CSR and financial performance using accounting based measure can be stated. There is a positive association between ROA and the control variables but only Capital structure has P value smaller than 0.05 and therefore has statistically significant impact on the outcome variable. The rest of the control variables have P value more than 0.05 and therefore have insignificant impact on the outcome variable.

The second equitation contains independent variable CSR and four control variables, which explains the second measured dependent variable EPS. In Appendix 9 the whole output of the second multiple regression is presented. The value of R (0.37) presents the multiple correlation between EPS, CSR and control variables. The value of $R2$ (0.137) shows that independent variables explain only 13.7% of dependent variable EPS. ANOVA table shows that for the model the F-ratio is 3.619. This model is also significant, because ($P < 0.05$) and it indicates that overall the regression model statistically significantly predicts the outcome variable. 0 for this model is 3.354 and the first regression coefficient is negative (−0.24) which indicates that the association between EPS and CSR is negative. Increasing the independent variable by one point, results the decrease of 0.24 on dependent variable. However, CSR variable has statistically insignificant impact on the outcome variable as P value is larger than 0.05. This means that the null hypothesis has to be accepted and stated that there is no statistically significant relationship between CSR and financial performance using market-based measure. There is a positive association between EPS and other control variables except the association between Risk and EPS is negative. Also every control variable has insignificant impact on the outcome variable.

4.4 Discussion

From the regression outputs it may be concluded that the accounting based measures showed that the independent variables explain 34.3% of dependent variable ROA in this model. The coefficient was found to be negative, which might indicate that CSR has negative impact on ROA. This was also found to be significant and shows that CSR variable has statistically significant negative impact on the outcome variable ROA. However, the results do not explain even the half of the whole sample. Therefore, any generalized conclusion cannot be made whether CSR results the decrease in

ROA and further research is needed, in order to define the other aspects that affect the outcome. Market-based measure showed that independent variables explain only 13.4% of dependent variable EPS and this model was also found to be significant. The first coefficient was found to be negative, which might indicate that CSR has negative impact on EPS. However, all coefficients in this model were found to be insignificant and therefore the independent variables cannot explain the impact on the outcome variable EPS.

The results might convey that the companies do not apply CSR practices only because they are profitable and also that applying CSR activities does not guarantee profitability. The findings may agree with the Carrol's construct (1991) which indicated that the aim is to meet the expectations of society rather than maximize profit and the other reason for this kind of findings might be similar to conclusion made by Lopez et al. (2007), who noted that when companies engage in CSR, this causes extra costs and therefore those companies perform lower on financial performance. It can be also noted that this research found similar results than previous research made by McGuire et al. (1988), that accounting based measure can predict the outcome better than market based measures.

Based on the findings of this research and previously made arguments, it might be concluded that corporate social responsibility has an impact on financial performance of Finnish companies, when using accounting based measure but not when applying market-based measure. Statistically significant evidence was found that CSR has a slight negative impact on company's financial performance when using accounting based model. When applying market based measure, CSR could not explain the impact on EPS. However, no direct conclusions can be drawn from these results and it cannot be said that CSR activities are the reason for weaker financial performance. The findings in future research may change significantly, when the obligation for CSR reporting comes into effect in Finland in 2018. This will result in more CSR data being available from Finnish companies.

As limitations of this study, the data covered four-year period only, which is relatively short time period for the research concerning CSR which it can be seen more as a long term investment. Also, the sample size of 30 companies was rather small due to lack of CSR data available, and the sample was only collected from Finnish companies, meaning that the results cannot be generalized to other countries. Last, this research only included two different financial variables, which cannot offer very wide scope of insight into how CSR affects different financial parameters. Therefore, extensive generalizations cannot be made based on the results of this research because it does not include all aspects that lead to company's financial gains in the long- and short-run.

5 Conclusion

The aim of this paper was to study the connection between corporate social relationship and financial performance of Finnish publicly listed companies. This topic has been universally studied with different methods but the mixed results have been found and no common ground has been obtained regarding the connection between corporate social responsibility and financial performance. This paper covered two different economic performance indicators: return on assets (ROA) and earnings per share (EPS) in order to cover accounting based measure and market-based measure of financial performance. Two multiple regression models were developed in order to answer to the research questions:

(1) Does corporate social responsibility have an impact on company's financial performance in Finland?
(2) What kind of impact corporate social responsibility has on company's financial performance in Finland?

According to this research, made from panel data from Finnish listed companies during 2013–2016, we can conclude as an answer to the first research question, that there is a statistically significant evidence that corporate social responsibility has an impact on company's financial performance when using an accounting based measure. When applying a market-based measure, corporate social responsibility also had an impact on company's financial performance, but the connection could not be established because the result was found to be insignificant. As an answer to the second research question, the results revealed that the impact is negative but weak when applying the accounting-based measure and there is a need to pay attention to the threat of collinearity, which might have affected the model. As mentioned earlier, when applying the market-based measurement, the result was insignificant and therefore the nature of connection in this case is not relevant. However, as the significant evidence found was not strong and the results varied between measurement methods, it cannot be generalized that corporate social responsibility has a negative impact on company's financial performance in Finland.

As mentioned previously, this research is only a small part of the larger entity and it does only give an insight to the subject. As the period in this study was from 2013 until 2016, it also might be that the impact of CSR is not visible yet in these years financial performance measurements, as CSR can be seen more like a long-term investment or the CSR might still be see as an "cost item" meaning that it does not affect the financial performance measurements positively yet,.

Therefore, further research on this topic is needed in order to make any generalizations. The suggestions for future research would be to apply the sample taken from a longer period of time, which would improve the research. Also different types of variables could be applied in order to generalize the conclusions based on the wider sample. The authors also believe that the financial performance measures are not enough to measure company's benefits of CSR engagement. A more holistic view is needed in order to see how different CSR activities impact the overall organization and its operations. For that purpose, several internal measures can be used, for example employee turnover, productivity, job satisfaction, or reputation or marketing-related measures would give a better picture of the correlation with CSR.

The results of this study also show the need to study the motives for engaging in CSR activities first before measuring its impact on financial performance. When the companies' reasons for CSR engagement are not determined, it is not certain, whether the purpose of companies is to achieve increased financial performance or possibly affect the other performance indicators, such as reputation or marketing based indicators.

Appendix 1: Model Summary ROA

Model	R	R Sq.	Adjusted R Sq.	Std. error of the estimate	Change statistics					Durbin-Watson
					R Sq. change	F change	df1	df2	Sig. F change	
1	586[a]	0.343	0.314	4.47353	0.343	11.915	5	114	0.000	2.123

[a]Predictors: (Constant), IND, RISK, CSR, SIZE, CAPSTRUC

Appendix 2: Anova Table ROA

ANOVA[a]						
Model		Sum of squares	df	Mean square	F	Sig.
1	Regression	1192.255	5	238.451	11.915	0.000[b]
	Residual	2281.423	114	20.012		
	Total	3473.678	119			

[a]Dependent variable: ROA
[b]Predictors: (Constant), IND, RISK, CSR, SIZE, CAPSTRUC

Appendix 3: Coefficients ROA

Model	Unstandardized coefficients		Standardized coefficients	t	Sig.	95.0% confidence interval for B		Correlations		
	B	Std. error	Beta			Lower bound	Upper bound	Zero-order	Partial	Part
1 (Constant)	−10.759	12.972		−0.829	0.409	−36.457	14.938			
CSR	−0.194	0.074	−0.209	−2.643	0.009	−0.340	−0.049	−0.095	−0.240	−0.201
CAPSTRUC	0.465	0.138	0.879	3.380	0.001	0.193	0.738	0.526	0.302	0.257
SIZE	5.327E−9	0.000	0.155	1.975	0.051	0.000	0.000	0.217	0.182	0.150
RISK	0.166	0.123	0.350	1.353	0.179	−0.77	0.409	−0.489	0.126	0.103
IND	6.181E−5	0.000	0.028	0.359	0.720	0.000	0.000	−0.028	0.034	0.027

[a]Dependent variable: ROA

References

Allouche J, Laroche P (2005) A meta-analytical investigation of the relationship between corporate social and financial performance. Revue de gestion des ressources humaines 57:18–41
Benn S, Bolton D (2011) Key concepts in corporate social responsibility, 1st edn. Sage Publication London, p 246
Brammer S, Brooks C, Pavelin S (2006) Corporate social performance and stock returns: UK evidence from disaggregate measures. Financ Manage 35:97–116
Campbell D, Craven B, Shrives P (2003) Voluntary social reporting in three FTSE sectors: a comment on perception and legitimacy. Acc Auditing Acc J 16:558–581
Carroll AB (1991) The pyramid of corporate social responsibility: toward the moral management of organizational stakeholders. Bus Horiz 34:39–48
Clarkson ME (1995) A stakeholder framework for analyzing and evaluating corporate social performance. Acad Manag Rev 20:92–117
Dahlsrud A (2008) How corporate social responsibility is defined: an analysis of 37 definitions. Corp Soc Responsib Environ Manag 15:1–13
Das S, Zhang H (2003) Rounding-up in reported EPS, behavioral thresholds, and earnings management. J Acc Econ 31–50
Davidson WN, Worrell DL (1990) A comparison and test of the use of accounting and stock market data in relating corporate social responsibility and financial performance. Akron Bus Econ Rev 21:7
Dechant K, Altman B (1994) Environmental leadership: from compliance to competitive advantage. Acad Manag Perspect 8:7–20
Deegan C (2006) Legitimacy theory. Methodol Issues Acc Res, p 161–183
Deegan C (2009) Financial accounting theory, 3rd edn. McGraw Hill, Sydney
Deegan C, Unerman J (2006) Financial accounting theory. European edition, McGraw-Hill Education
Dillard JF, Rigsby JT, Goodman C (2004) The making and remaking of organization context. Acc Auditing Acc J 17:506–542
Elliot AC, Woodward WA (2014) IBM SPSS by example: a practical guide to statistical data analysis. SAGE Publications, London
European Commission (2011) Communication from the Commission to the European Parliament, the Council, the European Economic and Social Committee and the Committee of the Regions. A renewed EU strategy 2011–2014 for Corporate Social Responsibility. EC, Brussels
FASB (2010) Accounting standards update (Update) No. 2010-20. The Financial Accounting Standards Board (FASB)
Fernando S, Lawrence S (2014) A theoretical framework for CSR practices: integrating legitimacy theory, stakeholder theory and institutional theory. J Theor Acc Res 10:149–178
Freeman RE (2001) A stakeholder theory of the modern corporation. In: Hartman LP (ed) Perspectives in business ethics. McGraw-Hill Boston, MA, pp 171–181
Gray R, Owen D, Adams C (1996) Accounting and accountability: changes and challenges in corporate social and environmental reporting. Prentice Hall, pp 1–2
Gray SJ, Linthicum CL, Street DL (2009) Have European and US GAAP measures of income and equity converged under IFRS? Evidence from European companies listed in the US. Acc Bus Res 39(5):431–447
GRI (2016) Management approach 2016. Global reporting initiative
Hackston D, Milne MJ (1996) Some determinants of social and environmental disclosures in New Zealand companies. Acc Auditing Acc J 9:77–108
IASB (1989) Framework for the preparation and presentation of financial statements
Karagiorgos T (2010) Corporate social responsibility and financial performance: an empirical analysis on Greek. Eur Res Stud XIII:1–24

Kooskora M, Vau K (2011) Strategic corporate responsibility: a key for surviving and succeeding thorugh turbulent times. In: Vadi M, Jaakson K, Kindsiko E (eds) Management theory and practise: synergy in organization pp 178–203
KPMG (2011) KPMG International survey of corporate responsibility reporting 2011 http://www.kpmg.com/PT/pt/IssuesAndInsights/Documents/corporateresponsibility2011.pdf
KPMG (2014) The KPMG survey of corporate responsibility reporting 2013 https://assets.kpmg.com/content/dam/kpmg/pdf/2015/08/kpmg-survey-of-corporate-responsibility-reporting-2013.pdf
Leppiniemi J, Leppiniemi R (2006) Tilinpäätöksen tulkinta, 4th edn. Juva, WS Bookwell Oy
Lindblom CK (1994) The implications of organizational legitimacy for corporate social performance and disclosure. Paper presented at the critical perspectives on accounting conference New York
Lopez MV, Garcia A, Rodriguez L (2007) Sustainable development and corporate performance: a study based on the Dow Jones sustainability index. J Bus Ethics 75:285–300
Maignan I, Ferrell OC (2000) Measuring corporate citizenship in two countries: the case of the United States and France. J Bus Ethics 23:283–297
Makni R, Francoeur C, Bellavance F (2009) Causality between corporate social performance and financial performance: evidence from Canadian firms. J Bus Ethics 89:409–422
Martinez-Ferrero J, Frías-Aceituno JV (2015) Relationship between sustainable development and financial performance: international empirical research. Bus Strategy Environ 24:20–39
McGuire JB, Sundgren A, Schneeweis T (1988) Corporate social responsibility and firm financial performance. Acad Manag J 31:854–872
McWilliams A, Siegel D (1997) Event studies in management research: theoretical and empirical issues. Acad Manag J 40:626–657
McWilliams A, Siegel D (2000) Corporate social responsibility and financial performance: correlation or misspecification? Strateg Manag J 21:603–609
McWilliams A, Siegel D (2001) Corporate social responsibility: a theory of the firm perspective. Acad Manag Rev 26:117–127
McWilliams A, Siegel DS, Wright PM (2006) Corporate social responsibility: strategic implications. J Manage Stud 43(1):1–18
Ministry of Economic Affairs and Employment (2017) https://tem.fi/en/
Moll J, Krueger F, Zahn R, Pardini M, deOliveira-Souza R, Grafman J (2006) Humanfronto-mesolimbic networks guide decisions about charitable donation. Proc Natl Acad Sci USA 103:15623–15628
Nelling E, Webb E (2009) Corporate social responsibility and financial performance: The "virtuous circle" revisited. Rev Quant Financ Acc 32(2):197–209
Oliver C (1991) Strategic responses to processes institutional. Acad Manag Rev 16(1):145–179
Pava ML, Krausz J (1996) The association between corporate social responsibility and financial performance: the paradox of social cost. J Bus Ethics 15(3):321–357
Peng C-W, Yang M-L (2014) The effect of corporate social performance on financial performance: the moderating effect of ownership concentration. J Bus Ethics 123:171–182
PwC (2013) Closer to business. Corporate responsibility Barometer https://www.pwc.fi/fi/publications/assets/corporate-responsibility-barometer-2013-pwc.pdf
Rahman (2011) Evaluation of definitions: ten dimensions of corporate social responsibility. World Rev Bus Res 1(1):166–176
Ross S (1973) The economic theory of agency: the principal's problem. Am Econ Rev 63:134–139
Scott W R (2014) Financial accounting theory, 7th edn. Pearson
Setyorini CT, Ishak Z (2012) Corporate social and environmental disclosure: a positive accounting theory view point. Int J Bus Soc Sci 3(9):152–164
Shnayder L, Van Rijnsoever FJ, Hekkert MP (2015) Putting your money where your mouth is: why sustainability reporting based on the triple bottom line can be misleading. PLoS ONE p 10
Simpson WG, Kohers T (2002) The link between corporate social and financial performance: evidence from the banking industry. J Bus Ethics 35:97–109
Simpson J, Taylor JW (2013) Corporate governance ethics and CSR. Kogan Page Publishers

Slaper T, Hall T (2011) The triple bottom line: what is it and how does it work? Indiana Bus Rev 86(1):4–8
Solomon RC, Hanson KR (1985) It's good business. Atheneum. Chicago, New York
Spicer BH (1978) Investors, corporate social performance and information disclosure: an empirical study. Acc Rev 53:94–111
Sprinkle GB, Maines LA (2010) The benefits and costs of corporate social responsibility. Bus Horiz 53(5):445–453
Tang Z, Hull CE, Rothenberg S (2012) How corporate social responsibility engagement strategy moderates the CSR-financial performance relationship. J Manage Stud 49:1274–1303
Thomson I (2007) Mapping the terrain of sustainability accounting. In: Unerman J, Bebbington J, O'Dwyer B (eds) Sustainability accounting and accountability. Routledge London pp 19–36
Turker D (2009) Measuring corporate social responsibility: a scale development study. J Bus Ethics 85:411–427
Valtioneuvosto (2011) Valtion omistajapolitiikkaa koskeva Valtioneuvoston periaatepäätös 3.11.2011. http://valtioneuvosto.fi/tiedostot/julkinen/periaatepaatokset/2011/omistajaohjaus/fi.pdf
van Marrewijk M (2013) Concepts and definitions of CSR and corporate sustainability: between agency and communion. Citation classics from J Bus Ethics (Celebrating the First Thirty Years of Publication) pp 641–655
Vance SC (1975) Are socially responsible corporations good investment risks? Manage Rev 64:18–24
Visser W (2006) Revisiting Carroll's CSR pyramid. Corporate citizenship in developing countries, pp 29–56
Waddock SA, Graves SB (1997) The corporate social performance-financial performance link. Strateg Manag J 18:303–319
Watts RL, Zimmerman JL (1978) Towards a positive theory of determination of accounting standards. Acc Rev 53:112–134
Watts RL, Zimmerman JL (1986) Positive accounting theory. Pearson
Weber M (2008) The business case for corporate social responsibility: a company- level measurement approach for CSR. Eur Manag J 26:247–261
Wu ML (2006) Corporate social performance, corporate financial performance, and firm size: a meta-analysis. J Am Acad Bus 8(1):163–171

Community Gardening: Integrating Social Responsibility and Sustainability in a Higher Education Setting—A Case Study from Australia

Johannes M. Luetz and Stephen Beaumont

Abstract Community gardening (CG) has been the subject of growing interest, both within and without the realm of academia. The reasons for this increase in interest are straightforward, given that CG typically offers benefits in at least three areas: (1) fostering a sense of community among contributing stakeholders; (2) promoting a sense of social responsibility; and (3) heightening awareness in areas of sustainability. As such CG is typically recognised as having the capacity to conjointly meet core human needs. This paper presents a case study that describes the inception and progressive implementation of a community garden project ("campus greening"), set within the university context of a private higher education (HE) provider in Brisbane, Australia. The paper charts progress made to date, highlights hurdles that have had to be overcome, distils relevant lessons learned, and extrapolates success factors for future similar projects. Capitalising on 'right timing' emerges as a critical success factor for incentivising, progressing and implementing CG projects. The case study analysis also culminates in a shortlist of tentative recommendations for different stakeholders: (1) soliciting input from alternative leaders; (2) building supportive interdepartmental coalitions; (3) building a broad stakeholder base; and (4) building momentum and support through unconventional means. Experiences and lessons gathered in this paper will be useful for education stakeholders who are interested to use CG to promote community, social responsibility, and sustainability.

Keywords Community gardening · Higher education · Social responsibility Sustainability · Sustainable development

---

J. M. Luetz (✉)
CHC Higher Education, Brisbane, Carindale, QLD 4152, Australia
e-mail: jluetz@chc.edu.au; j.luetz@unsw.edu.au

J. M. Luetz
University of New South Wales (UNSW), Sydney, Australia

S. Beaumont
School of Social Sciences, CHC Higher Education, Brisbane, Carindale, QLD 4152, Australia
e-mail: smbeaumont@bigpond.com

W. Leal Filho (ed.), *Social Responsibility and Sustainability*, World Sustainability Series, https://doi.org/10.1007/978-3-030-03562-4_26

1 Community Gardening, Social Responsibility and Sustainability

This case study discusses community gardening (CG) in a private higher education (HE) setting in Brisbane, Australia. The study was motivated by a desire to better understand some of the success factors involved in integrating social responsibility and sustainability within this context, including how to mobilise diverse stakeholder participation.

This introduction situates the case study within relevant literature by sketching both key conceptual understandings of CG and the diverse range of associated benefits (Sect. 1.1). It then concludes with reflections in respect of the intended paper contribution (Sect. 1.2).

1.1 Benefits and Understandings of Community Gardening (CG): A Short Literature Review

The manifold benefits of CG are quite well established in the literature (Davies 2012; Draper and Freedman 2010; Kitzman-Ulrich et al. 2013), and a number of conceptual understandings exist that conjoin ideas of 'gardening' with notions of 'community'. The American Community Gardening Association[1] has defined a community garden as "any plot of land that is gardened by a group of people" (cited in Kitzman-Ulrich et al. 2013, n.p.), and according to the University of California (UoC 2018),

> A community garden is any piece of land gardened by a group of people, utilizing either individual or shared plots on private or public land. The land may produce fruit, vegetables, and/or ornamentals. Community gardens may be found in neighborhoods, schools, connected to institutions such as hospitals, and on residential housing grounds. (para. 1)

The benefits of CG are broad and diverse and may include considerations of self-sufficiency (Candlin 2011), health-related benefits (Harris 2009; Zick et al. 2013), greening urban environments (Calverley 2017; Hodges Snyder et al. 2016), spatial area maximisation via vertical gardening (Fell 2011), innovation and business growth (Fisk 2010), biopsychosocial benefits (George 2013), individual, community, and environmental resilience (Okvat and Zautra 2011), "reductions in ethnocentrism" (Hoffman et al. 2010, p. 171), and sustainable business (Fetzer and Aaron 2010), among others (Draper and Freedman 2010; Kitzman-Ulrich et al. 2013). Further, at a time where democracy is conceived to be in retreat (Kurlantzick 2013), CG has also been described as having the potential for "cultivating deep democracy" (McIvor and Hale 2016, pp. 179–188).

[1]https://communitygarden.org—see also archived link: https://web.archive.org/web/20071204082111/http://www.communitygarden.org/learn.

In short, Kitzman-Ulrich et al. (2013) posit that "[c]ommunity gardens have been in existence since World War II but have gained recent popularity as a community-based strategy to improve access to food and to build healthier communities." (n.p.).

Set within education settings, CG additionally offers a range of benefits, including pedagogical (Guitart et al. 2014). Kraft and Kielsmeier (1995) have highlighted the manifold benefits of experiential learning, which include notions of "land-as-teacher" (Raffan 1995, p. 129), "learning to live more lightly on the earth" (Johnson 1995, p. 123), and creating "affective bonds to place" (Raffan 1995, p. 129). Further, the literature identifies significant benefits associated with outdoor and environmental education (Hammerman and Hammerman 1973; McRae 1990), including opportunities for 'pedagogy of place' in outdoor education (Wattchow and Brown 2011). Finally, there are also obvious connections to the formation of 'community', 'ownership' and 'education' inherent in CG (Australian Broadcasting Corporation [ABC] 2007), a point advocated by Bloom (2006) in *Creating a Classroom Community of Young Scientists*:

> […] discuss with your children the possibility of starting a community garden for growing vegetables. You can investigate the best conditions for growing vegetables, then proceed to preparing the garden and growing vegetables. […] From the inception, your students can determine what is required, the costs, the sources of materials, […] Taking such an approach not only captures children's imaginations and curiosity, but also provides them with a sense of ownership that is embedded in a meaningful and relevant context. Throughout our teaching, we need to constantly focus on (1) Stimulating and supporting curiosity; (2) Embedding all teaching activities in meaningful and relevant contexts; (3) Providing opportunities for children to take on a sense of ownership and control over the content, direction, and functioning of inquiry projects […] (pp. 137–138)

For the reasons mentioned, the cultivation of community project 'ownership' also emerges as a salient factor for appropriately informed environmental studies (Aplin 1998, pp. 101–104). This perceived need for 'ownership' in resource management situations has perhaps been most notably digested via the concept the 'tragedy of the commons' (Hardin 1968). This concept describes situations where resources that are owned by everyone (i.e., 'the commons') are in heightened danger of becoming degraded, despoiled and depleted by the unrestrained utilitarian self-interest of individual stakeholders (Hardin 1968). Thus, "[s]ome kind of regulation or self-imposed restraint often becomes necessary and, indeed, is frequently a feature of common-ownership regimes." (Aplin 1998, p. 103). This situation consequentially establishes a definitive nexus between CG and fostering social capital and social responsibility (Alaimo et al. 2010; Hoffman et al. 2010; Rodale 2006; The Garden Project 2018; cf. Henriques 2010).

In the literature, environmental sustainability and/or sustainable development are often linked to the well-being of future generations, wherefore CG also emerges as a compelling opportunity to promote sustainability education (Walid and Luetz 2018). For instance, the *Brundtland Report* (World Commission on Environment and Development [WCED] 1987) famously coined the following definition of sustainable development: "Sustainable development is development that meets the needs of the present without compromising the ability of future generations to meet their own needs." (pp. 16, 41). To date, this definition remains influential in respect of its widespread and enduring impact on social responsibility and sustainable development discourses (Luetz and Walid 2019). Furthermore, the following (non-exhaustive) overview gives an idea about some of the many approaches and models discussed in the literature that explore notions of sustainability education: "Sustainability Education" (SE), "Education for Sustainability" (EfS), "Education for Sustainable Development" (ESD), "Education for Environmental Sustainability" (EES), "Environmental Education" (ED), "Formal, Informal and Non-formal Education for Environmental Sustainability" (FINEES) (Huckle and Sterling 1996; Leal Filho 2018; Sustainable Development Goals [SDGs] Goal 4 n.d.; Walid and Luetz 2018).

For the manifold and diverse reasons discussed above, CG conjoins concerns for sustainability and social responsibility (Crane et al. 2009; Hardin 1968; Henriques 2010; Huckle and Sterling 1996), and has therefore become the object of growing research interest in Higher Education (HE) settings (Cheang et al. 2017; Laycock Pedersen and Robinson 2018; Moorhouse 2014; Tangwanichagapong et al. 2017; Wozniak et al. 2016). Such HE context provides the contextual backdrop for this case study.

1.2 Intended Paper Contribution

This paper features a case study that explores how CG may promote the integration of social responsibility and sustainability within a higher education (HE) setting. More specifically, this paper aims to extend previous research by exploring relevant success factors for the inception and implementation of campus gardens. It presents a case study that describes the inception and progressive implementation of a community garden project ("campus greening"), set within the university context of a private HE provider in Brisbane, Australia. The paper charts progress made to date, highlights hurdles that have had to be overcome, distils relevant lessons learned, and extrapolates success factors for future similar projects. The case study also culminates in some tentative hypotheses and offers a shortlist of recommendations for different stakeholders. In this sense, the study seeks to contribute to the field inductively through theory generation. Experiences and lessons gathered in this paper will be useful for education stakeholders interested in incentivising and promoting community, social responsibility, and sustainability.

2 Case Study Methodological Design Features and Approaches

For a range of reasons, which are delineated in this section, case study methodological designs offer important benefits for research into fields of investigation characterised by complex multidisciplinary and multicausal interrelationships (Bryman 2016, pp. 60–69; Johnson and Christensen 2017, pp. 433–441; Punch 2014, p. 124). The case study research design is therefore appropriate and well-suited to explore the CG implementation (Sect. 2.1). A synthesis of data analytical approaches, which were undertaken in this study, will conclude the discussion of methodological considerations (Sect. 2.2).

2.1 Case Study Methodological Appropriateness: A Short Introductory Overview

Expressed in simple language, "[t]he basic case study entails the detailed and intensive analysis of a single case." (Bryman 2016, p. 60). Relatedly and importantly, a case study "has a holistic focus, aiming to preserve and understand the wholeness and unity of the case." (Punch 2014, p. 120). Moreover, a case study may be especially useful in situations requiring in-depth evaluation, assessment or analysis (Stake 1995). According to Creswell (2014),

> Case studies are a design of inquiry found in many fields, especially evaluation, in which the researcher develops an in-depth analysis of a case, often a program, event, activity, process, or one or more individuals. Cases are bounded by time and activity, and researchers collect detailed information using a variety of data collection procedures over a sustained period of time. (p. 14; attributed to Stake 1995, and Yin 2009, 2012)

Further, cases may be comprised by myriad defining features that are conceived and studied within a bounded context and that may not be similarly captured, comprehended and described by alternative research methodological approaches. They are especially suited in situations where knowledge is shallow, fragmented, incomplete, non-existent:

> All data relevant to the case are gathered, and all available data are organised in terms of the case. The case study method gives a unitary character to the data being studied by interrelating a variety of facts to a single case. It also provides an opportunity for the intensive analysis of many specific details that are often overlooked with other methods. (Punch 2014, p. 121)

Even though case studies are open to the criticism that they are "only one case, so how can we generalise?" (Punch 2014, p. 122), their perceived benefits outweigh this potential weakness for at least five important reasons: (1) There are situations where generalisability may not constitute the primary study intention but "the case may be so important, interesting or misunderstood that it deserves study in its own right" (Punch 2014, p. 122). (2) Even so, dissimilarities should also not be overstated:

"Clearly, every case that can be studied is in some respects unique. But every case is also, in some respects, similar to other cases." (Punch 2014, p. 123); (3) Case studies are consistently employed as effective teaching tools (Bryman 2016; Johnson and Christensen 2017) in situations where "historical cases are studied in great detail and are used to train managers, doctors, lawyers, and so on" (Punch 2014, p. 124) to prepare them to deal with future scenarios. (4) Case studies enable both in-depth analyses of complex circumstances and contexts and are therefore well-suited to facilitate comparative analyses across "two or more cases" (Bryman 2016, p. 67) through "cross-case analysis" (Johnson and Christensen 2017, p. 437). (5) Finally, the capacity of the case study design to engage and manage complex research contexts surpasses the capacity of other research designs (Punch 2014), wherefore case studies are also an effective means for uncovering and conceptually describing opportunities for future research:

> [...] only the in-depth case study can provide understanding of the important aspects of a new and persistently problematic research area. This is particularly true when complex social behaviour is involved, as is the case in much social science research. Discovering the important features, developing an understanding of them and conceptualising them for further study, is often best achieved through the case study strategy. Following this line of argument, it may be that too much research has tried to go straight to measurement and quantitative mapping without a fuller understanding of the phenomena and processes involved that are best achieved by case studies. (Punch 2014, p. 124)

In short, "[s]ome of the best-known studies in sociology are based on this kind of design" (Bryman 2016, p. 60).

2.2 *Synthesis: Research Design, Case Study Dates, and Data Analytical Approaches*

This study used an 'exploratory design' paradigm (Creswell 2013, 2014), in the sense that the overall approach was weighted heavily on qualitative data analysis. It may also be generally situated "in a philosophical position which is broadly 'interpretivist' in the sense that it is concerned with how the social world is interpreted, understood, experienced, produced or constituted" (Mason 2002, p. 3).

The study commenced on 7 February 2017 with the inception of the community garden idea (elaborated in Sect. 3). Given that the CG project is ongoing, this case study reflects a snapshot of activities and critical assessments conducted over a 16-month period until 7 June 2018. Data analyses occurred iteratively throughout this time period. The case study was then prepared for peer review and publication during the months March to June 2018 after a level of "theoretical saturation" had been attained, meaning that "new data no longer suggest new theoretical insights or no longer suggest new dimensions of theoretical categories" (Bryman 2016, p. 412).

For the reasons mentioned in Sect. 2.1, the case study design seemed to be well-suited to engage and describe the multidimensional realities encountered in the course of the community garden project planning and implementation. In our case study approach, we move from description of project inception to planning and implementation, providing contextual and historical facts and figures, and then offer critical analyses and interpretations. On a descriptive level, we also sketch the community garden project progress to date and report specific planning and implementation hurdles and outcomes. We also explore to what extent the initial goals of the community garden were conceptually envisaged and reached. In our analyses, we report learnings, findings and success factors on a macro and micro level, including tentative interpretations and hypotheses. The last two steps will be undertaken in the discussion (Sect. 4), and distilled in the concluding synthesis (Sect. 5), which follow the descriptive production of results presented next (Sect. 3).

3 Community Gardening (CG): Inception and Implementation at CHC

CHC Higher Education/Christian Heritage College (CHC)[2] is a private higher education (HE) provider in Brisbane, Australia. Founded in 1986 in the State of Queensland, the second-largest and third-most populous state in the Commonwealth of Australia, and set in Carindale in the outskirts of Australia's third-largest city Brisbane, CHC currently has more than 750 students enrolled in HE courses that can be characterised by a Christian ethos and tradition (CHC 2018). Its mission statement is "Transforming People to Transform their World," and its self-stated ambition is to be recognised as "A Global Leader in Distinctively Christian Higher Education" (CHC 2015, p. 1). In terms of student experience and graduate employment, CHC's course offerings have been rated highly in surveys[3] conducted by the Australian Government Department of Education and Training. Overall results of its comparative analyses of HE providers[4] entitled Quality Indicators in Learning and Teaching (QILT n.d.), places CHC "among the best in national rankings" (CHC 2018, para. 1).

At the present time, CHC offers a total of 30 undergraduate and postgraduate degrees across five disciplines[5]: Business (6); Education (7); Liberal Arts (5); Ministries (5); and Social Sciences (7). Its diverse range of courses includes staple Bachelor and Master of Education degrees, leadership and management programs such as the Master of Business Administration or Master of Social Science Leadership, and leading-edge postgraduate specialisations such as the Graduate Certificate in Neuropsychotherapy, among others. Course offerings are accredited by the Tertiary Education Quality and Standards Agency (TEQSA),[6] which is the Australian Gov-

[2]http://www.chc.edu.au.

[3]Graduate Destination Survey (2015), Graduate Outcomes Survey (2016–17), Student Experience Survey (2014/15); see also CHC (2018).

[4]See, e.g., https://www.qilt.edu.au/institutions/list/institution/christian-heritage-college?ca=experience-of-newly-qualified-higher-education-undergraduates.

[5]http://www.chc.edu.au/future-students/course-finder.

[6]https://www.teqsa.gov.au.

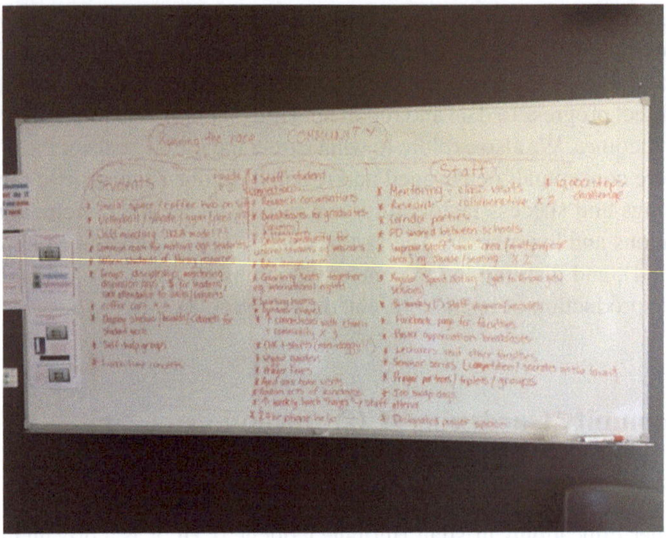

Fig. 1 The annual staff retreat on 7 February 2017 identified the community garden as one of several strategies to "build community" and strengthen "staff-student connections". *Photos* Authors

ernment's "independent national quality assurance and regulatory agency for higher education" (TEQSA 2017, para. 1). Officially, the institution is listed as Christian Heritage College under Provider Number 01016F in the Commonwealth Register of Institutions and Courses for Overseas Students (CRICOS),[7] which is administrated by the Australian Government Department of Education and Training.[8]

Set within this context, the idea of a community garden first emerged during a brainstorming exercise carried out on the occasion of CHC's annual staff retreat held on 7 February 2017. The idea was noted as one of several strategies with the potential to "build community" and strengthen "staff-student connections" (Fig. 1). Following its inception, the idea was then progressed by the co-author of this paper in an email dated 16 March 2017, which was circulated among staff at the School of Social Sciences and read: "Is anyone interested in starting a socsci community garden at the back of C block?" The tongue-in-cheek question drew an immediate and raucous response from members of the School, resulting in a spontaneous popcorn-style flurry of humorous email messages being exchanged and copied to all, which included spontaneous comments like:

> "Yes, I would like to be a part of that!"; "I am so for having chooks!!!"; "Great idea ☺"; "Yes – let's have chooks! ☺"; "So exciting! Chooks and all […]"; "The problem with chooks is they will attract foxes, and I will have to shoot them!"; "[…] Alpaca's are known as excellent guards against foxes […]"; "Goannas and pythons could also be a problem here ☹"; "And apparently foxes […]"; "Alpacas are very cute […]"; "All we have to do is have a really

[7]http://cricos.education.gov.au.

[8]https://www.education.gov.au.

> secure coop with a night light that comes on if a fox approaches. Believe me – we could do it. The alpacas could provide wool that [redacted by authors] could spin and knit us sweaters with!"; "I want to go shooting with the boss - I am sick of paintball. Let's have chooks."; (pers. comm.; on file with authors; selected de-identified comments are best taken with a "grain of salt")

At this stage in the conversation, and following 16 email exchanges (excerpts of which are featured above), it was realised by one email group recipient that a colleague from the School was actually missing from the list of email recipients, and that instead an external professional from a colleague institution had been accidentally cc'd into all the email exchanges. Following an immediate sense of embarrassment and humour-laden corridor conversations, the realisation resulted in the immediate removal of the incorrect email address from the list of recipients, and some more email banter that increasingly contained notions of sustainability:

> "I'm for solar panels and other resources to reduce our footprint on the planet. I will refrain from commenting on weapons."; "We could provide the manure and straw for the garden"; "Could we get a cow as well?"; "Hi fellow lovers of all things green, there are 10,000 litre water tanks for sale this month for $1595 with free water pump. Any ideas for some fundraising? Growfully yours, [redacted by authors]" (pers. comm.; on file with authors)

Regrettably, thereafter the community garden idea lost some momentum. The proposed water tank installation was declined by senior management in an email communiqué to the School on 22 March 2017. This was then followed by the commencement of the teaching semester, and activities and priorities shifted to lecturing and related academic commitments. Even so, there was a recognition on the part of the School as expressed by one staff member that "there will be students who would learn a lot from this experience and use it as a template for future practice. Our students will always benefit from 'seeing' things like this ☺" (pers. comm.; on file with authors).

This was the first instance where the matter of 'timing' emerged in the minds of the authors as an important success factor for the timely and unencumbered implementation of novel projects. Seeing that the community garden could no longer be implemented prior to the commencement of Semester 1, 2017 (and in view of the very short time window between Semesters 1 and 2), effectively put the project on hold until the very end of Semester 2, 2017. Therefore, having missed the implementation opportunity in March 2017 ultimately resulted in the community garden project becoming increasingly encumbered by the busyness and competing demands and constraints of mid-semester commitments. Hence implementation had to be temporarily suspended until teaching commitments eased towards the end of 2017. This implied a minimum delay of 7–8 months.

In November 2017 the project could finally be reinvigorated on the back of several concomitant developments: (1) On 16 November 2017 one of the authors of this case study proposed presenting a research paper on the community garden project at the World Symposium on Social Responsibility and Sustainability in Edinburgh, United Kingdom (27–29 June 2018),[9] thus raising research generation and reportable

[9]See https://www.haw-hamburg.de/en/ftz-nk/events/edinburgh2018.html.

publication output as a potential abetting factor for the reanimation and hoped-for implementation of the CG initiative; (2) On 7 December 2017 this idea was successful in securing some research grant funding via the institution's Research Committee, which accepted the rationale put forward in the funding application that presenting the case study would present a good opportunity to both showcase the CG, while concurrently raising the research profile of CHC; (3) The renewed momentum and funding support stimulated unexpected synergies. For instance, a colleague from the School of Education promptly applied for a $5000 grant through the Queensland Government. Overall there seemed to be a gradual resurgence of interest in the community garden idea. This was abetted by timing. In the Australian academic calendar the months November through February represent the long annual semester break, hence it is a period that is typically conducive to longer-term strategic and planning matters.

Discussing how best to build momentum towards community garden implementation, the authors of this paper (who were at this stage the principal leaders of the CG project) coordinated and strategized. They agreed that to send an unsolicited conventional 'catch-all' email to all staff of the entire institution would be unlikely to receive much notice. Hence a strategy of 'covert seeding' was devised, both in a metaphorical and literal sense. The idea was concocted, discussed and then implemented: Covertly coordinating with the security contractor in charge of locking up all building facilities at night, the plan was hatched to attach small packets of seeds (outside business hours) to the computer screens of all institution staff. The idea was that staff would then return to work the next day to find these seed packs attached to their screens without any further explanation provided, causing seed-receiving staff to ponder and wonder what the meaning of this might be. It was hoped that this approach might stimulate more imagination, discussion and momentum than attempting to 'drum up' interest and participation in the community garden project through conventional email messaging. Hence using building maps, each office was allocated to one of two rounds of 'covert seed-sowing'. The first-round targeted professionals deemed particularly 'vocal' and took place on 7 December 2017. The second round targeted all remaining staff and took place on 13 December 2017. In this way, all staff received at least one pack of seeds. In short, the secret plan proved highly successful in creating institution-wide 'buzz'. It generated some rather 'wild' and 'wide-eyed' speculations in corridor conversations as to who might be behind the activity, and why (Figs. 2, 3 and 4).

Following a short Christmas break, the next step in the process was to 'reap' the benefits of this secretive 'seed-sowing' activity. Hence a plan was devised to invite all staff to a community garden launch party event. For this purpose, a (community) garden party invitation letter was creatively designed by the young children of this paper's first author (Fig. 5). This invitation letter was then photocopied and prepared for institution-wide distribution with the help of another child who creatively attached seeds, leaves, and ferns to each invitation letter for maximum impact (Fig. 6). Thus, a final round of 'covert' distribution was arranged with the help of the institution's security contractor, who secretly distributed these individualised invitation letters to all staff (once more by attaching them to their computer screens) during the weekend

Fig. 2 Concocting a 'covert' distribution strategy, the community garden facilitators used conventional seed packs to unconventionally 'seed' the garden idea among institutional stakeholders. *Photos* Authors

Fig. 3 Covertly coordinating with the security contractor in charge of locking up all building facilities at night, garden facilitators aimed to heighten stakeholder suspense, imagination, discussion and momentum. *Photos* Authors

Fig. 4 A secret plan was hatched and implemented to attach small packets of seeds (outside business hours) to the computer screens of all institution staff, who then unsuspectingly found these the next morning. *Photos* Authors

20 and 21 January 2018. Once again, the 'secretive' distribution of invitation letters proved suspenseful and successful and resulted in a majority of staff attending a short morning (community) garden party on 24 January 2018, which also benefited from snacks and morning tea being provided by staff of the School of Social Sciences (Fig. 7). Of course, the garden party also (finally) revealed the identities of the heretofore covertly scheming organisers and 'seed sowers'. Importantly, the garden party resulted in 22 volunteers stepping forward to catalyse the implementation of the community garden (Fig. 8). Momentum also peaked when it was announced that on 19 January 2018 the above-mentioned grant funding application had been successful, and that $5000 had been received by the institution from the Queensland Government for the community garden project.

Even so, despite hopes running high to promptly establish the community garden within two weeks of the party, the actual in situ implementation was yet again delayed by institutional inertia, including the need to satisfy political processes, perform due diligence checks required by certain stakeholders, ensure building code and grounds compliance metrics, and so forth. This implied that three additional milestones had to be passed: (1) submission of additional site proposal documentation offering five possible community garden sites for consideration (21 February 2018); (2) an executive meeting officially recorded the decision for the preferred site (27 March 2018); (3) a final map was requested and submitted (18 April 2018) detailing the final proposed location of the community garden drawn to scale and in context (Figs. 9 and 10). In short, the opportunity to implement the garden before the commencement of the first semester 2018 was yet again missed. As had previously occurred in 2017,

Fig. 5 A community garden party invitation letter was creatively designed by the young children of this paper's first author, which was later photocopied and prepared for institution-wide distribution. *Photos* Authors

Fig. 6 The 'covertly' distributed party invitation letter was attached to the computer screens of all staff during the weekend 20–21 January 2018. Arriving back to work Monday, staff found personalised invitations. *Photos* Authors

delays implied that (once again) the project became encumbered by the busyness and competing demands and constraints of mid-semester academic commitments.

Fig. 7 The 'covert' invitation idea proved quite successful and resulted in a majority of staff attending the short morning 'garden party', which benefited from catering provided by the School of Social Sciences. *Photos* Authors

Fig. 8 The 'garden party' also resulted in 22 volunteers stepping forward to catalyse the implementation of the community garden. A further aim of the party was to 'stake out' a location for the garden. *Photos* Authors

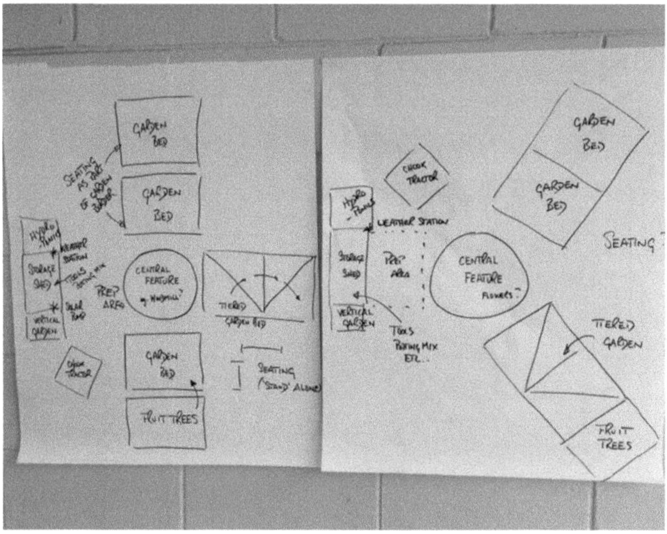

Fig. 9 Several plans were drawn up and presented to senior management in the process of identifying optional community garden concepts and favoured locations. *Photos* Authors

Notwithstanding these difficulties, a couple of 'working bees'[10] were organised on 19 and 26 April, and 24 May 2018, which saw at least some of the foundations laid for the community garden (Fig. 11). In summary, 'timing' and 'momentum' seemed to be reconfirmed in the minds of these authors as critical success factors for the unimpeded implementation of CG projects in HE institutional environments.

Seeing that community garden implementation could once again not be finalised prior to the commencement of Semester 1, 2018, the following discussion (Sect. 4) will now report our analyses, learnings, findings, and success factors on a macro and micro level, including tentative interpretations and hypotheses. Despite overall modest outputs achieved following 16 months of participatory CG project engagement, there is nonetheless a sense of "theoretical saturation" (Sect. 2.2), meaning that "new data no longer suggest new theoretical insights" (Bryman 2016, p. 412). Therefore, in terms of all data available to date, the following analyses seem to be sufficiently and appropriately informed.

[10] A "working bee" is a "voluntary group doing a job for charity" (*Collins English Dictionary* n.d.).

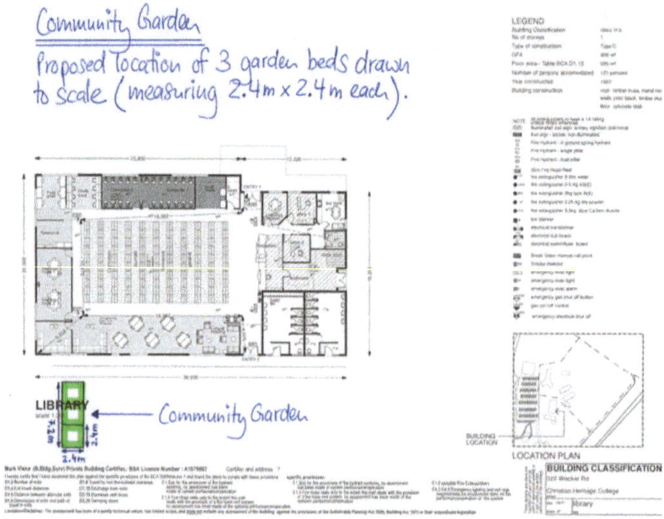

Fig. 10 The approval process also included the required submission of a map that was drawn to scale and presented the proposed garden location in context. Garden beds are highlighted in green. *Photos* Authors

Fig. 11 To date, the progress made to the establishment of the community garden is modest, if viewed in terms of outputs (e.g., garden beds laid, seeds sown and/or crops harvested). *Photos* Authors

4 Community, Sustainability and Social Responsibility: A Critical Discussion of Progress, Hurdles, Success Factors, and Lessons Learned

To date, the progress made to the establishment of the community garden is modest, if viewed in terms of outputs (e.g., garden beds laid, seeds sown and/or crops harvested) (Fig. 11). However, in terms of the broader goals of community building and social responsibility there have been several encouraging 'green shoots', challenges, and opportunities, which are critically analysed under four headlined subsections below.

4.1 *Resurgence in Community Gardens: The Unique Locus of Academic Contexts*

There appears to be a resurgence in community gardens worldwide. Commensurate with this resurgence is a growing amount of scholarly literature, including empirical research into analysing the benefits of establishing community gardens within a wide range of social and physical locations. One of these locations is within the context of academic institutions (Cheang et al. 2017; Laycock Pedersen and Robinson 2018; Tangwanichagapong et al. 2017). In addition to the broader societal and economic benefits identified in the literature for community gardens (Sect. 1.1), other more specific benefits are attributed to academic institutions (Moorhouse 2014; Wozniak et al. 2016). For example, Brown-Fraser and colleagues (2015) identify these as (1) influencing the communities to become "more interdependent and sufficient", and (2) "[i]ntergenerational connections made through community gardening can create lasting relationships between university students and surrounding communities, underscoring the importance of learning beyond classroom and serving that can make a difference" (p. 412).

An analysis of the current project to date suggests that there is some (albeit small) evidence of these perceived benefits becoming evident. Beginning with the first—"becoming more interdependent and sufficient" (Brown-Fraser et al. 2015, p. 412), the most observable facet is a growing sense of community among contributing stakeholders who share a sense of common ownership in the project. What is significant about this is that across the campus the level of interest and direct participation by disparate stakeholders from across different Schools has far exceeded the expectations of these authors and project facilitators. This was most clearly evidenced by the widespread and cross-departmental attendance of interested supporters and volunteers at the (community) garden party described above (Sect. 3). What was missing however, was attendance by the broader campus stakeholders (church, school, etc.), as well as potential civic stakeholders (e.g., local school community groups etc.).

Unfortunately, and not atypically for academic contexts, the institution is not immune to the twin evils of "silo mentality" (Cilliers and Greyvenstein 2012), and "organisational inertia" (Gilbert 2005), both of which often exist within educational organisations. While recognising the diverse theoretical perspectives regarding organisational inertia, the phenomenon observed in this institution appears to line up with Gilbert's propositions regarding threat perceptions and routine rigidity. Reduced level of experimentation is demonstrated by the dismissal of any new initiatives that may have the potential to introduce change and alter the status quo (Kotter 2011, 2012; Kotter and Cohen 2002). Focus on existing resources is demonstrated through tight budget restraints creating competition between Schools or Departments for scant resources. In the end, this organisational inertia has had some impact on thwarting collaborative initiatives, leaving Schools to operate for the most part, fiercely independent. This is further exacerbated by the potentially larger and unique set of stakeholders compared with non-academic public gardens (Scoggins 2010).

4.2 Building Momentum: Busting the Clay Soil of "Inertia" by Leading Creatively

So how did the concept of the community garden break through institutional inertia? Putting this into context it may be added that the idea of a community garden had been discussed even prior to the staff retreat mentioned above in Sect. 3. Although this paper has taken the staff retreat on 7 February 2017 as the official starting point for the garden inception (Sect. 3; Fig. 1), it would be more correct to suggest that the community garden idea predates that particular staff event, albeit without achieving any salient advancement. In regular fashion, the idea was floated again during the annual staff retreat 2017. So why did the 'seed' idea germinate at this juncture in time?

This can be attributed, at least in part, to the determination of the facilitators, who used strategic and unconventional means to build momentum (Sect. 3). Kerfoot (1998) describes leadership as "the leading of creativity which leads to creative change" (p. 181). Whilst the idea of a community garden was highly popular, further investigation discovered that while the project had been previously considered, even by senior leadership, steps towards implementation were not heretofore taken. What seemed to be needed was a creative and unconventional way to lead and progress the project. Within the copious literature on Innovation Theory there is significant emphasis on ideas having to be both 'promoted' within an organisation, and importantly, 'started' (MacFayden 2013). The reasons may be that organisations are not driven to action but rather towards "control, consolidation, preservation and conservatism" (Peters 1992, p. 14). Indeed, the manifold challenges of leading change in change resistant institutional environments are well-recognised in the literature (Kotter 2011, 2012; Kotter and Cohen 2002; Nelson and Luetz 2019). This was explored in a recent study entitled: "What Can We Learn from Pope Francis about Change Management for Environmental Sustainability? A Case Study on Success Factors for Leading Change in Change Resistant Institutional Environments." (Nelson and Luetz 2019). The study highlights the role of creative leadership and

> identifies the power of personal example as a key success factor for influencing change-resistant environments. This accords with the leadership maxim made famous by the French-German theologian, humanitarian, and Nobel Peace Prize winner Albert Schweitzer: 'Example is not the main thing in influencing others, it is the *only* thing.' (Schweitzer 1996, p. xviii, emphasis original). In the final analysis, it appears that a credible, authentic and authoritative messenger will ultimately supersede the significance of the message itself. Therefore, influencing change-resistant environments succeeds best if the messenger *owns, exemplifies, personifies* or even *becomes* the living message. (Nelson and Luetz 2019, p. 519; emphasis original)

In this CG project, a decision was made to step around management as the perceived sole sponsors and enablers of the initiative and to identify other key players from across the four Schools who evidenced passion for ideas without fear of failure. Hence getting the idea 'started' was paramount and required what is commonly known as 'managing upwards'. By managing the relationships with management, the facilitators were able to take control of their own actions and gradually step the garden towards mutually agreed common goals (Bawany 2014).

Beyond the key players, other staff (who the authors did not initially perceive as key stakeholders) became involved. This was also a surprise to the facilitators. Not that the authors did not expect to find interest, however some of the staff who ultimately supported the project were not those individuals that the facilitators had anticipated. Relatedly, the strategic transformation literature is rife with examples stressing the significance of building winning coalitions (Tattersall 2010). While these are typically conceived as leadership-led 'guiding' coalitions (Kotter 2012), in this case they came from 'alternative' leaders (Hensmans et al. 2012).

4.3 Select Opportunities: Pedagogy, Service-Learning, Community Engagement, Therapy

In terms of what specifically constitutes the vision for the garden, the facilitators see several specific opportunities. While the institution does not have a science department, which would naturally align with establishing agricultural and/or horticultural practices, it does have an Education Faculty that teaches science to pre-service teachers. This is a natural fit, and one of the facilitators is an experienced gardener who uses gardening for the kinds of pedagogical purposes highlighted under Sect. 1.1 (ABC 2007; Bloom 2006; Hammerman and Hammerman 1973; Kraft and Kielsmeier 1995; McRae 1990; Raffan 1995; Wattchow and Brown 2011).

Beyond these immediate pedagogical opportunities for the university students, there is also the promise and anticipation of wider community collaboration (Moorhouse 2014). The most significant opportunity is with the Prep to Year 12 School

Campus[11] that physically shares the same site. This 1680-pupil School has no community garden, wherefore there is much potential and high value to collaborate both inter-institutionally and inter-generationally, for example, by inviting both young children and aspiring teachers-in-training into the kinds of pedagogical opportunities elaborated above.

There is also a so-called 'Men's Shed'[12] ministry to men operating within 200 m of the garden, as well as a low-cost housing complex that typically provides accommodation for the elderly. Hence opportunities exist for intergenerational connections that may be made through CG that can create lasting relationships between university students and the surrounding community. It is envisaged that both groups will be approached to solicit their interest and participation in the garden project.

Lastly, in addition to the Prep to Year 12 School Campus, there is also a large nearby 'mega-church'[13] (Hey 2013), which owns both the school and land, but not the university. Including church attending individuals and families, and parents driving their children to school daily during the week, plus academics and students enrolled in courses at the university, the total population that may ultimately access the community garden may be estimated to be in excess of 5000 people. Although the site is not located within the immediate reach of any suburban neighbourhoods—aside from the housing complex adjacent to the garden beds mentioned above—there is a high potential for significant communitarian cooperation and engagement (see Harris 2009). This is in addition to opportunities to educate conservative Christian churchgoers about Environmental Sustainability and Creation Care (Luetz et al. 2018). Thus, the potential of the garden to spawn and support multidimensional, multipurpose and multistakeholder usage is overall promising and significant.

Moreover, the garden site is located next to the Social Sciences complex, which operates a Community Counselling Centre[14] that provides both student practicums and training, as well as professional counselling services. While it was never articulated during the CG planning stages, there is a recognised potential for the garden to be used by the centre for therapeutic purposes. In the U.S. at least, there is a significant prevalence of CG being affiliated with hospitals and academic community engagement (Wozniak et al. 2016). Notwithstanding, the precise ubiquity of healthcare-based gardens in the Australian context is difficult to ascertain. This site may prove to become popular as the therapists from the centre explore the therapeutic landscape of this community garden. While these ideas are not new (Parr 2007), empirical research is now being directed at evaluating the benefit of community gardens for specific therapeutic issues such as end-of-life care (Marsh and Spinaze 2016; Marsh et al. 2017), cognition and nutrition in older adults (Strout et al. 2017), and outpatient care (Milliron et al. 2017). The opportunity for our own therapists and counselling students to establish research-based, therapeutic interventions utilising the garden is encouraging, and has added to the momentum.

[11] http://www.brisbane.coc.edu.au.

[12] http://citipointechurch.com.au/brisbane/church-life/men.

[13] http://citipointechurch.com.

[14] http://www.chc.edu.au/community/counselling-centre/.

4.4 Confronting Challenges: Moving Forward

Whilst the facilitators remain positive and highly motivated, there are significant threats to the continued success of the community garden. The major ones are described above and centre around organisational culture and behaviour. The 'siloing' culture of the institution raises concerns over how well the community collaboration will harmonise with the disparity of resources across departmental Schools. The unique set of potential stakeholders complicates the issue further with no single garden manager appointed to manage the communication between the stakeholder levels. A decision has not been made whether the garden will be divided up into individual lots (e.g., across Schools), or communally, for example. There is danger that the garden may suffer from entropy.

Other challenges more broadly include those identified in the literature such as funding, participation, land and materials (Drake and Lawson 2015). This was demonstrated recently when practical assistance was sought from the affiliated church gardening and maintenance team, but was denied on the grounds that this was not a constituent part of the existing service contract. Further, since inception, funding for the enterprise has not been granted by senior management. To date funding and resources have been met by private donations, which have covered the purchase of tools and recycled materials. This includes seeds and other materials needed to popularise the garden project (Figs. 2, 4, 5, 6, 7 and 8), which were self-funded by the authors. To date, funding from the $5000 grant received from the Queensland Government (Sect. 3) has not yet been accessed. So far, it is uncertain whether the garden has actually promoted a sense of social responsibility both within and without the confines of the immediate community concerned, except by a few individuals. Further, the track record of environmentally responsible management at the institution is rather poor. Thus, in the end, it may be premature to judge whether the CG is a failure, or not.

Yet, in an interesting twist, the campus has recently made significant progress towards achieving carbon neutrality through the large-scale installation of solar panels (Fig. 12). This was however, not based on strategic planning by management but resulted from the initiative of an external benefactor. Even so, it is hoped that the visibility of solar panels will further raise awareness of the institution's social and environmental responsibility. At the micro-level the campus does not (yet) recycle water, food, or other common products, although it is hoped by the authors that relevant steps may eventually be taken. Composting food waste for the garden may soon get underway.

Fig. 12 Recent large-scale installations of solar panels have enabled progress towards carbon neutrality. It is hoped that this may further raise awareness of the institution's social and environmental responsibility. *Photos* Authors

5 Concluding Synthesis: Success Factors, Lessons Learned, Future Prospects

This paper features a case study that explores how community gardening (CG) may promote the integration of social responsibility and sustainability within an Australian private higher education (HE) setting. The paper charts progress made to date, highlights hurdles that have had to be overcome, distils relevant lessons learned, and extrapolates success factors for future similar projects. Experiences and lessons gathered in this paper are useful for education stakeholders seeking to incentivise CG as a means of promoting community, social responsibility, and sustainability.

This research is deemed important as to date, while the literature on CG in HE is emerging, there is little published information documenting the implementation and prevalence of CG in Australian academic settings. This case study highlights some of the formidable challenges facing growing initiatives in private HE settings. These challenges have not been given sufficient attention in the literature. Because of the general bias in the literature against reporting 'failed' growing initiatives, there is opportunity for new knowledge to be generated and documented that focusses on preventing failure. This is one of the three key literature gaps identified by Laycock Pedersen and Robinson (2018) who report that "[f]ailure of CG projects is under-explored." (p. 664) Hence this case study contributes to the literature by digesting some of the diverse challenges encountered during this CG initiative. Notwithstanding these difficulties, given that the project is ongoing (Sect. 2.2), these authors do not deem the CG initiative to date as a 'failure'.

As discussed above, while there is considerable potential for the benefits of CG particularly in the areas of sustainability and community building, tertiary settings are problematic at many levels. Suggested future action-based research could provide an effective methodology to specifically address these shortcomings, through piloting new models of implementation. Further, research could attempt to quantify the perceived potential environmental sustainability benefits from CG in HE as these are not well documented in the literature. There is also considerable scope to explore more fully the additional potential benefits the growing initiatives can bring to the university itself outside of the other stakeholders.

The perceived major threats to the viability of this project identified in this paper also warrant further investigation. In particular, qualitative research that can provide a more salient and richer understanding of the institutional blockages, such as organisational inertia, could assist in the design of more effective implementation strategies. This would be further enhanced by research designs establishing any direct causal links between failure and these perceived blockages.

To conclude, to date the progress made to the establishment of the community garden is modest, if viewed in terms of outputs (e.g., garden beds laid, seeds sown and/or crops harvested) (Fig. 11). However, in terms of the broader goals of community building and social responsibility there have been several encouraging 'green shoots' (Figs. 1, 2, 3, 4, 5, 6, 7, 8, 9 and 10). A key lesson learned is the realisation that seizing opportunities (and successfully implementing novel projects) hinges on 'timing'. Therefore, capitalising on 'right timing' emerges as a critical success factor for incentivising, progressing and implementing CG projects. Seeing that for two consecutive years the community garden could not be implemented prior to the commencement of Semester 1 (2017 and 2018), and in view of the very short time window between Semesters 1 and 2 (Sect. 3), ultimately implied delays that were caused by academic calendar rhythms and arising competing mid-semester commitments. Finally, distinctive success factors for the community garden may be expressed as a shortlist of recommendations for different stakeholders: (1) soliciting input from alternative leaders; (2) building supportive interdepartmental coalitions; (3) building a broader stakeholder base from within and without the campus; and (4) building momentum and support through unconventional means and creative and sustained resourcefulness.

Acknowledgements The authors wish to thank Paul Willis for his enthusiastic support of the community garden, Kyano Maddock for his practical logistical help with successive 'secretive' seed pack and invitation letter distributions, Noah, Daniel and Aurora Lütz Barrón, and Leila Margus, for their creative invitation letter artwork and design, Kirsty Andersen for her copy-editorial support, Stephen Jones and Jesse Keech for their volunteering spirit during 'working bees', and the School of Social Sciences for the constructive input and enduring support throughout the community garden design phase.

References

Alaimo K, Reischl TM, Allen JO (2010) Community gardening, neighborhood meetings, and social capital. J Commun Psychol 38(4):497–514
Aplin G (1998) Australians and their environment: an introduction to environmental studies. Oxford University Press, Oxford
Australian Broadcasting Corporation (ABC 2007) Gardening Australia: creating a school garden. Feature length documentary: 30 mins. ABC Commercial. Marcom Projects
Bawany S (2014) Managing your boss: the two-way conversation of management. Leadersh Excellence Essentials 31(10):47. https://www.hr.com/en/topleaders/all_articles/managing-your-boss-the-two-way-conversation-of-man_i2nhsliy.html. Accessed 7 June 2018
Bloom JW (2006) Creating a classroom community of young scientists, 2nd edn. Routledge, New York
Brown-Fraser S, Forrester I, Rowel R, Richardson A, Nicole Spence A (2015) Development of a community organic vegetable garden in Baltimore, Maryland: a student service-learning approach to community engagement. J Hunger Environ Nutr 10(3):409–436. https://doi.org/10.1080/19320248.2014.962778
Bryman A (2016) Social research methods, 5th edn. Oxford University Press, Oxford
Calverley J (2017) The urban farmer: how to create a productive garden in any space. Harper Collins, Sydney
Candlin A (2011) The self-sufficiency manual: a complete, practical guide to living off the land. Murdoch Books, Millers Point
Cheang CC, So WW, Zhan Y, Tsoi KH (2017) Education for sustainability using a campus eco-garden as a learning environment. Int J Sustain High Educ 18(2):242–262. https://doi.org/10.1108/IJSHE-10-2015-0174
Christian Heritage College/CHC Higher Education (CHC 2015) CHC strategic plan—towards 2020: 'Raising the standard' 2015–2019. Resource document. Author. http://www.chc.edu.au/wp-content/uploads/2015/11/Strategic-Plan-Latest_WEB.pdf. Accessed 28 Apr 2018
Christian Heritage College/CHC Higher Education (CHC 2018) CHC rates among the best in national rankings. General CHC news. Online news article. http://www.chc.edu.au/chc-rates-among-the-best-in-national-rankings. Accessed 29 Apr 2018
Cilliers F, Greyvenstein H (2012) The impact of silo mentality on team identity: An organisational case study. SA J Indus Psychol/SA Tydskrif vir Bedryfsielkunde 38(2), Art. #993. http://dx.doi.org/10.4102/sajip.v38i2.993. Accessed 7 June 2018
Collins English Dictionary (n.d.). Working bee. Online dictionary entry. https://www.collinsdictionary.com/dictionary/english/working-bee. Accessed 7 June 2018
Crane A, Matten D, McWilliams A, Moon J, Siegel DS (2009) The Oxford handbook of corporate social responsibility. Oxford University Press, Oxford
Creswell JW (2013) Qualitative Inquiry and research design: choosing among five approaches, 3rd edn. Sage, Thousand Oaks
Creswell JW (2014) Research design: qualitative, quantitative and mixed methods approaches. Sage, Thousand Oaks
Davies A (2012) Enterprising communities: grassroots sustainability innovations. Advances in ecopolitics vol 9. Emerald, Bingley, UK
Drake L, Lawson LJ (2015) Results of a US and Canada community garden survey: shared challenges in garden management amid diverse geographical and organizational contexts. Agric Hum Values 32(2):241–254
Draper C, Freedman D (2010) Review and analysis of the benefits, purposes, and motivations associated with community gardening in the United States. J Commun Pract 18(4):458–492
Fell D (2011) Vertical gardening: grow up, not out, for more vegetables and flowers in much less space. Rodale, New York
Fetzer AV, Aaron S (2010) Climb the green ladder: make your company and career more sustainable. Wiley, Chichester

Fisk P (2010) People, planet, profit: how to embrace sustainability for innovation and business growth. Kogan Page, London
George DR (2013) Harvesting the biopsychosocial benefits of community gardens. Am J Public Health 103(8):e6. https://doi.org/10.2105/ajph.2013.301435. https://www.ncbi.nlm.nih.gov/pmc/articles/PMC4007866/. Accessed 7 June 2018
Gilbert CG (2005) Understanding the structure of Inertia: resources versus routine Rigidity. Acad Manag J 48(5):741–763
Guitart DA, Pickering CM, Byrne JA (2014) Color me healthy: food diversity in school community gardens in two rapidly urbanising Australian cities. Health Place 26:110–117
Hammerman DR, Hammerman WM (1973, eds) Outdoor education: A book of readings, 2nd ed. Burgess, Minneapolis
Hardin G (1968) The tragedy of the commons. Science 162(3859):1243–1248. https://doi.org/10.1126/science.162.3859.1243
Harris E (2009) The role of community gardens in creating healthy communities. Aust Planner 46(2):24–27. https://doi.org/10.1080/07293682.2009.9995307
Henriques A (2010) Corporate impact: measuring and managing your social footprint. Earthscan, London
Hensmans M, Johnson G, Yip G (2012) Strategic transformation: changing while winning. Palgrave MacMillan, New York
Hey S (2013) Mega churches: origins, ministry and prospects. Mosaic, Preston, Vic
Hodges Snyder E, McIvor K, Brown S (2016) Sowing seeds in the city: human dimensions. Springer Science & Business Media, Netherlands
Hoffman AJ, Wallach J, Sanchez E (2010) Rediscovering community: Interethnic relationships and community gardening. Gateways: Int J Commun Res Engagem 3:171–184
Huckle J, Sterling S (eds) (1996) Education for sustainability. Earthscan/Routledge, New York
Johnson B (1995) Earth education: learning to live more lightly on the earth. In: Kraft RJ, Kielsmeier J (eds) Experiential learning: In schools and higher education. Association for Educational Learning, Kendall/Hunt, Boulder, CO, pp 123–127
Johnson RB, Christensen L (2017) Educational research: quantitative, qualitative and mixed methods approaches, 6th edn. Sage, Thousand Oaks
Kerfoot K (1998) Leading change is leading creativity. Pediatr Nurs 24(2):180–181
Kitzman-Ulrich H, Momoh J, Martin A, DeHaven M (2013) Community gardens. Oxford University Press, Oxford
Kotter JP (2011) Change management vs. change leadership—what's the difference? Forbes. Online article. https://www.forbes.com/sites/johnkotter/2011/07/12/change-management-vs-change-leadership-whats-the-difference/#4764c6e44cc6. Accessed 14 Feb 2018
Kotter JP (2012) Leading change. Harvard Business Review Press, Boston
Kotter JP, Cohen DS (2002) The heart of change: real-life stories of how people change their organizations. Harvard Business Review Press, Boston
Kraft RJ, Kielsmeier J (1995) Experiential learning: In schools and higher education. Association for Educational Learning, Kendall/Hunt, Boulder
Kurlantzick J (2013) Democracy in retreat: the revolt of the middle class and the worldwide decline of representative government. Yale University Press, New Haven
Laycock Pedersen R, Robinson Z (2018) Reviewing university community gardens for sustainability: taking stock, comparisons with urban community gardens and mapping research opportunities. Local Environ 23(6):652–671
Leal Filho W (ed) (2018) Handbook of sustainability science and research. Springer, Cham
Luetz JM, Walid M (2019) Social responsibility versus sustainable development in United nations policy documents: a study of key terms in human development reports. In Leal Filho W (ed) Social responsibility and sustainability: how businesses and organizations can operate in a sustainable and socially responsible way. Springer Nature, Switzerland. https://doi.org/10.1007/978-3-030-03562-4_16

Luetz JM, Buxton G, Bangert K (2018) Christian theological, hermeneutical and eschatological perspectives on environmental sustainability and creation care: the role of holistic education. In Luetz JM, Dowden T, Norsworthy B (eds) Reimagining Christian education: cultivating transformative approaches, (Ch 4). Springer, Singapore, pp 51–73. https://doi.org/10.1007/978-981-13-0851-2_4

MacFayden JS (2013) Culture of innovation. Holist Nurs Pract 27(4):196–198

Marsh P, Spinaze A (2016) Community gardens as sites of solace and end-of-life support: a literature review. Int J Palliat Nurs 22(5):214–219

Marsh P, Gartell G, Egg G, Cross M (2017) End-of-life care in a community garden: findings from a participatory action research project in regional Australia. Health Place 45:110–116. https://doi.org/10.1016/j.healthplace.2017.03.006

Mason J (2002) Qualitative researching, 2nd edn. Sage, London

McIvor DW, Hale J (2016) Common roots: urban agriculture's potential for cultivating deep democracy. In: Hodges Snyder E, McIvor K, Brown S (eds) Sowing seeds in the city: human dimensions. Springer Science & Business Media, Netherlands, pp 179–188

McRae K (1990) Outdoor and environmental education: diverse purposes and practices. Macmillan, South Melbourne

Milliron BJ, Vitolins MZ, Gamble E, Jones R, Chenault MC, Tooze JA (2017) Process evaluation of a community garden at an urban outpatient clinic. J Commun Health 42(4):639–648. https://doi.org/10.1007/s10900-016-0299-y

Moorhouse PA (2014) Dialogic landscape: reciprocity between a campus garden and a university community. Northern Arizona University, unpublished thesis, on file with authors

Nelson W, Luetz JM (2019) What can we learn from Pope Francis about change management for environmental sustainability? A case study on success factors for leading change in change resistant institutional environments. In: Leal Filho W, Consorte-McCrea A (eds) Handbook of sustainability and humanities. Springer, Cham, Switzerland, pp 503–524. https://doi.org/10.1007/978-3-319-95336-6_29

Okvat HA, Zautra AJ (2011) Community gardening: a parsimonious path to individual, community, and environmental resilience. Am J Commun Psychol 47(3–4):374–387

Parr H (2007) Mental health, nature work, and social inclusion. Environ Plann D: Soc Space 25(3):537–561

Peters J (1992) Total strategy. Manag Decis 30(8):12–21

Punch KF (2014) Introduction to social research: quantitative and qualitative approaches, 3rd edn. Sage, Thousand Oaks

Quality Indicators in Learning and Teaching (QILT n.d.) Compare student experience and graduate employment: find a study area or institution. Online resource. Author https://www.qilt.edu.au. Accessed 28 Apr 2018

Raffan J (1995) The experience of place: exploring land as teacher. In: Kraft RJ, Kielsmeier J (eds) Experiential learning: in schools and higher education. Association for Educational Learning, Kendall/Hunt, Boulder, pp 128–136

Rodale A (2006) Precious garden of hope. Prevention 58(8):204

Schweitzer A (1996) Brothers in spirit: the correspondence of Albert Schweitzer and William Larimer Mellon, Jr. Syracuse University Press, New York, NY

Scoggins HL (2010) University garden stakeholders: student, industry, and community connections. HortTechnology 20(3):528–529

Stake RE (1995) The art of case study research. Sage, Thousand Oaks

Strout K, Jemison J, O'Brien L, Wihry D, Waterman T (2017) GROW: green organic vegetable gardens to promote older adult wellness: a feasibility study. J Commun Health Nurs 34(3):115–125

Sustainable Development Goals (SDGs n.d.) Goal 4: ensure inclusive and quality education for all and promote lifelong learning. Online resource. United Nations. https://www.un.org/sustainabledevelopment/education/. Accessed 6 June 2018

Tangwanichagapong S, Nitivattananon V, Mohanty B, Visvanathan C (2017) Greening of a campus through waste management initiatives: experience from a higher education institution in Thailand. Int J Sustain High Educ 18(2):203–217

Tattersall A (2010) Power in coalition: strategies for strong unions and social change. Allen & Unwin, Crows Nest

Tertiary Education Quality and Standards Agency (TEQSA 2017) Australian Government. https://www.teqsa.gov.au/. Accessed 7 June 2018

The Garden Project (2018) We don't just grow plants, we grow people too. Online resource. Author. https://www.gardenproject.org. Accessed 26 Apr 2018

University of California (UoC 2018) Community gardens: what is a community garden? Online resource. Author. http://ucanr.edu/sites/MarinMG/Great_Gardening_Information/Marin_Community_Gardens. Accessed 22 Apr 2018

Walid M, Luetz JM (2018) From education for sustainable development to education for environmental sustainability: reconnecting the disconnected SDGs. In: Leal Filho W (ed) Handbook of sustainability science and research. Springer International, Berlin, Germany, pp 803–826

Wattchow B, Brown M (2011) A pedagogy of place: outdoor education for a changing world. Monash University, Clayton

World Commission on Environment and Development (WCED 1987) Our common future. Brundtland Report. Oxford University Press, Oxford. http://www.un-documents.net/our-common-future.pdf. Accessed 5 Mar 2018

Wozniak JR, Bellah J, Riley JM (2016) Building a community garden: a collaborative cross-disciplinary academic community engagement project. J Bus Strat 33(2):95–115

Yin RK (2009) Case study research: design and methods, 4th edn. Sage, Thousand Oaks

Yin RK (2012) Applications of case study research, 3rd edn. Sage, Thousand Oaks

Zick CD, Smith KR, Kowalski-Jones L, Uno C, Merrill BJ (2013) Harvesting more than vegetables: the potential weight control benefits of community gardening. Am J Public Health 103(6):1110–1115

Dr. Johannes M. Luetz is Senior Lecturer, Postgraduate Coordinator and Research Chair at CHC Higher Education in Brisbane, Australia, and Adjunct Lecturer at the University of New South Wales (UNSW) in Sydney in the School of Social Sciences, where he also earned his Ph.D. in Environmental Policy and Management. Dr. Luetz has worked extensively with World Vision International on research projects raising awareness of the growing effects of climate change on poor and vulnerable communities in Asia, Africa and Latin America, and the need to meaningfully address vulnerabilities through praxis-informed education approaches that work in the real world. He has travelled extensively and has a passion for holistic education and human development, which previously found expression in his former role as Lecturer in Development Studies at UNSW, Sydney.

Dr. Stephen Beaumont is Dean of Social Sciences, CHC Higher Education in Brisbane. Dr. Beaumont has worked extensively in the not-for-profit sector, specifically in the design and implementation of early intervention programs for at-risk youth. He holds several board positions on various social enterprises and charities. His current research interests include youth engagement in the not-for-profit sector, and mentoring for non-traditional students.

MIX
Papier aus verantwortungsvollen Quellen
Paper from responsible sources
FSC® C105338
FSC
www.fsc.org

If you have any concerns about our products,
you can contact us on
ProductSafety@springernature.com

In case Publisher is established outside the EU,
the EU authorized representative is:
Springer Nature Customer Service Center GmbH
Europaplatz 3, 69115 Heidelberg, Germany

Printed by Libri Plureos GmbH
in Hamburg, Germany